SHOOTING THE FRONT

ALLIED AERIAL RECONNAISSANCE IN THE FIRST WORLD WAR

Colonel **Terrence J. Finnegan** served forty years within the Department of Defense in a broad array of assignments that covered the globe. Key postings included NORAD/US Space Command, Defense Intelligence Agency, National Security Agency, European Command, NATO, Pacific Command, Air War College, and Central Command during Operation *Desert Storm*. He spent the past ten years supporting the US National Guard on domestic response missions. Terry has written several articles on the role of military intelligence and aviation in the First World War, including 'Studies in Intelligence', 'Over the Top' and 'Over the Front'. His latest book, *'A Delicate Affair' on the Western Front* is scheduled to be published by The History Press in 2014.

SHOOTING THE FRONT

ALLIED AERIAL RECONNAISSANCE IN THE FIRST WORLD WAR

TERRENCE J. FINNEGAN

I dedicate this book to my parents, Theodore J. and Elizabeth M. Finnegan, and the loving family they created.

Above: The aerial observer demonstrates how the exposure plunger on the 'L' Type camera works. The pilot could also acquire aerial photographs through the Bowden wire release attached to the fuselage. See page 272.

Frontispiece: Farman Experimental (FE) 2d from No. 20 Squadron assigned to the Fighting Reconnaissance mission on the Western Front. The aircrew check the aeroplane and 'L' Type aerial camera before the mission. The FE 2d 225-hp Rolls-Royce Eagle V-12 engine had nearly twice the horse power of the FE 2b's six-cylinder Beardmore. See page 286. (Nesbit, *Eyes of the RAF*)

First published 2011
This paperback edition published 2014
by Spellmount, an imprint of

The History Press
The Mill, Brimscombe Port
Stroud, Gloucestershire, GL5 2QG
www.thehistorypress.co.uk

British Library Cataloguing in Publication Data.
A catalogue record for this book is available from the British Library.

ISBN 978 0 7524 9954 3

Typesetting and origination by The History Press
Printed in India

Contents

Acknowledgements

Aerial reconnaissance in the Great War is a fascinating subject. I have been immersed in it since 2002 and have never been bored. You stumble across a nugget or two that promotes understanding of what the reality was back then – survival, time-sensitive priorities, and immersion in strategy and tactics. In this edition, exploration of the role of Sir David Henderson in particular showed how the usual capability and technology interpretation did not tell the whole story. The legendary 'Red Baron', Manfred von Richthofen, summed up the process: 'Often a photographic plate is more important than shooting down an enemy machine.' That plate influenced the thinking of all combatants, particularly at the command level. My desire to update the first edition was also prompted by a surge of recent outstanding scholarship on the era. Particularly exciting was SHD's elaboration of French *Militaire Aviation* escadrilles and the *Flight* web archives. I only wish I had a lifetime and an unlimited travel budget to explore the Triple Alliance aerial reconnaissance experience as well as the Eastern and Southern fronts.

Getting published in one's lifetime is a wonderful feeling. Having your book issued by an international publisher is even better. My deep gratitude goes to the staff of Spellmount at The History Press for making the commitment to bring *Shooting the Front* to an international readership. I am greatly indebted to Nicholas Watkis (of 'Thackers' fame) for his initiative in introducing The History Press to the project. Thanks to Nicholas, I have had the pleasure of working with Shaun 'b rgds' Barrington in getting the book ready.

Along with Nicholas, Mike Mockford and Chris Halsall, Trustees of the Medmenham Collection, were magnanimous in promoting the publication to today's photographic interpretation experts and carrying the torch for the book's significance to military enthusiasts.

My thanks also to all those at the National Defense Intelligence College. In particular, I want to thank again John Rowland, Denis Clift, Jim Lightfoot, Ric McQuiston, and Russ Swenson for their work on the first edition. John's vision for my role at the college made all this possible. Jim was always magnanimous in making sure any archival data from his desk files was accessible.

My continued access to the United States Air Force Academy seniors made the entire experience particularly gratifying. Colonel John Abbatiello was instrumental in acquainting me with relevant research. Steve Maffeo, Ed Scott and Mary Elizabeth Ruwell kept me energized by revealing new acquisitions in the library for subsequent research and discussions. I am extremely grateful to Air Chief Marshal Sir Stuart Peach for providing the foreword at a time of intense activity, in April 2011.

A global network of aviation history enthusiasts remain close friends and astute critics. Jim Streckfuss, Steve Suddaby, Aaron Weaver, Alan Toelle, Jim Davilla, Dennis Showalter, Daniel Hary, and Marcellin Hodier never failed in to respond quickly when I needed confirmation about a discovery. New-found friends (this is beginning to sound like a social media discussion) that were generous with time, analysis and data include Mike Hanlon, Carl Bobrow, Walter Pieters, Jack Herris, Birger Stichelbaut, Andy Renwick, Ian Passingham, Tony Langley, Chuck Thomas, Eric Ash, Steve Erskine, Colin Huston, Basilio Di Martino and David Zabecki.

Within my close friend and family circle, I continue to remain blessed with the right mix of diversified opinions. Bill Kelly helped me to use the correct idioms in the English language. At home, Susan, Willie Lincoln and Teddy provide the right loving atmosphere to make the best use of my time. I rededicate *Shooting the Front* to the two people who have given me the most inspiration in my passion for the study of history, my parents and closest friends: Theodore J. and Elizabeth M. Finnegan.

Abbreviations and Acronyms

AEF	American Expeditionary Force
AIR	Air Ministry Record, Public Records Office
ALGP	Heavy Artillery of Great Power (*artillerie lourde à grande puissance*)
BEF	British Expeditionary Force
BIO	Branch Intelligence Office or Branch Intelligence Officer
CA	*Corps d'Armée*
CAS	Chief of Air Staff
EVK	*Eskadra Vozdushnykh Korblei*
G-2	Assistant Chief of Staff for Intelligence
GDE	*Group des divisions d'entraînement*
GHQ	General Headquarters
GQC	*Grand Quartier Général*
GPO	Government Printing Office (US)
HQ	Headquarters
IWM	Imperial War Museum
MA&E	Musée de l'Air et de l'Espace
NARA	National Archives and Records Administration, Washington DC
NASM	National Air and Space Museum, Washington DC
OHL	Oberste Heeresleitung
PC	*Poste de commandement*
RAF	Royal Air Force
RFC	Royal Flying Corps
RG	Records Group
RNAS	Royal Naval Air Service
SHD	*Service Historique de la Défense*
SPAé	*Section photo-aérienne*
SRA	*Service des renseignements de l'artillerie*
SROT	*Service des renseignements de l'observation du terrain*
SS	British Army Publications and Stationery Office
TNA:PRO	The National Archives: Public Record Office (UK)
TPS	*Télégraphie par le sol*
TSF	*Télégraphie sans fil*
USAMHI	United States Army Military History Institute
WO	War Office Papers (UK)

West Point Atlas Map Legend

BASIC SYMBOLS:

Regiment	III	Infantry	⊠
Brigade	X	Cavalry	
Division	XX	Cavalry covering force	
Corps	XXX	Trains	
Army	XXXX	Artillery	
Army Group	XXXXX		

Examples of Combinations of Basic Symbols:

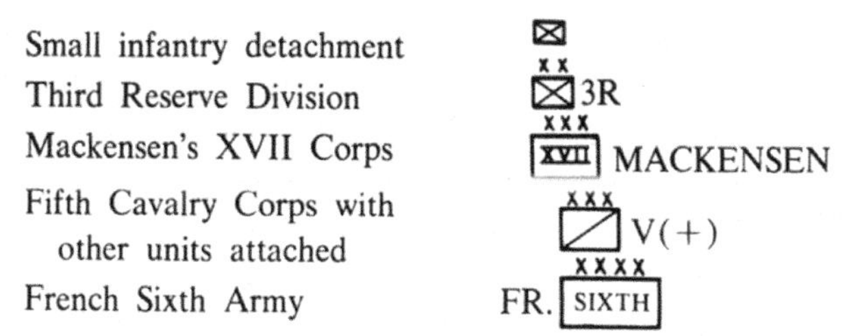

OTHER SYMBOLS:

	Actual location	Prior location
Troops on the march		
Troops in position		
Troops in bivouac or reserve		

	Occupied	Unoccupied
Field works		
Strong prepared defenses		

Route of march

Boundary between units — XXX — (Appropriate basic symbol)

Troops displacing and direction — or

Troops in position under attack — or

Fortified Areas

Foreword

Interest in World War One remains, rightly, high. *Shooting the Front* highlights an important and neglected element of the story of air power: the rapid development of aerial photography.

Colonel Finnegan tells the story in a text packed with data, information and – mostly – meticulous research from the archives. The book shows not just how important aerial reconnaissance was to the war effort but also how effort in the First World War laid the foundation for the exploitation of imagery and geospatial intelligence to this day.

War is a human activity. People make change happen. This story of photography demonstrates very well how small bands of determined people make a difference. In the case of the UK it was John Moore-Brabazon (later Lord Brabazon of Tara), a true aviation pioneer in every sense of the word. And the story is full of people who made a difference at all levels: technicians, mechanics, photographic interpreters; all developed the system that enabled the daring and professional risk of pilots and observers to be turned into intelligence product which made a serious difference to the war effort.

Throughout history, making sense of intelligence is like trying to complete an enormous jigsaw without most of the pieces! By the end of the First World War the system of aerial reconnaissance developed by all protagonists (but particularly by the French) was very impressive. This book tells that story.

In this edition of Colonel Finnegan's important, if deeply specialist, work, the story is told in great detail and with enough primary research that it is unlikely to be surpassed.

So I commend this book to anyone interested in the history of intelligence – which should mean anyone interested in military history!

Air Chief Marshal S. W. Peach KCB CBE BA MPhil DTech Dlitt FRAeS, RAF
Chief of Joint Operations

(Sir Stuart's operational service has included duty in Germany, Turkey, Iraq, Saudi Arabia, Italy, Kosovo and the USA. His long and distinguished career in the Royal Air Force began with a tour in Canberra aircraft in the photographic and strategic reconnaissance role; which makes the Air Chief Marshal's foreword to this book even more appropriate.)

Preface

The first edition of *Shooting the Front* was published to enlighten enthusiasts of the Great War on the role of intelligence collection, analysis and dissemination and how they served as a catalyst in spurring aviation's development and application over those tumultuous years. Aerial reconnaissance has been neglected in the majority of histories which have focussed instead on the exploits of chivalrous combatants in aerial battle thousands of feet above the trenches. Aeroplanes committed to reconnaissance were, for the most part, unequipped for engaging in battle, thus generating little excitement among general readers. The roles of the aerial observer and the photographic interpreter have been virtually ignored over the past century. This oversight left a rich vein of documents for my research and enabled me to discover an aspect of the war that few have known. The real value of aviation, it is now evident, was in the information that was acquired by aerial reconnaissance to better support the combatant on the ground. Everything else at the time, including the exploits of the fighter pilot, was ancillary.

The second edition retains this core thesis. The Great War saw intelligence evolve into a form familiar to present-day practitioners. Positional warfare – a euphemism for trench warfare – resulted in the development of a fairly sophisticated intelligence architecture that utilised all facets of available and emerging technologies, including the aeroplane and photography, although no new photographic principles were discovered. Combatants, from field command posts to headquarters, quickly came to understand what an aerial photograph provided and to incorporate those findings into their operational plans to shape campaigns. Behind every battlefield decision lay aerial (and other) intelligence sources that encouraged or precluded the next tactical or strategic step. What was missing was the strategically important ability to transition quickly from prevailing doctrine, including the intelligence techniques developed for the static battlefield, to find a means to break out of the siege state and achieve victory. This finally began to appear in the more mobile battlefield of the final weeks of the war. The strategic application of intelligence was by no means a failure unique to the Great War; it has been a persistent problem for operations and intelligence practitioners through the ages.

The story in this second edition is enhanced by a surge of recent scholarship on aviation's role and makeup, particularly with regard to French aviation. Of the Allies, the French military clearly paid the most attention to aerial reconnaissance and photographic interpretation. The present research shows a greater commitment to aerial camera use, backed by French industry, which held the competitive edge in producing optics. Their British counterparts made tremendous strides in establishing an important role for aerial reconnaissance despite limited camera production and a disconcerting reliance on airframes lacking manoeuvrability. Aerial photographic interpretation was codified by the French, and British military seniors discovered early in the war that the French were more committed to the aerial photograph in planning and conducting operations. The sharing of ideas and procedures benefited the Allied armies and provided the newly arriving Americans with a short course in the realities of warfare involving a continent-wide siege.

Only recently in academic circles has there been an awareness of the extent of the relations between the French and Americans in the final year of the war. My research illuminates the degree of interaction that existed between these two Allies in their aviation and aerial photographic interpreter communities. There nevertheless remains an assumption among many observers that the Anglo-American relationship, based on a common language and culture, created a natural conduit for providing the American Expeditionary Force (AEF) with the most current war fighting techniques. The archival record clearly proves otherwise. Despite the recommendations of American general officers for intelligence officials to model their program after the British, it was French and American pilots, observers, photo interpreters, and technicians who formed their own *entente cordiale* at the working level. This entente was the foundation for meeting AEF wartime requirements, and it laid the conceptual framework for aerial reconnaissance and photographic interpretation conducted throughout the 20th century.

Allied aerial reconnaissance also included the laudatory accomplishments of the Russian *Eskadra*

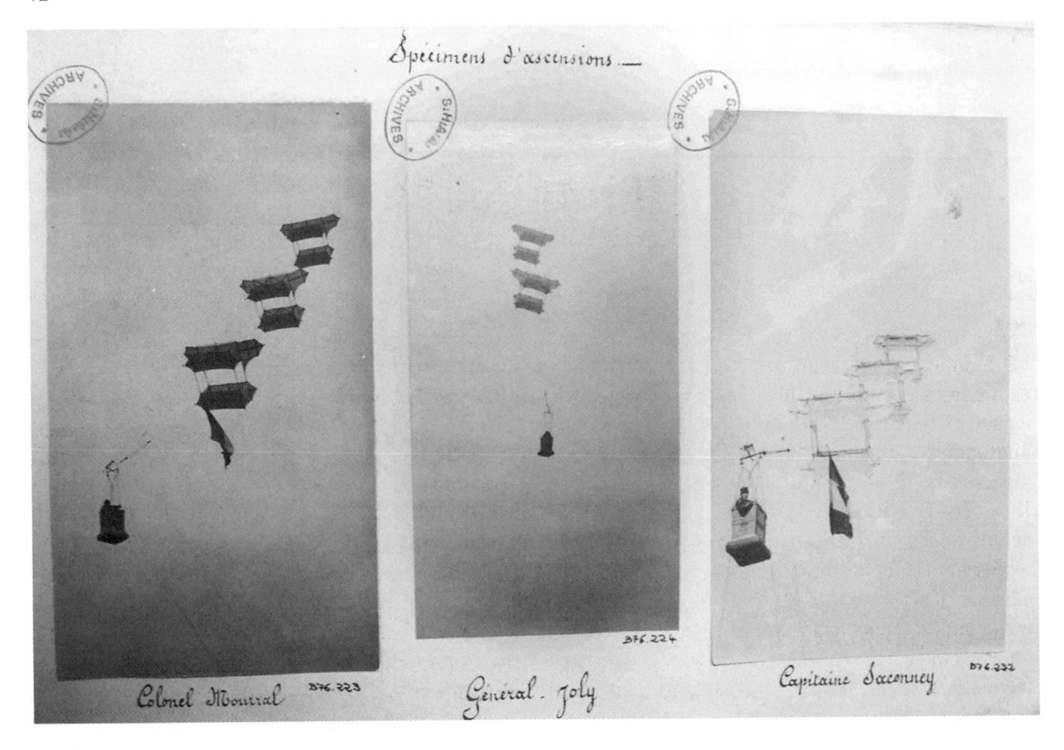

Three French aerial reconnaissance pioneers demonstrate the potential of man-lifting kites. Colonel Mourral (left), *Chef de Section Technique du Genie* (Technical Engineering Section) responsible for studying new warfare technologies; Général Joly (centre), responsible for evolving aviation applications such as the French Aerial Photographic Service, observes a 1910 military exercise from a Saconney-designed kite; and Capitaine Jacques Theodore Saconney, the greatest *aérostier* of the war and aerial reconniassance visionary, takes to the air in the same year. (SHD, B76.244)

Vozdushnykh Korablei (EVK) with its impressive Il'ya Muromets four-engine aeroplane (see page 208), *Aviation Militaire Belge* (Belgian) and *Servizio Aeronautico* (Italian). An effort has been made here to encompass as much of the total Allied effort on this subject as possible but there was a conscious decision to focus on original research. Colleagues have covered Russian, Belgian and Italian roles in aerial reconnaissance in greater detail elsewhere. They have been kind enough to provide imagery for this book.

Whenever possible, I made a conscious decision to use terminology contemporaneous to the period, ranging from military ranks and organisational designations to the use of 'aeroplane' instead of 'aircraft' or 'airplane'. While the word 'aircraft' appears in some reports, it was not common. Therefore, I refer to a British and American aircraft as an 'aeroplane' and to a French '*aéroplane*' as appropriate. Likewise, various descriptions of the battleground incorporated into this study include French nomenclature to reacquaint the reader with the language that the Allied combatants shared.

The fundamental premise of French influence on aviation remains central to the discussion. However, the second edition also brings to light the amazing scholarly neglect over the past century regarding the roles of the first British titans of military aviation, particularly Lieutenant-General Sir David Henderson and Major-General Sir Frederick Sykes. Both shaped the Royal Flying Corps into an important force arm for the main British Expeditionary Force during the critical first weeks of the war. Without their leadership in defining and employing aerial reconnaissance roles, the German advance on Mons and subsequently Paris would have resulted in drastically different outcomes – possibly defeat. It was in the nature of these two, particularly Henderson, to provide maximum support with minimum publicity. Their characters were quietly subordinated to more marketable themes – until now.

Introduction

'The aerial photograph is born of trench warfare.'[1] Simple and yet profound, this pronouncement expresses an important but frequently overlooked legacy of the Great War. Aerial reconnaissance and photographic interpretation reinvented the way that modern battle was envisioned, planned and executed. They became the catalyst for expanding intelligence acquisition, exploitation, and dissemination in the modern era. The advantages of an elevated platform capable of covering broad areas of enemy territory made photographic images an invaluable commodity for assessing enemy intentions. Acceptance of this new technology and procedure was sometimes challenged. The senior command was initially sceptical of the new intelligence source, but quickly comprehending its value, became its strongest advocate. Aerial reconnaissance soon became the primary information source behind most battlefield decisions on the Western Front and it provided a permanent photographic record of the results of those decisions. Without it, the First World War would have depended less on artillery, the most destructive of the force arms, since targeting accuracy would have remained limited to the practices of the previous century. Furthermore, the evolving role of aeroplane aerial bombardment deep behind enemy lines never could have achieved a high degree of sophistication, since validation of damage inflicted would have been limited to whatever could be acquired through eyewitness reporting. In many ways, trench warfare generated what the world knows today as military intelligence. Thanks to an aerial platform and a photographic medium, operational intelligence quickly evolved into a staple of the modern battlefield.

Photography had established itself as a fixture of nineteenth-century society. Its capabilities were well recognised in literary and intellectual circles, not only as a way of capturing and immortalising a moment in time, but also as a new expression of knowledge and emotion.[2] As the second decade of the twentieth century unfolded, photography had developed from being merely a technology for capturing light; it became a recognised art medium alongside canvas and paint. Thanks to a host of commercially produced, mass-market box cameras in Europe and the Americas, photography was a medium that everyone could recognise and understand. What had been missing was a reasonable aerial platform to create a new view of the world.

'The pride of France was in her aeroplanes.' Blériot Escadrille XI at Belfort prior to the war. (*Service Historique de la Défense* (SHD), B83.932)

Equally intriguing is the legacy of aerial reconnaissance and photographic interpretation. There has been little attention paid to the role of these functions in the First World War and their impact on intelligence. Shortly after the war ended, however, Theodore Knappen published an account acknowledging that not all aviation in the war was oriented toward a personal tally of victims:

> It is intriguing to reflect upon the numerous post-war reviews of aviation that provided a perspective that was well known at the time, but clearly forgotten today. The thrilling duels and other combats of the air which have marked this war as being conducted in a third element which was not the medium of attack or defence in preceding wars have centered public attention on aerial combat as the end of military aircraft ... While military men admit the great value of airplane offensives on account of their effect on the morale of enemy troops and civilian population, the prevailing view among them seems to be that the chief military use of the airplanes is as a means of securing information – as the modern scout.[3]

Aviation in the First World War provided information to the ground commander and combatants. This was true on both sides of the front lines. Allies and adversaries were trapped in a strategic standoff that extended for hundreds of miles from the Swiss border through northern France, up to the Belgian coast and over to the Dutch border. Given the standoff, prosecution of the war demanded any information that helped gain the upper hand. In today's military parlance, information and intelligence advantage is a force multiplier. Aviation provided the platform, and photography the source.

Even though aerial reconnaissance and photographic interpretation were important influences on all combatants along the Western Front, this book focuses on three key Allies: France, Great Britain, and the United States. The reason for this is simple: documentation on the thinking behind the Allied evolution in aerial reconnaissance was accessible and readily translatable. The Germans and their allies went through a parallel evolution. German aeroplanes were superb creations demonstrating the leading edge of flight technology. German cameras were the best in the world. The world's leading optical lens producer was Zeiss, located at Jena, Germany. In spite of German technological superiority, the Allies established their own priorities for aerial reconnaissance and through perseverance created an impressive legacy.

The first six chapters describe the evolution chronologically. The sequence of events shows how aerial observation justified the aeroplane's role in battle during those first months that featured manoeuvring armies. The subsequent attrition resulting from the paralysed front lines increased dependency on aerial reconnaissance and the evolving role of photographic interpretation. Prevailing doctrine and strategy contributed most to the blind application of armies against a well defended adversary. Information access improved thanks to aeroplanes capable of flying at higher altitudes fitted with cameras possessing powerful focal arrays. By war's end intelligence derived from aerial reconnaissance and photographic interpretation was as integral to the battle as the weapons that required targeting to be effective. Those who disregarded the role of information and intelligence suffered more casualties. Decisions based on intelligence created an air of greater confidence for the front-line combatant in what often seemed a fruitless and never-ending endeavour.

The second section describes how Allied aerial reconnaissance matured along with photographic interpretation to create a modern-day intelligence discipline. Each Ally established its own methodology to ensure aerial photography contributed to the enemy's ultimate defeat. It served as the baseline for analysing each relevant target in the Great War. Describing the genesis for modern-day intelligence was made more difficult by the pronounced absence of personal reminiscences covering the intelligence experience of this era. Lacking this insight required an in-depth study of official writings. Specific entries culled from surviving manuals and written lectures provided a better appreciation for how the intelligence medium transformed itself to serve military objectives through photographic production, cartography, and analysis. Beyond the front line, strategic analysis came into its own, focussing on aerial photographic signatures of rear-echelon activity to confirm German strategic intentions. The first three years of combat also helped modernise the Allied intelligence process so that the existing technologies of the day were employed in a variety of ways to ascertain the enemy's situation and intentions. When the Americans arrived, they adopted both the technology and its means of exploitation. For all Allies on the Western Front, the role of information acquisition, processing and analysis, refined in the Great War, became the foundation for intelligence for the remainder of the twentieth century.

The third section describes the challenges to aerial reconnaissance and photographic interpretation development based on lessons learned from the battlefield, the focus of intense Allied scrutiny for four years. The No-Man's Land baseline shaped the tactical and strategical missions as well as the technology that became the foundation for modern aerial reconnaissance. It had an impact on camera design, aeroplane design and the discussions among the Allies concerning critical resources. There was a parallel evolution of the critical role and application of camouflage and deception. The combatants fought this war in an environment based on a photographic dimension; only to be challenged by camouflage and illusion designed to confuse and contradict the analytical process. Art mirrored science in the conflict between military reasoning and illusion.

The final section focuses on the technological and human legacy behind aerial reconnaissance and photographic interpretation. It is a legacy rich in accomplishment and worthy of greater attention. Aerial cameras and reconnaissance aeroplanes were overshadowed by the drama of aviation's bellicose potential, yet they legitimised aviation's role in the conflict. The biggest gap in analysis and perhaps the most intriguing aspect for the author is the absence of any evidence of aerial photographic interpretation reporting on a given target, other than line drawings. The extant reporting consists of short summations of daily reports. All that exists are dry descriptions of images. There were attempts to publicise the process through the media outlets of the day. The public's knowledge was derived from exaggerated accounts describing an aerial reconnaissance and photographic interpretation ability to transform basic imagery into a vast storehouse of intimate files describing the adversary's units and cognitive process. The story relies on the reflections of a group of pioneers who worked to develop aerial reconnaissance into the intelligence medium that exists today.

One of the first French pioneers in the art of photographic interpretation, *Capitaine* Jean De Bissy noted that intelligence in the First World War was originally designed 'to follow the destructive work of our Artillery and to register the victorious advance of our Infantry'.[4] This early assessment clearly understates the use and value of intelligence derived from aerial reconnaissance as it evolved during the war. From the horrors of trench warfare evolved one of the most important sources of military information today.

Maurice Farman MF 11bis and crew with 120 cm camera. As Air Chief Marshal Peach points out in the foreword, the French were quickest amongst the Allies to comprehend the potential of, and then develop, aerial photographic reconnaissance. (From Charles Christienne and Pierre Lissarague, *A History of French Military Aviation)*

Aerial Reconnaissance and Photographic Interpretation During the War: A Chronological Overview

CHAPTER 1

Aerial Reconnaissance at the Start of the War

CONCEPT BECOMES REALITY

Aeronautical reconnaissance from the first five months of war clearly demonstrated the acceptance of aerial observation as an integral part of modern warfare. Both sides made significant advances through reconnaissance, thanks to the array of uniquely designed aeroplanes that provided the battlefield commander with a view from the 'higher ground'. The opening weeks of combat established patterns for one of the most important military roles of aviation in the twentieth century. Information became an integral part of aeronautics. Observation became both science and art as aeroplanes manoeuvred with ground forces, contributing data to commanders that helped reveal enemy intentions. The ability to detect the movement of forces meant that whatever was on the ground was now accessible to reconnaissance. These lessons were quickly learned by aviators and commanders alike. Confidence in the aeronautical arm was established. However, the initial salvos of the first months of combat also demonstrated the need for technology to reinforce the aerial observation.

Aerial reconnaissance for military purposes begins in the eighteenth century. Then the French demonstrated that a balloon was not merely a novelty and could support military objectives. With each trial the aerial proponent discovered that an aeronautical advantage correlated to a combat advantage. The powers of Europe had the resources to leverage aeronautics for their standing military. Pioneer British aerial historian Sir Walter A. Raleigh observed in 1922, 'The pride of Germany was in her airships, and the pride of France was in her aeroplanes.'[5] Britain dabbled with both, albeit not to the extent of being competitive. She was preoccupied with the naval race against Germany. However, she was not blind to the advantages achieved through military aviation. British assessments in 1911 concerning French aviation acknowledged it equated to national power projection. As Raleigh stated, 'There is no doubt at all but that the Germans have suddenly realised that the French Army since the general employment of aeroplanes with troops has improved its fighting efficiency by at least twenty per cent.'[6]

Pondering aerial reconnaissance potential was not only done by the French or German military. Colonel [eventually promoted to Lieutenant-General] David Henderson laid the groundwork for intelligence thinking that influenced the British Army for the first 30 years of the twentieth century. In 1897 he joined the staff of the Intelligence Department. He gained notice for his bravery and intellect in South Africa during the Second Boer War, prompting the British commander, Lord Horatio Herbert Kitchener, to appoint him as Director of Military Intelligence (DMI). In that capacity, Henderson led development of the tactical intelligence system for the theatre of operations. Henderson established precise requirements for intelligence officers at different levels resulting in a more structured daily and weekly reporting process. Intelligence under his direction not only served senior headquarters, it was also disseminated to subordinate units.[7] Following duty in South Africa, Henderson wrote the fundamental treatise for British Military Intelligence, *Field Intelligence, Its Principles and Practices* in 1904 under the sponsorship of the British General Staff. The work was an up-to-date guide for a field intelligence officer that captured most of the tactical intelligence lessons of the second Boer War. *Field Intelligence* was the key text for the British Army for several decades. It was a concise, well organised field manual consisting of three chapters that addressed methods of collection followed by five chapters on personnel and organisation, the interpretation of battlefield information, the provision and handling of guides, the preparation and dissemination of reports, and counterintelligence.[8] 'The successful Intelligence officer must be cool, courageous, and adroit, patient and imperturbable, discreet and trustworthy. He must understand the handling of troops and have a knowledge of the art of war.'[9] Henderson also wrote *Regulations for Intelligence Duties in the Field* in 1904 providing another seminal

NOVEMBER 9, 1912.

FLIGHT

MEN OF MOMENT IN THE WORLD OF FLIGHT.

"Flight" Copyright.

Director of Military Training. Royal Flying Corps: Brigadier-General D. C. B. HENDERSON, D.S.O., C.B.

1015 C 2

Brigadier-General Sir David Henderson, the first Commander of the Royal Flying Corps, who with his military intelligence background addressed reconnaissance in a modern-day context. His significant role in aviation history is acknowledged under the banner, 'Men of moment in the World of flight'. (*Flight*, November 9, 1912)

source for British Army intelligence. Topics not covered in *Field Intelligence* such as the minimum establishment of the General Staff required for field intelligence duties in wartime were addressed in *Regulations*.[10] Both texts were still applicable when Britain went to war in 1914, most notably in the areas of reconnaissance and communications.[11]

Henderson's primary argument regarding the role of reconnaissance comprised his book *The Art of Reconnaissance*, first published in in 1907. Henderson examined reconnaissance potential to influence the battle. His third edition published just prior to the outbreak of the First World War drew attention to aerial reconnaissance potential. In his preface Henderson wrote: 'A book on reconnaissance, in which the possibilities of aerial scouting are not considered, must be classed as obsolete, and yet, in this new field, experience is so limited, and progress is so rapid, that any dogmatic pronouncement on military aeronautics would, at this stage, have no permanence and little value.'[12]

The British weekly *Flight* immediately highlighted the potential of aerial reconnaissance by printing aerial photography of known landmarks on the covers. For a visionary like Henderson, such capability represented a strategic leap forward for army manoeuvres because it meant the high ground was adaptable and immediately accessible. The traditional scouting serviced by cavalry was now supplemented, if not replaced by, aerial technology. Though aerial reconnaissance as a mission was still evolving.

Early aerial reconnaissance introduced a unique technology. Fragile pre-war aerial platforms were incapable of sustaining meaningful sorties of long duration. Yet the opportunity to alert infantry and artillery to the advancing enemy provided the impetus to integrate aviation within the standing forces. The French were keen on this idea: 'The aeronautic mission was initially fixed, strategic reconnaissance against a rapidly advancing enemy.'[13] Aerial photography at this moment was an experiment, considered by many in

Flight, March 13th, 1909.

Flight

A Journal devoted to the Interests, Practice, and Progress of Aerial Locomotion and Transport.

OFFICIAL ORGAN OF THE AERO CLUB OF THE UNITED KINGDOM.

No. 11. Vol. I.] MARCH 13TH, 1909. [Registered at the G.P.O. as a Newspaper.] [Weekly, Price 1d. Post Free, 1½d.

PARIS, AS SEEN FROM AN AIRSHIP.—Remarkable photograph taken from an airship, at a height of about 800 metres, of the Place de l'Etoile and the Arc de Triomphe, showing the radiating boulevards, all leading up to the famous archway.

B 2

Public attention to aerial photography was introduced via popular journals such as the Aero Weekly *Flight*. The cover shows the Arc de Triomphe from an altitude of 800m taken from an airship. (*Flight*, 13 March 1909)

Capitaine Saconney, the French Aviation Militaire's visionary for aerial observation platforms. (From Charles Christienne and Pierre Lissarague, *A History of French Military Aviation*)

the military to be another minor novelty of technology. However, photography was a well established part of the culture of the day and was soon recognised as a medium that could contribute to the conduct of a campaign. Advances in aeronautics over the 60 years since the American Civil War had demonstrated observation roles for lighter-than-air balloons and the later-designed dirigibles. In France, the means to capture the aeronautical dimension for military purposes was accomplished through ongoing experimentation with man-carrying kites for observation.

During this vibrant period of experimentation, the roots of aerial photographic reconnaissance took hold. Public awareness of aviation's potential for acquiring information was demonstrated in unique ways prior to powered flight. Innovative aerial platforms for photography went beyond balloon units. In 1896, the public was entertained by the potential of aerial photography through publications describing successful coverage from parakites and pigeons armed with miniature cameras.[14] Throughout the latter half of the nineteenth century, photography and aerial platforms proved their military usefulness. As early as the late 1860s, the Prussian General Staff organised a corps trained in photographic methods to conduct aerial military surveys, aiding its planning for military operations in the Franco-Prussian War of 1870.[15] Campaigns were employing aerial platforms to commence initial planning. Innovations by balloonists in the US Civil War, the Franco-Prussian War of 1870, and the Spanish-American War confirmed the utility of balloon-hosted aerial observation and photography. The British also experimented with photographs taken from balloons during the Boer War in South Africa at the turn of the century.

Experimentation blossomed as existing technologies were applied to aviation. In 1906, a British inventor, Samuel Franklin Cody, built a man-carrying kite that was accepted by the British Army. A year later, Cody built the first British-designed aeroplane, which was subsequently identified as Army Aeroplane No. 1.[16] In 1910 Robert Loraine, in a Bristol aeroplane fitted with a transmitting apparatus, succeeded in sending wireless messages from a distance of a quarter of a mile.[17] The French Aerial Photographic Service conducted research in both aerial photography and restitution (analysis of photographic images applied to developing and enhancing information on a map) under the command of *Capitaine* Jacques Saconney.[18] French military trials in February 1910 used man-bearing kites to support artillery observation and attempt aerial photography. Saconney developed and employed a man-bearing kite to an altitude of 560 metres. The three-kite configuration allowed various passengers to both observe and photograph the area. Despite the apparent risk, Capitaine Saconney's commander *Général* Joly went aloft in the kite.[19]

However, despite the progress through experimentation, aviation's potential was not fully understood by future military commanders. While attending an air race in eastern France in 1910, Général Ferdinand Foch commented, 'Flying, you must understand is merely a sport, like any other; from the military point of view it has no value whatever.'[20] His attitude was echoed by another future senior commander of the coming conflict, General Sir Douglas Haig. Haig spent his career in the cavalry, and as 'the Apostle of Cavalry,' he expressed his bias in 1911 on the technology that was being introduced. 'Tell Sykes [Captain Frederick H. Sykes, soon to become the first British chief of staff for the pre-war Military Wing, Royal Flying Corps (RFC) and eventually served as the Royal Air Force (RAF) Chief of Air Staff in 1918] he is wasting his time, flying can never be of any use to the Army.'[21] As the war proceeded, both became advocates of aviation's potential.

Aerial reconnaissance ceased being a concept and became reality when the Italians successfully employed aéroplanes and airships in North Africa during the Turco-Italian War (1911–1912). The conflict generated a series of firsts for military aviation, to include the first wartime use of aéroplanes for aerial reconnaissance, directing naval gun fire, aerial bombardment, night reconnaissance and the first aerial photographic reconnaissance sortie in a combat environment. The Italians arrived on 15 October with nine aéroplanes: two Blériot Type XI, three Nieuport IV G, two Etrich Taube, and two Farman biplanes. After a week of getting oriented to the area, the leading Italian aviator, Captain

NOVEMBER 11, 1911.

AEROPLANES AT TRIPOLI.

THE AEROPLANE IN THE ITALIAN-TURKISH WAR.—Members of the Italian Army Aviation Corps at the front. From left to right, Lieut. Poggi, Lieut. de Roda, Capt. Piazza (Commander of the Corps), Lieut. Falchi, Lieut. de Tondo (behind on horseback), and Lieut. de Marro.)

MR. QUINTO POGGIOLI, who will be remembered by our readers as having taken his pilot's certificate in England under the Royal Aero Club's regulations, sends us some interesting details of the practical work being carried out in Tripoli in connection with the Italian-Turkish War. Mr. Poggioli writes :—

"On the 25th Oct. Capt. Piazza with his Blériot, and Capt. Moizo on his Nieuport, observed three advancing columns of Turks and Arabs of about 6,000 men. The Italians, after receiving this information, could successfully calculate distances and arrange for their defence.

"On the day following, the 26th Oct., the battle of Sciara-Sciat took place, resulting in the loss to the Turkish Army of 3,000 men. During the battle two aeroplanes, Lieut. Gavotti with his Etrich and Capt. Piazza, were circling the air. The flights took place above the line of fire, so as to be able to direct the firing of the big guns from the battleship 'Carlo Alberto,' and also of the mountain artillery. The aeroplanes were often shot at by the guns of the enemy, but with no result. The only difficulty they had was caused by the currents of air caused by the firing of the big guns.

"Previously, on the 22nd Oct., Capt. Moizo when reconnoitring passed over an oasis, and, in order to observe better the movements of the enemy, descended to an altitude of about 200 metres, and in consequence the wings of his machine were pierced by bullets in six or seven places, and also a rib was broken.

"On November 1st Lieut. Gavotti (Etrich) flew over the enemy, carrying four bombs, carried in a leather bag; the detonator he had in his pocket.

"When above the Turkish camp, he took a bomb on his knees, prepared it and let it drop. He could observe the disastrous results. He returned and circled over the camp, until he had thrown the remaining three bombs. The length of his flight was altogether about 100 kiloms.

"The bombs used contained picrato of potassa, type Cipelli."

THE first *official* communication by one of the belligerents, in regard to the use of aeroplanes in actual warfare, has been issued by the Italian authorities, dated November 5th, from Tripoli. As a matter of historical record we reproduce the text in extenso as follows :—

"Yesterday Captains Moizo, Piazza, and De Rada carried out an aeroplane reconnaissance, De Rada successfully trying a new Farman military biplane. Moizo, after having located the position of the enemy's battery, flew over Ain Zara, and dropped two bombs into the Arab encampment. He found that the enemy were much diminished in numbers since he saw them last time. Piazza dropped two bombs on the enemy with effect. The object of the reconnaissance was to discover the headquarters of the Arabs and Turkish troops, which is at Sok-el-Djama."

⊛ ⊛ ⊛ ⊛

A New Machine at Havre.

A BIPLANE, or, rather, double monoplane, as the inventor, Capt. Hayot, prefers to call it, has just been seen in the air at Havre, piloted by Molon. The main planes are of 10·5 metres span, and have a surface of 48 square metres. The top plane is placed a good deal in advance of the lower plane. There is a monoplane elevator tail, with a vertical rudder. The machine may also be steered when rolling along the ground, as the landing wheels are operated by a lever. The engine is a 70-h.p. Gnome, driving a 2·5 metre propeller.

989

Italian pilots led by Captain Piazza flying the Blériot Type XI aeroplane proved aviation's potential to execute aerial reconnaissance successfully in wartime during the Turco-Italian War. (*Flight*, 11 November 1911)

Carlo Maria Piazza from the Military Aviation School of Udine, flew the first aerial reconnaissance sortie in a Blériot XI on 23 October 1911.[22] The *New York Times* reported that Piazzi [sic] successfully located Turkish infantry entrenched in the Zanzur Oasis 20 miles from Tripoli. His flying created 'an extraordinary sensation among the native populace'.[23] On 23 February 1912, Captain Piazza flying the Blériot brought aerial reconnaissance to the next level by taking an aerial photograph with a Bebè Zeiss hand-held camera. During the course of the thirteen-month campaign aviation also became involved in combat directly. Two Italian aviators were wounded from ground fire. A third acquired the distinction of becoming the first aviator prisoner-of-war when his aéroplane landed behind Turkish lines and was captured.[24]

NOVELTY TURNS INTO PART OF THE OPERATIONAL SCHEME

In France the command staff of the French *armée*, the *Grand Quartier Général* (GQG), supported aviation's role in several instances. They first demonstrated aerial reconnaissance's advantage on 18 September 1911. During a military exercise, Capitaine Eteve and Capitaine Pichot-Duclas flew a 42-km track in a Maurice Farman north from Verdun to Étraye and west to Romagne at an altitude of 1,000 metres, providing in-depth observations of activity.[25] The ongoing success of Italian aviation against the Turks also motivated the French. When their armée deployed in 1912 to Morocco, they brought four Blériot Type XI aéroplanes along to provide aerial reconnaissance support. These aéroplanes were now part of the operational scheme.[26] At the same time, the French established a Blériot Cavalry (BLC) unit with three Blériot aéroplanes.

Prior to the war, French aviation actively exercised aerial reconnaissance in military operations and exercises. Here a Blériot pilot debriefs the reconnaissance to an intelligence officer. (Claude Grahame-White, *The Aeroplane in War*)

Though critics were wary, influential proponents saw potential value, and aviation became more legitimised. On 19 March 1912, the *Aviation Militaire* (also referred to as *Aéronautique Militaire*) was recognised by the GQG as an integral part of the French armée. [27] Ten days later, the French Aviation Militaire units were formed under Permanent Inspector of Military Aeronautics Général Auguste-Edouard Hirshauer comprising an aerial force strength of five *escadrilles*. [28] In 1913 the French Aviation Militaire was placed under the command of French *Corps d'Armée* (CA) commanders, or local administrators, to diminish the authority of the permanent inspector for aviation. Military manoeuvres that year showed reconnaissance information from Aviation Militaire was extremely useful. On 4 April 1914, Aviation Militaire became a separate department of the Ministry of War, making it an independent French military service. [29]

Military staffs also started to refine the routing of information from their aeronautical sources. French officers were encouraged to become familiar with aviation and its reconnaissance potential. Staff officers were expected to make passenger flights, and some were even detached for a period as observers. Absolute familiarity with the manoeuvres and formations of all arms was considered essential. The French realised early on the potential for aéroplanes to observe artillery fire. A British intelligence estimate described the process:

> Two white lines are marked on the ground to show the observer where the battery is in position. The aeroplane ascends behind the battery and rises to a height sufficient to be clear of the trajectory of the shells. The battery commander then fires his bracket. The observer is provided with a sheet of paper across which two parallel lines are drawn indicating the two extremities of the bracket. Having observed the first rounds he marks on the paper the position of the rounds with reference to the space between the two lines representing the extremities of the bracket. He then drops the paper near the battery commander's position.[30]

In another manoeuvre, a suitable landing place for the dropped information was selected by marking it with a large white sheet. A party of two non-commissioned officers and eight troopers were permanently on duty so as to be ready to take messages to the proper quarter.[31] Additionally, the French started realigning their command centres to areas where aéroplanes operated. Messengers were stationed at designated landing zones to transfer reports to the various units. Over time the

January 10, 1914. FLIGHT

FIFTH ARTICLE.

MORANE-SAULNIER.

The Morane-Saulnier exhibit is one of the most artistically-arranged, as well as being very interesting, from a technical point of view. The three machines shown are a single-seater similar to the machine flown by M. Garros across the Mediterranean, a tandem two-seater military monoplane, and the "Parasol." Interesting as the two first-mentioned machines are, though they are standard types, we think that the "Parasol" is the most interesting on this stand, as it represents radical alterations in design and construction. It is fitted with a 9-cyl. Gnome *monosoupape* engine mounted on overhung bearings in the nose of the *fuselage*. A neat aluminium cowl of the usual Morane type prevents oil thrown out by the engine from being blown back in the pilot's face.

The *fuselage* is similar to those of the machines now flying at Hendon, and is built up of four *longerons* of ash converging towards a horizontal knife's edge at the stern. The struts and cross-members are made of ash in the front portion of the *fuselage*, where the greatest strength is required, and of spruce in the rear part. The chassis is of the ordinary Morane type, which is already known to our readers through descriptions in the columns of FLIGHT. The most striking departure from standard design is in the wings, or rather in the mounting of the wings. Instead of bolting these to the sides of the *fuselage*, they have been raised a couple of feet above the *fuselage*, and are mounted on a structure of steel tubes, which converge to form a three-cornered *cabane* above the wings

'Flight' Copyright.

The Morane-Saulnier "Parasol," and a view of the chassis and engine mounting.

31

The 1913 Paris Aero Salon forecast aerial technology design to accommodate photographic aerial reconnaissance with a Morane-Saulnier 'Parasol' monoplane with an internally mounted aerial camera. ('The Paris Aero Salon 1913', *Flight*, 10 January 1914)

landing zones became the rendezvous point for military commanders to disseminate their own information.[32]

The French proceeded to develop an aerial camera, commencing under strict conditions of secrecy.[33] However, during the 1913 Paris Aero Salon French engineering publicly demonstrated the future of aerial photography with their display of a uniquely configured Morane-Saulnier 'Parasol' monoplane. The Aero Weekly *Flight* reported 'behind the passenger's seat is a specially constructed camera, the lens of which is pointed vertically downwards. By simply pulling a cord the observer exposes a plate, and when the cord is released the camera automatically changes the exposed negative and substitutes in its stead an unexposed plate.'[34] By August 1914, the *laboratoire des recherches aéronautique* (Aeronautical Research Centre) at Chalais-Meudon had three wood-frame prototype cameras for *aérostat* observation. They employed 100cm, 60cm, and 50cm focal lengths. However, despite their progress, the laboratory encountered a lack of interest as their cameras were demonstrated to artillery staffs.[35] By the time the French mobilised for war, they had established three photographic sections to work with the aérostat units. The Paris section was led by Capitaine Georges Bellenger (*chef d'avion VIème Armée*). He was an artillery officer whose aviation career had an auspicious beginning. He served as advisor on aeronautical matters to Senator Emile Reymond, a leading French parliamentarian and aviator in his own right.[36] Bellenger became the catalyst for creating the aerial observer's role in the upcoming first weeks of the war. The Verdun section was led by *Lieutenant* Maurice Marie Eugène Grout, another French artillery officer by training who proved to be the most successful initiator for the French military's adaptation of aerial photography.[37] Lieutenant Grout was a brilliant and extremely hard-working staff officer. Educated at France's elite *École Polytechnique*, he was a key member of the staff at Chalais Meudon, developing aerial telegraphy and photography before the war started.[38] The third photographic section supported the eastern French region headquarters at Toul.[39]

Both the French and British military developed coherent aviation organisational structures. The French promoted homogeneity of aircraft types for the various *escadrilles* to save on maintenance while promoting a diverse collection of craft primarily used in observation and reconnaissance missions. Along with infrastructure came development of a specific air doctrine for observation and reconnaissance missions. By 1913 Aviation Militaire comprised eight companies for *aérostat* (*ballon/dirigeable*) work and ten sections for aviation work.[40] As for the RFC, aviation was organised into squadrons comprised of a mix of aeroplanes to support the British Army. The Royal Navy deployed both aeroplanes and a dirigible fleet for naval missions.[41]

French armées and aviation assets commenced mobilisation when the war started. 'At the time of mobilisation military aeronautics was still in its earliest infancy. We may say it was as the fighting went on that this science was evolved and developed.'[42] French *aéronautic* units were governed under Plan XVII (14 February 1914) comprising both *aérostation* and aviation.[43] The French aviation section now comprised 21 flights of six aéroplanes. Two BLC flights of four aéroplanes were added at the start of the war. The mainstay Blériot XI was capable of flying for two hours at 1,500 feet, providing a mobile aerial platform that manoeuvred over the countryside and acquired vital glimpses of enemy forces. Immediately after mobilisation, new flights were created from aéroplanes delivered or salvaged behind the lines. Aviation personnel numbered about 3,500, including 480 officers and noncommissioned officers.[44] Increased awareness of the value of 'higher ground' reporting resulted in a support element to convey critical information. Each escadrille had a 'fast car' and a motorcyclist assigned to rapidly disseminate airborne-acquired reports to the respective ground commander.[45]

French intelligence, known as the *2ème (Deuxième) Bureau*, actively monitored the advancing German forces with all the resources at its disposal. The fledgling radio intercept function was effective in view of the capabilities of the time. Established in 1909, the French radio intercept specialists were recognised as experts in this field. 2ème Bureau also conducted cryptological analysis that served the highest levels of the government.[46] The French had committed their intercept service in full, even before the beginning of the war, and were following German Army traffic attentively. After a few days of combat, they possessed a perfectly clear picture of the operational structure of the German Army in the west as it marched through Belgium in the direction of Paris.[47]

A REDUCED BUT MOTIVATED BRITISH PRESENCE

The British also commenced structuring their aeronautical force. The RFC was constituted by a Royal Warrant on 13 April 1912. It is perhaps by design rather than coincidence that the first RFC commander was the former army intelligence officer, Brigadier-General Sir David Henderson. Henderson was competently assisted by Major Frederick Sykes from the War Office Directorate. Besides acquiring his aviator's certificate in 1911, Sykes became the point man for observing flying operations underway in Europe. His 'Notes on Aviation in France' alerted British military seniors to the dismal state of their aviation and strengthened Sykes' reputation as one of the most knowledgeable officers. When the RFC was established under Henderson's command (Henderson's official title became the Director General of Military Aeronautics (DGMA)), Sykes was promoted to Lt-Col. and assumed command of the RFC's Military Wing.[48]

Brigadier-General Henderson's deep understanding of aviation's potential made him the right person to serve the British Army in this important chapter of military evolution. Henderson advocated that two types of aeroplane, a fighting machine and a scouting machine, would be essential. He envisioned fighting aeroplanes overcoming an enemy air force to allow deep penetrations by scouting aeroplanes to gain accurate information on the dispositions of enemy troops. Fighting aeroplanes were simply necessary to block 'the enemy's endeavours to gain information'. As for the scouting aeroplane serving in a reconnaissance role, preliminary reconnaissance gave the commander the ability to chart the existence of large bodies of troops and observe changes in disposition. An ancillary role for scouting aeroplanes would include observing changes and movements of enemy aeroplanes.[49] Discussion

Capitaine Bellenger, an early French aviation pioneer, directed the French Armée photographic section for the Paris sector at the outbreak of the war. (SHD, B84.3023)

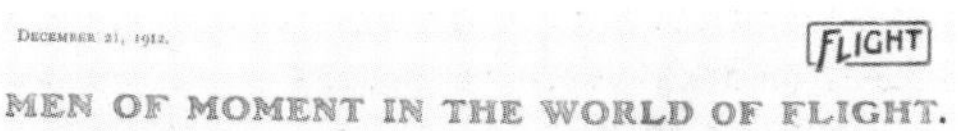

December 21, 1912. FLIGHT

MEN OF MOMENT IN THE WORLD OF FLIGHT.

Major F. H. SYKES, 15th Hussars, Officer Commandant in charge of Records, Royal Flying Corps, Military Wing.

1189

Major Frederick Sykes, Deputy to Brigadier-General Henderson as Commander of the RFC Military Wing follows his boss a month later as one of *Flight's* 'Men of Moment in the World of Flight'. (*Flight*, December 21, 1912)

of aerial reconnaissance during this pre-war period was characterised by analogy to other foundations of military reason, including knowledge gleaned from battlefield experience, human nature in a conflict environment, mechanical knowledge applied to the capabilities of flight at that time and Henderson's most important consideration in prediction: common sense.[50]

Thanks to Henderson and staff, British public interest in military aviation surged. Lt-Col. Sykes aggressively sought to prove that the RFC was competent and innovative in supporting British Army operations. In the first year of its existence, the RFC dabbled with taking photographs from the air (airships, balloons, and kites). Public awareness increased thanks to exhibitions

Lieutenant Frederick Charles Victor Laws was the Royal Flying Corps' expert on aerial photography for most of its existence. (Courtesy of Michael Mockford and the Medmenham Collection)

that featured photographs from aeroplanes taken by a 'Panros' (Pan Ross) press camera with a 6-inch lens producing images on 5- by 4-inch glass plates. The RFC aerial photographs were the main feature of the 1912 Annual Exhibition of the Royal Photographic Society at Farnborough.[51] These attempts to combine photography and aeroplanes remained experimental well into the first months of the war.

Growing interest in aerial photography generated a call for volunteers to join the Military Wing of the RFC. One volunteer was Frederick Charles Victor Laws. He transferred from the 3rd Battalion the Coldstream Guards and became a First Class Mechanic Air Photographer. Laws improved the photographic laboratory at the Farnborough test grounds shortly after his arrival. He was subsequently promoted to sergeant and became the RFC's first non-commissioned officer in charge of photography.[52] Laws made RFC history by serving as aerial photographer using the first British-designed aerial camera, the Watson Air Camera. In 1913, he flew on a dirigible and took overlapping aerial photographs of the Basingstoke Canal, demonstrating a capability that enhanced the reputation of aerial reconnaissance. It was an auspicious start. Laws became the leader of many British initiatives in developing aerial photography for the remainder of his military career.[53]

RFC No. 1 Squadron became the initial aeroplane unit to explore aerial photography. In 1913, the British transferred all dirigibles from the RFC to the Royal Naval Air Service (RNAS). No. 1 Squadron served as an 'Aircraft Park' to experiment on new aerial capabilities.[54] While assigned to No. 1 Squadron, Laws proved the capability for developing a photograph in flight. In early 1914, RFC No. 3 Squadron took the lead in developing British aerial photography. The officers purchased their own cameras and commenced to learn the art of photography on their own. They even learned Laws' technique for developing negatives in the air and had them ready for printing upon landing. On their own initiative they photographed the defences of the Isle of Wight and the Solent at an altitude of 5,000 feet.[55]

Two months prior to the commencement of conflict the culmination of British efforts to clearly demonstrate aviation's military purpose and significance – to a diverse crowd of senior government, media, public, as well as a host of foreign attachés – took place at Netheravon under the designation Concentration Camp of the Military Wing of the RFC. The aerospace weekly publication *Flight* covered in detail the various demonstrations to include an aerial review with as many as 30 machines flying at one time. Over 700 officers and men of the Headquarters Flight, Aircraft Park, Nos. 2, 3, 4, 5, and 6 Squadrons, and a detachment of the Kite Section were present.[56] The idea of bringing the squadrons together seems to have originated with Sykes, whose arrangements were admirable in their detailed forethought and completeness. New-found capabilities for British military aviation were revealed as aeroplanes armed with machine guns, rifles, and bombs demonstrated aerial operations against targets on the ground. This was combined with insight into experimental work being conducted in wireless

An aerial photograph acquired during the RFC's Concentration Camp at Netheravon and published in *Flight* shows the configuration of the camp. Note the identification of items of interests ('targets') on the photograph. (*Flight*, 14 July 1914)

telegraphy. It was mentioned that messages as far as 20 miles away were being sent and received by wireless officers while flying in the aeroplanes.[57] Aerial photography was also demonstrated. Flight-Sergeant Laws flew as aerial observer in a Henri Farman (HF 7, 'Longhorn') with the Watson Air Camera mounted on the nose. The sortie ended when the engine motor cut off and the plane crashed. Not to be dissuaded, two days later Laws took off as observer in a Maurice Farman (MF 11, 'Shorthorn') and photographed a military review and parade in progress. What caught Laws' attention as the photographs were developed was the difference in light reflection on the adjoining grass from personnel walking through the area. It spurred him to further explore the potential that aerial photographic interpretation could offer.[58]

Final pre-war feelings towards British aviators varied within certain social groups. Despite the amazing effort underway to validate aviation's role with capabilities such as aerial photography, the prevailing attitude among military circles was mixed: 'If that young fool likes to get himself killed, let him do so.'[59] With each demonstration of aerial photography, 'those who came to scoff remained to pray.'[60] *Flight* commented that RFC morale was high. The article went so far as to claim that 'One is apt to measure military strength by numbers; but those who are acquainted with the art of war are well aware that morale is of far greater importance than numerical superiority, and organisation than either of these factors.'[61] The conflict in the coming weeks provided ample opportunity to test the veracity of the journal's statement.

CHAPTER 2

August to Late 1914: The Evolution of Aerial Reconnaissance as a Force Multiplier

FIRST AERIAL RECONNAISSANCE SORTIES ON THE CONTINENT

French aviation was put to the test as the war commenced. The French used their aéroplanes for daytime reconnaissance and dirigible fleet for night-time observation. Each armée included a flight of aéroplanes. Order of battle for Aviation Militaire comprised Ième Armée (six escadrilles), IIème Armée (four escadrilles), IIIème Armée (four escadrilles), IVème Armée (two escadrilles), Vème Armée (five escadrilles) and two BLC escadrilles.[62] The first French wartime reconnaissance mission was accomplished by a Maurice Farman (MF) 7 from escadrille MF 2.[63]

Like the French Aviation Militaire, the British Military Wing prior to the commencement of the war was designed for aerial reconnaissance, with the squadron assuming the role of the basic British aviation unit.[64] Five squadrons had been established prior to the war. On 31 July 1914, total operational aircraft from RFC Squadrons No. 2, 3, and 4 numbered 22 BE 2a, five Blériot XI monoplanes, and six Henri Farman F.20. An additional 24 aeroplanes were to be added prior to deployment to France.[65] The RFC inventory of cameras comprised five 'Panros' press cameras fitted with 6 Homocentric lenses.[66] It was indicative of the way aviation evolved between the French and British that a common airframe was used in the first years of the war. Both Blériot XI and HF 20/F.20 were flown by French and British pilots. Of the two, Blériot XI was a more recognised airframe thanks to the first crossing of the English Channel in 1909. However, as an observation platform, the Blériot design was not ideal. The broad wings were an obstacle to observers.[67]

The British had yet to adopt homogeneous airframe types for each squadron. It was not yet practical due to the embryonic condition of the British aircraft industry. Their most enduring aeroplane, the BE 2a, had a maximum speed of 65mph and took approximately 35 minutes from take off to reach 7,000 feet.[68] However, despite the diversity of aeroplanes that complicated ground maintenance, squadron operations were effectively maintained. Sorties did not include formation flying. The standard procedure called for each aeroplane to act independently.[69]

The German adversary was no stranger to the potential of aerial reconnaissance. This resource had strong advocates within the senior headquarters, the German High Command (*Oberste Heeresleitung* (OHL)). During the first month of combat, successful aerial reconnaissance gave the German armies an advantage in their initial operations on the Eastern Front. Each active German corps and headquarters (*8. Armee*) was assigned a six-aeroplane *Feld Flieger Abteilung*, comprised of monoplanes (*Tauben*) and biplanes to provide timely information to German battlefield commanders. Priority for the mission and command of the two-seater aeroplane went to the observer, not the pilot. The Germans also employed Zeppelin craft for longer-range strategical reconnaissance.[70] German aerial reconnaissance gained favour in the war's first major battle at Tannenberg in eastern Poland. The OHL was personally briefed by Flieger Abteilung crewmembers as the Russians commenced the attack. German reconnaissance provided locations of manoeuvring Russian units and aided planning for appropriate countermoves.[71] Intelligence information proved critical in this first major battle of the war. Success came to the Germans when they intercepted Russian radio transmissions containing exact force disposition and locations. Aerial reconnaissance had reinforced German command decisions, but not as decisively as the radio intercept. Tannenberg became the first battle in history where interception of enemy radio traffic played a decisive role.[72] Intercepts included Russian operational orders and, more significantly, the organisation and destination of the Russian Second Army. The Russian manoeuvres were successfully countered.[73] Modern intelligence technology had validated itself on the battlefield.

During the first days of combat on the Western Front, the Germans employed dirigibles for aerial

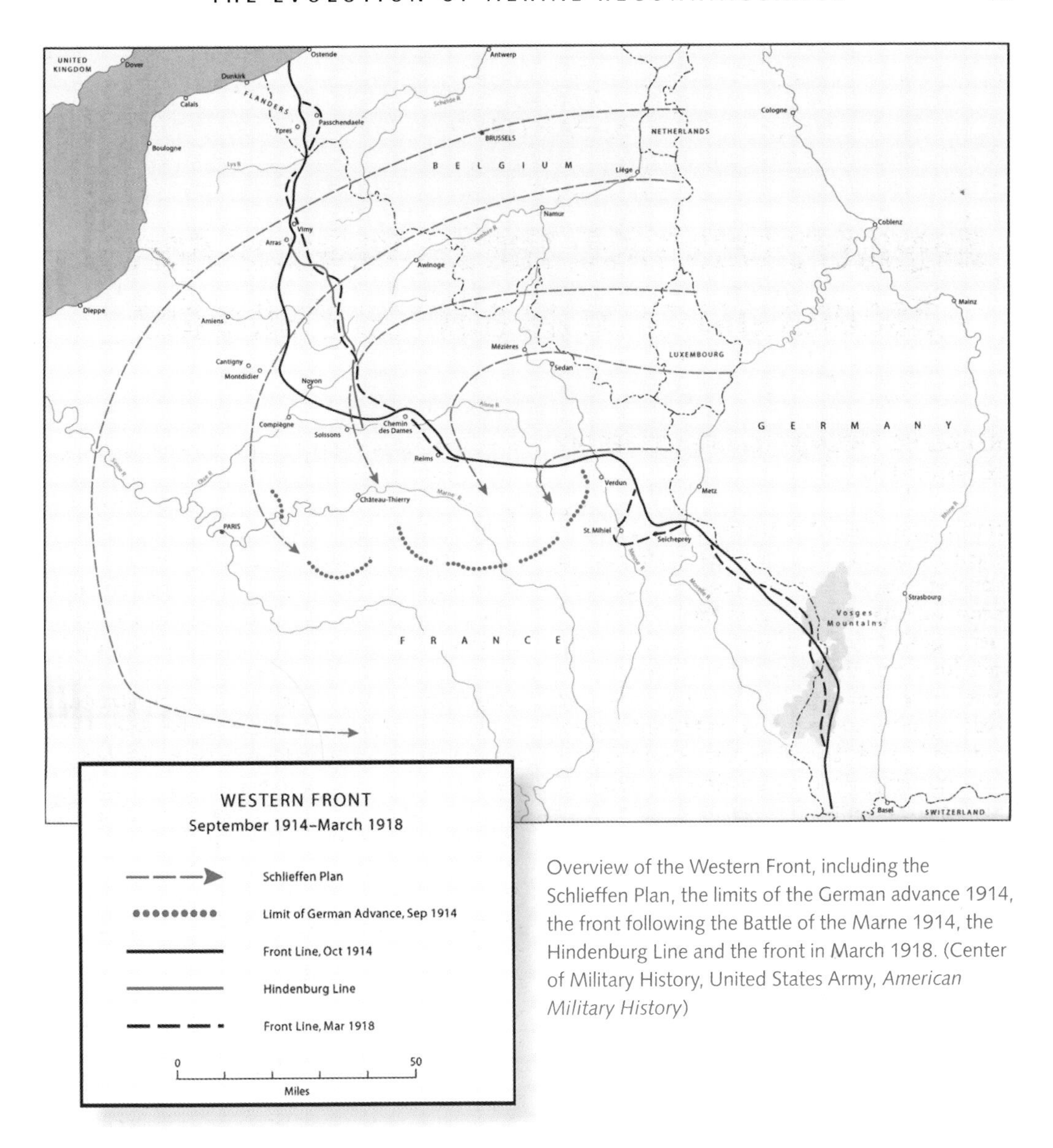

Overview of the Western Front, including the Schlieffen Plan, the limits of the German advance 1914, the front following the Battle of the Marne 1914, the Hindenburg Line and the front in March 1918. (Center of Military History, United States Army, *American Military History*)

reconnaissance and bombardment. Their first aerial bombardment sortie targeted Liège on 6 August 1914. During this sortie, a dirigible was hit by cannon and returned to the German base with damage. Another dirigible was hit over Alsace while on a reconnaissance mission. A third, dirigible, ZVIII, was shot down by French artillery.[74]

On 13 August, the RFC under Brigadier-General Sir David Henderson (Command of the RFC in the Field) deployed to France in support of the British Expeditionary Force (BEF). The first unit to make the crossing was No. 2 Squadron, with the first aeroplane departing early that morning and arriving at Amiens approximately two hours later. The pilots flew in line astern at two-minute intervals across the English Channel. Each was fitted with an inflated bicycle tyre for a life preserver should the aeroplane had to ditch.[75] The departure of other RFC aeroplanes that day was not so fortunate. Several aircraft of No. 4 Squadron were damaged while following their leader in a forced landing on a ploughed field.[76] The following RFC squadrons and aeroplanes that flew to France on 13–15 August 1914 were No. 2 Squadron, 12 BE 2a's; No. 3 Squadron, 6 Blériot XI's, a Blériot Parasol, 4 F.20s; No. 4 Squadron, 11 BE 2a's; No. 5 Squadron, 4 F.20s and 3 Avro 504s.[77] Those were the aircraft that arrived; others were lost on the way. Two days later, a British pilot met an unusual welcoming committee while landing near Boulogne-la-Grasse. Lieut. R.M. Vaughan of No. 5 Squadron was arrested by the French and held in prison

The first RFC aerial reconnaissance in wartime was flown by Captain Philip Joubert de la Ferté. (Sir Philip Joubert de la Ferté, *The Fated Sky: An Autobiography*)

for three days while local officials tried to determine his true nationality and purpose. Lieut. Vaughan was able to rejoin his squadron on the eve of the Battle of Mons.[78]

Initial RFC attempts to commence aerial reconnaissance operations against the Germans were precluded by weather. The heat of summer brought thunderstorms, mists and haze. Flight logs described these conditions as 'unsuitable for reconnaissance'.[79] Aerial observation did not operate from airfields as understood in today's terms. Aeroplanes and support were totally mobile, operating from any suitable area, known as the 'landing ground'. Aerial reconnaissance operations got underway on 19 August with the first RFC aerial reconnaissance flight flown by Captain Philip Joubert de la Ferté, No. 3 Squadron, and Lieut. Gilbert 'Gib' Mapplebeck from No. 4 Squadron. Joubert de la Ferté flew a Blériot XI-2 (without observer) while Mapplebeck took off in a BE 2a from Maubeuge (on the France-Belgium border, 15km south of Mons) that morning. Mapplebeck flew to the north while Joubert de la Ferté was ordered to inspect the Belgian country west of Brussels and report on any evidence of enemy troops. Shortly after takeoff, the pilots struggled with cloudy weather and a general unfamiliarity with the region, resulting in their getting lost. Both considered it 'rather bad form to come down and ask people the way', but such diffidence was soon overcome.[80] Joubert de la Ferté recalled:

> I wandered round Western Belgium for some time and then seeing a large town over which the Belgian flag seemed to be flying. I landed on the parade ground. I discovered that this was Tournai, still in Belgian hands, and I was given a most excellent lunch by the commandant of the garrison. Leaving an hour later, I lost myself again very quickly. Finally I recognised the Bruges Ship Canal, flew south from there, and landed near Courtrai. Here my reception was not at all cordial. The local police threatened to put me in jail. I was saved by the kindness of a little North of Ireland linen manufacturer, who was visiting Courtrai, where there are a lot of spinning mills. He placed a Union Jack on my aircraft, and the tone immediately improved. I got what I wanted, which was petrol, and the directions how to find my way back to Maubeuge, and flew off very thankfully, completed my reconnaissance, and landed back at Maubeuge at 5.30.[81]

As for Lieut. Mapplebeck in the BE 2a, he flew over Brussels itself without recognising it. The result of both reconnaissance missions was negative. Joubert de la Ferté and Mapplebeck reported with assurance on where the Germans were not, but had nothing to say about where they were.[82] The time had come for aerial observers to be included on reconnaissance sorties.[83] Gib Mapplebeck's aerial adventures included a narrow escape in 1915. While conducting aerial reconnaissance over German lines near Lille, he experienced engine problems, landed, and two German soldiers literally grabbed one of the wings. Mapplebeck escaped into Lille wearing peasant clothing that he found on the way. He then persuaded a French businessman to cash a London cheque that provided him with French notes bearing a German stamp. Mapplebeck proceeded to buy a new suit and walked from Lille through Belgium to the Dutch border. He obtained passage to London and reported in at Farnborough. The Air Ministry provided him with a new aeroplane to fly to France. When he showed up at the squadron in France, Mapplebeck reported for duty 'just as though nothing unusual had happened'.[84]

That same day, the French committed their *dirigeable* (dirigible balloon) forces to reconnaissance missions. The French dirigeable *Fleurus* became the first Allied airship

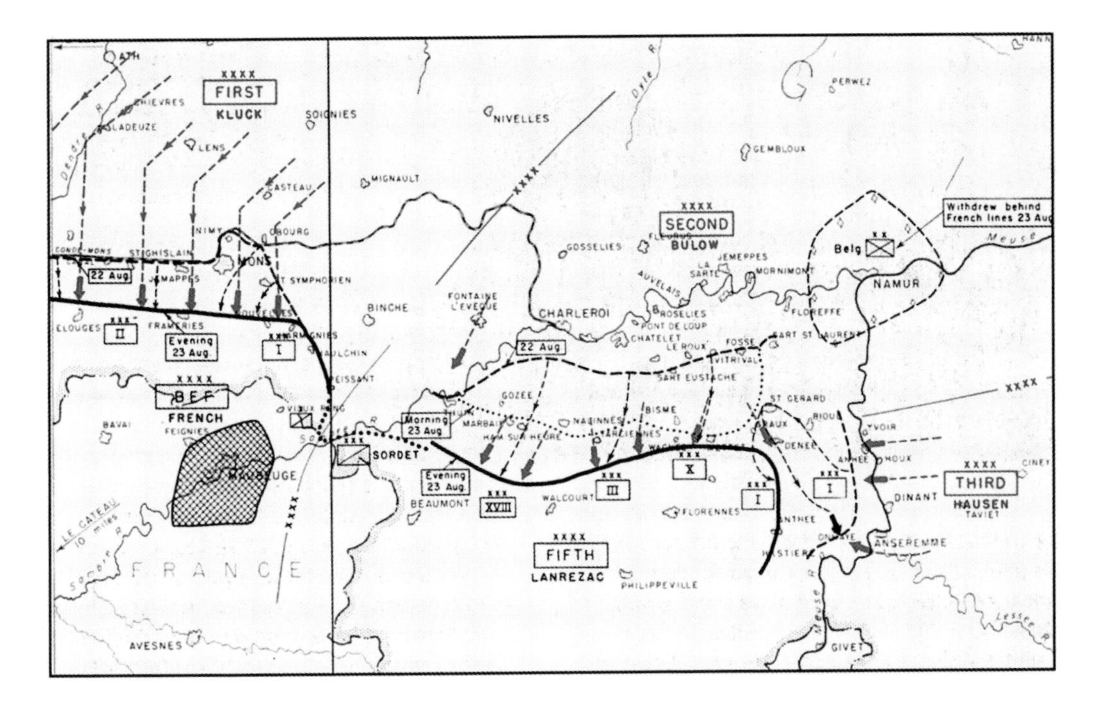

The German advance on Paris 22–23 August 1914. (*West Point Atlas*, courtesy Department of History, US Military Academy)

to fly over Germany, successfully flying over the Saar and Trèves while on a night reconnaissance sortie.[85] Early French dirigeable flights suffered more from friendly fire than from German anti-aircraft artillery. Two dirigeables were damaged. One French dirigeable, the *Dupuy de Lome*, suffered casualties (the pilot was killed) as well as being shot up fairly severely while heading toward the French city of Rheims. Following these mishaps, the French delayed dirigeable operations until April 1915.[86] For the remainder of the war, French dirigeables played a secondary role serving armée objectives in 1916 and were eventually transferred to the navy.[87]

MONS

On 20 August British cavalry were on the move in Belgium. They pushed forward as far as Binche (16km east of Mons) without encountering the enemy. At the same time, RFC aerial reconnaissance was ordered to find the German army, pinpoint its position, and estimate its strength. The RFC discovered elements of the German army heading through Louvain (approximately 25km southeast of Brussels). Aerial observation estimates of strength proved too difficult since German army columns went beyond the visual horizon. That day the German army moved into Brussels. The main echelon was heading toward France.[88] On the morning of 21 August, aerial reconnaissance was incapable of operating owing to ground mist. Reconnoitring squadrons and patrols were pushed out toward Soignies and Nivelles in southern Belgium.[89] Once the weather cleared, RFC reconnaissance commenced and quickly acquired significant observations of German cavalry and artillery manoeuvring southeast of Nivelles.[90]

Perhaps the most significant day in RFC aerial reconnaissance history was 22 August. British cavalry engaged the Germans for the first time near Soignies.[91] That day the RFC flew twelve reconnaissance sorties from Maubeuge and reported on extensive German manoeuvres and the presence of large masses of troops. Field-Marshal French's Director for Intelligence, Colonel George Mark Watson Macdonogh, recalled in 1922:

> He [Henderson] consequently sent out a reconnaissance on the 22nd, whether one or more machines I do not remember, but he certainly let me know that day that a long column, correctly estimated at one corps, had been seen moving along the Brussels–Ninhove road, which on reaching Ninhove had bent south-westwards towards Grammont. Putting two and two together we came to the conclusion that this was the II Corps and the report showed very clearly that our position on the Mons Canal was likely to be outflanked.[92]

Lieutenant Edward L. Spears, the British liaison officer at the French Vème Armée headquarters echoed the drama of the moment: 'It was bound to outflank us. Now we knew. No possible doubt could subsist. The German manoeuvre stood fully revealed.'[93] This critical issue was personally brought to the attention of Field-Marshal French by Brigadier-General Henderson and Lieut-Col. Sykes. Field-Marshal French was initially sceptical until he received further verification from his French counterparts.[94] The RFC's aerial reconnaissance that came back with this information was probably the most fruitful of the whole war.[95] The information gleaned enabled the British forces to keep ahead of the German manoeuvre and averted a catastrophe.[96]

The first losses from aerial combat then occurred. An Avro 504 from No. 5 Squadron was shot down while performing aerial observation. The downed aeroplane provided the Germans with their first positive confirmation that British forces were now engaging them.[97] The significance of the occasion was recalled by German commander General Alexander von Kluck, observing that a British reconnaissance aeroplane from Maubeuge had been shot down.[98] Aerial combat now commenced in earnest. One RFC pilot, Lieut. Louis A. Strange, fitted a Lewis gun on his F.20 and took off in pursuit of a German aircraft discovered in the area. However, it was to no avail since the German simply out-climbed the Farman.[99] Additional opportunities for the first aerial combat occurred when two British BE 2as armed with Lewis guns chased an Albatros biplane for 45 minutes.[100]

With the campaign now underway, the benefits of aerial observation were beginning to gain appreciation among senior commanders. In his post-war reminiscence, Field-Marshal French described his view of aerial reconnaissance during the German advance on Mons:

> The opening phases of the Battle of Mons did not commence until the morning of Saturday, 22 August. The intelligence reports which constantly arrived, with the results of cavalry and aircraft reconnaissance, only confirmed the previous appreciation of the situation, and left no doubt as to the direction of the German advance; but nothing came to hand which led us to foresee the crushing superiority of strength which actually confronted us on Sunday, 23 August. This was our first practical experience in the use of aircraft for reconnaissance purposes. It cannot be said that in these early days of the fighting the cavalry entirely abandoned that role. On the contrary they furnished me with much useful information. The number of our aeroplanes was then limited and their power of observation were not as developed or accurate as they afterward became. Nevertheless, they kept close touch with the enemy, and their reports proved of the greatest value. Whilst at this time, as I have said, aircraft did not altogether replace cavalry as regards the gaining and collection of information, yet, by working together as they did, the two arms gained much more accurate and voluminous knowledge of the situation. It was, indeed, the timely warning they gave which chiefly enabled me to make speedy dispositions to avert danger and disaster. There can be no doubt indeed that, even then, the presence and cooperation of aircraft saved the very frequent use of small cavalry patrols and detached supports. This enabled the latter arm to save horse flesh and concentrate their power more on actual combat and fighting, and to this is greatly due the marked success which attended the operations of the cavalry during the Battle of Mons and the subsequent retreat.[101]

French was a career cavalryman. However, his next comments reflect the military intelligence transformation: 'At the time I am writing, however, it would appear that the duty of collecting information and maintaining touch with an enemy in the field will in future fall entirely upon the air service, which will set the cavalry free for different but equally important work.'[102]

The Battle of Mons took place on Sunday, 23 August. The weather early that morning was misty

Aircraft recognition proved challenging for both sides in the first days of the war, resulting in incidents of friendly fire. A Maurice Farman (MF) 11 is observed flying low over infantry. (SHD, from Christienne and Lissarague, *A History of French Military Aviation*)

and raining but gave way prior to noon to fair weather for flying.[103] The German attack commenced against British positions at Mons while pilots and observers flew over and behind the battlefield looking for enemy movements and locating artillery batteries.[104] A furious battle ensued all along the filthy Mons-Condé Canal. The German artillery mercilessly battered the BEF's lines with shell and shrapnel. The British II Corps under General Horace Smith-Dorrien fired round after round from their Lee-Enfield rifles into the advancing German infantry in close formation.[105] Meanwhile the RFC headquarters initiated their departure from Maubeuge south to Le Cateau.

That afternoon the RFC made further reconnaissance toward Charleroi (about 40km south of Brussels) and ascertained that at least two German army corps (Guard Corps and the Guard Cavalry Division) were attacking the French Vème Armée on the line of the Sambre. When the observers returned in the evening, they presented grave news that the French armée centre had been driven back.[106] Field-Marshal French learned in the evening when he returned to his headquarters that the French forces on his left flank were pulling back. Since the BEF was still advancing, this news, coupled with further ominous reports from RFC assets that German movements were putting his forces at risk, made it clear to Sir John French that the BEF must fall back as well.[107] The British now commenced the famous 'retreat' from Mons and headed south.[108]

The RFC observers drafted their in-flight reports on enemy movements and forwarded information to commanders immediately upon landing. Confusion reigned as the battle heated up. Captain Joubert de la Ferté recalled, 'We were rather sorry they [the British soldiers arriving] had come because up till that moment we had only been fired on by the French whenever we flew. Now we were fired on by French and English. To this day I can remember the roar of musketry that greeted two of our machines as they left the aerodrome and crossed the main Maubeuge-Mons road, along which a British column was proceeding.'[109]

Aeroplane recognition, or rather lack of it, posed a significant risk to the early aviator. Infantry on both sides were not accustomed to aeroplanes and their role. Establishing and assigning distinguishing marks of nationality became a necessity to safely operate at lower altitudes. 'Union Jacks have been put on the under surfaces of the lower planes. They are not large enough and larger ones are necessary. The French have a blue spot, surrounded at an interval, by a red ring. The Germans have a black cross on some of their machines.'[110] Confusion over identity eventually led to the British adopting the French-style roundel with a blue outer circle and red spot.

THE GERMAN ADVANCE ON PARIS

When the RFC departed from Le Cateau on 24 August and eventually arrived at Melun (near Paris) on 4 September, the headquarters and aircrews had made no less than ten moves.[111] RFC ground support involved during this operation was fondly remembered by Major-General Sir Sefton Brancker (Henderson's deputy in the Military Aeronautics Directorate for controlling all RFC units operating in the United Kingdom) for the diversity of available kit:

> The line of march of the R.F.C. looked like a circus. Except for a fair sprinkling of 30-cwt. Leylands and the Daimlers every vehicle was different and still

RFC support vehicles during the retreat from Mons. (Sir Frederick Sykes, *From Many Angles: An Autobiography*)

> wore trade markings. Among them were two Maple furniture vans and a lorry with a quaint steel body for liquid refuse which belonged to No. 2 Squadron. But they were all dominated by a large red van with 'THE WORLD'S APPETISER' written on it in huge gold letters and which belonged to No. 5. As can be imagined, the march discipline and maintenance of such a mixed lot were no easy job.[112]

Equally challenging was a peripatetic aerial operation – as described by Lyn Macdonald:

> For the past week the Flying Corps, like the Army, had been moving gypsy-like round France sleeping wherever chance had led them, in barns, in châteaux, in lorries, under hedgerows, even occasionally in hotels. Every day they had spent many hours in the air searching for the enemy and half as long again searching to find their own base, for there was no guarantee that the makeshift airfield the patrols had left would still be in the same place when they returned. Standing orders were not much help – 'Fly approximately twenty miles south and search for aircraft on the ground.'[113]

Despite the confusion, the RFC was demonstrating its potential. On 25 August Major Hubert Harvey-Kelly and two other pilots from No. 2 Squadron forced a Rumpler Taube to land near Le Cateau.[114] Lieutenants Borton and Small from No. 5 Squadron flying a Henri Farman demonstrated aviation's support to the British command in the field during a hair-raising confrontation with German forces. In an effort to deliver a critical message to General Sir Douglas Haig, Borton and Small landed between the firing lines of the ensuing battle and found a cavalry patrol that took their message to Haig. Upon returning to the aeroplane, they started up the engine and flew away with two German Uhlan cavalrymen bearing down on them. The quickening pace of battle created a greater challenge for successful communication within the British Corps. Every possible means of liaison, including the creation of new links with aviation assets, was being put to a critical test.[115]

On Friday 28 August, Général Commandant en Chef des Armées, Général Joseph Jacques Césaire Joffre, asked Field-Marshal French to undertake air reconnaissance on the western flank of the Allied forces covering French territory.[116] While Joffre went to the RFC for answers, the *Vlème Armée* 2ème Bureau commanding officer, Commandant Dutilleul, had acquired a copy of the directive to the German commander, General von

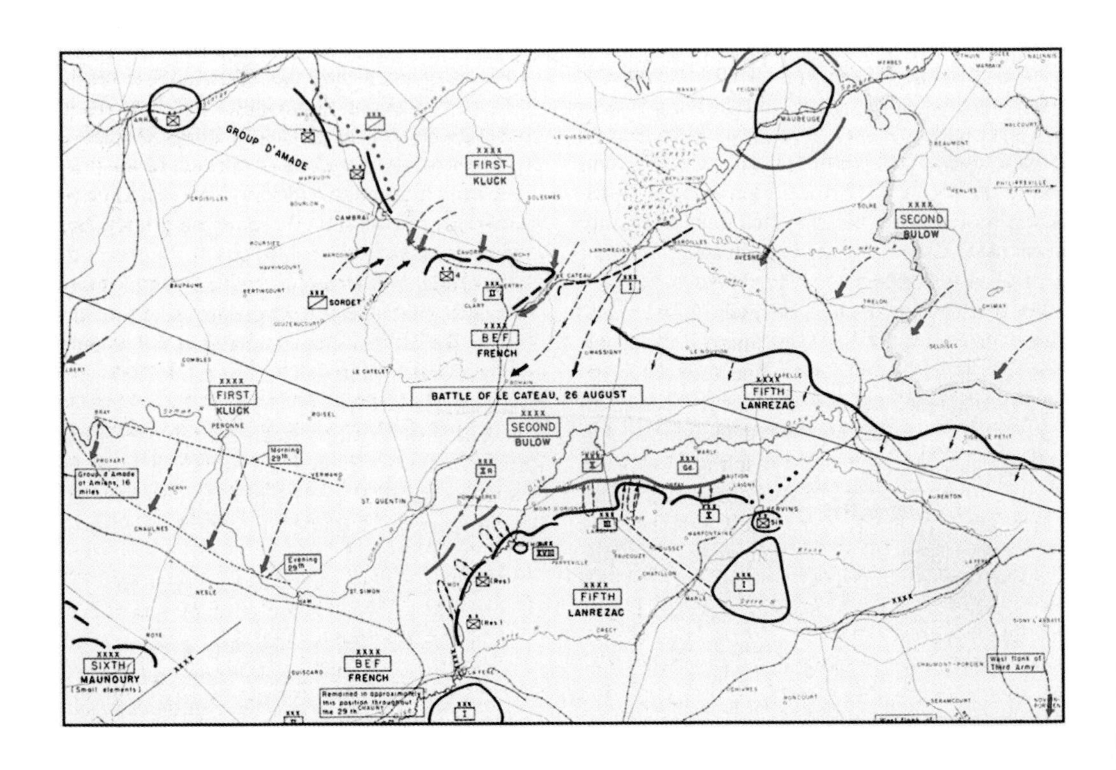

Battle of Le Cateau and Battle of Guise, 26–29 August 1914. (*West Point Atlas*, courtesy Department of History, US Military Academy)

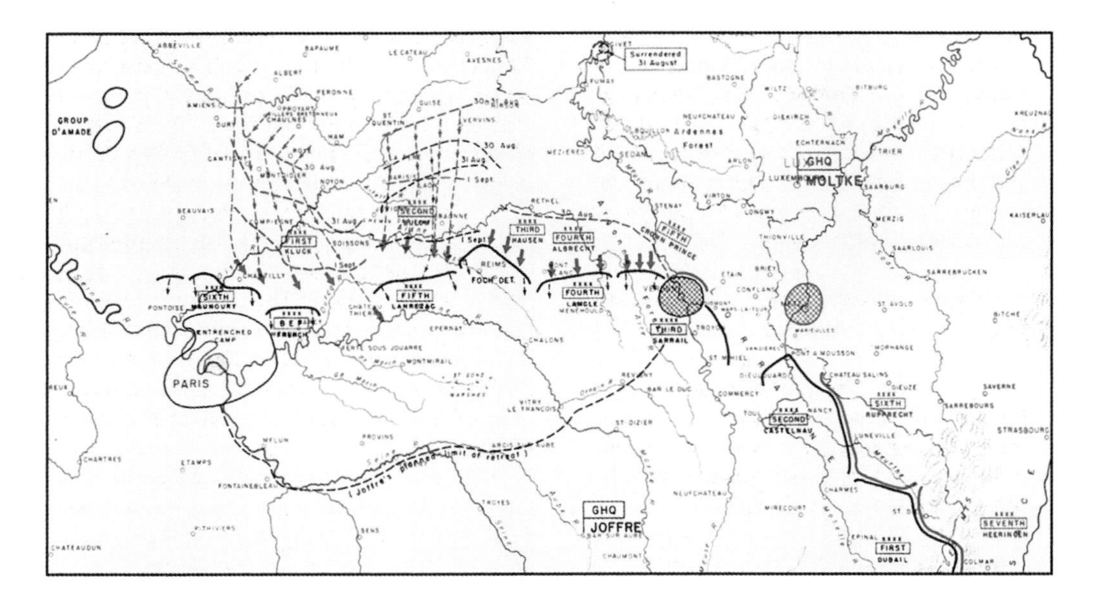

30 August–2 September 1914, Battle of the Marne. (*West Point Atlas*, courtesy Department of History, US Military Academy)

Kluck, to commence his manoeuvre. Despite access to this incredible source, Dutilleul remained sceptical about reports on von Kluck's manoeuvre from Capitaine Bellenger's REP 15 and MF 16 escadrilles.[117] Meanwhile, RFC aerial reconnaissance during the day showed German columns sweeping southward over the Somme between Ham and Peronne, advancing on the French VIème Armée and between the Oise and Somme west of Guise near St Quentin. Aerial observation reported several sightings of devastation as the Germans continued south.[118] In the meantime, von Kluck continued to employ his own aerial reconnaissance near Albert-Doullens-Amiens, but his pilots were unsuccessful in locating the retreating British forces.[119]

At this moment, unbeknownst to the Allied commanders, General von Kluck deviated from the original strategy of advancing directly on Paris and decided to go after the retreating Allied armies. Von Kluck's solution was to annihilate the French armies in the field without enveloping Paris. On the evening of 30 August, a message from the overall German commander, General Karl von Bülow, ordered von Kluck's manoeuvre to help von Bülow 'gain the full advantage of victory' over the retreating French Vème Armée. Von Bülow's request complemented von Kluck's perspective and the decision was made to move to the southeast toward Noyon and Compiègne.[120] The next day RFC aerial observation began to see signs of the move. Despite the constant flux of RFC operations as a result of continued mobility, RFC sorties over the Compiègne region managed to detect signs that the Germans were changing direction.[121] On the afternoon of 31 August, RFC pilot Captain E.W. Furse discovered the lead German cavalry corps spearheading the German army in a new direction. A subsequent sortie from RFC No. 4 Squadron flown by Captain Pitcher and Lieut. Soames confirmed the manoeuvre.[122] This, combined with reports from French cavalry, alerted the Allied commanders that something significant was in progress. The Germans had reached the limit of their western advance and were now heading southeast.[123] The French Vème Armée commander, Général Charles Lanrezac, and VIème Armée commander, Général Michel-Joseph Maunoury, were provided the information at the most critical time, saving their forces from a potentially decisive German manoeuvre.[124] The RFC was achieving tremendous credibility at this climactic moment.

Throughout the campaign, the French were equally energetic in maintaining aerial observation on the advancing German forces. Unfortunately for their aviators, they not only had to find the Germans, but also had to convince their seniors of what was observed. The French GQG had a variety of intelligence sources to draw from in determining the location and direction of the German advance. French agents had reported that the German campaign plan called for marching from Brussels towards southwestern Belgium, eventually heading to the Oise River. However, by 1 September the direction of the German advance did not validate that intelligence source. Aerial reconnaissance kept confirming the southeast advance. At this critical moment, French Armée headquarters staff became reticent, not following up on the new information,

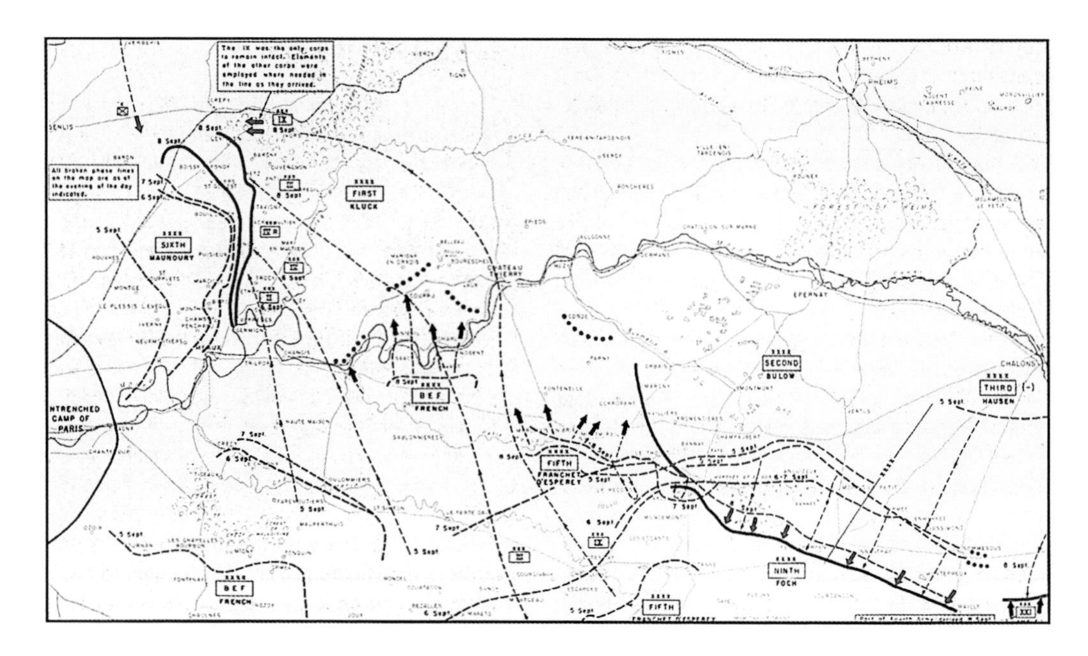

5–9 September 1914, Battle of the Marne. (*West Point Atlas*, courtesy Department of History, US Military Academy)

apprehensive because it contradicted the prevailing opinion of the GQG. With indecision, critical time was lost. French staff began withholding information, afraid to admit possible mistakes. Capitaine Bellenger, commanding the air section of the French VIème Armée, became increasingly alarmed at this state of affairs. He quickly sent one of his observers, Lieutenant Wateau (a lawyer by training), to the GQG to make the case for their aerial observation sighting of the German manoeuvre in progress. Wateau successfully explained the case and at the same time took credit for the discovery.[125]

The Allies' aerial observation reporting on the German manoeuvre received further substantiation from a map retrieved from a German officer.[126] Armed with this more convincing information, the Allies gained an advantage. German forces were now surprised when they came into contact with British forces in unsuspected locations, such as Compiègne. With the increase in fighting, the British manoeuvred to a more strategic and defensible position.[127] Meanwhile, the RFC demonstrated a new role in the campaign with initial attempts at aerial bombardment. They dropped two bombs on two columns of enemy troops at a crossroads, causing confusion and a stampede.[128]

Allied aerial coverage continued to monitor the enemy's southeastward march. Aviator confirmation of German movement toward Noyon alerted the headquarters that the enemy was heading towards Soissons on the Aisne. The BEF GHQ set up north of Paris at Dammartin-en-Goële received progress reports on several German columns from RFC patrols.[129] The French were equally successful in their observations. On the evening of 2 September, French aerial reconnaissance confirmed the presence of two large German encampments (German *1. Armee*) between the Oise and Ourcq Rivers.[130] It was a brief respite for an increasingly fatigued German force. General von Kluck ordered a further march hoping to envelop the BEF. Again, the British were able to escape the trap, crossing the Marne just in time.[131]

Corporal Louis Breguet, an early pioneer in aéroplane design, accomplished one of the key observations of the day. Flying his own prototype Breguet AG 4, Breguet spotted the German army moving to the east of Paris. The reconnaissance suggested von Kluck was moving southeast to eliminate the British corps and surprise the French forces moving to the rear. Subsequent aerial observation by escadrilles REP 15 and MF 16 confirmed the information. The French VIème Armée was quickly dispatched to attack the German's flank.[132] Breguet's report further confirmed General von Kluck's intentions were shifting. Capitaine Bellenger complemented the reports by personally flying in the sector. His challenge continued to be with the GQG. This time the intelligence assessment did not confirm the aerial observation. French 2ème Bureau sources showed von Kluck was working according to the Schlieffen Plan and was manoeuvring to encircle Paris.[133] Bellenger was now forced to call on additional aviators to testify to the ongoing observations, while trying to maintain an aggressive aerial observation sortie rate.[134] The Germans under von Kluck also continued their aerial reconnaissance effort. On 2 September, they were ordered to monitor Allied forces across the Marne.[135]

Capitaine Bellenger's luck changed when he made contact with the British Third Army. He finally gained an audience that listened to what he had acquired from the aerial observations. He subsequently went to Général Joseph Simon Gallieni, commander of French forces within the region of Paris, to emphasise further the meaning of observations from aerial reconnaissance at this critical time. Gallieni proved invaluable in accepting Bellenger's recommendations and taking the time to persuade Général Joffre of their value.[136] Bellenger always retained a high regard for Général Gallieni: 'He is one of the rare generals that believe in aviation; he is a man of decision and is not afraid of responsibilities.'[137] Général Gallieni's success with Général Joffre helped to direct French manoeuvres to counter the Germans at the Marne. This significant event sounded the final death knell of the German Schlieffen Plan of 1905.[138]

RFC aerial reconnaissance of 3 September also substantiated French observations on the German manoeuvres. The BEF Intelligence Director, Colonel Macdonogh, was profuse in his praise: 'A magnificent air report was received disclosing the movements of all the Corps of the German *1. Armee* diagonally South East across the map toward the Marne.' The resultant battle between the French VIème Armée under Général Maunoury and the German army under General von Kluck helped thwart the German offensive strategy. Key to the battle was the effective position of the BEF, which permitted exploitation of a gap between the German armies in the region, eventually forcing a German retreat.[139]

Despite the positive response to aerial photography, other intelligence sources performed an even more key role. Allied radio intercepts of German transmissions describing their manoeuvres were critical in the overall analysis. This was partially down to German 'lack of discipline in radio operation'. By 4 September French intelligence confirmed: 'The German First Army, neglecting Paris and our Sixth Army, before which nothing has shown itself, continues its march toward the Marne.' German fatigue and logistical shortfalls were also discerned. General von Kluck's order to withdraw was picked up by the French radio intercept analysts.[140] Général Joffre personally learned of the Russians' defeat at Tannenberg through a radio intercept of a German transmission that described the destruction of three Russian corps, the capture of two corps commanders and 70,000 other prisoners, while announcing triumphantly, 'The Russian Second Army no longer exists.'[141]

THE BATTLE OF THE MARNE

At the Marne, the Germans attempted to swing around the western end of the Allied line and envelop it. Concurrently, they planned to break through the Allied centre, forcing the remainder of the western half of the line back to Paris, thereby completing the process of envelopment and creating a second Sedan of 1870 on a grand scale. The French had a similar plan – flank around the west end of the German line, break through the centre, and split the German forces. The two flank attacks began on the plain north of Paris, while the two attempts to break the enemy's centre were staged on the low plain of Champagne.[142]

A French aerial observation report reflected extensive activity and a concern for France's future:

> Came down at Tours-sur-Marne to report to Général Foch; then at Heitz-le-Maurupt, to the staff of Langle de Cary. Second reconnaissance on Somme-Suippe and Cuperly; landed at Saint-Dizier. Today's was the maximum speed of retreat attained in one day. How far are we to retreat? To the Loire, some say, to the Central Massif, according to others. It is terrible![143]

Meanwhile, French radio intercept units were providing extensive reporting. In the course of 14 days during the German advance on the Marne, the French radio intercept service picked up approximately 350 radiograms from the cavalry corps under General Johannes Georg von der Marwitz alone. The French radio intercept service did not fail to note the movement of the German First Army towards the north in order to avoid being outflanked by French Général Maunoury's VIème Armée.[144]

RFC aerial observation was also accelerating by keeping track of every German corps of the First Army. Information was forwarded to the GQG and marked on the battle map. The whole of the German right wing had now thrust itself into a vast arc, with the French VIème Armée on one side of it, the IIIème and IVème Armées on the other, the Vème and IXème Armées and BEF at the base. The time to attack, Joffre decided, had come.[145] The next day French reconnaissance kept seniors informed amidst a flurry of activity:

> Saint-Dizier, Reims, Fismes, Bergeres-les-Vertus, where we descended to report to Général Foch, who is in command, it is said, of three army corps, forming the IXème Armée which is from this time to come between us and the army of Franchet d'Esperey on our left. We saw four German army corps today marching in order of battle across the camp of Châlons and the neighbourhood of Reims.

> What feelings it aroused! But what a splendid spectacle it was! I dropped two bombs on a large bivouac and saw dense smoke rising from the very centre of it. The bombs had gone home. We then went on to Saint-Remy-en-Bouzemont near Vitry-le-François, where we descended; we lunched off a crust of bread, then went on to Brienne-le-Château. I heard afterwards that my bombs had wounded a captain and three men.[146]

Early on, escadrilles were often shifted around the battlefield to support the ongoing battle. Amongst the initial Allied aerial inventory, the Blériot XI was a favoured airframe because it could be rapidly disassembled, placed in closed containers and reassembled for transport to a new sector. In a two-month period (September to October 1914) escadrille BL 10 moved to eleven locations along the front. It was not without hazard. On 5 September escadrilles supporting Général Foch's IXème Armée were ordered to Mailly-Champagne (near Reims) to support operations in the sector. At seven in the morning, escadrilles BL 3 and BL 10 assigned to the IXème Armée were operating from the aerodrome at Belfort in eastern France. The 12 Blériot XI's at Belfort had a support unit of motor-drawn trailers readied to provide necessary transport should the aéroplane encounter maintenance problems. The fragile aéroplanes were wearing down and required delicate handling. Exposure to the cold night air and rain caused structural problems as the airframe wood expanded and canvas stretched.[147] However, the urgency of the moment required that the escadrille reach Mailly (approximately 250km to the northwest) in the shortest time possible.[148] The Blériot XI escadrille station commander, Capitaine Constantine Zarapoff, set off from Belfort by road with supplies and equipment while his pilots took off with their machines for the aerodrome at Mailly. One pilot was killed while taking off. Two other pilots then stayed behind to bury him. Of the twelve aéroplanes in the escadrille, nine eventually landed at Mailly.[149] The stresses on the older airframes were beginning to take their toll. New airframes were ordered to handle the demands of both mobility and endurance. Conversion to more effective aircraft such as Morane-Saulnier Ls, Caudron G.3s and G.4s would take place in late 1914. For the moment, they had to rely on fragile aéroplanes enduring greater stresses.[150]

On 6 September the British reorganised their aerial reconnaissance resources to reflect the current battle situation. Sir John French reallocated three of the RFC aeroplanes directly to the British I and II Corps for tactical reconnaissance – the first time that aircraft were decentralised under Corps authority. The realignment enabled the British commanders to observe the retirement of the German II Cavalry Corps on Coulommiers (due east of Paris) and established that the French Vème and VIème Armées were engaged in battle.[151]

Meanwhile, French reconnaissance of the Marne region reported that the Germans were occupying Châlons and the district of Ay due east of Paris.[152] Upon hearing this, Général Gallieni ordered the French *7ème Division* to move up to the front line. Rail could only move half of the division. The French had only 250 Army motor vehicles available, resulting in the famous employment of Paris taxis to move the troops of the XIVème Brigade of the 7ème Division that evening to Nanteuil-le-Haudouin.[153] In all, they transported five battalions of 4,000 infantry a distance of 30 miles.[154]

The Germans were now being pressed by Allied manoeuvres. Fighting created a gap between the First and Second German Army, forcing them to stabilise their position. The cavalry corps under General von der Marwitz was ordered to create a screen between the two armies. However, the deception unravelled owing to the lack of discipline amongst the German radio operators who transmitted in in clear and provided a picture of the situation while being monitored by French radio intercept units. This intelligence breakthrough provided French and British forces the opportunity they needed

Escadrille BL 10 commander Capitaine Zarapoff in the cockpit of a Blériot XI. (SHD, B90.2389)

to counter the German advance at its weakest point.[155] On 7 September aerial reconnaissance confirmed the general impression that the Germans were withdrawing two corps to the north, despite sightings of cavalry and infantry in the same general area of the battleground.[156]

Allied aerial reconnaissance on 8 September monitored the German withdrawal. A large number of infantry were observed waiting their turn to cross the river near La Ferté-sous-Jouarre.[157] French aerial reconnaissance provided a broad overview of the situation.

> Reconnaissance on Vitry and Châlons. We were told to watch the arrival of enemy reinforcements on the Châlons highroads; other machines were to watch the east of Vitry-le-François. The battle was raging below, following a line decidedly east and west. To the right this line seemed to incline toward the northeast of Verdun. To the west, it passed by Sompuis and inclined toward the Marshes of Saint-Gond. A second flight with twenty-three shells: I dropped them on the heavy batteries to the east of Vitry. Then I disturbed an artillery brigade (echelons of *ravitaillement* [resupply]), which scattered in every sense, but reassembled as soon as I had gone. I discovered the positions of twenty-four guns on the line to the west of Vitry. I marked them on the map at 1/80,000 and informed the corps concerned.[158]

The next morning's reconnaissance (9 September) reported that 'large hostile forces' were marching north of Château-Thierry.[159] A noon report placed the Germans about four miles north of Château-Thierry with additional small columns on roads to the east, all going north. In this moment of high drama, RFC aeroplanes were employed to discover their own BEF positions as well as provide aerial observations of the British flanks. At the same time, RFC aeroplanes were keeping watch on the French units in the area. The nearest troops of the French Vème Armée were seen that morning near Viels-Maisons (24km south of Château-Thierry) moving northeast away from the BEF.[160] Both French and British aerial reconnaissance remained active with a French liaison to GHQ who passed on the first indications that German movements were underway. The reporting was further substantiated by British aerial reconnaissance. At 1535 hours on 9 September, Captain Grey and Captain Boger of RFC No. 5 Squadron took off in an Avro to reconnoitre the area in front of the II Army Corps. They landed at approximately 1700 hours and reported indications of the withdrawal of von Kluck's left from opposite the French VIème Armée West of the Ourcq River. Additional afternoon aerial reconnaissance discovered an eight-mile-long column marching northward from Lizy (northwest of Soissons on river Aisne) at the back of the Ourcq battlefield and then other columns of transport and troops retiring north-northeast on La Ferté-Milon (southwest of Soissons on the river Ourcq). The ground on the left flank of the BEF in the angle between the Marne and the Ourcq was reported clear.[161] With each report greater attention focussed on accuracy. Debate arose over 'how it [aerial reconnaissance] will be employed tactically, for the benefit of those commanders that have to use it.'[162] The lack of headquarters staff and intelligence personnel familiar with the limits of aerial observation was showing. Inflated reporting was becoming an issue of concern.[163] German aerial reconnaissance kept General von Kluck appraised of the situation, reporting on the advance of four long Allied columns manoeuvring towards the Marne with advance troops on the line at the river Ourcq at Nanteuil-Citry-Pavant.[164]

On 10 September the RFC reported that near Troesnes (on the river Ourcq above La Ferté-Milon) German artillery and cavalry were retreating at a rapid pace. Villers-Cotterêts was blocked with trains and all the rail sidings were very full. Later that evening additional reporting observed numerous bivouacs around Soissons, where the troops coming from the south were assembling.[165] French reports confirmed the fact that the Germans continued to withdraw:

> On our left [IVème Armée], toward the camp of Mailly, there were no forces opposing us. There was a gap, a space between the two German armies, with only some cavalry a long way behind. Our division of cavalry was in the camp at Mailly. The obvious thing was for us to drive in a division or a corps if possible. Général Dubois was charged with this task. We watched closely for the moment when this movement could be effected without being noticed by the enemy, and we flew over Sompuis, Mailly, Sommesous, Lenharree, Coole [flight west-northwest in an area 20km southwest of Châlons-sur-Marne] in spite of heavy clouds. And we never saw any German reserves. Second reconnaissance with Capitaine Thiron. We alighted beside the XXIst corps, informing them as to the positions of batteries and troops, telling them that our troops were too far in the rear, that our guns did not carry far enough, and that there were no troops on our left. While we were at Saint-Ouen [Saint-Ouen-en-Brie, approximately 20km east-southeast of Paris], the Général sent us a message that the cavalry had signalled an enemy division in front of him. We told him that was not the case. But we proposed to make another reconnaissance. We set off, saw a French infantry division which was marching slowly forward, sustained by the artillery.

> We also saw the French cavalry division in the camp at Mailly. We flew low down, for it was late. No Germans. We returned to carry our information to Saint-Ouen. We had made three reconnaissances. Returned to Brienne [Brienne-le-Château] for the night.[166]

That evening Field-Marshal French estimated that the Battle of the Marne had concluded. Now the Battle of the Aisne was to be fought.[167] It was a significant moment. The German strategy to defeat the Allies quickly, based on years of thorough planning and major commitment of resources, was thwarted. In the annals of history it is regarded as one of the most decisive battles.[168]

Aerial observation at the Marne had secured for itself an integral role in battle. Daily observation defined the battlefield. Both French and RFC sorties had been successful in reporting the movements of the manoeuvring German advance in a timely manner. The observer cadre made the difference, with inputs covering enemy forces sighted, status of bridges and road networks, and other information critical to the command's battlefield awareness. The aerial observation role also had another benefit. British aerial reconnaissance not only covered German positions; it also clarified locations of the British vanguard. The Allied commanders were impressed. Général Joffre in an emotional discussion with the BEF command staff on the eve of the battle of the First Marne heaped praise on the RFC, saying, 'Please express most particularly to Marshal French my thanks for the services rendered to us every day by the English Flying Corps. The precision, exactitude, and regularity of the news brought in by them are evidence of their perfect organisation and also of the perfect training of pilots and observers.'[169] For the British, the first weeks of conflict had an additional impact. Those engaged in the early campaign became the strongest proponents of aerial reconnaissance for the remainder of the war.[170]

THE BATTLE OF THE AISNE

After the conclusion of the Marne campaign, Field-Marshal French and his staff remained optimistic, persisting in their plans to push across the Aisne. French wired Lord Kitchener, the British Secretary of State for War, 'Very heavy rain falling to-night, making roads most difficult. Should much hamper enemy's retreat. Pursuit continues at daybreak.'[171] The greatest impediment to aerial reconnaissance at this critical time, in early September 1914, were torrential rains that made aerial operations impossible. GHQ was desperate to learn where the Germans were digging in. All they could do was hope for an improvement in the weather so that flights could resume. One storm was so violent that British aeroplanes were tossed into the air and wrecked. Unfortunately for RFC squadrons arriving at Saponay (about 20km north of Château-Thierry) landing ground, the storm wrecked half of the unit. The No. 3 Squadron diary recalled:

> Before anything could be done to make the machines more secure, the wind shifted, and about half of the total number of machines were over on their backs. One Henri Farman went up about 30 feet in the air and crashed on top of another Henri Farman in a hopeless tangle. BEs of No. 2 squadron were blowing across the aerodrome and when daylight arrived and the storm abated, the aerodrome presented a pitiful sight. The Royal Flying Corps in the field could not have had more than ten machines serviceable that morning.[172]

Aerial reconnaissance now transitioned from recording manoeuvres to pinpointing targets for the artillery. At the beginning of the Battle of the Aisne, on the morning of 13 September, the BEF attempted to continue the pursuit of the German forces over the Aisne and succeeded in taking a bridge at Venisel (directly east of Soissons on the river Aisne). However, the Germans had entrenched several hundred yards behind the crest of the ridge north of the Aisne. By mid-morning, it was clear to British commanders that considerable artillery support was required. Additional problems occurred when initial rounds were fratricidal. That night, a British attack could not break through German positions that were being reinforced with more troops. The British were suddenly at a disadvantage since the Germans were staying put in terrain that gave them dominance of the area. Artillery directed at German lines proved ineffectual owing to the topography of Aisne valley, which provided excellent cover for German machine guns.[173]

Since the Battle of Mons, the Germans had experimented with aerial observation to increase artillery effectiveness. Now the British realised that they needed any available resource to increase their own. At the Aisne, they turned to the aeroplane to meet that urgent need. The British Army had tried aerial spotting for artillery prior to the war, but had not followed up on enhancing the capability.[174] However, there were not enough Allied aeroplanes to meet the immediate demand. The stalemate at the Aisne portended the battle to come. The RFC was still recovering from damage suffered during the recent storm. The critical lack of aerial and cavalry reconnaissance prompted the GHQ to work with mere assumptions about the threat location instead of relying on actual information.[175]

That night Field-Marshal French and staff were under the impression that the German army was retiring. GHQ issued orders the following day: 'the Army will continue the pursuit tomorrow at six a.m. and act vigorously against the retreating enemy.'[176] The next morning, the strongly entrenched German forces were successful in holding back the BEF attack and conducting a counterattack.[177]

15 September was frustrating for Field-Marshal French and the BEF. They made no progress on the battlefield and had to endure an intense German barrage. French remained convinced that the German withdrawal was, in fact, over. He wired Lord Kitchener that evening, 'We have been opposed not merely by rearguards but by considerable forces of the enemy which are making a determined stand in a position splendidly adapted to defence.'[178] At 2030 hours Operation Order 26 finally brought the advance to a close, stating 'The Commander-in-Chief wishes the line now held by the Army to be strongly entrenched, and it is his intention to assume the general offensive at the first opportunity.'[179] The British military historian Brigadier-General James E. Edmonds noted that Operation Order 26 'proved to be the official notification of the commencement of trench warfare'. For the rest of September, the BEF endured daily artillery bombardments as well as sporadic infantry attacks.[180]

One of the great innovators of aerial targeting support for artillery was Captain Douglas Swain Lewis. His recommendations revolutionised British artillery operations through the squared map and clock code. (Peter Mead, *The Eye in the Air*)

AVIATION INITIATIVES

The demands on aerial observation to locate artillery targets during the Battle of the Aisne now assumed greater prominence. The requirement soon transformed cartography. Artillery targeting required detail that prewar maps lacked. It took the initiative of several RFC members to meet the challenge. Major Geoff Salmond, Commander of No. 3 Squadron, was approached by Captain Douglas Swain Lewis concerning a role for the wireless radio. Lewis was convinced that wireless communication could increase the reconnaissance advantage and better serve the needs of artillery units. Geoff Salmond, an experienced artilleryman prior to his transfer to aviation, understood that artillery fire demanded accuracy. The two set up a wireless experiment with the artillery that was an immediate success. They created two maps lined with 400-yard squares. One map was provided to the battery commander involved in the experiment, the other to Lewis. The successful demonstration initiated a cartographic revolution for the British. From then on, maps were reconfigured to serve aerial-field artillery wireless coordination.[181] The 'squared map', as it was called, revolutionised British maps issued to artillery and aviators. The numbered and lettered squares allowed aerial reconnaissance to effectively identify the target area and relay information to the battery. The process of finding the target was called pinpointing. Soon the entire battle area was covered by a series of contiguous sheets.[182]

The experiment was then continued with British forces at the front. On 15 September the British III Corps assigned its aeroplanes to the divisional heavy and howitzer batteries. Later that month the British 3rd Division reported to its higher headquarters at II Corps that an RFC wireless-equipped aeroplane had successfully supported the artillery in achieving direct hits on several previously hidden German positions. By early October the following procedure described by the staff of the I Corps prevailed throughout the BEF: 'The hostile batteries are first located as accurately as possible by aeroplane; these results are then communicated to divisions and an hour at which fire is to be opened is fixed. At that hour the aeroplane again goes up, observes the fire, and signals any necessary corrections by wireless.'[183] However, despite these significant successes, the equipment was a liability over time. A few weeks later when the British advanced to the Ypres sector, the wireless equipment began to

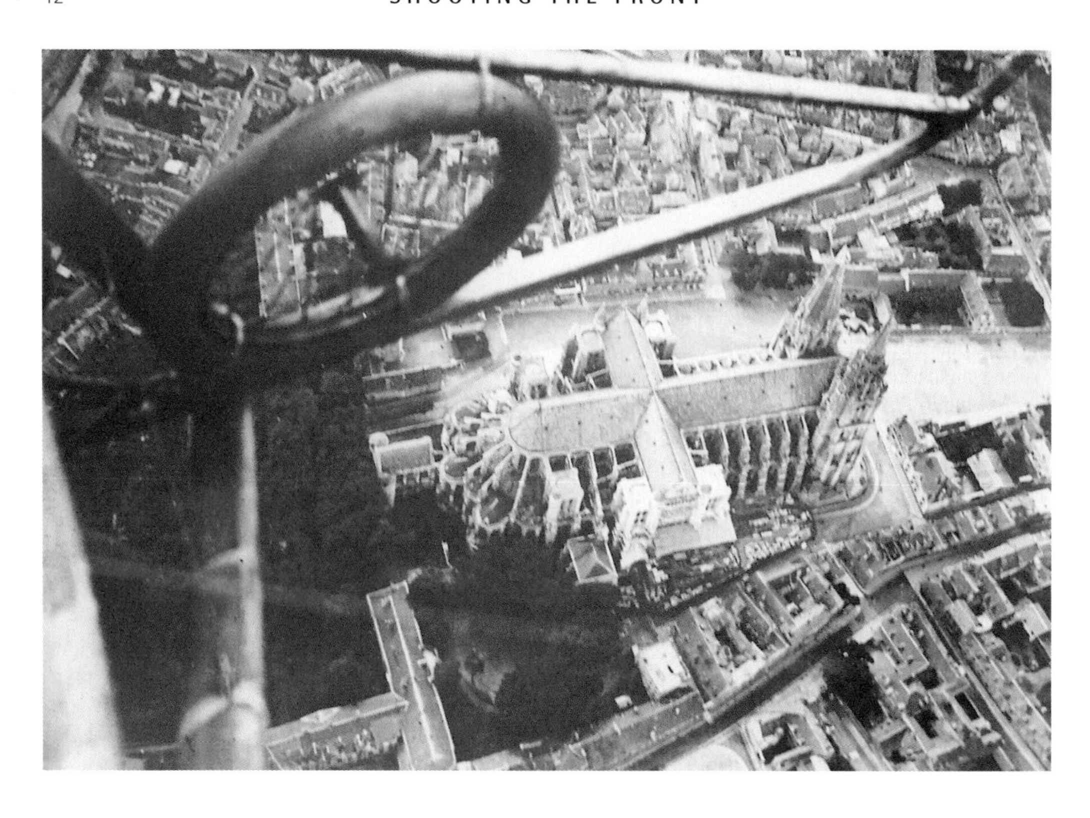

Interest in aerial photography grew as aviators acquired photographic souvenirs for loved ones such as this shot of the famous Chartres Cathedral. (SHD)

malfunction. It required extensive maintenance with each sortie.[184]

The genesis of aerial photography within French aviation circles was defined by a mixture of innovation, mission necessity, and personal motives. The first aerial photographers were amateurs. Their intent was not to confirm or compliment aerial observation in progress. It was to generate souvenirs for loved ones showing off the 'pretty landscapes' of the front lines. Such acts of bravado soon caught the attention of artillery staff, who saw the opportunity of using photography as a means of 'investigating German batteries'.[185] French aviation also applied their aerial resources to artillery spotting. By mid-September, as the French faced a standoff against German forces at the Reims-Vosges sector in eastern France, French artillery personnel demonstrated an important initiative. Artillery personnel flew with the escadrilles to gain a better idea of the German artillery positions. It was during these sorties that the French artillery observers carried their own personal cameras to detect and locate the batteries. The results were beneficial, alerting the French GQG that aerial reconnaissance was better served with a camera on board.[186]

While success was demonstrated with artillerymen on board the aéroplane, the actual liaison with the artillery unit was challenging. The pilot tried conducting various manoeuvres with the aéroplane to

Capitaine Barès was the first chef de l'aeronautique à GQG. However, his animosity toward aerial photography at the beginning of the war did not aid Bellenger's attempts to make aerial reconnaissance more effective. (SHD, from Christienne and Lissarague *A History of French Military Aviation*)

link up with the battery. That proved confusing at first. Subsequent means of controlling the rounds included using a specific brand of tracer round to allow the aerial observer the ability to track the shell toward impact. However, attention to the challenge soon met with success.[187] Still in September 1914, French pilots succeeded in spotting the artillery of a German army corps in manoeuvre. They identified the location for the French artillery at Thiaucourt (Thiaucourt-Regniéville), whose subsequent barrage destroyed half of the corps, a feat recognised and cited by Général Joffre.[188]

The stalemate at the Aisne set the stage for significant evolution of aerial reconnaissance. Taking aerial photographs was on the minds of several RFC aviators. On 15 September 1914 an RFC No. 3 Squadron pilot, Lieut. George F. Pretyman, took five photographs of German gun positions on the Aisne. Photo quality at this early stage was not impressive, but the experiment demonstrated that aerial photographic technology was possible in a combat environment.[189] Major Robert Moore 'Henry' Brooke-Popham, commander of the RFC No. 3 squadron, was the driving force for his squadron's efforts to employ photography. His prewar photographic trials laid the groundwork for the first aerial photographs of the German forces.[190] However, existing camera technology did not function well in the new combat environment. The 'press' type cameras of the Pan-Ross pattern became available later that year.[191]

French acceptance of aerial photographic reconnaissance commenced after British recognition of its value. Despite the notable success of the Marne observations, Capitaine Bellenger still met resistance from GQG staff. He continued his struggle to institutionalise aerial photographic reconnaissance. In October 1914, he attempted to create a new photo service, promising to provide maps of the enemy. To some, aerial photography represented a fad, not an integral part of the military. His counterpart at GQG, Capitaine Edouard Barès, *chef de l'aéronautique a GQG,* responded to one query with 'use a Kodak that you can purchase from the local shop. Don't ask the government to pay for this.'[192] Bellenger continued to find cameras, organise a lab, create a team of photographic interpreters, and teach aerial observers the techniques of aerial photography. All were essential, but advocacy for his efforts was still absent.[193] In post-war reminiscences, Bellenger recalled his confrontations with GQG. In December 1914, in an attempt to gain favour with Général Henri Berthelot, Général Joffre's chief of staff, by showing aerial photos from various missions, Bellenger was told sarcastically that 'he already had a map.' Berthelot concluded the conversation with 'I don't care about your pictures.'[194]

The solution to GQG's intransigent attitude was found by Bellenger's counterpart, Lieutenant Grout of the French Artillery. Grout developed the right contacts at GQG and proceeded to establish the seminal French aerial reconnaissance program. While flying as an observer on a reconnaissance mission on 7 October 1914, Lieutenant Grout photographed German batteries at Fort Douaumont in the Verdun sector. The photographs aided French artillery in subsequent targeting. Grout also examined methods for transferring the information onto a map. The success of the mission resulted in Grout being placed in charge of an elite element to establish aerial photography's role. He proceeded to test cameras on *aérostat* platforms. One such design evolved into the mainstay 120cm camera. Another long-term legacy from this advance group was the introduction to aerial photography of one of the greatest aerial photographic interpreters of the war, Lieutenant Eugène Pépin.[195] Grout's persistence in proving aerial photographic reconnaissance continued into the next year, providing a successful example for not only the French staff, but an increasingly curious British staff.[196]

Lieutenant Maurice Marie Eugène Grout, the leading French pioneer of aerial photography, acquired command trust early in the war, resulting in his establishing the French Section Photo-aérienne (SPAé). (SHD 05/0676)

As the Marne campaign came to its conclusion, aerial reconnaissance was also on the minds of the German General Staff. General von Kluck blamed the campaign's failure on several issues including the lack of critical information from aerial reconnaissance. 'Both cavalry and air reconnaissance have failed during the past few days.'[197] Von Kluck was relieved of command along with the senior German commander, General Helmuth

Johannes Ludwig von Moltke. Dismissal did not result in a transfer for these senior officers. Von Moltke was ordered to stay in the background while General Erich von Falkenhayn commenced command of the Aisne campaign.[198]

The stalemate at the Aisne continued into late September with British commanders resolved to transfer the BEF to the left of the Allied line. If the war was to go in favour of the Germans, they wanted to be in place to best defend British interests. Their choice for the remainder of the war was to defend the coast and critical ports. On the afternoon of 1 October, GHQ directed the Army to move north to Flanders, ordering three British corps to withdraw in succession.[199] In preparation for the move, Field-Marshal French directed Brigadier-General Henderson to establish aerial reconnaissance toward Antwerp. Field-Marshal French then expedited the move of all the British forces to the northern theatre.[200]

The First Battle of Ypres signified the termination of mobile warfare on the Western Front for the next four years. The rapidly evolving battlefield demonstrated to commanders on both sides that mobility was a secondary consideration to firepower. As the Germans headed north in the 'Race to the Sea' toward the British lifeline of the Channel ports, they were stopped through an effective combination of artillery and infantry fire. Artillery had assumed the dominant position in warfare.[201]

The German General Staff had envisioned the Flanders plain as ideal territory for enveloping the Allied armies. The plain consisted of an abundant network of roads, railways, and canals that could transport and supply a great army.[202] However, geological and topological factors were the ultimate gauge of how the battle would be fought. A clay formation typified the Flemish landscape, which when wet became heavy, sticky mud.[203] For the next four years, Flanders became the stereotype of all the horror associated with the Great War. The German attempt to take Ypres led to a series of catastrophic battles fought over the worst imaginable terrain. Damage from artillery fire was greatly reduced when the shells exploded in the mud. Shell holes filled with water that did not drain away, turning the battlefield into an almost impassable morass that prevented any advance prepared for by the ceaseless bombardment. Flanders was a logistical nightmare. Munitions and other supplies could not be moved in time to support an advance properly. Reinforcements floundered in the mire, while attacking forces were forced to relinquish hard-fought positions. Artillery suffered from an inability to operate off the roads. Light field artillery required macadam-like material to convert the shell-torn route to one meeting the weight and bulk demands for its movement. A British artillery officer, reflecting on the continual challenge remarked, 'I am carrying forward my guns and ammunition, the material for making my road as I go along, and the material for fortifying my new position.... I am half expecting orders to bring along an acre of ground with me, too.'[204] 'Stationary warfare now the rule' summed up the condition for most of the Western Front.[205] In light of the topological horror, aerial reconnaissance assumed prominence, and with it the aerial photograph.

As positional war established itself with networks of trenches from the Belgian coast to the Swiss border, aerial observation assumed a permanency with employment of 'captive' balloon aérostats. The French had ceased their captive balloon (*ballon captif)* program in 1911 in favour of developing their aéroplane inventory. The results of the first months of the war led to rethinking the role of aerial observation resources. The existing fleet of aérostat dirigibles were relegated to the background and soon disappeared from the scene. As the immediate post-war French assessment stated, 'The science of aeronautics was reduced to that of aviation.'[206] Once the mobile battlefield ceased the French realised their mistake and quickly established an operational ballon captif aérostat program in October 1914. The innovative aerial pioneer Capitaine Saconney not only led the design of the French ballon observateur program but also configured its operational employment, determining suitable missions such as observation, artillery ranging, verification of demolitions, and sector reconnaissance.[207] By early 1916 the French had 75 aérostat companies supporting all armées along the front line.[208] The British Army followed suit with the employment of captive balloons at the front in May 1915, with the deployment of their first kite balloon at Poperinghe, Belgium.[209] The ballon captif aérostat had a distinct advantage over the aéroplane thanks to being directly linked via telephone to the headquarters or artillery battery. The ballon captif employed a new technique for *aérostier* survivability, the parachute. Aerial observation was now a permanent feature with the aérostat platform. The ultimate ballon captif, the Caquot, (named after the designer Capitaine Albert Caquot) became the standard design for the second half of the war. Testament to its success was that the Germans produced their own fleet (designated the Type Ae 800) after capturing Caquots that had broken from their moorings and drifted east.[210]

With the decline and eventual termination of mobile warfare, observation by aéroplane assumed a more ubiquitous role. The French Aviation Militaire expanded from 31 flights to 65, of which most served the reconnaissance mission. Each French flight consisted of approximately six aéroplanes each. By March 1915, the French had 53 flights of aéroplanes.[211] The Aviation Militaire now comprised 16 army reconnaissance and

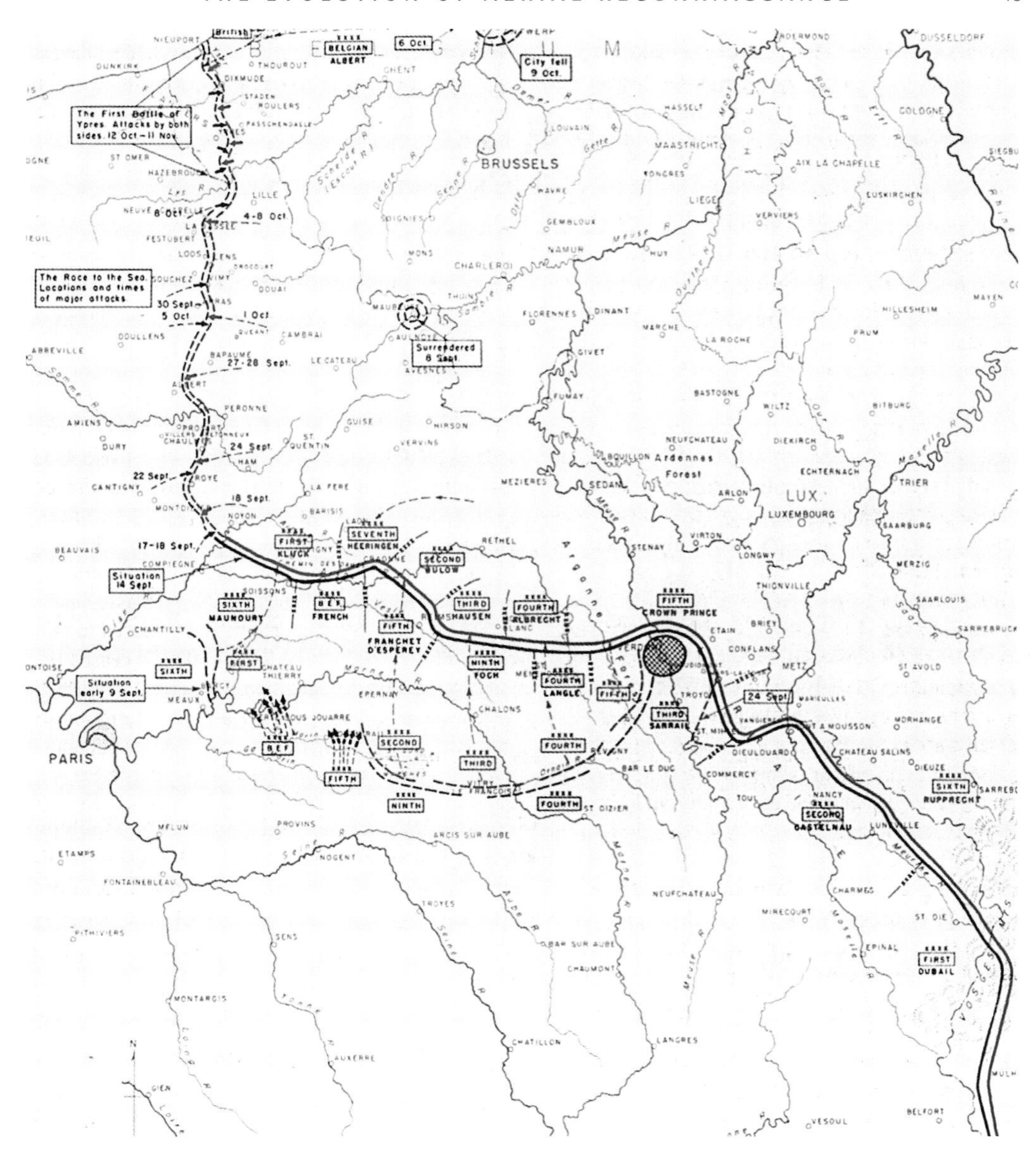

Race to the sea, September–October 1914. (*West Point Atlas*, courtesy Department of History, US Military Academy)

pursuit flights, 16 bomber flights, and 30 army corps observation flights.[212] British attitudes towards aviation also became more realistic. Lord Kitchener spoke for the prevailing view that more aeroplanes were needed to support the rapidly expanding British Army. When the Directorate for Aeronautics forwarded a proposal for 50 more squadrons in the spring of 1915, Kitchener wrote on the memorandum 'Double this. K.'[213]

The last phase of mobile warfare cemented a relationship that continued for the remainder of the war. As the forces adopted positional warfare, there was an impetus for greater cooperation between aerial observation and the traditional infantry and artillery forces. The principle that 'Photography is the basis of good artillery' started to evolve.[214] After the war Air Vice Marshal (AVM) Geoff Salmond stated that aerial photography owed its existence to wireless and artillery.[215] His seminal efforts with Captain Lewis established a base for subsequent aerial observation efforts to support artillery. However, officers from the artillery arm were still slow to accept the critical role of aerial observation, despite the successes at the Aisne. Greater liaison was required to communicate aerial reconnaissance capabilities to both artillery and infantry senior officers and staff.[216] The pioneer in

British aerial photography, John Theodore Cuthbert Moore-Brabazon, reflected later in life that the artillery culture felt that its ballistics were an exact science and did not require assistance from the aviation community. Moore-Brabazon suggested that it took one of the first British aerial celebrities, Major Harvey-Kelly, to drive home the value of aerial reconnaissance for the artillery. One day during the First Battle of Ypres, Harvey-Kelly transmitted via wireless a 'very curious message'. The British had moved several large mortars, the pride of the artillery, to the front. Near the Ypres battleground was a large lake (Bellewaarde Lake) well within RFC aerial coverage. During a mortar barrage, Harvey-Kelly commented on the accuracy of the mortar fire: 'If anybody is firing on the middle of Bellewaarde Lake, he is hitting it.' Moore-Brabazon stated that this dry assessment did more to convince the British Army to rely on British aviation to help direct artillery fire than any other effort.[217]

Wartime intelligence operations evolved along with aviation operations thanks to the aerial reconnaissance legacy of the opening salvos and subsequent battle of the First Marne. Despite the significant and speedy advances, military intelligence was still regarded with suspicion. By the time the First Battle of Ypres had commenced, it was clear amongst the members of the staff that intelligence and operations had to learn to coexist. Colonel Macdonogh, the lead British intelligence officer for the BEF who at the conclusion of the war was promoted to Lieutenant-General, was one of the most brilliant intelligence officers to serve in the forces. When Field-Marshal French was informed that the Germans had reinforced their position at Ypres with three new reserve corps, he flew into a rage at Macdonogh: 'How do you expect me to carry out my plans if you will bring up these bloody divisions!'[218] A testament to changing attitudes was that Sir John French now had Macdonogh brief him twice daily, at 0830 hours every morning and at an appointed time in the afternoon. Joining Macdonogh in the daily discussions were French's chief of staff and Henderson, RFC Commander. In Henderson's absence, Sykes substituted and provided aerial reconnaissance updates.[219] It took technological advances in aerial photography to lend credibility to the assessments and eventually convince the operators to accept the information provided.

In early 1915 the US military attaché to Great Britain who later commanded the US Army Signal Corps for the remainder of the war, Colonel George O. Squier, commented in his report on the changing face of battle:

> It is strange, this modern war, when a [modern] strong fortress like Antwerp falls in ten days and then a simple trench dug in the open field can completely stop the German Army at its best. At Ypres 4 Corps stopped something like 8 or 10 corps, including the Prussian Guards. What are we going to do about fortresses in the future? One thing is clear, concealment comes first, and protection is secondary.[220]

Squier's comments were prophetic. The evolution of warfare in the coming year brought aerial reconnaissance to the point of threatening the prevailing military mindset. The battlefield had been transformed.

CHAPTER 3

Late 1914 to October 1915: Aerial Photographic Reconnaissance Takes the Stage

EVOLUTION OF BRITISH AERIAL RECONNAISSANCE OPERATIONS

Both French and British military thinking after the race to the sea campaign was in a quandary. Lack of battlefield mobility had become the norm. Defining battleground success in a modern-day positional combat environment required intimate knowledge of the immediate battlefield terrain now known as No-Man's Land. With trench warfare a routine was established and carried on until the final months of the war. The front lines of 1914–1918 gauged combat in terms of measured destruction. Aerial reconnaissance and photographic interpretation served commanders as a recorder, measuring progress through the medium of the aerial photographic print.

The British Army underwent an evolution in thinking during this first year of the war. Despite the fact that they were seasoned veterans from fighting in South Africa and in a variety of military operations around the world, the British had to come to terms with major combat involving large formations against an equivalent adversary. They became merged with the conventional military thinking of the early twentieth century as defined by the French and Germans. When the BEF commenced its campaign in Belgium and France the initial manoeuvres transitioned to holding ground at all costs. Departing significantly from their experiences of the previous military campaigns, where the British commander and staff were in the middle of the fight, their leadership and thinking was now away from the battlefield in a protected sanctuary. Positional warfare confirmed this dislocation. The access to information to make the necessary decisions now came through evolving aerial observation technology, to include the aerial photograph.[221]

The disarray that presided over RFC operations since the late-August retreat from Mons finally concluded on 12 October when Henderson's RFC staff settled in St Omer, their headquarters location for the next two years. Despite the trial of managing aerial operations 'on the run' the RFC had done a remarkable job in launching sorties and acquiring relevant information. At St Omer, Henderson and Sykes deliberated on what aviation role best served the BEF on a static front line. With Sykes providing the details, Henderson decided to decentralise the RFC by splitting it into Wing organisations attached to the BEF's army corps.[222] Each Wing, commanded by a Lieutenant-Colonel, was charged with the formation and training of squadrons capable of reinforcing the air units already operating in the field. The institution of RFC wings now provided a structure to support the British Army with both reconnaissance and offensive mission capability. The British corps commanders were the responsible echelon to task the wing for tactical reconnaissance.[223]

Henderson's was determined to ensure RFC integrity and not have it become controlled by BEF corps commanders who lacked an appreciation of air power capabilities or the risks of flying.[224] The best way to counter ignorance was through dissemination of relevant information. To gain an appreciation of the extent of the operation, the RFC directed that each unit keep a 'War Diary' on missions flown and other significant events of the day. The Wing also provided any information on landing ground locations should the squadrons be required to move.[225] Additional attention was given to demonstrating the aeroplane's ability to go beyond the front lines and gather strategic information necessary to define the enemy's intentions. The initial five operational RFC squadrons (Numbers 2, 3, 4, 5, 6) that had deployed to France in August were now tasked to search for rolling stock at all railway stations in the region so that commanders could have a possible indication of offensive build up in the Ypres sector.[226] This function demonstrated the strategic purpose that remained a priority for aerial observation.

Refining the aerial observation operations also called for a greater understanding of the army echelons that required information. In December 1914, Major-General Henderson and Colonel Sykes imposed rules of operation so that the missions comprised both artillery and tactical artillery reconnaissance, with artillery support preeminent. The squadron landing ground was to be as close to the artillery headquarters as possible. The RFC also encouraged division counterparts to locate their headquarters as close to the landing ground as possible. Aerial reconnaissance communication with British divisions was accomplished through the RFC liaison at the location, using a variety of communication devices to include dropping a message, communicating by wireless or lamp, or having the aeroplane fly vertically over the target and drop a smoke ball.[227] The RFC liaison was assigned to the advance guard and provided with a light tender (truck) and two motorcyclists. The officer reconnoitred landing grounds at intervals of five miles over roads along which the division was to advance. When a landing ground was selected, a 'T' marker was placed and a report of the location forwarded to the division headquarters.[228]

At year's end, Major-General Henderson staffed the plan that provided direction to the RFC for years to come. Now that the RFC was accomplishing tactical reconnaissance for the Wings, it was recognised by the British Army as an autonomous service element. On 15 January 1915, the 'Organisation of the Royal Flying Corps in the Field' was published under the signature of Lieutenant-General Sir Archibald J. Murray, the Chief of the Imperial General Staff.[229] Sykes authored the definitive 'Notes on Aerial Reconnaissance' portion of the document, which outlined the long-distance strategical reconnaissance functions assigned to units attached to GHQ and the tactical reconnaissance functions assigned to units attached to the armies. The RFC organisation now prevented overlapping of tasking and a had systematic process to cover the ground.[230] One of the key points was that aerial reconnaissance was not to be requested if not vitally necessary. Equally important, aeroplanes were to be flown only on missions suited to their type. Henderson and Sykes laid the foundation for the RFC's maintaining control of its mission while the size of the armies increased at the front.[231]

FRENCH ADVANCES IN RECONNAISSANCE

The French also took steps to organise their aviation resources to suit the changing parameters of warfare. Aviation Militaire supporting the GQG now had a permanently assigned chief (*chef*) of the *Service Aéronautique*. The senior aviation presence was further established with each armée acquiring a chief of the Service Aéronautique on its staff.[232] Each armée acquired two strategic reconnaissance escadrilles equipped with either MF 11s or Morane-Saulnier Ls.[233] An armée cooperation escadrille was also assigned to each CA along with a company of ballon observateur.[234] The French would later expand escadrille assignment to support artillery regiments with Caudron G.3 and G.4 aéroplanes.[235] As the campaigns of 1915 progressed, they realigned aerial resources to provide fighter support. In an interesting twist, a few multi-seat Caudron G.4 and Nieuport 11 aéroplanes were assigned to a new unit, the *Groupe de Chasse de Malzéville*, to defend other escadrilles from German bomber attacks.[236] As the year progressed, the existing line of aerial reconnaissance platforms fell victim to improved German fighters. French industry was producing an early type aerial reconnaissance platform; designs for multi-seat long-range reconnaissance aéroplanes such as the Caudron R.4 and G.6 were not produced until 1916. As 1915 concluded, this situation proved deadly for the units on the line flying older airframes.[237]

The French initiated their operational aerial photographic infrastructure on 23 October 1914. The success of Capitaine Grout's efforts and the inclusion of French artillerymen serving on aeroplanes with cameras soon caught the attention of senior French commanders.[238] Général Beliu, Director of the Service Aéronautique, approved a directive establishing *Section photo-aérienne du Group des divisions d'entraînement* (SPAé du GDE), one assigned to each armée. It advised that personnel with a background in photography should be considered for this assignment.[239] Grout established the initial SPAé architecture and commenced operations on 1 December 1914. The SPAé brought significant change to the production of photographic products for military operations and and clearly demonstrated the potential of aerial photographic reconnaissance for senior officers and staffs.[240] Manning for each SPAé consisted of an officer, a draughtsman, three photographers and a driver.[241] With Capitaine Grout's SPAé now in operation, several initiatives became reality. The French hand-held 26cm camera (13 x 18cm photo plate) became operational after several months of development. The 26cm was followed by aerial cameras of greater capability that became the mainstay of French (and later US) aerial photography – the 50cm camera and 120cm camera. The French developed a mobile photographic laboratory housed in a truck. By the start of 1915, the SPAé concept was taking hold within the French armées as well as acting as a prompt to both the British RFC and the Belgian *Aviation Militaire Belge*.[242]

The British became aware that the French were collecting information on trenches from aerial photographs and were transposing sketches from aerial observation onto their tactical maps. When trench warfare commenced, the first maps of the battleground were crude imitations of French and Belgian maps that captured topographical and tactical detail. Later, aerial

Allied reconnaissance acquired a new generation of aeroplanes such as the Caudron G.3 to support artillery targeting and battle planning. (James Davilla, MA 35860)

An ideal design for early aerial reconnaissance was the Henri Farman (HF) 20. (James Davilla, MA 26005)

photography contributed details to the map features.[243] As a courtesy, the French started to forward their aerial photographs to GHQ, labelled as 'aerial maps'. They showed the British that they were earnest about aerial photography and were making major progress in its support of military operations. As a result, Colonel Sykes sent the innovative Major Geoff Salmond to visit the French unit responsible for the aerial coverage.[244] Besides observing French progress in camera technology, Major Salmond quickly recognised that the French were resolute about integrating aerial photography into their battle planning.[245] Grout's seminal French SPAé program successfully demonstrated that aerial photography served the Allied effort. The British were spurred to develop their own program.

The French realised that their battlefield maps were inadequate to keep up with the significant changes underway in the field. Sketches from aerial observers were insufficient. Accuracy was called for and the memory of the air crew was limited; they could not enhance the results from the primitive hand-held cameras some used. The need for more information on the German artillery increased the demand to improve aerial reconnaissance. French staffs came up with a solution. They created the *Plan Directeur*, a master map that became the mainstay of French intelligence and operational targeting for the remainder of the war.[246] With the Plan Directeur, battlefield information now provided a framework for operations. As post-war French assessments of this period reflected, '... aerial combat became more and more general but the escadrilles were not yet specialised in pursuit, bombing, or observation.'[247] French aerial observation now included hand-held cameras. Artillery spotting became more refined. French air crews initiated the use of either flares (*fusées*) or coloured flags to signal the French artillery units to correct their aim. Also, French wireless, *télégraphie sans fil* (TSF), commenced operation to provide the air crews with a means to rapidly signal any necessary alterations of fire. TSF became the primary means to conduct artillery spotting for the remainder of the war. TSF experienced problems like any new technology, with unintelligible or incomplete transmissions during the sortie.[248]

Procedures institutionalised in 1915 increased the effectiveness of the SPAé's ability to interpret aerial photography and disseminate the information quickly. Planners for aerial photographic missions strove to achieve a common map scale to allow information from the print to be quickly transferred to the combat map. The scale of choice became the 1/10,000. By standardising Plan Directeur formats, French target planning became consistent across the front.[249]

The mainstay French aéroplanes employed for aerial observation were the Henri Farman (HF) 20 and Maurice Farman (MF) 11. Both were ideal designs for aerial reconnaissance and artillery spotting, with the cockpit forward of the wings, yet vulnerable through limited armament and slow operating speed.[250] Both aéroplanes exemplified the state of aerodynamic art. Their airspeed was relatively slow (HF 20: 165km/hr/102mph; MF 11: 118km/hr/73mph). However, they both possessed respectable mission endurance for observation, each being able to operate for around three hours.[251] Barring any threat, aerial observation could be maintained for extended periods of time. Examples of French aerial photographic work in early 1915 reflected the manner in which the escadrille functioned. On 30 January 1915, an HF 20 escadrille flew numerous reconnaissance sorties over enemy batteries at Harville, took aerial photographs over Eparges and Saint Rely, and used a TSF-equipped machine to direct artillery fire over the Ième Armée sector.[252] In a single day in February 1915, one MF 11 aéroplane flew two reconnaissance missions, three artillery spotting missions, and took several photographs of enemy camps.[253]

THE BRITISH ADOPT AERIAL PHOTOGRAPHY

French military seniors were quick to share their aerial photographic experience with their British counterparts. French advances in aerial photography would prompt the British, as traditional rivals, to keep up and demonstrate an equal proficiency. The genesis of the British aerial photographic program stemmed from the accomplishments of two 'civilians' who had come on board with the British military at the first call. Lieut. J.T.C. Moore-Brabazon from the RFC and Lieut. Charles Duncan Miles (C.D.M.) Campbell from the Intelligence Corps, GHQ. Moore-Brabazon was a leader in the British aviation community, winning the 1909 *Daily Mail* prize for the first all-British machine that could fly a circular mile; Campbell was a photographic pioneer, later responsible for helping the US Air Service establish a photographic capability.[254] Following the race to the sea campaign both Moore-Brabazon and Campbell were assigned to the RFC's No. 9 Squadron at the St Omer aerodrome. No. 9 Squadron was dedicated to technological experiments applicable to combat aviation such as 'wireless'. Also arriving at No. 9 Squadron to apply his proven aerial photographic expertise was Flight-Sergeant Laws.[255]

Moore-Brabazon and Campbell's first challenge was to allay the scepticism of the British military establishment. It required expert knowledge on the subject, an ability to educate an obstinate culture of military professionals, and the constant pressure of proving success every step of the way. Moore-Brabazon: 'Considerable difficulty was experienced at once in getting the Army to realise the full possibilities of photography, but slowly they came round to understand the advantages that were being put before them for a trench warfare proposition.'[256] His subsequent memoirs provided further observations on that military mindset:

Charles Duncan Miles (C.D.M.) Campbell was the genius behind the British development of aerial photography. Together with J.T.C. Moore-Brabazon and Victor Laws, he was able to establish a viable British aerial photographic reconnaissance program in early 1915. (Roy Nesbit, *Eyes of the RAF*)

> The reluctance of the older Services to adopt new devices and inventions sprang from fundamental principles that were little understood, but were not unworthy. First of all, officers are up-to-date, progressive and technically minded; but they look upon the struggle with an enemy quite objectively, with no personal dislike at all, even though they might lose their lives at any moment. They take the greatest exception to an enemy who refuses to play according to the rules as understood or who, in their opinion, indulges in a dirty trick.... Aerial photography was a thing that had not been done before, and to expose the whole set-up behind the enemy's lines was to invade a privacy that had always been accorded the enemy, never mind that it happened because nobody could do anything about it.[257]

The air crew eye was effective only up to a certain altitude. The ultimate gauge of effectiveness for the observation mission was the ability of the observer to detect and remember what was observed. In the early days pilots recorded their information by sketching on cavalry sketch-boards attached to the forearm or on kneeboards. British aircrews referred to these 'trench-sketching' sorties rather optimistically as 'close reconnaissance and photography'.[258] Army demands for strategical and tactical information by far surpassed the initial capabilities of aerial observation. It was time to turn to available technology.

The impetus for aerial photography rested with the junior officers. Thanks to the pre-war initiative promoted by Major Brooke-Popham, No. 3 Squadron was recognised within the British Army for its role in demonstrating the utility of the aerial photograph. Lieut. Pretyman's first attempt at photographing German artillery at the Aisne proved to the British that it could be done in a war environment.[259] With the coming of positional trench warfare, further refinement of aerial photography was accomplished. One of the No. 3 Squadron's aerial observers, Lieut. Charles C. Darley, had acquired an Aeroplex Camera (with a 12-inch lens) and proceeded to take photographs from his aeroplane. His initial efforts were poor due to fogged plates and surface scratches on the film surface caused by metal plate holders. Darley persevered despite the poor results. What impressed everyone was his outstanding initiative, accomplishing the entire process on his own. Not only did Darley take his own photographs, he also established his own photographic laboratory to produce the prints from his sorties. Darley purchased his own chemicals in Bethune, developed the slides in a dark room converted from the stables of the château where he and several squadron personnel were billeted. He had one assistant, 'who knew nothing about photography'. When Darley finished developing his own photographs, he completed his own interpretation of the prints. Darley persevered in his efforts to get aerial photography recognised by spending every evening personally showing the results to staff members at the nearby Corps Headquarters or Divisional Headquarters.[260] Darley later recalled that his photography improved immensely thanks to Campbell's loaning him a Pan-Ross camera with a Mackenzie Wishart plate holder.[261]

Over the winter of 1914–1915, Colonel Sykes assumed command of the RFC on several occasions. Major-General Henderson departed the RFC to assume command of the First Infantry Division. Henderson's transfer was terminated by Lord Kitchener a few weeks later. Upon his return, Henderson was afflicted by a medical condition that limited his ability to serve the requirements of the office. Sykes proved a worthy substitute. He took the initiative to establish the first British Photo Section using the talent resident at No. 9 Squadron.[262] Major Geoff Salmond, a General Staff Officer (GSO) at RFC Headquarters, was tasked by Sykes to learn how the French were organising aerial photography. Salmond proceeded to select Moore-Brabazon, Campbell, Flight-Sergeant Laws, and 2nd Air Mechanic W.D. Corse to build a photo section equivalent to what the French had established under Grout.[263] They were instructed to work out with the Wings the best form of organisation and camera for air photography.

One of the first objectives for the British photo section was deciding on the best type of camera for aerial operations. Moore-Brabazon, Campbell, and Laws were ordered to the RFC 1st Wing under the command of Lieut.-Col. Hugh Trenchard on 16 January 1915.[264] The initial plan was to have the team work with the 1st Wing for a month, then proceed to the 2nd Wing for another month. Following the two visits, they were to put forward a scheme to RFC Headquarters on the best course of action for aerial photography implementation. The team accomplished their purpose early while at 1st Wing. Based on the recommendations of Laws, the three produced an assessment that each RFC headquarters should include a Photo Section assigned at the Wing. The Photo Section was to be equipped with a newly configured photographic lorry. The success of Lieut. Darley helped influence the team to determine that the responsibility for taking aerial photographs was to be given to the aerial observers. Each RFC squadron was to have a camera available at all times for supporting ongoing operations as well as acquiring photos of any targets of opportunity. The RFC flight echelon acquired two cameras for ready aerial reconnaissance missions. After each sortie, all the photographs taken by squadrons were to be sent to the Photo Section at Wing Headquarters for development, fixing, printing and distribution.[265] The question of where photographic

Offcicial portrait of Major-General 'Boom' Trenchard as General Officer Commanding RFC in the Field. Trenchard continued Henderson's advocay of aerial reconnaissance after assuming his command. (Courtesy Andrew Renwick, RAF Museum)

plates were to be developed prompted the decision of what echelon should serve as the photographic hub. Based on discussions between Moore-Brabazon, Campbell and Darley, the final decision was made by Lieut.-Col. Trenchard that the RFC-manned wing (not squadrons) Photo Sections were to support each British Army.[266]

As aerial photography popularity increased, it became necessary to expand production by adding a second photo section under the direction of Moore-Brabazon at the RFC's 2nd Wing. Campbell was transferred back to London to establish an aerial photographic staff role within the Ministry's War Office. An RFC training program for aerial photography was required alongside

Colonel Frederick Sykes (seen here in 1918 as Major-General Chief of Air Staff for the Royal Air Force) was Trenchard's adversary. Regrettably, though both were leaders in advocating aerial reconnaissance's contribution to warfare, the conflict between the two clouded their accomplishments. (Eric Ash)

a definition of personnel requirements for photographic officers. It was hard work for Campbell, for he was forced to operate without an effective sponsor within the War Office. However, his credible staff role in London could now better serve BEF demands for aerial photographic support.[267]

THE AERIAL PHOTOGRAPHIC BREAKTHROUGH

Throughout this period the RFC aggressively carried out reconnaissance and artillery spotting, despite unfavourable winter weather. Hand-held photographs covered the front lines to a depth of some 1,500 yards. It was a period of transition for the GHQ. Sceptics still were unconvinced that aerial photographs could generate a detailed picture of German trenches and barbed wire. It was Allied planning for an early offensive in 1915 that spurred acceptance of aerial photography. The RFC's No. 3 Squadron, commanded by Major John Maitland Salmond, and including the innovative aerial observer, Lieut. Darley, started receiving queries on the layout of German trench networks opposite the British sector. It fell on Darley to accomplish this objective. He constantly updated his aerial map with each additional sortie. In the first days of February, Darley's photographs were provided to French and British staff officers planning an operation north of the town of Neuve

Initial aerial reconnaissance mapping at Neuve Chapelle was accomplished by aeroplanes such as these observed at the RFC No. 4 Squadron's St. Omer aerodrome. (Peter Liddle, *The Airman's War, 1914–18*)

Chapelle. The battleground comprised the brickfields and railway triangle south of La Bassée canal.[268] A significant finding was construction of a new German trench that would evidently serve as a launch point for a subsequent attack. The discovery created a major impression, resulting in the senior staff modifying the plan of attack. The photographs also significantly aided the planning by determining the precise junction between the attacking Allied troops. On 6 February, the Allied attack commenced and achieved their objectives in capturing important ground and taking a considerable number of prisoners.[269]

The success of Lieut. Darley's photographs at the La Bassée Canal sector prompted a vigorous effort by the RFC to employ aerial hand-held cameras to acquire as much aerial coverage of the battleground as possible. Following the initial proselytising over the potential of the aerial photograph, the GHQ staff became the strongest advocates, maintaining 'an unswerving loyalty to Air Photography, and ... ever pressing the R.F.C. for more and more photographs.'[270] By the second week of February, Darley was acquiring excellent photos with his hand-held camera of the Neuve Chapelle area and fusing them together to develop an 'aerial map' mosaic. Darley's format was immediately accepted by Major John Salmond and taken to the sector army headquarters (Darley recalled it was the Indian Corps Headquarters) where it generated considerable interest. Lieut.-Col. Trenchard from the 1st Wing now got involved. Moore-Brabazon recalled Trenchard 'going about with these photographs in his pocket, trying to make people use them.'[271] Trenchard personally escorted Darley to a conference of British and French generals deliberating on Neuve Chapelle. Général Joffre was so impressed with the photographic array that he requested that he be included in future distribution of photographs taken by the RFC.[272] Darley's initiative accomplished one of the most significant intelligence production feats of the war. He clearly demonstrated to a British staff lacking training in aerial photography that an aerial map created from a mosaic of photographs did an excellent job in portraying the battlefield and significantly improved battle planning. The aerial photographs contributed to a last-minute alteration of the Allied attack plan and spurred both French and British to collaborate on aerial photographic techniques. Not only did the hand-held photographs show the extent of German preparations; they also prompted the construction of the first British trench map. There was another result attributable to Darley's initiative. It moved the British command to task Moore-Brabazon and his team of photographic experts to build the prototype 'A' Type camera that became the British technological breakthrough.[273]

NEUVE CHAPELLE – THE FIRST IMAGERY-PLANNED BATTLE

After 180 days of conflict, the Allies started to institutionalise their preparation of the battlefield with photographic coverage. The initial photographic successes by the British design team were timely. Field-Marshal French recognised that a 1915 spring offensive in a location where the British had a tactical advantage was necessary. The decision was made to conduct the offensive at Neuve Chapelle, just south of the Belgian-French border.[274] Field-Marshal French appointed General Douglas Haig to command the British forces. His senior intelligence officer, Brigadier-General John Charteris, played a major role in the BEF's acceptance of aerial photography. Charteris was enthused by the aerial photographic source. His 24 February 1915 diary entry mentioned that intelligence preparation for the upcoming battle at Neuve Chapelle now included this significant resource. 'My table is covered with photographs taken from aeroplanes. We have just started this method of reconnaissance, which will I think develop into something very important.'[275] When Major-General Henderson was initially briefed on the planning he immediately attached Moore-Brabazon, Laws and Corse to Trenchard's 1st Wing to commence operations.[276]

The aerial photographic trench map (mosaic) covered the German trench network to a depth of 700 to 1,500 yards facing the British First Army.[277] The 'big picture' planning base became a precedent for combat operations in the war.[278] Moore-Brabazon recalled, 'At first, the R.F.C., in order to show the real value, made actual maps of the trenches from their photographs, and sent them out in large numbers. Map-making not, however, being the work of the R.F.C., this was eventually handed over to the Army, and we supplied the photographs from which they marked in the trenches.'[279] Constructing the trench map also included the traditional map-making element of the British Army, the Royal Engineers. Their field surveys soon complemented the aerial coverage.[280] Mosaics were carefully traced on 1/8,000 scale maps that detailed the plan of attack. The trenches were eventually mapped to a 1/5,000 scale with 1,500 copies distributed to each corps. The process proved a masterful stroke for intelligence. Analysts could study the battleground and make assessments on the potential main line of approach of a German attack. For the first time in the history of the British military, the British Army went into action with a total picture of the hidden intricacies of the enemy defences. When the first assault commenced, British bombing parties were able to make their way without loss of time to their separate objectives.[281]

The stabilised front at Neuve Chapelle and Ypres. (*West Point Atlas*, courtesy Department of History, US Military Academy)

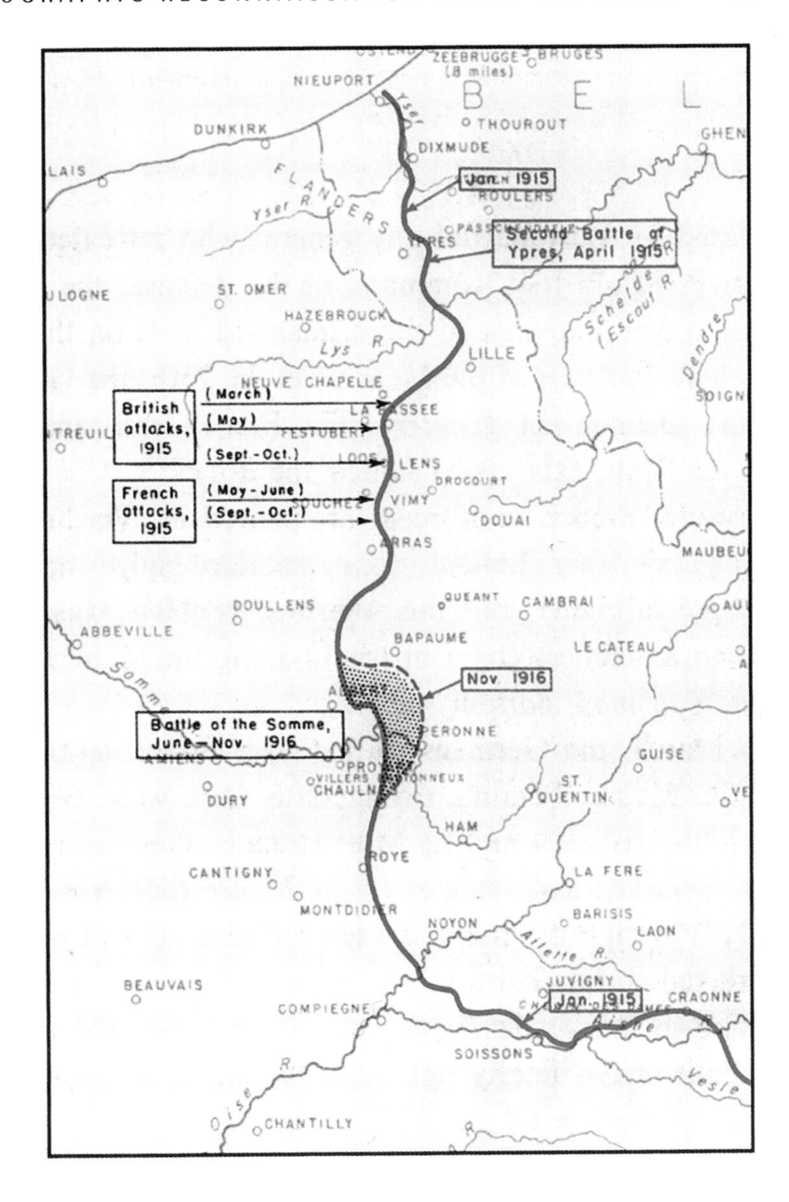

The first day of fighting showed the benefits of the careful planning based on the mosaic. Charteris commented on the Neuve Chapelle battle that, 'Our Intelligence show was successful, in that we found the Germans exactly as we had located them, and their reinforcements arrived to the exact hour that we had predicted they would.'[282] For the first time in the war, a British artillery barrage was employed to help the infantry advance, resulting in the immediate objective being attained with minimal losses. At a critical juncture of the battle, an RFC airman made a very useful report on the location of the advancing British infantry, which enabled the forces to penetrate deeper into enemy territory.[283] Despite the significant gains in planning made from employment of aerial photography, victory was not secured. Failure at Neuve Chapelle was blamed on obstacles remaining despite the heavy artillery barrage. It led the British commanders to consider a strategy of annihilation: destroying everything in the path of the attacking infantry regardless of damage to the sector terrain or the absence of surprise.[284] Artillery was now the key weapon in the arsenal for subsequent campaigns and aerial reconnaissance became an integral partner in this.

Despite the disappointment at Neuve Chapelle, British intelligence gained a positive reputation. Charteris mentioned for the first time the growing influence of photographic interpretation on the evolving battlefield,

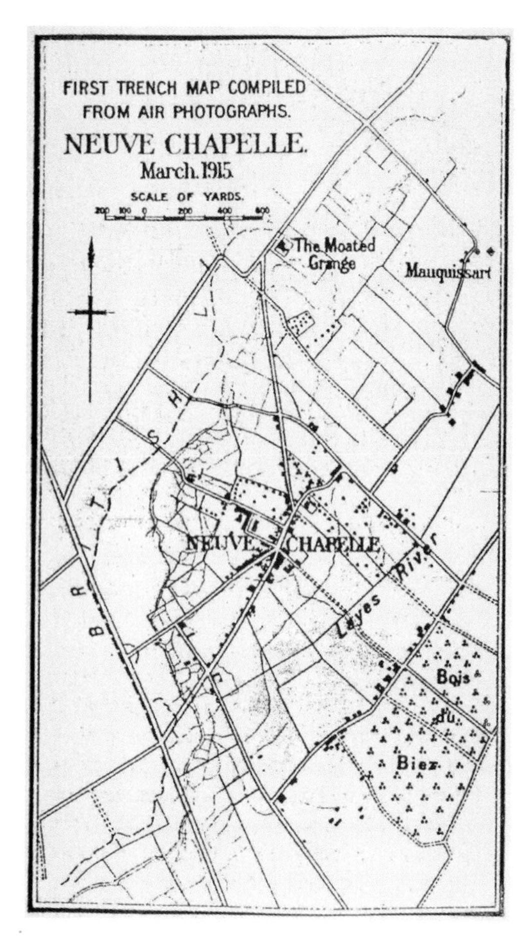

The success of the 'A' Type camera provided the British with their first trench map derived from aerial photography. (H. A. Jones, *The War in the Air, Being the Story of the Part Played in the Great War by the Royal Air Force, II*)

'I have now two regular majors under me, and three temporary officers – a barrister, a diplomat, and a stockbroker. They are all good linguists, keen and clever, so my work grows easier – only it never stops. The barrister's especial job is studying air photographs, at which he is getting extraordinarily expert; finding out all manner of things, some very important, from them.'[285] Neuve Chapelle also initiated a standard report that carried through the remainder of the war. The British and their Canadian Commonwealth counterparts crafted their first Intelligence Summary to support operations at the front. This was the first time that a British Division generated a regular daily intelligence summary for their combatants on the line.[286]

The British improved their production and dissemination of intelligence to serve the increasing demand. British aerial reconnaissance now started producing panoramic photographs that enabled the cartographer to draw out the current features of the landscape. Once that was accomplished, the British produced lithograph templates. Colonel Squier, US attaché, reported that the daily cycle had aerial photographic coverage arriving around 2130 hours. During the night the staff completed a master copy and readied 100 copies for distribution by 0600 hours the following morning.[287] The distribution was extensive and included the lowest echelons of the British forces. In the British sector during the battle of Givenchy (Givenchy-lès-la-Bassée), every platoon commander and all staffs from battalions rearward had copies of an intelligence-produced panoramic sketch of the battleground.[288]

THE SECOND BATTLE OF YPRES

Following failure at Neuve Chapelle, the next major battle took place to the north at Ypres in April 1915. This battleground witnessed some of the cruellest fighting of the war. Ypres took warfare to another dimension, to inflict mass casualties through technology. Ypres became the first battleground to witness the widespread use of chemical warfare on the Western Front. To the total surprise of the Allied combatants, the Germans had released heavy, asphyxiating chlorine gas from their own trenches. This innovation in warfare was first discovered from the air by Captain Louis A. Strange from the No. 6 Squadron based at Poperinghe (Belgium). He reported a bank of yellow-green cloud coming from the German trenches and moving towards the French lines (held by a territorial division and African troops). However, the observations were of no consequence.[289] What intelligence had been collected on the chemical threat had not benefitted the defenders. Despite the chaos caused by the gas attack among the Allies, German forces were unable to conduct a breakthrough. Chemical munitions became another sad fact that contributed to the overall misery that constituted positional war. Only now it meant increased casualties without an appreciable gain in territory.

It was at Ypres that intelligence at the front was put to the test. When the Germans did advance under the chemical cloud, intelligence was reduced to a flurry of hurried telephonic messages trying to figure out what was occurring. It took reliance on aerial photography and map distribution to provide structure for intelligence under these extreme conditions. British intelligence officers at Battalion, Brigade, and Division were now sensitive to acquiring as much information as possible not only to prepare for the battle but also to ensure that there was enough intelligence to be ready for any disruption caused by new weapons such

Royal Naval Air Service aerial reconnaissance was accomplished by floatplanes such as the Short 184. Hand-held cameras such as the 'A' Type were acquired in 1915. (Nesbit, *Eyes of the RAF*)

as chemical munitions.[290] It was intriguing to note senior British staff confidence in locating chemical munitions through aerial photographic coverage. On 28 April 1915, Charteris remarked:

> We have had to take over some more of the French line, and have not yet regained all the ground that was lost. It will take some considerable time before we get things straight again in that part of the line. It does not, however, immediately concern our own army – except that we must now expect to have the same methods used against us; that is mainly a medical job. As far as I am concerned it involves only another thing to look for in air photos – and fortunately preparations are easily distinguishable in these photos – and a careful record of wind currents, for this gas depends entirely upon a favourable wind. We shall of course now have to use gas ourselves, as soon as we can get it going.[291]

Ironically, the surprise attack at Ypres eroded the goodwill between the RFC photographic section and Army Staffs that had been established from preplanning experience at Neuve Chapelle. Survey officers from the Topographical Sub-Section of the General Staff, also known as Section I(c) or 'Maps GHQ,' were competing with aerial photographs for accurate replication of the front.[292] Moore-Brabazon reflected on the strife within the staff:

> After the first gas attack on Ypres, the line was pretty fluid. The British photo team accomplished a map of the enemy trenches. It took an extensive effort, aided by Captain Wyllie [son of the great marine artist] who took an enormous risk to his own safety by flying very low over the trenches in order to identify the uniforms of the occupants. A map of the German lines was prepared. The photo team then started to circulate the map, causing extensive commotion among the army staffs. [Trenchard told Moore-Brabazon to] 'stop butting into affairs' I had nothing to do with. Maps were to be done by Maps GHQ and nobody else had any right to go making them, and it was a disgraceful thing, and I ought to be ashamed of myself. Obviously he had been told by GHQ to 'strafe' me, and that he had done. However, Trenchard followed with a positive burst, 'That's the spirit; I am too delighted for words at what you have done – magnificent! But you know what the Army is, so now do you mind getting in touch with GHQ and going hand in hand together?[293]

Positional warfare confounded the prevailing military mindset. Being stuck in a front line sector for extended periods of time was not something they had experienced over the past quarter century. Their information regarding location and possible enemy intentions came traditionally from sources such as cavalry. Now technology was supplanting the human resource for information. Evolving strategies toward containing the enemy turned to using aerial reconnaissance as routine insurance against surprise.[294] The combatants now engaged in active deception to counter the risk posed by aerial reconnaissance. It didn't take long for both sides to develop their own security policy for strategic cover. A German 'Order of the Day' commented:

> According to the report of a squadron of Airplane Observers, our troops are very easy to mark in fighting, in spite of their grey uniforms, because of the density of their formation, while the French know apparently how to protect themselves perfectly against aerial reconnaissance. During a fight it is necessary that our troops should make the

> task of aerial reconnaissance more difficult by more careful use of the country – making use of narrow files along trees, edges of villages, shelter of houses, avoiding mass formations.... At the approach of an aeroplane all movements ought to cease.... Upon the approach of an enemy aeroplane there should be no firing, for the flash of the gun betrays the position from afar.[295]

THE ROYAL NAVAL AIR SERVICE

British forces maintained two separate photographic programs in place until the formation of the RAF in 1918. In 1915, RNAS aeroplanes equipped with a variety of hand-held cameras, to include the 'A' Type camera, operated out of Dunkirk and photographed significant targets along the coastline to a depth of four miles.[296] They maintained aerial reconnaissance on all known German defence systems from Nieuport to the Dutch frontier, to include determining detailed characteristics of the coast and inland batteries, the general topography, and information addressing slopes of beaches, locations and heights of sea walls. Aerial photographs provided the necessary detail to map the coastal region. Analysis of enemy dispositions along the Belgian coast was forwarded to the Admiralty's Intelligence Department of the Naval Staff.[297]

The relations between the RFC and RNAS concerning photography were positive. Captain Campbell worked very closely and harmoniously with his RNAS leading photographic officer counterpart, Lieutenant Charles W. Gamble. Both realised that cooperation was essential, saving on duplication of work, time and expense.[298] By the autumn of 1915, close cooperation between the RFC and the RNAS had become the order of the day.[299] Both RFC and RNAS treated aerial photography as a sensitive program, requiring the utmost security. One memorandum from the RNAS Station commander working with the RFC Photographic Branch at Chingford, England provided a glimpse of the security policy. He stipulated that all officers and men attached to the Photographic branch at Chingford operate totally at a strictly confidential level. One officer was to be on duty at all times, 'never to leave the building on any pretext whatever'.[300]

ADVANCES IN ARTILLERY SPOTTING

Positional warfare settled down to a routine in the spring of 1915. All the major elements of aerial observation comprising aeroplane and ballon observateur resources were now operational and employed throughout the front. The emphasis on aerial observation covering the German artillery batteries started to have a major impact in 1915. Artillery batteries became concerned by their vulnerability and proceeded to employ extensive camouflage materials and procedures to cover themselves from aerial observation.[301] In addition to aerial observation from aeroplanes, aerostat activity increased throughout the front. In May 1915, during the French operations in the Artois sector, ballon observateur was employed on a large scale for the first time. One of the first aerial observation successes from the ballon observateur was achieved by Sergeant Toutay, a pioneer with French *aérostiers*. During one ascent, Toutay reported to his assigned battery that he observed German convoys. Following his report, French artillery reacted with a barrage that successfully destroyed the convoy. Another French pioneer aérostier, Warrant Officer Arondel, accomplished 25 artillery rangings in the course of five and a half hours, an exceptional feat in the course of this war.[302]

While the British were fighting at Neuve Chapelle, French SPAé under Capitaine Grout advanced the role of aerial photography for targeting artillery. In April 1915, Grout and his subordinate, Lieutenant Eugène Pépin, attached the Douhet-Zollinger film camera to a Voisin 5 and flew a sortie over the Verdun sector. The mission was a major success. It acquired the most comprehensive coverage to date of a battle area from one aerial sortie. The coverage proved deadly for German artillery in the Verdun sector for the photographs provided accurate target locations for French counterbattery operations. Subsequent missions showed the German artillery suffered extensive losses from the barrage. It prompted new courses of action involving extensive camouflage and greater attention to deception techniques.[303] The initial success experienced by the Douhet-Zollinger camera was soon eclipsed by its operational limitations. The camera's limited focal length made aerial observation more vulnerable to ground and anti-aircraft fire. A few more sorties were attempted that year with the Douhet-Zollinger camera in mapping the Verdun and Champagne sectors. However, the camera was soon replaced by the sturdier and more effective range of plate cameras, the French 26cm, 50cm and 120cm.[304]

The British also started emphasising aerial photographic coverage against artillery as a means to detect German camouflage.[305] Charteris made this bold claim:

> The Germans can quite easily cover up gun positions and other defences, so that the observers in aeroplanes cannot detect them. It is next to impossible to conceal them from the camera. There is the negative result also, that these air photos teach us how to conceal our own gun positions, though as far as we know at present the Germans are not

Above: The EVK's Il'ya Muromets 'Kievskiy' strategical aerial reconnaissance aeroplane literally dwarfs all airframes in the first half of the Great War.

Left: Russian aerial observer Warrant Officer Yegorev with Ulyanin 20cm aerial camera.

Below: Potte aerial film camera. This was employed extensively throughout the entire war on the Eastern Front. (All courtesy Marat Khairulin)

A 1916 pose in front of their Morane Parasol by 16 kao (*Korpusnoi Aviatsionniy Otryad* – Corps Aviation Department) members, including the aerial observer with Ulyanin aerial camera, mechanics and the mascot. (Marat Khairulin)

The Voisin 5 served as the aeronautical testbed for the Douhet-Zollinger camera. (SHD, from *Vues d'en Haut 1914–1918*)

> using air photographs. The plans for the battle were all worked out on maps, brought up-to-date from air photographs for the first time in the war.[306]

It was then that individual initiative within the RFC came through with a more accurate means of guiding artillery towards the intended target.[307] Again, it was the innovative thinking of Captain D. S. Lewis from RFC No. 3 Squadron. Lewis, having initiated the squared map during the Battle of the Aisne, further refined targeting through the 'clock code'.[308] The 'clock code' took the target as being the centre of a clock face, with 12 o'clock pointing true north. Imaginary circles with the target as centre at radial distances of 10, 25, 50, 100, 200, 300, 400 and 500 yards were lettered Y, Z, A, B, C, D, E, and F respectively. The RFC pilot conducted the 'shoot', noting shell bursts with reference to the imaginary circles and clock hours, and signalling the result by giving the letter of the smallest circle within which the shot fell, followed by the hour of the clock indicating its direction from the target. Thus a shell which fell 90 yards west of the target was reported as B9 – that is less than 100 yards from the target and due west of it. The 'clock code' proved to be a highly successful tool for aviation and artillery throughout the war.[309]

> Working with the wireless required a standard procedure be put in place. Having reached a height of some 1,500 feet, one would let the aerial down, all the approximately 100 feet of it, and start calling up the home W/T station on the aerodrome and check the strength of one's signals. One would then make sure that the little sheet of mica or celluloid, bearing the six concentric circles scratched and inked upon it (in a surrounding bolder circle representing a clock-dial, with the hour numbers indicated), was properly positioned securely, with its centre-mark over the first target to be dealt with, and the '12 o'clock' mark pointing due north, before sitting back and relaxing. Each of the inner circles represented actual distances in yards from the centre, the innermost circle indicating 50 yards and the next larger circle 100 yards, and so on.[310]

Once the preparation for the targeting process was in place, the observer commenced observation procedures:

> Flying at 3,000–4,000 feet, the observer approached the line. The observer would go through the preamble and the battery being supported would be ready to fire a sighting shot. In ranging a battery on a target, it was very necessary for the observer to be so placed as to be able to watch the entire cycle of operations undisturbed by any change of direction on the part of the aeroplane, and it behooved him and his pilot to agree upon a course that would give him a direct view of battery and target up to the moment of sending down corrections. The pilot would usually steer an oval course, so arranged that both battery and target would be in full view during the aeroplanes' advance along one leg of the oval, thus allowing time for the sending of corrections and for the battery-crew to make the necessary sighting adjustments called for while the artillery was returning along the other leg. Needless to say, once a shoot had begun, it was highly desirable not to deviate from this course, or alter the rhythm adopted. The occupants of the observing aeroplane were at the mercy of those enemy anti-aircraft gunners who were gathered around their guns waiting for the operation to start.[311]

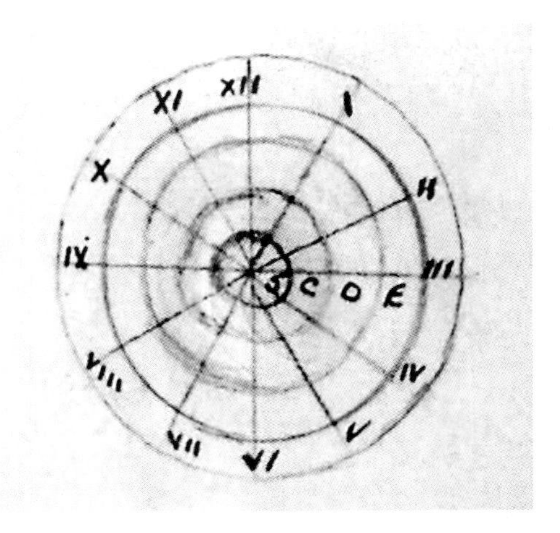

Captain D.S. Lewis in his January 1915 correspondence to a colleague drew a 'clock code' describing how aerial targeting support for artillery was accomplished. (TNA, AIR 1/834/204/4/240)

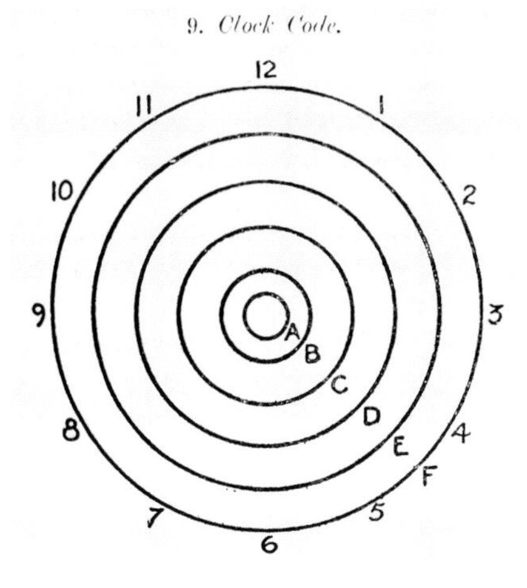

The clock code was a remarkably simple device. It correlated the location of the intended target to the center of the clock and allowed the aviator to radio the actual impact point to the superimposed clock face. (TNA, AIR 1/834/204/4/240)

Captain Lewis did not survive the war. Later that year, while flying over the front lines to show a Guards officer the view of the front, his Morane Parasol aeroplane was destroyed by a direct hit from 'Archie'.[312]

The French did not employ the 'clock code.' Their artillery spotting process involved

> ...corrections being given as such and such a distance long, short, right or left (or in the event of a battery's position being unknown, North, East, South or West of the target). For corrections in hundreds of metres the unit figure is always given the same as the hundred figure (that is, '200 in right') would be sent down as 202 'in right' in order to check any possible error of transmission of the hundreds. For corrections in tens a the-figure number beginning with 0 is used to avoid confusion with code signals which are two figured. Moreover the tens and units are given the same. For example '60 metres left' would be sent as '066 metres left.'[313]

RFC MODIFICATIONS AND REORGANISATION

Major-General Henderson maintained direct control until 19 August 1915 when his RFC command was passed to Colonel Trenchard. Prior to that reassignment, Colonel Sykes had been transferred to the Mediterranean to support the Dardanelles campaign on 26 May 1915. Lieut-Col. H.R.M Brooke-Popham became the new deputy commander for the RFC.[314] Sykes' legacy for the first three years of the RFC remains a debatable topic. The extent that he led operations in those first weeks of turbulent progression from the retreat from Mons through the battle of the First Marne and arranging aerial reconnaissance for positional war has been understated over the years.[315] Key to the success of British aerial reconnaissance throughout this period was the fact that the British senior command kept Henderson in charge over strong personalities directly responsible for RFC evolution, such as Sykes and Trenchard. In the aerial reconnaissance arena, it was Henderson's proven abilities in understanding and applying intelligence and reconnaissance resources that provided the base for keeping the RFC relevant to the British Army in the first year of conflict.

By the summer of 1915, the RFC assigned their Wings to each British Army. General Haig's First Army now had the services of the 1st Wing commanded by Lieut-Col. E.B. 'Splash' Ashmore. 1st Wing comprised four squadrons (RFC Squadron Numbers 2, 3, 10, and 16) and two Kite Balloon Sections (Numbers 6 and 8). The primary British aerial observation aeroplane was the ubiquitous BE 2c. Allocation of aeroplanes showed emphasis on the artillery observation mission with all possessing a wireless transmission capability. Artillery spotting procedures were refined to include distinct use of wireless to prevent jamming between aeroplanes. All aeroplanes cooperated with the ground forces within the squadron's assigned sector. Aerial photography was now a regular feature, particularly to assist artillery commanders in their choice of targets. Additionally, each squadron included one single-seater to provide stand-by pursuit or intercept duty.[316] RFC Squadrons'

No. 2 and No. 3 each had two flights detailed for counter-battery work and one for observing trench bombardment. No. 10 Squadron had one flight for long distance reconnaissance and two for artillery observation. No. 16 Squadron had one flight for tactical reconnaissance, and two for artillery observation.[317]

Along with the ongoing evolution of the squadrons, the experiences of the RFC Photo Sections over the first six months now provided seniors with an appreciation of aerial photography's potential. During July 1915, the RFC conducted a conference at the War Office in London to discuss improving aerial photography. Attendees included the representative from the General Staff, Lieut-Col. C.C. Marindin, Captain J.T.C. Moore-Brabazon, and Captain Campbell. The summary of the conference highlighted the need for extensive modifications:

> Photographic work in the Royal Flying Corps is at present in a somewhat chaotic state ... it is difficult to initiate or develop useful experimental work at home unless the requirements of the Expeditionary Force are more fully known. The best means of co-ordinating photographic work would be through (1) appointment of an equipment officer for photography at the headquarters; (2) appointment of a photographic expert at the War Office; (3) appointment of an assistant equipment officer in charge of a photographic experimental section; (4) standardisation should be tried for as far as possible; (5) all officers at home, both pilots and observers, should be taught something of photography [their notes as to the conditions under which photographs are taken would then be of great assistance to the Photographic Section]; and (6) establish the requirement for testing all apparatus and material in the Field before its final adoption.[318]

The discussion at the War Office had an impact. The RFC's Wings commenced assigning photographic officers to the headquarters staff. As the British armies increased their demand for aerial photography in the subsequent months, emphasis was laid on providing Photo Sections for all squadrons assigned at corps echelon. This led to the establishment of the Army Reconnaissance squadron, the most specifically assigned unit established in the war for strategic coverage. As for the issues raised in the conference, the desire to field a fully aware aerial observation force would remain in committee. The British aerial culture soon gravitated toward greater emphasis on the combat role.[319]

At the front, the British and Commonwealth forces were refining their intelligence infrastructure to accommodate the new sources of intelligence such as aerial photography. A direct channel of information was now established to ensure reports passing from the Battalion Intelligence Officer through brigade to the division. The British increased intelligence manning at the Corps to improve coordination and support.[320] The addition of personnel to the intelligence staffs now provided each echelon with an improved capability to handle the increasing volume of aerial photographs. A new functionary was added to the intelligence ranks known as the draughtsman. This position helped in the interpretation of the photograph as well as tracing key information for the command maps.[321]

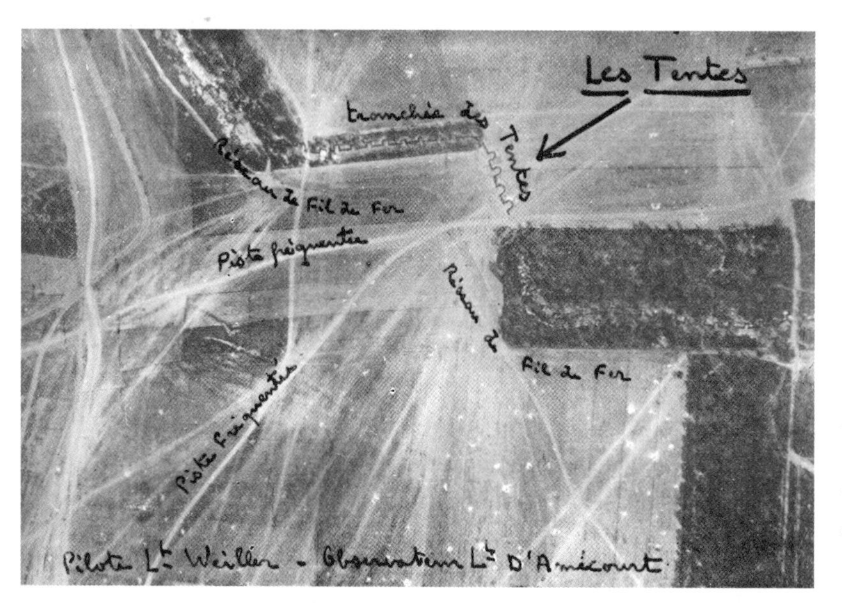

French aerial reconnaissance won converts by confirming to a sceptical senior staff that the French infantry occupied the sector known as the Tranchée des Tentes. (Capitaine Paul-Lewis Weiller from *L'Aeronautique pendant la Guerre Mondiale, 1914–1918*)

LESSONS FROM THE LOOS (ARTOIS) BATTLE

Throughout 1915 the Allies refined their entire aerial observation program to accommodate the lessons learned from the initial campaigns. RFC strength had expanded to meet the demands for aerial observation and photography. The three RFC Wings now comprised 160 aeroplanes, backed by 40 aeroplanes in reserve at two aircraft parks. Each Photo Section supporting the Wings now sported mobile darkroom vans on lorries. These additions complemented the darkrooms at various buildings in the area and provided direct service to aeroplanes arriving at the landing grounds.[322] The Battle of Loos (Artois) commenced on 25 September 1915. The French also commenced an offensive along the Champagne front. Loos was fought on a larger front sector than that of the March 1915 Neuve Chapelle offensive. However, the number of British artillery pieces covering the wider sector was approximately the same as that employed earlier. To achieve the appropriate level of destruction as envisioned by the planners, a longer period of fire was conducted. The four days of artillery barrage accomplished an appropriate level of damage, but surprise was abandoned. The British artillery also applied a new artillery tactic called the 'lifting barrage' against the Germans. Artillery fire concentrated on one trench at a time. In the ongoing 1915 Artois campaign, 16 reconnaissance escadrilles were assigned to cover a 35-km front.[323] However, despite the use of aerial observation, the attack was effectively countered by the German forces. Thanks to a lack of surprise at the commencement of the offensive, the Germans were able to regroup their reserves and mass against the attacking Allies.[324]

The British employed gas for the first time in the attack. Lieut. J.G. Selby, an observer for No. 3 Squadron, recalled the moment seeing the gas floating towards the German lines. Now the Germans were to experience the terror. However, Selby noticed the wind shifted and the gas reversed course towards the British lines. The sight of that prompted the pilot to turn around and shout, 'Thank God we're in the Flying Corps, old boy!'[325]

The role of French aerial observation, reconnaissance, and artillery spotting was practically codified after the Loos. The experience gave Aviation Militaire the chance to clearly review what aerial reconnaissance provided to positional war. It was concluded that military aviation was most useful to armée commanders when providing tactical information such as location of enemy troops and regulation of artillery fire. Also, aerial reconnaissance reports on location of French troops were clearly demonstrated and strongly endorsed. The basic unit for aerial operations would be the armée, cooperating with the escadrille assigned to the CA. Escadrilles supporting heavy artillery were to closely coordinate their operations with CA escadrilles. Finally, Aviation Militaire decided that each CA would require one aéroplane for photo reconnaissance, two for army cooperation duties, and one for artillery regulation (*avion de réglage*).[326]

Aerial photographic reconnaissance was now integral to the evolving battleground with a modern equipment suite, to include cameras, plate photographs and mobile field laboratories. Air-to-ground liaison was expanding with an array of capabilities to include télégraphie sans fil (TSF) with one-way Morse code transmission, cloth panels, and new procedures for dropping messages. Despite improvements in components, French airframe and power plant design did not develop. Their hesitation to advance aeronautical design translated into new vulnerabilities against a rapidly improving German pursuit force fielded in the coming months.[327]

The defining moment for French aerial reconnaissance occurred during the Loos offensive. It showed that scepticism could have lethal consequences for the French military. On 25 September, a French company had succeeded in taking a German trench network known as *Tranchée des Tentes* (Tent Trench). A German counterattack forced most of the French troops to withdraw, but the forward unit, which had no warning of the retreat, stubbornly held on. The French command, unaware of the French forward unit's presence, commenced general artillery fire on the position. Throughout this barrage, French aerial observation began to detect the ground signals laid out in desperate fashion by the forward unit. The aerial observation proceeded to notify the CA post of command (PC) that French soldiers were still in place in the sector. Tragically, the PC did not believe the reports. For the next three days, the French artillery bombarded the position. Only after a French officer of the unit managed to safely escape and make it back to the French lines to personally relay the fratricidal horror taking place was the French command convinced of the forward unit's location. The incident made the case that aerial observation was a credible source for gauging the battle. For French aerial observation, the Tranchée des Tentes silenced French military sceptics for the duration.[328]

AN ENDURING LEGACY OF FRENCH AERIAL OBSERVATION

Several French masters matched the demonstrated initiative of British aviators in evolving the role of aerial reconnaissance and photographic reconnaissance; most notably, Capitaine Paul-Louis Weiller. Like the RFC's Captain Lewis, Weiller came from an artillery background. In the first year of the war, he was instrumental in exploring aerial photographic

Sous-lieutenant Paul-Louis Weiller, legend of French aerial reconnaissance in the First World War, is decorated with the Croix de Guerre in 1915. (Mousseau, *Le siècle de Paul-Louis Weiller 1893–1993*)

The RFC expanded its aerial reconnaissance role in 1915 through deployment of a new line of British-produced aeroplanes such as the Vickers F.B.5. The photograph shows the observer receiving the hand-held 'A' Type camera for the sortie. (Nesbit, *Eyes of the RAF*)

reconnaissance possibilities. Flying as an MF 11 observer, *sous-lieutenant* Weiller took numerous aerial photographs with his personal 'vest pocket' camera (a Kodak). Weiller was one of the leading proponents behind establishing the French military's restitution process that transferred important information from the aerial photograph to the map. His efforts encouraged the French staff to rethink the role of aerial photography in updating maps. Whilst assigned to the Voisin aerial bombardment escadrille, V 21, Weiller worked with engineers in establishing a method of updating maps that became the standard for the French *Service Géographique*.[329] In August 1915, Weiller acquired a pilot's licence and returned to V 21. He established the process for achieving the longer range armée reconnaissance mission profile, determining specific altitudes and ranges for the existing French aéroplane inventory. The lessons of 1915 honed his aerial observer skills for the remainder of the war, establishing an example of aerial reconnaissance excellence that became recognised at the highest levels of the French military.[330]

AERIAL COMBAT BEGINS

The ascendancy of aerial observation coincided with the introduction of the pursuit fighter. Aerial combat now became a common fixture of the war, affecting all facets of aerial reconnaissance. During the late summer and autumn of 1915, British aerial reconnaissance beyond the front lines now started to come into contact with a new type of German aeroplane. RFC thinking up to that time had not placed emphasis on aerial combat. It was not until the battle of Loos that it became a serious and permanent part of the aerial scheme.[331] In November 1915, the British encountered a Fokker *Eindecker* armed with a synchronised machine gun. A Vickers aeroplane was undertaking army reconnaissance deep inside the enemy's lines. A brief fight ensued, but no casualties came from the engagement.[332] Aerial combat transformed aeronautics for the remainder of the war and established aviation's primary military purpose for the twentieth century. While aerial combat became refined in 1916, aerial observation now came to its greatest prominence. It took the horrors of two cataclysmic battles, Verdun and the Somme, to determine the limits of aerial observation for refining positional war's potential. For the remainder of the war, the impact of aerial reconnaissance became dependent on the seniors who employed the information.

CHAPTER 4

October 1915 to October 1916: Fortifications Across the Front Line

The year 1916 is remembered for the climactic battles that culminated in catastrophic loss of life with no discernible gain. Verdun symbolised the cost of war for nearly a century to the French nation. Likewise, the Somme was to be Britain's benchmark for military planning gone appallingly wrong. Aerial photography was a novelty in its first year. By 1916, aerial reconnaissance had become a standard fixture of the operational mission. Decisions were made based on aerial photographic validation. However, the dominance of a failing doctrine overshadowed the potential that aerial reconnaissance provided. The application of mass against well established defences proved catastrophic throughout the second year of conflict.

VERDUN

Verdun was 'the crucible in which the aviation of the French was fired.'[333] Prior to that battle, French

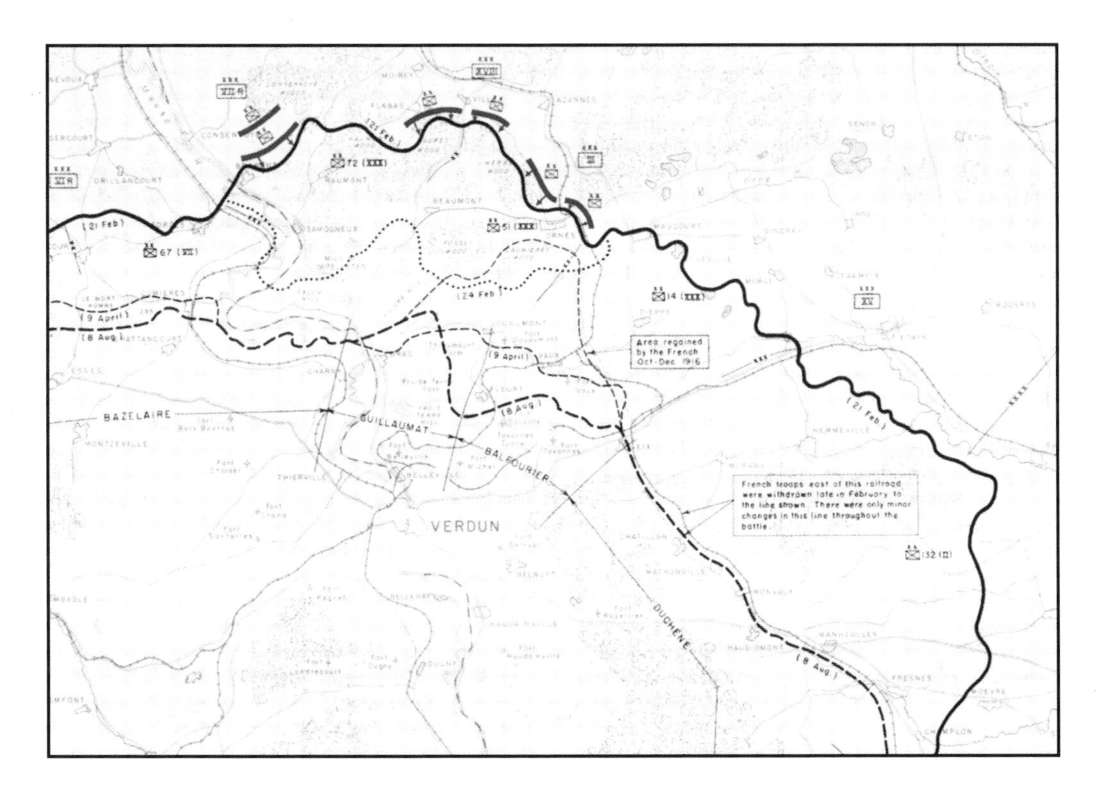

Verdun. (*West Point Atlas*, courtesy Department of History, US Military Academy)

military leadership maintained a very narrow view of aviation's role in operations. Fighter/pursuit aircraft were employed to provide a protective escort for an aerial observation, but aerial operations over the Verdun sector were concentrated in a small area. The consequence was that this allowed German fighters to marshal their pursuit forces sufficiently to gain air superiority.[334] At the end of 1915, German pursuit aviation led by the Fokker Eindecker transformed aerial operations over the battleground. The Allies had to contend with the realisation that German pursuit aviation was dominant. The existing pusher aeroplane configuration, ideal for aerial observation, was no match for the more manoeuvrable tractor (aeroplane engine in front of the fuselage) fighter aircraft with synchronised machine guns. The French aerial observation aéroplane performance was mediocre at best. Its average engine power plant could only achieve a ceiling of 3,000 metres carrying a small load. By the end of 1915, several French prototypes were in the testing stage, but replacements were slow in coming.[335]

The battle of Verdun commenced on 21 February 1916. German preparations leading up to the first salvos were extremely effective. Warning signs varied with each intelligence source. However, German deception effectively masked critical signatures prior to the offensive. The combination of bad weather over the battleground and German aerial supremacy from Fokker Eindeckers denied the French aerial access to establish the more expansive strategic aerial picture needed to determine indications of an attack. Aerial photography was now integral to German battle planning. Coverage went beyond Verdun to strategic French locations 25 miles deep in the rear echelon.[336] The Germans recognised that an offensive in the Verdun sector was a major challenge. The attack plan had to contend with a maze of ridges that allowed the French to command practically every avenue of attack. The other significant topographical feature was the plain of the river Woevre and its dominating plateau scarp. It forced the Germans to attack from the north. In the end, the geographical obstacle proved one of the final determinants of French victory.[337]

German planning at Verdun established an effective camouflage and deception program that brought into question the credibility of the entire French aerial reconnaissance and photographic interpretation effort. French aircraft three days before the offensive commenced were only reporting German forces far from the Verdun front lines. Their short-range tactical coverage did not pick up any unusual activity commensurate with the offensive, such as indications

A panoramic northeast view of the Verdun battlefield taken from an altitude of 800m. (SHD)

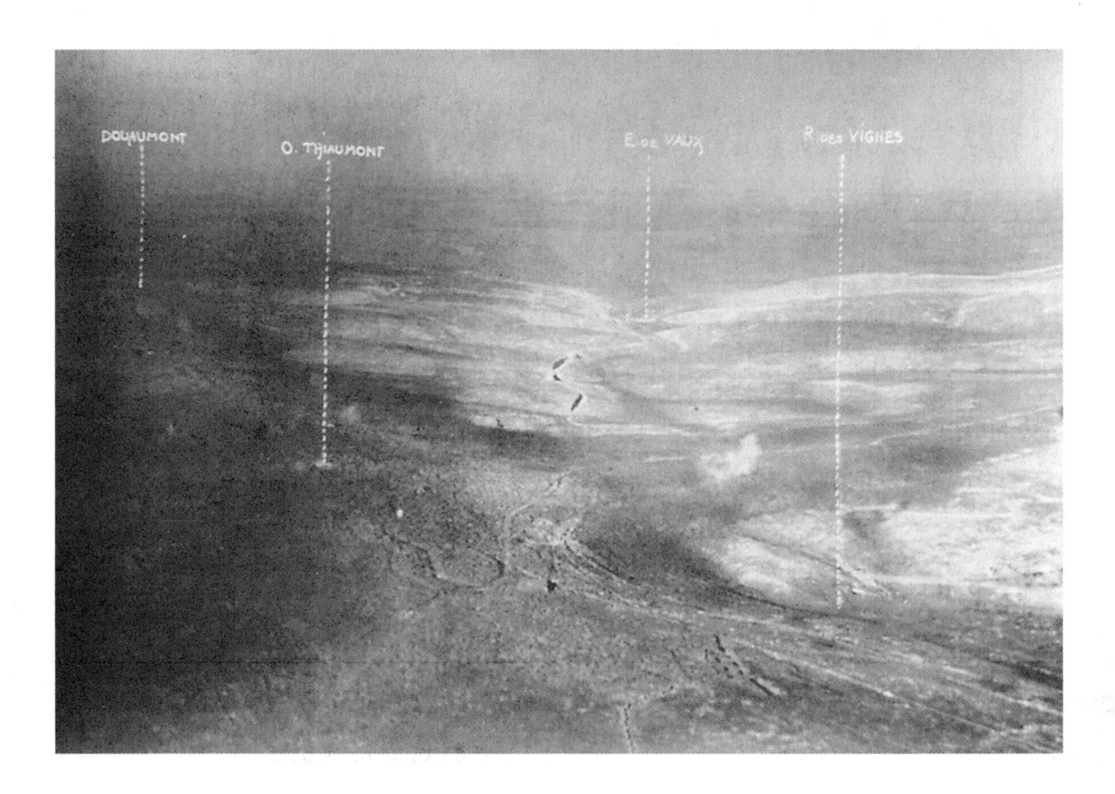

Entrance to a *Stollen* tunnel housing the first wave of German infantry prior to the 21 February commencement of the battle of Verdun. (*Bibliotek fur Zeitgeshichte*)

of increasing supply dumps or logistical activity. It took the cross-referencing of prisoner interrogations with subsequent aerial coverage to discover German construction of *Stollen*, or tunnels, which secretly conveyed the troops to the jump-off point.[338] Initial French intelligence assessments of Stollen were faulty, assuming the activity represented an upgrade in defensive works. German strategic camouflage was now playing a major role, masking construction activity. The Germans worked at night and effectively disposed of the earth material. When the Verdun battle had commenced, the Stollen had served their purpose. The numbers of German shock troops positioned in the front lines far exceeded anything the French believed possible. The advance stockpiles were masked, and contained new weaponry, such as flame-throwers.[339]

The German chief of staff at Verdun, General Erich von Falkenhayn, also intended to blind the French artillery from the start by depriving them of their ballon observateur and CA aéroplanes. His tactic for eliminating the French ballon observateur was to move heavy artillery up to the front lines as close as possible under the cover of darkness and extensive camouflage. When the barrage started, German long-range guns were fired as a pair, one firing time-fused shells at a ballon observateur already aloft and the other firing percussion shells at the winch on the ground. French aérostiers were trained in the classical evasive movement called the 'cartesian diver'. Upon observing the fire, the aérostier counted on a lead time of about 45 seconds, whereupon the winch operators were contacted via telephone to raise or lower the ballon observateur. The descending rate averaged five metres a second. Given a successful reaction, the shell burst either too high or too low from the ballon observateur. However, the German tactic at Verdun cut the effective response time. The opening German salvos were unanswered. The French desperately needed to regain air superiority so that they could reclaim the artillery spotting advantage.[340]

In a simultaneous show of aerial force, von Falkenhayn created an effective aerial blockade against French aviation through a concentration of Fokker Eindecker fighters. The Germans flew parallel to the front lines, countering or destroying any French reconnaissance aeroplanes. Additionally, the German fighters prevented French planes from taking off from nearby landing grounds. Those French aéroplanes that did arrive were met with formations of three or four German aeroplane patrols. Verdun's first day culminated in two significant gains for the German advance. They possessed air superiority for the first time and countered French attempts to gain the 'high ground' for aerial observation.[341] French post-war recollections of their aviation losses were to the point: 'They paid dearly for this disregard of economising of effectives, and in spite of their heroism they were wiped out by the German air force at the beginning of the battle of Verdun.'[342] Verdun's lessons reinforced critical facets of military aviation in general. Pursuit needed to be advanced to protect aerial observation missions. If artillery was to remain effective in targeting, it required closer liaison with an Aviation Militaire that possessed greater authority to designate objectives and regulate fire. At the same time, Aviation Militaire needed a more effective liaison with the infantry to provide a greater awareness of the battleground. Aéroplanes operating over enemy lines were now required to place greater emphasis on formation flying. Finally, French headquarters needed more aviation-experienced personnel on the staff to support planning as well as judiciously apply aerial resources to CA operations.[343]

With the serious setback of the first days of the

German offensive, the French military realised they had to react immediately and decisively. The chef de l'aéronatique at GQG at this time, Commandant Paul-Fernand du Peuty, stressed a more aggressive combat role with greater presence of pursuit fighters in protection of aerial reconnaissance. He also stipulated that aerial observation was not to be flown solo over the front lines. The French also changed their aviation organisation to be more responsive to the ground commander and the threat. In each CA, an aerial element, the *Secteur Aéronautique* was established to centralise the use of aeronautic units working missions supporting infantry and artillery.[344] In turn, Général Joffre called upon Commandant Tricornot de Rose, escadrille commander of Moraine Saulnier (MS) 12, and gave him authority to take whatever steps necessary to regain air superiority and reestablish aerial observation. De Rose quickly responded by assembling 15 flights of elite pursuit escadrilles (*groupes de chasse*), tasked with 'sweeping the sky.' In the weeks that followed, the elite pursuit units proved successful, regaining air superiority to the point that German aviation was incapable of penetrating the French sector. Aerial observation resumed with the renewal of French air superiority.[345]

The French also made a significant change in supporting heavy artillery targeting. Elite reconnaissance aircrews were assigned to *premières escadrilles d'Artillerie lourde* (AL) to detect and photograph significant targets for the most powerful artillery units.[346] The French numerically designated the escadrilles from 201 through 228. The aerial reconnaissance specialists were experts, forwarding essential information to longer range heavy artillery units capable of delivering more lethal firepower.[347] Such was the case on 2 April 1916 when one French aerial observer reported Germans rapidly advancing in a certain sector. His timely updates on the exact positions contributed to a massive French artillery barrage that was quickly followed by a ground attack. German prisoners taken reported the French barrage 'littered the railroad track with German cadavers'.[348] The concerted effort of the entire French military to do whatever it took to halt the German advance at Verdun finally saw its efforts rewarded. The German advance halted.

As the Verdun campaign entered its third month, the French military made demands on the government to produce a new line of more effective aéroplanes to counter any future German aviation dominance. Général Joffre personally requested the government halt production of obsolete aéroplanes such as the HF 7 and MF 11 and replace them with aéroplanes with the engine in front to better accommodate rear defences.[349] The French aeronautic industry was in a quandary. A revolution in design was called for to both keep ahead of any future German fighter planes as well as supply the numbers necessary to maintain superiority. Verdun became the catalyst for all of France to do whatever was necessary to stop the German advance and regain lost territory. The pervasive influence of the battle culminated in a significant upgrade for French aviation.[350]

The French adopted a new intelligence strategy that modified aerial photography's role to be more readily available and have sufficient quantities of aéroplanes to meet any demand. The battle cry for intelligence became, 'Follow the Attack,' supporting the infantry and artillery with large numbers of aerial photographs.[351] Further supporting the realignment of reconnaissance expertise was the addition of new SPAé. This increased the numbers of intelligence personnel supporting the aeronautical sectors, in addition to those supporting the armée.[352] Mobile SPAé were set up at landing grounds near CA headquarters allowing the aircrew to photograph the target, land, quickly download and develop the photographic plates, produce prints and personally distribute the intelligence products to the Corps commander within an hour. Other recipients included artillery batteries and infantry companies. The surge was illustrated when one French photo station was credited with producing 5,000 prints in one day.[353]

The French took steps to increase the timeliness of their aerial reconnaissance force by assigning dedicated additional aéroplanes to respective units. Three photographic reconnaissance aéroplanes were attached to the division to photograph the German trench system and enemy batteries. Determining the extent of destructive fire required a minimum of eight aéroplanes (fitted with radio) and two ballon observateur (the famous Caquot ballon captif was introduced at this time) to provide constant updates. A 'sentry' aéroplane was assigned to search for rear echelon German reinforcement activity and reported via TSF any indications that the enemy was preparing a counterattack. Each French counter-battery artillery was assigned two radio-equipped aéroplanes. Any information of significant priority acquired through the aerial reconnaissance was telephoned immediately to the respective centres upon landing.[354]

Another factor that improved French aerial observation was the emphasis on 'continuity' by sector. Verdun was now divided into air sectors, each covered continually by the same observers. This enabled the observation officers to have an exact knowledge of the sector, an essential ingredient for maintaining an intelligence baseline to better support the ongoing Verdun campaign.[355] By the second month of fighting, the French aerial campaign under Commandant de Rose was regaining supremacy. French aviation in the sector improved, through pursuit to counter the Fokker Eindecker threat, and through aerial observation to supply more targeting information to support the aggressive artillery employment behind the counterattacks. The fighting was still bitter, but now it was in German-held territory. In May 1916, after months of intensive combat, the French key to the Verdun campaign, Fort Douaumont, was retaken and the German advance was finally halted.[356] Regaining Fort Douaumont also included a significant achievement for aerial photography. On 22 May 1916, 40 minutes prior to the French attack, aerial photographic interpretation discovered German troops at a position critical to the French attack.[357] The IVème Armée commander, Général Charles Marie Emmanuel Mangin, acquired prints that had been quickly processed and flown to his advanced headquarters for review. With this information, he took decisive action, quickly modifying the attack plan and launching a successful assault on the sector. Mangin acknowledgeded aerial observation as his 'eyes' throughout the attack.[358] Mangin's confidence was soon curtailed. His attack ultimately failed through an inability to sustain the offence. A fierce German rearguard action from underground passages into the fort ultimately brought defeat to the French operation.[359] Despite the setback, aerial reconnaissance and photographic interpretation had successfully served its purpose. Commanders such as Mangin trusted the intelligence they saw.

German regimental histories of Verdun recalled the aggressive French reconnaissance over the front lines. Both Prussian Guards and Bavarian forces recalled having to stay undercover as the French observation aéroplanes circled overhead, calling in artillery against any observed movement. 'The French observation and artillery fire direction from the especially low, swaying aircraft were so accurate and skilful that throughout the course of the afternoon of May 28, their fire crept toward our positions.... The impact of three heavy shells – direct hits – annihilated the battalion staff.'[360] A few days later, the Royal Bavarian 20th Infantry Regiment history recalled:

> As so often happened – especially at Verdun – the brilliantly carried out attack was not the worst. Already by the evening of June 1, the presence of a circling French flyer over the captured ground warned us that the enemy was not about to take the loss lightly. In the early morning of June 2, the retaliation fire began. This quickly swelled and grew into a powerful bombardment of annihilation.[361]

GÉNÉRAL HENRI PHILIPPE PÉTAIN

Général Henri Philippe Pétain is one of history's greatest military enigmas. His successes in the First World War are clouded by his later capitulation to Nazi Germany and his position as head of the Vichy government. Despite that disgrace, Pétain was a positive force behind the French victory in the First World War. His military legacy was forever established thanks to his leadership at Verdun. He reconstituted the French Armée to withstand the Verdun horror and achieve the ultimate

The destruction of Fort Douaumont was symbolic of Verdun's horror. The devastation is clearly discernible in late 1918. The dark area in the foreground is all that remains of the fort. (US National Air and Space Museum, Garber Facility Archives)

Maréchal Pétain, a post-war portrait of the Allied commander who used aerial photography to great advantage in the war. (Henri Philippe Pétain, *Verdun*)

victory, rightfully earning him the soubriquet the Hero of Verdun. In the following year, Pétain again rose to the occasion following the debacle of the failed Nivelle offensive. On the strength of his reputation, he kept the French Armée in check during a wave of mutinies and desertions, achieving a force balance while awaiting the arrival of the Americans. As to aerial photography, it was Pétain's understanding and recognition of its intelligence value that influenced the senior leadership and staff to take action based on a photograph.

Général Pétain personified the best of military leaders, committed to the mission without needlessly sacrificing troops in the process. His brilliance lent credibility to aerial photographic reconnaissance attempts to influence decision-making in an environment beset with outdated doctrinal procedures. He led French development of the art and applied it to the operations for which he was responsible. In a post-war reflection on aerial reconnaissance, Pétain made it clear that structure gave credibility to the entire process, a credibility that both saved lives and achieved victory:

> Observation planes of the Farman and Caudron types now guaranteed with the greatest exactness the operation of liaison with the infantry, of the photographic service, and of the timing of the artillery. They were constantly fighting the pursuit planes of the enemy, were hampered many times by unfavourable atmospheric conditions under the lowering gloomy sky of Verdun, as they skimmed over the pockmarked and tumbled crests of the hills, and suffered heavy losses, but never failed in their mission.[362]

The 'greatest exactness' became the standard by which French aerial reconnaissance proceeded for the remainder of the war.

Pétain had a passion for detail that impressed all of his counterparts.[363] He was methodical in his leadership. He personally visited with all echelons to better gauge the potential of the unit as well as to demonstrate a deep personal commitment to his troops. He questioned regimental officers and NCOs to make sure they knew they were clear as to their individual roles.[364] This attention to detail provided the impetus for legitimising aerial photographic reconnaissance. Soon Pétain

A new generation of longer-range aeroplanes such as the Caudron G.6 entered service to enhance aerial reconnaissance with increased firepower. This aeroplane was assigned to escadrille C 11. (SHD, B83.5656)

Aviation Militaire Belge contributed substantial aerial reconnaissance to the northern sector of the Western Front. Capitaine d'Hendecourt and Adjudant Rondeau in front of a flamboyant Farman F.40 appropriately named *Brownie*. (Courtesy of Walter Pieters; *The Belgian Air Service in the First World War*)

established a select group of aerial reconnaissance experts to provide him with a better understanding of the battleground as well as educate the other staffs in this new art. One of the leading figures in this effort was Capitaine Grout's deputy for the SPAé, Lieutenant Eugène Pépin. Pétain organised an elite group of brilliant and proven staffs to introduce new methods of conducting warfare. Known as the *groupe d'armées du centre* (Army Group Centre (GAC)), the staffs became key to Pétain's objective of successfully integrating new technologies.[365] Pépin commenced an extensive training program, publishing *Études de Photographie Aérienne*, one of the first of many manuals on the art of photographic interpretation. Pépin required training of the staffs and unit commanders to gain a complete appreciation of this technique. Pépin also served as an advisor to Pétain on acquiring the maximum benefit from analysis of the aerial photograph. It was a professional relationship that benefited aerial reconnaissance for the remainder of the war.[366]

TRANSFORMATION WITHIN THE ROYAL FLYING CORPS

Like the French, the British also had to contend with the threat posed by the introduction of the Fokker Eindecker. On 14 January 1916, RFC headquarters mandated that any aeroplane working aerial reconnaissance had to be escorted by at least three fighting machines in close formation. Shortly afterwards they extended formation flying to bombing missions. British aerial combat was now becoming specialised, profiting from the French experiences at Verdun. It became an established policy that, in order to win freedom of action for the three key missions of aviation, (reconnaissance, artillery observation and bombing), the enemy had to be denied freedom to operate in the air. The British now committed to conducting a relentless aerial offensive.[367] Brigadier-General Trenchard, RFC commander, reinforced this policy: 'It is now necessary to provide protection in the form of patrols for machines doing artillery work. The demands of Corps for artillery work, photography and close reconnaissance are now such that they occupy the entire services of a squadron. The aeroplanes of artillery corps squadrons are therefore not available for other duties, without impeding the work for which they are now intended.[368]

More aircraft were being committed to controlling the sector, with the 12 squadrons assigned in October 1915 increased to 27 squadrons at the commencement of the Somme battle. The squadrons were better equipped, better armed and better trained.[369] Trenchard advocated that the aeroplane establishment of artillery squadrons be increased to 18, providing three single-seat fighters with engines of the same type. Once the artillery squadrons had received their compliment of 18, an equivalent number of aeroplanes was assigned to newly established Army and GHQ squadrons.[370] In April 1916, British RFC commanders made a critical shift in controlling the airspace of the battlefield. All fighting machines at corps squadrons were now reassigned to Army squadrons. Throughout the remainder of the war, air fighting was the duty of specific squadrons of higher echelon wings subordinate to the British Army.[371]

The demands on aerial reconnaissance increased with new artillery tactics being applied at the front. The characteristic straight barrage of 1915 was now replaced with the 'piled-up barrage', one that employed parallel lines of fire which shifted to firing at irregular intervals along the line. The advancing barrage would then 'pile up' on the German position simultaneously. The tactic had its disadvantages. The infantry would not advance until the entire German line was engaged by fire. Those with the greatest area to advance had to leave the trench first. The desired effect was to have British forces reach the German trenches at the same time. The idea of mass concentration had its perceived benefits. However, British forces were thus more vulnerable to unidentified enemy positions. Dependence on intelligence from aerial reconnaissance was now even greater.[372]

Existing limitations on the accuracy of artillery fire were self-evident. Communications between aeroplane and receiving station were short-range and sporadic. Captain Lewis' clock code was now actively employed to improve targeting accuracy and photographic maps were in greater demand. However, camera distortion from motor/engine vibration led to mapping inaccuracies. The lack of precise coordinates from the aerial maps prompted artillery units to demand continued registration from the aerial platforms. Sighting aerial observation platforms at work gave the enemy indications of an impending attack, allowing more time to prepare defences.[373] The increased use of aerial observation exposed the lack of experienced aerial observers coming into the RFC. Despite thes shortcomings, more information than ever was being forwarded to the artillery. British tactics now focussed on a massive bombardment to 'crush all resistance'. Artillery would restore infantry mobility by annihilating any opposition. It was time to apply the new strategy and achieve the breakthrough.[374]

Far from being challenged like the French at Verdun, the British RFC dominated its sector. The RFC enhanced its support to the British armies by establishing the Brigade echelon. Each RFC Brigade was better organised to meet the increasing demands for close reconnaissance, photography, and artillery cooperation on its own front. The RFC also met the increasing need for longer range reconnaissance beyond the corps area of responsibility. The Brigade comprised a 'corps wing' to accomplish short-distance reconnaissance, photography, and artillery cooperation missions within five miles of the front line. An 'army wing' was also created to better support the Army commander beyond the first five miles of front lines with long-distance reconnaissance, air fighting, and photography. Since Army Wing responsibilities required an extended radius of action, British aeroplanes that possessed the greatest performance and firepower to better protect themselves were assigned. The RFC also established a new wing, the 9th Wing, allotted to the GHQ with air assets to conduct strategic work and a host of special missions that included destruction of kite balloons and close-in photography. Finally, the RFC headquarters staff was strengthened and reorganised on the model of a corps staff.[375]

Close cooperation and constant communication between aerial observers and artillery battery commanders became normal practice. British radio technology was refined to better support the ground combatant as well as counter German electronic jamming. British transmissions now included a 'clapper break,' which varied the pitch or tone of the note sent out by the aeroplane transmitter. The clapper break allowed the ground receiver to distinguish one aeroplane's signal from another and increased the number of aeroplanes that could operate in a given sector without jamming each other's transmissions.[376]

By the spring of 1916, British army and artillery demand for aerial photographs became so enormous that the traditional photographic plate processing and print development centres at the wing were overburdened, resulting in critical delays in print dissemination. Moore-Brabazon recalled:

> The demands for photographs increased out of all proportions to the possibilities of being able to deal with it from the point of view of hand-held and hand-changing cameras, so that a semi automatic device was got out, whereby the camera was fixed to the machine, and by a simple movement of the hand the plate was changed, the shutter wound and the photograph taken. This increased the number of photographs enormously, and although the development, printing and identification of them were at the time done at Wing Headquarters, the proposition soon become so big that it was decided, during 1916, to departmentalise down photography and make it a Squadron proposition.[377]

This was especially true for artillery, which required a constant update of photographs to supplement its firing maps. Increased aerial coverage was necessary for ongoing and future planning, especially in preparation for the upcoming offensive at the Somme. With the increase in aerial photographic support came an increase in manning establishment for British intelligence. Now every squadron with an observation and photographic reconnaissance mission possessed an intelligence section (manning included one intelligence officer, two draughtsmen, and one clerk). A photo section was added to each RFC corps and army reconnaissance squadrons were augmented with a photo section manned by an NCO and two draughtsmen.[378] Brigadier-General Charteris commented on the expansion underway, 'On the Aisne (1914) each corps only had one officer for 'I'

Brigadier-General Charteris was in charge of 'I' work for Field-Marshal Haig until late 1917. (John Charteris, *At G.H.Q.*)

work. Now there is one with each brigade and division, and altogether seventeen at Army Headquarters and every corps is asking for a larger staff.'[379]

Back in England, Captain Campbell worked with the Ministry to keep pace with the demand for photographic expertise. The RFC School of Photography would train a complete photo section (one Officer, one NCO and five men). The personnel were busy loading magazines with unexposed plates, uploading and downloading aerial cameras to the aeroplanes prior to and after the flight, and working on the photographic negatives (developing, washing, drying, and plotting). Several men were constantly engaged in the enlarging room, exposing and developing as many as 100 prints in an hour.[380] The measure of success was speed, maintained by constant work at high pressure. However, Campbell's greatest challenge was not training a team of aerial photographic experts for interpretation and the photographic laboratory. It was the attitude of the pilots and observers to aerial photography. He observed that pilots and observers were not inclined to take instruction from anyone, particularly an NCO. His prior experience with squadron aircrew told him they even sometimes took unofficial photos, 'happy snaps', resulting in a waste of material. Campbell concluded that in light of aerial photography's increased importance, 'the flying officer should undergo a practical examination to include taking a good photograph of a given area. If this suggestion be adopted, a good deal of our present difficulties in getting work done at home would vanish.'[381] Schooling pilots and observers in the correct process became an obsession for Campbell and his Air Ministry counterparts. Campbell made it clear in his report to the Ministry:

> In conclusion, the training of photographic personnel is in good working order, but it is regretted that despite the efforts of the photographic Officers at Wings, the training of Pilots and Observers in Aerial Photography still leaves much to be desired. Very few photographs are taken, and consequently success is not always attained when the squadrons or drafts arrive overseas.[382]

The report went through the Ministry, prompting the Deputy Director of Military Aeronautics, Lieut.-Col. C.C. Marindin, to echo, 'I am directed to call your attention to the necessity of training all pilots in the handling of cameras and in taking aerial photographs before they proceed abroad, and to request you to take such steps as will ensure this being carried out in all service units under your command.'[383] The concern lingered on throughout the remainder of the war. The American photographic expert Major Edward Steichen commented on the issue in his post-war critique:

> The Photographic Section was rarely asked to make assemblages except for office ornaments. Stereoscopic prints were only occasionally called for. Not once was a large assemblage requested of an entire sector. Initiative in these directions on the part of the Photographic Sections was discouraged and there was a general atmosphere concerning these things which suggested that it was none of our affair and did not conform to Army Intelligence Regulations.[384]

THE SOMME

The awfulness of the battle of the Somme included the greatest number of casualties ever experienced by the British army in one day of fighting. The Somme battleground represented extensive open country with unobstructed vision and broad fields of fire. Geology favoured the defenders, with extensive areas of solid chalk reinforcing the trench lines more than the soil on most sectors of the Western Front. Additionally, the region's dense clusters of houses formed veritable

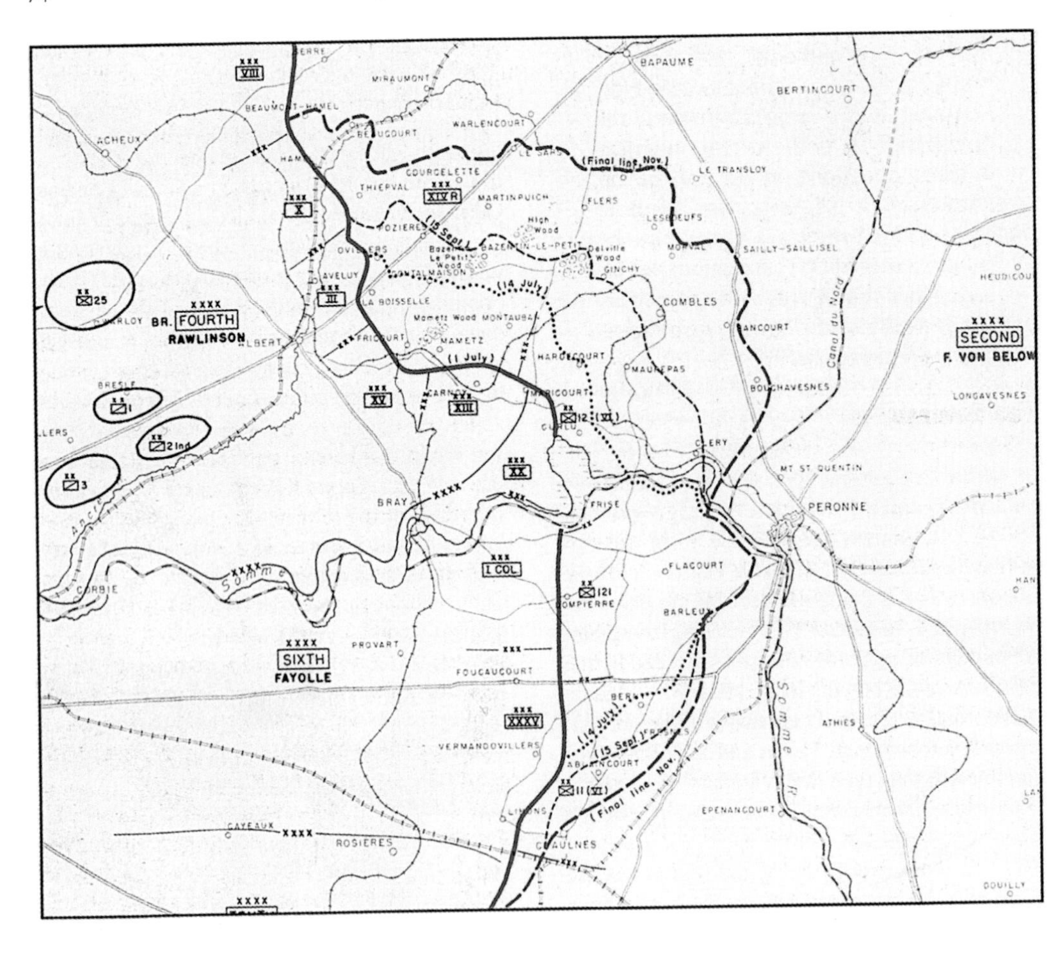

The Somme. (*West Point Atlas*, courtesy Department of History, US Military Academy)

forts, which became the strong points for the German defensive system. The battlefield was a monotonous succession of low, rolling plains. The British had commenced planning for the Somme offensive as early as 1915, acquiring aerial photographs that detailed the first and second line networks. The total concentration of the RFC on the Somme was evidenced by its complement of 185 aeroplanes. In the opening phase of the battle, the German *Fliegertruppen* was considerably outnumbered and remained so until the middle of July. During the buildup and first days of the offensive, the RFC reinforced protection of corps reconnaissance through a continuous line of patrolling pursuit aeroplanes. Aerial activity increased that spring with more units augmenting the sector, to include the RFC's 9th Wing, configured for special operations. The 9th comprised four squadrons (Numbers 21, 27, 60, and 70). In addition to more strategic reconnaissance of the region, the RFC increased their attacks against German aviation and conducted bombardment of communication networks.[385] The RFC preparation strategy for the Somme offensive worked. In the preparation for the battle, not one German aeroplane came over the Fourth Army zone either to bomb the British marshalling areas, direct enemy fire, or observe or photograph the preparations.[386]

The RFC had honed its aerial reconnaissance skills right up to the kickoff of the offensive. The effort to achieve timely photography for the commander was clearly demonstrated through a 'speed test'. The RFC's I Brigade had three squadrons from 1st Wing acquire coverage of German targets with amazing results. A BE 2 from No. 9 Squadron took an aerial photograph over the front on 10 June at 0750 hours, landed 20 minutes later and had the first print dispatched to 1st Corps via motorcycle at 0832 hours. Seven minutes later the prints arrived. Total time from aerial photograph to receipt of coverage at the Corps was 49 minutes. On 16 June, FE 2bs from No. 18 Squadron conducted a similar test. The photograph was taken at 1706 hours. The FE 2b landed with the camera quickly downloaded and the first print developed 21 minutes later, an incredible accomplishment in a war environment. Ten prints were handed to the motorcyclist and arrived at IV Corps Headquarters approximately 37 minutes after producing the first print. Finally, a BE 2c from No. 10 Squadron on 18 June photographed the German target, arrived at the landing ground, developed a print and delivered it to XI Corps Headquarters in an astonishing 30 minutes.[387] The speed tests showed the staff that aerial photographic reconnaissance could be depended on to provide the battlefield commander with timely information as the campaign progressed. Such preparations contributed to a growing confidence in the upcoming battle.

The RFC took excellent photographs daily, which were in the hands of the artillery commanders the same night. This greatly helped in determining the results of the fire.[388] The aerial reconnaissance reported new and more formidable defences being built, including the discovery of a third German defence line. The British soon acquired aerial coverage of the entire German trench network in the Somme region facing the British Fourth Army. Ten days prior to the actual offensive, British aerial photography was again acquired of the entire sector. The preparation had gone well. Aerial coverage had reinforced British confidence to proceed as planned. The British commander-in-chief, General Haig, notified Général Joffre the British would 'continue the battle with such vigour as will force the enemy to abandon his attacks on Verdun'.[389]

During the last six days of June, the Allies continuously bombarded the German positions. Aerial reconnaissance was in constant demand. On 26 June, the bombardment was stopped in order to allow the RFC to photograph the area and obtain an exact record of fire for the artillery. Aerial photographs not only showed how the heavy shells had fallen, but also provided indications of the state of the defensive wire entanglements around the German trench networks. The coverage showed the artillery was accomplishing its objective.[390] One of the Corps commanders remarked, 'The aeroplane photographs showed admirably the effect of the bombardment both on the wire and on the trenches and were of the greatest value.'[391] The weather in the region hampered aerial reconnaissance coverage for the next two days. Despite the interruption, aerial observation continued whenever there was a break in the clouds.[392]

On the eve of the battle, Charteris predicted, 'We do not expect any great advance, or any great place of arms to fall to us now. We are fighting primarily to wear down the German armies and the German nation, interfere with their plans, gain some valuable position and generally to prepare for the great decisive offensive which must come sooner or later, if not this year or even next year. The casualty list will be big.'[393] On 1 July, the British army suffered a catastrophic 60,000 men killed and wounded.[394]

Throughout that first day on the Somme campaign, the RFC retained command of the air. Although 110 pilots were in the air for a total of 408 hours, there were only nine recorded aerial combat engagements with the Germans. Infantry contact sorties tried their best to report the progress of the infantry. Ongoing artillery spotting missions searched for targets. However, the sheer volume of artillery marred the ability to associate impact with accuracy. Strategic aerial reconnaissance discovered very little activity on the roads and railways behind the German lines. A fair number of aerial photographs were taken, including some of the German third line of defence.[395] After a week, the Fourth Army had yet to secure the important preliminary Somme campaign objectives of Trones Wood, Mametz Wood and Contalmaison. The British Fourth Army commander, General Sir Henry Seymour Rawlinson, issued a preparatory order for the attack against the German second position, to be launched on 10 July. The preparations were similar to those before the first major assault on 1 July with aerial reconnaissance acquiring coverage of new objectives, to include the newly discovered German third line.[396]

The resultant Somme campaign did not lack for intelligence from aerial photographs. In the five months of battle, the RFC took more than 19,000 air photographs and generated 430,000 prints.[397] British planning for the14 July attack was aided by aerial photographs that revealed additional enemy positions. However, quality of analysis was now becoming suspect owing to unforeseen enemy resistance. Aerial reconnaissance reports were losing credibility; claims of limited enemy resistance not being believed. The situation did not improve. Charteris commented in a letter to Major-General Macdonogh at the Air Ministry, 28 days after the Somme campaign commenced, 'Air reconnaissance as a battle means of information has failed us. Aeroplanes have to go too high now to do much good, and in any case the number of machines is very limited and the information they give meagre and amateurish.'[398] It is interesting to note that German recollections reflected just the opposite view. Allied reconnaissance, in their view, *was* effective. One German veteran of the Eastern front remarked on his time during the August campaign at the Somme:

> What I have been through during this time surpassed in horror all my previous experiences during the second year of the war ... how the English, with the aid of their airmen, who are often 1,500 feet above the position, and their captive balloons, have exactly located every one of our

A BE 2c variant like the aeroplane pictured here successfully delivered intelligence to the headquarters within 30 minutes of target acquisition. (Courtesy of Les Rogers via Aaron Weaver)

The FE 2b variant with its pusher-propeller configuration also demonstrated the ability to deliver intelligence to the headquarters fast, in this case reportedly within 37 minutes of target acquisition. (Gorrell Report, NARA)

A detailed oblique aerial photograph of French troops on the attack at the Somme on 10 October 1916. (SHD, from Christienne and Lissarague, *A History of French Military Aviation*)

> batteries and have so smashed them up with long-distance guns of every calibre that the artillery here has had unusually heavy losses both of men and material.[399]

After the war, the *Generalquartiermeister der OHL General der Infanterie* Erich Friedrich Wilhelm Ludendorff reflected that Allied aeroplanes made 'the troops feel that they had been discovered in places which heretofore they had thought afforded safe cover.'[400]

The Somme campaign continued for months. Prior to the British assault of 15 September, Army reconnaissance of the rear German sector revealed no abnormal activity upon the roads and railways in the area. The RFC monitored key cities behind the German lines for signs of troop buildup and reinforcement, but did not discover any sign of activity. From the opening of the Somme offensive, British Intelligence employed all intelligence sources to determine any change in the German order of battle. British analysis showed 32 enemy divisions had been fighting since the commencement of the campaign on 1 July. By September, 19 enemy divisions had departed to other fronts.[401] When the British started their 15 September offensive, the RFC flew more hours and experienced more fighting than on any day since the war began. Artillery spotting in support of the Fourth Army front revealed the location of 150 batteries, 70 of these being engaged. At the same time the RFC's Army Squadron sorties over German rear echelons did not detect any abnormal movement.[402]

The British Army fully accepted the need for extensive aerial coverage to conduct its campaign. The demand resulted in innovative ways to acquire aerial photographs. Major Robert Smith-Barry, commander of No. 60 Squadron, and one of the original pre-war RFC pilots of the Military Wing, encouraged use of the newly acquired Nieuport 17 for aerial photography. It foreshadowed future aerial reconnaissance. Not only was the Nieuport 17 very manoeuvrable, it was faster than most aeroplanes in the operational inventory of the time. The Nieuport 17 was configured with a 'C' Type camera to the rear of the cockpit. The improvement came with a cost. When one errant pilot safely landed behind German lines, Nieuport 17 pilots were suddenly aware that 'their guns paid that much more attention to us.'[403]

The demand also had an impact on the entire aerial photographic community. British photographic training was well in place, producing photo section staff necessary for the campaign. The pioneers of British aerial reconnaissance, to include Moore-Brabazon, introduced measures to guide staffs and combatants in photographic interpretation with the first British manual on aerial photographic reconnaissance, *Notes on the Interpretation of Aeroplane Photographs.*[404] Improvement in camera and mount technology was accompanied by a greater awareness of the potential of aerial coverage. Lieut. Victor Laws was in the process of developing the prototype 'L' Type camera, the best British aerial camera of the war.[405] Better designs for mounting aerial cameras had been developed, eliminating many problems associated with vibration from the aeroplane motor.[406] Innovation on the battlefield was on the rise as the combatants entered into the third year of the conflict. Stabilisation was addressed and solutions were found. Captain Campbell had developed a new mounting device for the aerial camera, using rubber sponges inside the fuselage to reduce vibration. He had even purchased the entire stock of rubber sponges in St Omer for this purpose.[407]

HONING HIS CRAFT

Lieutenant Paul-Lewis Weiller continued to learn all aspects of aviation during the third year of the war. In his first full year of combat as a pilot, he endured a near fatal crash on 7 August while flying a Caudron G4 near Verdun. He subsequently recovered in hospital. On 15 October, Weiller was tasked to ferry a Nieuport 17 from Le Bourget to his home base. He ventured near the front lines and discovered an Albatros observation aeroplane. Weiller attacked the Albatros successfully and witnessed it crashing. He subsequently received his fifth citation from General Fayolle, praising his boldness, aggressiveness, passion, and ignorance of fear.[408] This was the first of four German aeroplanes that Weiller shot down in the war. He later achieved the distinction of shooting down two German aeroplanes in one day (9 July 1918) flying a Breguet 14 A 2 during the French Xème Armée offensive in the Ypres sector at Mont Kemmel.[409]

Weiller's enthusiastic advocacy of aerial photographs proved to be his greatest legacy. He recognised the prevailing scepticism that came with aerial observation, reminding colleagues that military seniors 'were always incredulous when given simple oral reports'. Weiller's operational experience proved invaluable in assisting the French to design one of the most important aerial cameras of the war, the 120cm. The impetus for the longer focal length design was a camera that could satisfy Général Mangin, IVème Armée commander, who wanted 'to see some Boches' on aerial photographs.'[410] Weiller personally tested the 120cm by flying solo over the German lines in his Caudron G4 with the camera installed in the observer's cockpit. The experience exemplified the motto he lived by: '*Fais ce que dois; advienne que pourra.*'[411]

CONTEMPORARY ANALYSIS OF THE WESTERN FRONT – A SUMMARY OF 1915–1916

Allied understanding of German intentions based on the evolving nature of positional war at this time was summed up in published manuals for aerial photographic interpretation. The most in-depth discussion was contained in Étude et Exploitation des Photographies Aériennes.[412] (See Appendix E.) The Germans had effectively fortified their entire front in the course of the first year and a half of conflict. Thanks to the lessons learned from the Loos Offensive, the Germans had shifted to economise their forces, emphasise defence to reduce losses, and exercise a logical use of terrain. Economising force was done through echeloning the troops in depth. The first position facing No-Man's Land was no longer composed of two lines of trenches. The Germans constructed three lines with the first line containing troops that provided initial defence against attack. The second trench line contained the counter-attacking force and supporting elements. The third trench line comprised additional counter attack elements as required. Additional security provided by dugouts was planned according to the terrain that contained the front. A trench was positioned in front of a crest followed by a second trench on the reverse slope. The Germans also considered deepening and widening trenches and traverses. Shelters were reinforced or dug deeper to resemble mine-galleries. Front line logistics became better served thanks to a network of narrow-gauge railways leading up to the forward trenches. Finally, reinforced gun emplacements now substituted for the older earthwork fortifications that were built initially to provide standing position defence. Allied analysis recognized that the terrain of the front line was critical for maintaining the best 'obstinate defensive' posture. German intentions were to make an Allied breakthrough impossible. In the worst case an Allied breakthrough would be limited only to the first lines - allowing German forces to quickly reinforce the sector and keep their rear echelons safe.

THE US MILITARY COMMENCES AERIAL OBSERVATION

While conflict continued in Europe, Americans were studying aerial reconnaissance. The War College Division sponsored studies in 1915 to 'consider various aeronautical appliances in regard to their practical value in campaign'.[413] The aerial reconnaissance role earned primary consideration. In the early stages of the conflict, assessment of photography's role paralleled cartographic construction. In the American staff assessment of 1915, photography was 'utilised to the greatest extent possible in aerial reconnaissance'.[414] The Americans recognised that timely development was essential to the value of the aerial photograph, and that institutionalising the interpretation process was necessary: 'The plates or films thus made are rapidly developed and are thrown on a screen by means of a stereopticon, when all details are magnified to any extent desired and details invisible to the naked eye are brought out plainly. These details are then entered on the maps of the officers concerned.'[415] Cartography also benefitted from the coverage. 'As the height at which an aeroplane is flying can be taken from the barograph, and as the focal angle of the lens of the camera is known, a scale can easily be worked out and the views form good maps of the terrain photographed.'[416]

Military employment of American forces on the Mexican border following the 1916 attack by Pancho Villa on Columbus, New Mexico, exposed an acute lack of US military preparation. However, the subsequent campaign against Pancho Villa highlighted aviation's potential for information gathering. Major Benjamin D. Foulois and the First Aero Squadron provided the limited aerial support in the campaign. Photographic expertise was available to support the fledgeling effort. James Bagley, a world leader in photographic mapping and exploration, was assigned to Pershing's staff to share his expertise. The ongoing campaign proved a challenge for any reconnaissance effort. Pershing's forces operated in a 19,000m^2 area.[417] Despite the experience, the US military was not prepared for the demands associated with modern warfare awaiting it on entering the European war in 1917. The Americans lacked the knowledge to exploit aviation, particularly in aerial photography. America's entry in the war exposed that woeful lack of preparation, forcing it to depend upon Allies who were already strained beyond capacity.[418]

CHAPTER 5

October 1916 to December 1917: Positional Warfare at its Zenith

THE AERIAL PHOTOGRAPHIC INSTITUTION

During the third year of the Great War, aerial reconnaissance and photographic interpretation fully integrated with operational planning and execution. There was no longer a debate with the military staff and combatants on capabilities and potential. The past year had proven that aerial observation, validated by the photograph and transformed into a map, was the most useful source for defining the battlefield. With that process fully institutionalised, the Allied aviation establishment now gravitated to developing pursuit resources to maintain air supremacy. Experienced aerial reconnaissance and photographic interpretation personnel were now in place at all echelons. Aerial reconnaissance and photographic interpretation's evolution differed in two major respects in the final two years of conflict. The quantity and quality of photographic production enjoyed substantial increases. The second key improvement was in aeroplane design. With fuselage, propeller, engine, and other design changes, reconnaissance aeroplanes could operate at higher altitudes, above the range of air defence cannons and enemy pursuit.

Aerial observation remained a critical priority for both sides, but emphasis was shifting toward achieving air superiority. The German General Fritz von Below of the General Staff described the German aerial posture against the Allies in his January 1917 memorandum: 'The main object of fighting in the air is to enable our photographic registration and photographic reconnaissance to be carried out, and at the same time prevent that of the enemy. All other tasks, such as bombing raids, machine-gun attacks on troops, and even distance reconnaissance in trench warfare must be secondary to this main object.'[419] Von Below's comments governing the offensive accurately prophesied things to come: 'It is this principle that the enemies have now realised and are now acting upon. What we need to defeat their application of it is a sufficient quantity of efficient weapons; that is to say, a sufficient quantity of fast fighting machines.'[420]

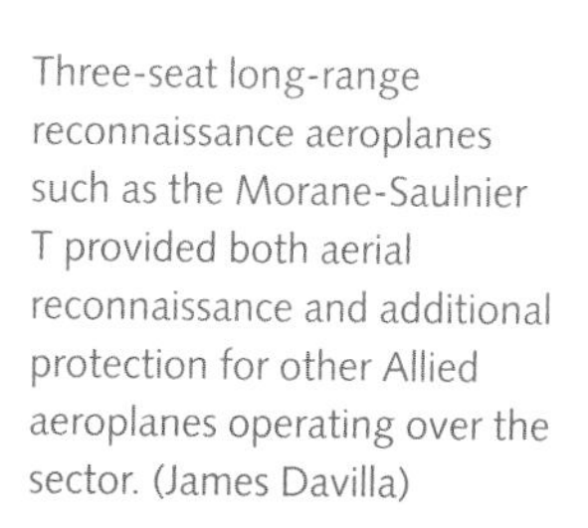

Three-seat long-range reconnaissance aeroplanes such as the Morane-Saulnier T provided both aerial reconnaissance and additional protection for other Allied aeroplanes operating over the sector. (James Davilla)

The GN5, Belgian-designed aerial reconnaissance aeroplane was developed from the Farman F.40. Two highly decorated Belgian aircrew, Capitaine Jules Jaumotte and his observer Adjudant Jean Wouters pose with a 50 cm camera. (Courtesy of Walter Pieters; *The Belgian Air Service in the First World War*)

The GQG vision for Aviation Militaire at the commencement of 1917 was to increase their aeronautical strength to 187 escadrilles, each with a nominal count of 10 aéroplanes. The emphasis was on the reconnaissance mission protected by pursuit.[421] In addition to this realignment in thinking, Aviation Militaire came to terms with the ever more inferior quality of their aerial reconnaissance force. New enemy fighter aeroplanes were being introduced and raising havoc. France's existing fleet of Sopwith 1 A2 (French variant of the British Sopwith 1½ Strutter, two-seater reconnaissance and bombardment aeroplane), A.R.1, Caudron G.6, Salmson-Moineau S.M.1s and Morane-Saulnier Ts suffered from technical limitations. The GQG decided to reorganise aerial resources. A 16 March 1917 memorandum directed that many three-seat, long-range reconnaissance aeroplanes (Caudron G.6, R.4, Salmson-Moineau S.M.1s, and Morane-Saulnier Ts) disperse across Aviation Militaire units because of 'serious inconveniences' being imposed on escadrille personnel. Maintenance time and landing ground overcrowding prompted the new arrangement. Rearranging long-range reconnaissance throughout the escadrilles of the Militaire Aviation had a consequence for aerial reconnaissance. The mission was integral to all of French aviation and remained in place to the conclusion of the war.[422]

In 1917 British aerial reconnaissance enjoyed a resurgence caused by the arrival of a new generation of aeroplanes that served until the end of the war. Key to aerial reconnaissance was the Reconnaissance Experimental (RE) 8 and Airco DeHavilland (DH) 4. One of the most famous of the RFC (and RAF) units was No. 55 Squadron. They were the first to acquire the DH 4 in March 1917 and achieved an impressive strategical bombardment and long-range reconnaissance record.[423]

The reconnaissance mission permeated the standard makeup of RFC squadrons. By the early part of 1917 every aerodrome at the landing ground had its map-room, the walls of which were adorned with sections of maps and photographic mosaics. The maps and mosaics enabled aircrews to quickly study the terrain and identify critical landmarks.[424] With the increased demand for aerial reconnaissance came the need for more intelligence personnel to interpret what was observed and photographed. This led the British to increase intelligence training at their schools in the UK and within the BEF. The Somme campaign generated large volumes of information without much benefit, resulting in the British turning to their French counterparts for another view of interpreting and processing information. The French experience in 1916 had clearly demonstrated the benefits of integrating intelligence personnel at the escadrilles. A British study of the French intelligence process entitled *Notes on the French System of Intelligence, 23rd September 1916*, outlined the French method of acquiring and disseminating aerial reconnaissance and photographic interpretation information during a major campaign. The analysis was published and disseminated to British staffs.[425] The next step was to learn from the French experience and modify to suit what worked best for the evolving British program. In October 1916, Major-General Trenchard proposed that British Intelligence Sections be established in certain squadrons and wings where the Intelligence Officer could be in intimate touch with the flying and photographic personnel.[426] This was approved and instructions issued to the British forces in December

1916. In a step taken to retain intelligence personnel, the RFC established its own Branch Intelligence Sections at each Corps Squadron and Army Wing. Trenchard preferred that the intelligence units remain integral to the RFC, but he was overruled by seniors at GHQ who decided that intelligence must remain under the Army Intelligence Chief.[427]

In the fall of 1916, senior factions of the Royal Artillery and Royal Flying Corps debated the role of the aviator. The Royal Artillery made demands on the British RFC to allocate more aerial observer resources to their artillery-spotting mission. British artillery was upgrading with greater numbers of heavy artillery pieces and howitzers to fire larger and more lethal munitions against an increasingly fortified German adversary.[428] The premise was simple: increases and improvements in artillery required more dedicated artillery spotting aerial platforms. Improving radio capabilities also added to the allure of acquiring more aeroplanes. British aerial priorities had committed twice the number of aeroplanes to artillery spotting over the previous year.[429] With these trends, the British high command in October 1916 proposed that the control of the Corps Squadrons should pass to the Artillery. General Rawlinson made his move to reorganise the aviation assets, 'To ensure perfection, we require [that] the observers must be highly skilled. Aeroplane observation for artillery purposes is skilled work of a very high order. With an unskilled observer the best trained and best equipped battery is useless.'[430] Trenchard fought Rawlinson's suggestion with a dose of aviation reality, 'Artillery work is not the entire duty of the Corps Squadron, which is also charged with contact patrol work, trench reconnaissance, and trench photography. Nor are Corps Squadrons at present equipped with machines of a suitable type to do all the work required by the artillery, such as photography, at a distance behind the lines which must be done by fighting machines.'[431] Trenchard won this battle. The allocation of aeroplanes remained within the jurisdiction of the RFC.

THE *SIEGFRIED STELLUNG* (HINDENBURG LINE)

The lessons of Verdun and the Somme left an indelible mark on both sides. The human casualties from

Siegfried Stellung at Bellencourt. 'Held by an Army of unimpaired morale, it would have been well-nigh impregnable.' (Great Britain General Staff (Intelligence), *Illustrations to Accompany Notes on the Interpretation of Aeroplane Photographs (S.S. 550a)*)

positional warfare greatly exceeded every known estimate. British and French leaders considered their tactics of destruction successful, despite the incredible cost. They assumed that sufficient artillery could win any limited objective. Planning a successful campaign required longer-range artillery and more effective counter-battery fire. These objectives depended more than ever on aerial reconnaissance.[432]

By the close of 1916 the German OHL had become deeply disturbed by the impact of the Allied offensives in the Somme region. General Ludendorff feared the German army might not stand up to a similar offensive in the coming spring, so the decision was made to transform the front to the most intricate defensive system ever devised. The Germans called their master network the *Siegfried Stellung*, but to the Allies it was known as the Hindenburg Line. Charteris viewed the obstruction with awe, 'Held by an Army of unimpaired morale, it would have been well-nigh impregnable.'[433] The Siegfried Stellung prompted many improvements in Germany's approach to positional warfare, to include more elaborate tank manoeuvres, unregistered artillery bombardments, elastic defences, and the use of storm troops to infiltrate enemy lines.

The Siegfried Stellung reestablished the configuration of the Western Front in a more strategic posture of elaborately built networks of defence. The German OHL decided to build five such strategic systems. The *Flandern* Line ran from the Belgian coast via the Passchendaele ridge through the Messines salient to Lille. The *Wotan* Line extended from Lille to the rear of the Loos-Arras-Vimy and Somme battlefields. The *Siegfried* Line began at Arras and proceeded west of St-Quentin and Laon to the River Aisne east of Soissons. The *Hunding* Line began near Peronne on the Somme and ran to Étain on the Meuse and then north of Verdun. The fifth line, the *Michael* Line (The Germans called it the *Michael I* Zone) covered Étain to Pont-à-Mousson on the River Moselle.[434] The Siegfried Stellung also helped the Germans' strategic objectives in the other theatres. The Siegfried Stellung allowed German staffs to effectively transfer thirteen German divisions and 50 batteries of heavy artillery to the Austrian southern front opposite the Italians without experiencing a major shortfall in manning.[435]

Within the La Malmaison sector near the city of Reims, the Germans' Hunding line shows the rough symmetry of the traverse network. (*Étude et Exploitation des Photographies Aériennes*)

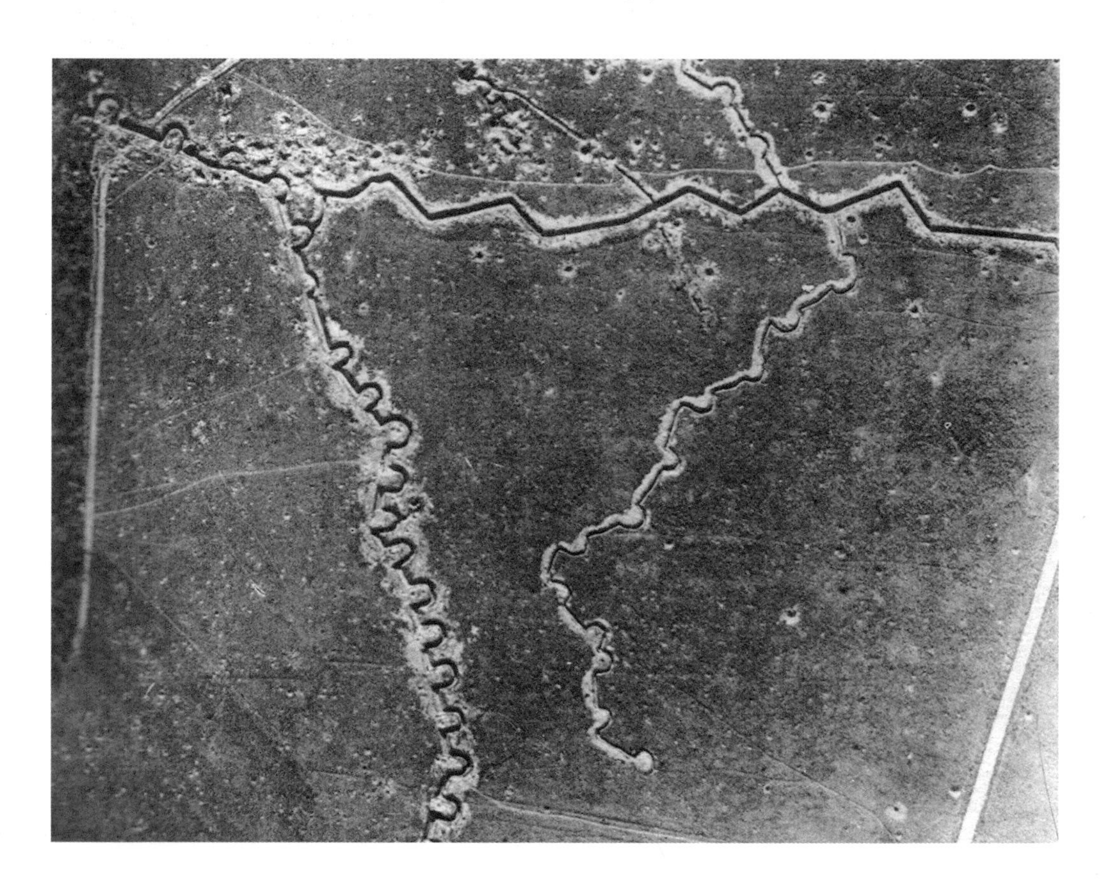

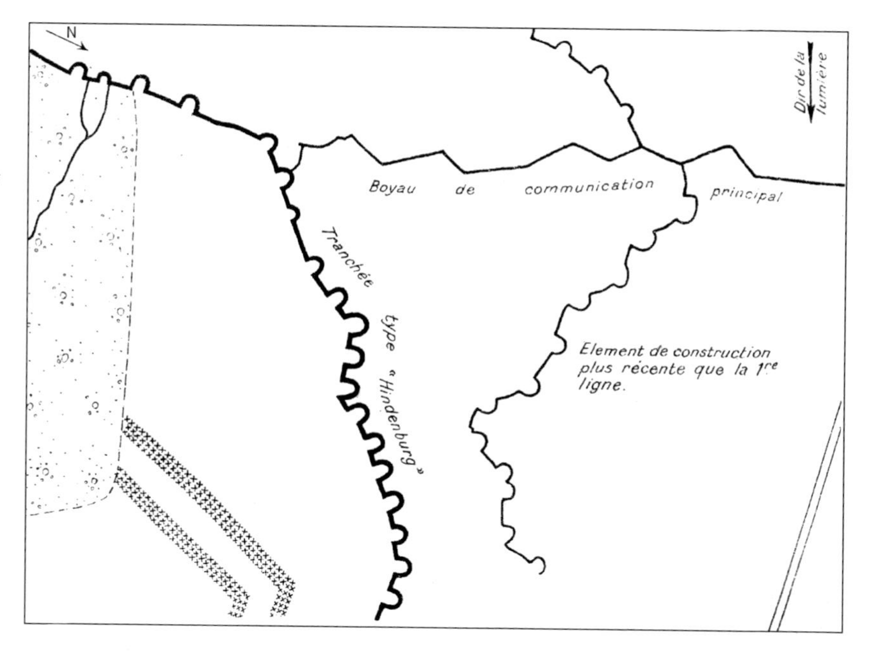

Companion line drawing of the traverse network. (*Étude et Exploitation des Photographies Aériennes*)

The first indication of construction was acquired by British aerial reconnaissance toward the end of October 1916. Fresh earth was discovered near Quéant (12 miles south-east of Arras and 14.5 miles east of the front line at Monchy au Bois). Suspicions were raised any time defensive network construction was discovered, especially since it was observed well behind the existing front line. A special reconnaissance was ordered. On 9 November 1916, the first aerial reconnaissance report was generated, describing trench construction through Quéant and Bullecourt, crossing the Sensée and the Cojeul and linking up to the Germans' rear third line east of Neuville Vitasse.[436] Despite noticeable construction activity in several sectors, the Allies were slow in recognising the overall strategic implications. When construction commenced, the large number of gaps in construction did not suggest an overall defence network. One December aerial reconnaissance of the eventual location of the Siegfried Stellung near the French town of Marcoing reported, 'observed no unusual activity.'[437] The Allies did share the little they knew. On 2 January 1917, the French commander-in-chief, Général Robert Georges Nivelle, prompted by reports from agents and repatriated civilians, instructed his aviation services to cooperate with the British in reconnaissance of the defences. Later that month, the British GHQ Intelligence Summary mentioned the report of French 'rapatriés' that the Germans were constructing a line of defence 'from Arras to Laon'. Although the Siegfried Stellung did not follow that very same course, it did show extensive activity in the strategic rear areas.[438] Additional reports from prisoner interrogations were treated as unconfirmed rumours at Field-Marshal Haig's GHQ. Analysis continued, but elicited a consistent 'Confirmation is required.'[439] Général Nivelle received more reports from agents and repeated his order for more reconnaissance.[440] Through the remainder of the winter, the British monitored the construction while their armies slowly advanced on the Somme front. However, aerial superiority was now turning in favour of the Germans.

The 'very combative' attitude of German aircraft that prevented both aerial reconnaissance and photography contributed to the inability to gauge the extent of construction effectively. The lack of aerial reconnaissance validation of human intelligence in January resulted in the French 2e Bureau still deliberating over the Siegfried Stellung's ultimate purpose. Effective reconnaissance did not resume until the Germans had departed their forward areas for the Siegfried Stellung in mid-March. French post-war analysis later confirmed their frustration: 'The General-in-Chief was thus not informed of the situation of the Hindenburg Line.'[441] German General von Below's remarks of January 1917 reveal a sensitivity to the effectiveness of the Allied reconnaissance and a determination to neutralise their efforts:

> The beginning and the first weeks of the battle of the Somme were marked by a complete inferiority of our own air forces. The enemy's aeroplanes enjoyed complete freedom in carrying out distant reconnaissances. With the aid of aeroplane observation the hostile artillery neutralised

> our guns, and was able to range with the most extreme accuracy on the trenches occupied by our infantry; the required data for this were provided by undisturbed trench reconnaissance and photography.[442]

Aerial reconnaissance provided a piecemeal picture of the target. On 2 February, the British reported from one sortie south of Quéant:

> There appears to be a complete system of trenches around the town. On the south east corner of the town there is a short line about 3/4–1 mile running up through Itancourt. On the north side, running east of the canal, is a fairly continuous line with strongpoints at about 1/4 mile intervals. Between Bellicourt and Bellenglise the line of trenches is very broken.[443]

Strategic analysis of the Siegfried Stellung tested the analytical skill of Allied aerial photographic interpreters. They eventually confirmed that the Siegfried Stellung was a masterful work of defence. The Siegfried Stellung defences, though they varied in different sections, represented the application of mass production to fortification. Comparatively shallow dugouts of ferro-concrete (steel-reinforced concrete) were constructed according to a fixed pattern beneath the parapet for each squad. Mined dugouts were spaced throughout. The Siegfried Stellung's formidable barbed-wire obstacles included three belts, each ten to fifteen yards in depth and five yards or more apart and in front of the fire trench, configured in a zig-zag pattern. This allowed machine guns to sweep the sides.[444] There were a number of small forts and strong points along the line.[445] In each sector, trenches were configured for transverse stopping positions (*Querstellungen*). Narrow gauge railways; flanking organisations and shelters; observatories; defences for PCs; munitions depots and drainage works were constructed in elaborate configurations. Defences were irregular and dispersed in several bands. The non-continuous entanglement allowed for any requirement for a German counter-attack and more effective placement of machine guns.[446] These 'centres of resistance' provided places to which the garrisons could retire and hold out until reinforcements arrived. They were skilfully built and sited, designed deliberately for use after the enemy broke through the adjacent lines. Once past, the Germans could attack from the rear with machine gun and rifle fire.[447] Many trenches were widened to entrap tanks. Tank pits were dug in the road and covered with camouflage to provide additional obstacles. Opposite the British, the average Siegfried Stellung trench measured between 3.20 and 3.40 metres in width. The German trench mortars (*minenwerfer*) were placed outside of the firing trenches. Finally, a network of observation stations was placed throughout, echeloned in depth to handle the vast contingencies of an experienced army, well versed in positional warfare.[448]

The wide trenches of the Siegfried Stellung served as obstacles against a tank attack. Most of these 3.20–3.40m wide trenches were in sectors opposite the British. (*Étude et Exploitation des Photographies Aériennes*)

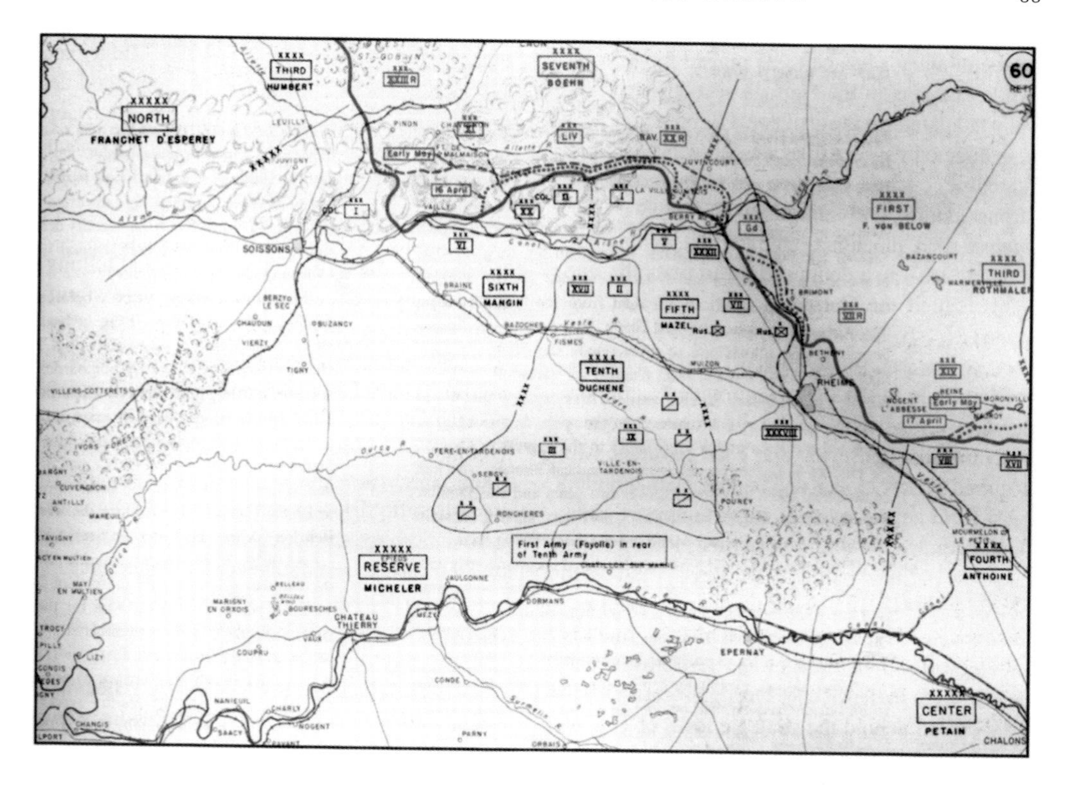

The Nivelle Offensive. (*West Point Atlas*, courtesy Department of History, US Military Academy)

The German withdrawal back to the Siegfried Stellung challenged aerial photography's support to artillery targeting. Additional detailed maps were required of the new sector. Triangulation targeting through flash-spotting and sound-ranging had to be moved forward and accurately positioned. Long-term reliance on cartographic data suddenly required a major overhaul.[449] The demand on aerial photography increased as analysts scrambled to acquire detailed information that would explain the complexities of the new front line to a host of consumers, to include artillery and infantry. The RFC repeatedly paid the ultimate price throughout this transition. The German opposition became the source of aerial legends with the sector being the domain of Germany's top fighter pilots. The elite German ace, Baron Manfred von Richthofen, operated his *Jagstaffel* out of Douai. On 15 April alone, Richthofen and his unit shot down six reconnaissance aeroplanes on photographic missions over the Siegfried Stellung.

By early February, the Siegfried Stellung was ready for occupancy. Under the code name Operation *Alberich*, the German forces commenced their withdrawal from 9 February 1917, moving both their own forces and the French civilians out of the front areas. They commenced a scorched earth policy, abandoning the former front line sector, leaving the Allies with an unusable countryside. The Germans methodically destroyed 264 villages and 38,000 houses in the region. Trees were sawed to the base, wells were clogged up, and landing grounds for aeroplanes were ploughed up and mined. The transformation was accomplished by forced labour from French civilians. Aerial reconnaissance failed to successfully determine the extent of the withdrawal. By the end of February, the German Armee withdrew their front line on the Somme battlefield 20 miles to the east and occupied the Siegfried Stellung. They had totally disengaged from the English Fifth Army and moved east. Two weeks later, the sector opposite the French IIIème Armée was evacuated. The remaining lines did not finalise until the last withdrawal of German troops took place on 8 April 1917.[450] The significance of the withdrawal had even greater strategic implications. The German move to the east complicated Anglo-French plans for a spring offensive.[451]

THE NIVELLE OFFENSIVE

The enduring legacy of the French armée in 1917 was the failed Nivelle Offensive. Général Nivelle enthusiastically supported his own master strategy to defeat the Germans through a large offensive with the objective 'to beat and, if possible destroy the enemy army'.[452] Convinced that a breakthrough would outrun the reconnaissance, Nivelle did not value aerial photography like Pétain. When Nivelle's offensive commenced on 16 April at the Chemin des Dames, the Aviation Militaire was assigned the spearhead role to destroy German aerial reconnaissance and bombardment over the advancing sectors. At this time the objective proved difficult because the Germans dominated the air over the Aisne sector. French aerial observers were not successful in adjusting the massive artillery barrage to the best possible effect. Thanks to the German's present policy to position their forces on the rear slope away from direct fire and ground observation the infantry was able to successfully weather the incessant pounding and stifle the French advance.[453] Under a newly established *Groupement de Combat de la G.A.R.*, approximately 100 fighters were assigned to the IVème Armée to protect the reconnaissance. Despite the realignment, the French armée cooperation escadrilles suffered extensive losses.[454] When the offensive was halted on 25 April, the French had suffered 134,000 casualties. The French government made the only decisive move that proved beneficial to the armées by appointing Pétain as Nivelle's *chef d'état-major* (Chief of Staff). The Nivelle offensive in the Chemin des Dames sector not only broke the spirit of the French Armée, but became the catalyst for mutinies throughout the French military. Général Pétain's assumption of command proved providential – it stemmed the surge of discontented manpower throughout France.[455]

French armée revival required better use of existing resources. Pétain increased aerial reconnaissance and aerial photography to better gauge areas where the armée could maintain a presence without a major commitment to battle. Increased reconnaissance provided a source of effective control during this critical time. Intelligence from aerial reconnaissance reinforced the operational picture, eliminating the fruitless assaults against a clearly defined German defence.[456] Pétain also decentralised the groupes de chasse to better support missions such as aerial reconnaissance. For Aviation Militaire, reinforcement of traditional missions supplanted aerial supremacy for the remainder of the year.[457] Pétain re-emphasised the *Groupe des Armées du Centre* (GAC) role in encouraging the use of aerial photography in battle planning. He ordered Général Jean-Baptiste Eugène Estienne, commander of France's newest combat arm, *les chars d'assaut* (tank force), to work directly with Lieutenant Pépin to ensure aerial photographic interpretation became an integral part of chars d'assaut planning.[458]

The remaining problem for the French Armée in 1917 was exhaustion. There was little energy left in the force to maintain any offensive action. The only positive note was that America's declaration of war meant a fresh new army would soon arrive. The most intriguing feature at this stage is the almost incredible oversight by German intelligence in failing to recognise the precarious state of the French Armée. In the end, this oversight proved fatal.[459]

VIMY RIDGE

One of the most decisively successful battles for the Allies was the Canadian offensive at the strategic heights of Vimy Ridge. A model of planning at the time, it involved all facets of intelligence and operations. Planning commenced in January for acquiring in-depth aerial photography of the ridge and German rear echelons.[460] German defences on the ridge and rear echelons were completely photographed weeks before the opening of the bombardment, creating an accurate map of the entire sector. The map not only formed the basis of the artillery preparation, but also enabled each stage of the infantry advance to be rehearsed in detail. All key targets in the German sector were carefully tabulated for the artillery. The bombardment was directed especially against trench junctions, dugouts, concrete machine-gun emplacements, strong points and entrances to tunnels; also against road junctions, ammunition dumps and light railways to a depth of four to five thousand yards behind the German front.[461] The RFC assigned No. 16 Squadron to the Canadian Corps for reconnaissance operations. The No. 2 Squadron, attached to the I Corps, provided additional aerial coverage. The RFC units were exemplary in their efforts, ensuring that good coverage was acquired.[462] Low clouds and bad weather curtailed most RFC aerial coverage opportunities. Sorties flew whenever visibility permitted. Photographs taken on 24 March and four days in the first week of April disclosed the obliteration of the front-line system of defence, the destruction of long stretches of communication trenches, and the cutting of lanes in dense wire entanglements both in front of, and behind, the strategic town of Thelus. Aerial photographic interpretation detected the wire entanglements after a snowstorm brought out the features. Aerial photographs of the region were exemplary in detecting active German battery positions. Some 190 German batteries were identified for subsequent targeting.[463] Final details were applied to the plasticine model used by officers and non-commissioned officers for extensive study of operational roles. During this battle and the preparation that preceded it, the systematic recording of targets for destruction and coordination of all arms for destructive shoots were both successfully accomplished. This resulted in a corresponding reduction in waste by elimination of overlapping tasks.[464]

Vimy Ridge demonstrated that 'destruction' still achieved results. The downside was that the devastated terrain and road system were unable to support subsequent exploitation of the victory. Despite the ability of the Germans to counter any deep penetrations, the Allies still clung to the perception that 'destruction' was the best course. This was the case with the failed Nivelle offensive, Vimy Ridge and subsequent campaigns.[465] With the continued dependency on artillery to achieve this goal, aerial observation required continued improvement and refinement.

As was the case with the Siegfried Stellung, RFC aerial reconnaissance losses were significant in trying to cover Vimy Ridge. The German aviation resurgence culminated in 'Bloody April' where the Allies suffered major losses of aeroplanes of all types. The state-of-the art German aircraft such as the Albatros (DIII) and Pfalz pursuit aeroplanes wrought havoc on Allied aircraft. 'Bloody April' became the critical motivation for Allied aircraft industries to design and field more effective aircraft, both pursuit and reconnaissance.

MESSINES

The first half of 1917 held few redeeming qualities for the Allies. The eastern front was a disaster as the Russian military disintegrated into chaos. The Nivelle offensive had crippled France's armée. British Commonwealth successes at Vimy Ridge demonstrated in-depth planning could achieve positive results. The most important development was not military, but political.

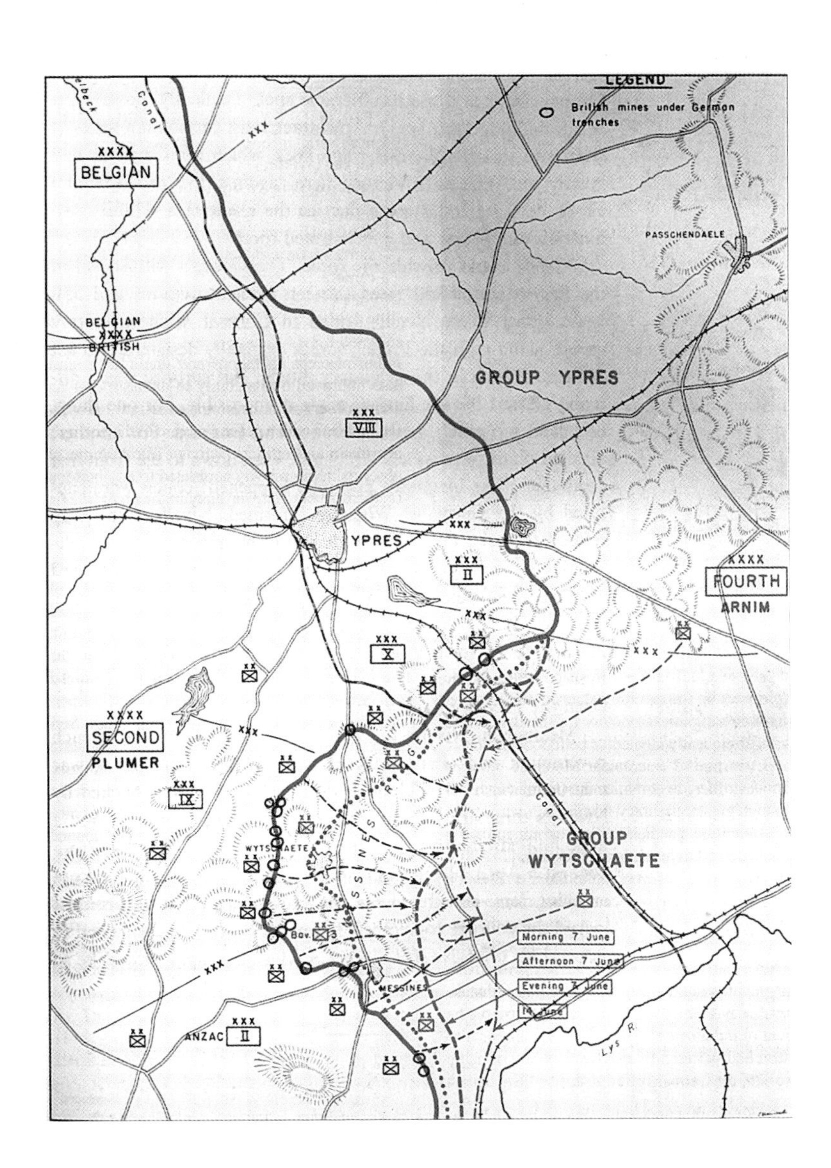

Messines. (*West Point Atlas*, courtesy Department of History, US Military Academy)

A unique piece of international cooperation in aerial reconnaissance observed for the first time in 1917. An American pilot from the illustrious Lafayette Escadrille, *Sergent* Harold Buckley Willis (later promoted to Lieutenant), flew an aerial reconnaissance sortie in a French Spad VII fighter with a British-made 'C' Type Camera. (Charles Wooley, via Alan Toelle)

The United States had declared war on Germany and her allies Austria-Hungary and Turkey on 6 April 1917. Now the Allies could draw on the tremendous American reserves. However, almost a year passed before American forces were noticeable on the front lines. Besides, the Americans had to be trained in the realities of positional warfare. That drew further resources from the already strapped British and French staffs.

The British realised that this was a critical time to consolidate their position on the continent while awaiting reinforcement from the US. They decided to concentrate on further protecting the Belgian coastline, the one area that presented the greatest risk to their strategic lifeline. The British decided to attack the Messines-Wytschaete Ridge south of Ypres, one of the enemy's most important strongholds on the Western Front. The ridge at Messines completely overlooked the British positions in the Ypres sector. The Germans' Messines-Wytschaete network of defences exemplified positional war at its zenith. They held the higher ground reinforced with artillery and infantry. The German forward defences on the Ridge consisted of an elaborate and intricate system of well-wired trenches and strong points forming a defensive belt over 2,000 yards in depth.[466] Defences included an array of concrete bunkers providing cover for machine-gun emplacements, observation posts, and large dugouts. Ongoing RFC aerial photography confirmed many of the bunkers were resisting British artillery shellfire.[467] The burden was on the RFC to ensure that their artillery-spotting mission dominated the information battleground. Success at Messines required the British artillery to dominate the German batteries through greater accuracy.

All intelligence elements were put to the test. The first known use of signals deception on a large scale occurred at Messines. GHQ manipulated the patterns of their radio traffic, 'leading the enemy to believe that Corps which were on any given front, now unimportant, are still there, and also that no increase has been made in the No. of Corps on the front of importance.' The British signals security effort had successfully masked the order of battle work of the German traffic analysts.[468] The British radio intercept operations and intelligence staff also improved in their ability to monitor German aerial observation radio transmissions, providing the RFC with a general picture of the enemy's aeroplane and artillery activity. The volume of observed radio activity correlated to any buildup or decrease of enemy artillery in the sector.[469]

The improvements in British signals intelligence at Messines also engendered the creation of Compass Stations. Compass Stations provided indications of German aerial observation within the respective centre of operation.[470] When the Americans arrived in theatre, they also exploited a similar network to track German aerial operations. Known as the Airplane Information Service, it comprised a network of radio detection and monitoring units called goniometric stations. When the enemy aeroplane commenced radio transmissions for artillery spotting, the goniometric units intercepted the message, telephoned the unit that was to be targeted, and contacted Allied artillery to commence counter-battery fire. It performed an important function during positional war. However, when mobile war commenced in the final weeks, the process was discontinued.[471]

The British commenced the attack with systematic trench bombardment and wire cutting on 21 May 1917. RFC aerial artillery spotting was active supporting British 18-pounders and 6-inch howitzers. During the final phase of prepping the battlefield, the British conducted extensive artillery night firing. Billets, headquarters, and villages in back areas were shelled by long-range pieces and a continuous barrage was maintained on the enemy's communications.[472] Messines is remembered most for the dramatic initial British offensive. On 7 June 1917, 19 deep mines containing an aggregate of 937,000 pounds of explosive were detonated under the German trenches. The massive explosion blew up large portions of the enemy's front and support line.[473] Mining operations were an integral feature of positional warfare. Aerial reconnaissance was continually looking for any signs of excavation that indicated this activity. The British had nine tunnelling companies in 1915, expanding to a total of 25 from 1916 to 1918. In 1916, 48 of 128 km of the front held by the BEF were protected by underground galleries, in several instances at more than one level. But mining was effective only against a strongly held front line, and by the close of 1917 it became a recognised principle to hold the front by firepower rather than manpower. Troops were deployed in depth and protected from bombardment by dugouts, subways, and other subsurface excavations. Despite the increase in tunnelling companies, offensive mining declined, so the companies were increasingly diverted to the construction of dugouts.[474]

British artillery was active from 21 May to 6 June. With supporting aeroplane observation they carried out 297 trench bombardments. Wire was cut along front and support lines by trench mortars and 18-pounder batteries, and on rear lines by 6-inch howitzers with aeroplane observation.[475] On 5 June 1917, the British artillery engaged in intense counter-battery work. Every available howitzer was to be used for intense and destructive counter-battery work. 'Maximum value from aeroplane and balloon observation' was to be arranged. Gas shelling was planned against the German batteries during the three to four nights before the assault in sufficient quantity to ensure the German artillery personnel were to be forced to use their masks and be kept awake and become exhausted.[476]

Prior to the actual attack, the RFC's No. 53 Squadron was tasked to keep three planes in the air throughout the hours of daylight to support the artillery. Normally, two planes were employed on counter-battery work. The third plane carried out observation of the effect of trench bombardments and wire cutting as well as providing registration in support of the heavy artillery. The RFC was required to photograph the entire trench area within the flank limits of the British IX Corps once a day. No. 53 Squadron was responsible for delivering proof copies to IX Corps heavy artillery one

Prepping the Messines battlefield was accomplished with aerial coverage one day before the 7 June 1917 battle commenced. (*Artillery Operations of the Ninth British Corps at Messines, June, 1917*)

hour after the aerial reconnaissance aircraft landed, with secondary distribution going to the divisions and IX Corps headquarters. Aerial photographs of hostile battery positions were to be taken as soon as possible after the completion of a destructive artillery barrage, with copies delivered to the Counter-Battery Office. Every evening the counter-battery staff officer issued the program of targets for the next day. An effective liaison network was accomplished when No. 53 Squadron assigned the same aircrew to support a designated artillery battery throughout the campaign. An example of artillery spotting work by No. 53 Squadron on 18 May 1917 shows the extensive effort: 'Battery 405th Siege firing on target Ochre Trench, O 26 c 11.42 to 05.55. Successful. About 60 rounds observed. 2 OK's, 6 Y's, 8 Z's. Battery shooting well when machine left. Several rounds fell very close to trench, one causing a fire, smoke of which lasted 10 minutes.'[477]

The RFC's II Brigade provided vital support throughout the campaign. Most of the German batteries were out of sight of the ground artillery observers. The Brigade's 300 serviceable aeroplanes successfully carried out the essential artillery observation and aerial photographic missions, acquiring daily coverage of German defences during the bombardment period as well as coverage every other day of the counter-battery area.[478] The RFC at Messines averaged a daily flying rate of 612 hours, accomplishing a total of 716 zone calls against German targets.[479] Observation balloon operations monitoring the German rear areas at the Messines Ridge were connected via telephone to most of the heavy artillery groups of the British Ninth Corps. These resources worked sector observation reporting intensity and extent of enemy barrages, updating progress of Allied barrages, and locating any tanks in the sector.[480]

The British conducted systematic aerial photography to determine the effect of artillery fire. The demand for increased coverage was levied on all flights. This was backed by the proliferation of photographic sections assigned at the squadrons. The schedule of coverage was continuous to satisfy Corps requirements. British Corps increased their demands, prompting the squadrons to continue to provide coverage from the front line to 5,000 yards beyond. To ensure timeliness, the squadrons were responsible for distribution. Photographs of lower priority targets were forwarded to the Army Printing Section at Hazebrouck for additional printing.[481] Army reconnaissance covered the Germans' main railway communications farther to the east, to include Bruges, Ghent, Grammont, and Ath. Particular attention was paid to the rail lines converging on Courtrai and Menin. DH 4 aeroplanes of No. 55 Squadron, fitted with cameras and flying at high altitudes between 16,000 and 20,000 feet, usually attempted the more distant reconnaissance. The closer and more detailed strategic reconnaissance was made by formations of Sopwith 1½ Strutters of No. 70 Squadron, escorted by a flight of pursuit fighters.[482]

Messines was a major success story for British military operations, 'a tactical masterpiece of its kind'.[483] The carefully planned offensive resulted in a significant victory for the British and was subsequently labelled 'a perfect operation of siege warfare'.[484] The British Second Army advanced 2½ miles in a single day, capturing the objective. The cooperation between the aerial observer and battery was recognised as outstanding, allowing the artillery to effectively target and destroy the enemy artillery. They did experience problems with fleeting targets, a premonition of future challenges as the battleground transitioned to mobile warfare. However, the relationship between aviator and artillery reflected the growing sophistication of communication technology, resulting in effective results.[485]

JUNE 1917

The French-English 1917 spring offensive led the German General Staff to publish, on 10 June 1917, a Supplement to the Regulations for Stationary Warfare, in which they summarised the most important principles in those regulations. This supplement was annotated and enlarged by General Friedrich Sixt von Armin (commander of the Fourth Army) in an army order dated 30 June 1917. This order fell into Allied hands, and some essential extracts include:

> Necessity of camouflage – 'Strength in a defensive battle depends essentially on the precautions taken to conceal from the eyes of the enemy all our means of fighting. These fighting means (trenches, dugouts, machine-gun and battery emplacements) will surely be destroyed, if they appear on the enemy's aerial photographs...
>
> Depth organisation – 'There must be substituted for the old system of position, which may be plotted and destroyed by the enemy, a defensive zone organised in depth...
>
> Utilisation of shell holes by the fighting line – 'During a battle it is no longer necessary to have continuous trenches in the front line. They may be replaced by shell holes held by bunches of men and isolated machine guns, arranged in checkerboard...
>
> 'In front of the first row of shell holes will be placed an irregular barbed-wire entanglement, as continuous as possible. It is also advisable to protect with barbed wire the holes in the front of the first line, to prevent their occupation by the enemy...
>
> 'Farther back, it is preferable to organise isolated defensive works, by surrounding the shell holes

with auxiliary defences in such a way as to leave passages for counterattacking troops...

Placing supporting troops and reserves – 'A large part of the supporting troops and of the reserves will be sheltered in the open field, in shell holes, woods, ravines, anywhere to avoid aerial observation. (Villages must be avoided, as they draw the enemy's fire.) These troops assist in the formation of a continuous line, avoiding so far as possible the view of the enemy. Thus a supporting line is established for the defence troops placed at successive intervals in front of it...

First-line defences – 'These must include several lines of trenches, protected by a strong barbed-wire entanglement, with passages for assaulting troops. Deep dugouts will be constructed only on the second and third lines. The first line will have only small dugouts, capable of holding about one-sixth of the occupants of this line...

New-line defence – 'The organisation of new lines back of the first will be governed by the same principles as above. If possible, the construction of dugouts for the garrison must be undertaken in succession for the whole depth of the zone. (For one-sixth of the men in the first trench, two-sixths in the second trench, three-sixths back of the second trench as far as the line of artillery defence.)...

By the creation of numerous shelters in the rear of the lines, preparations are made for the opening of the defensive battle, the orderly retreat of the troops in the trenches, and their redistribution on the intermediate terrain...

Order of urgency of the works – Staking out the works, placing auxiliary defences, bomb-proof shelters, digging trenches...

Machine-gun emplacements – Machine-gun locations should not be chosen in front positions, but on the sides of hills or in depressions, at points convenient for flanking. The enemy should come upon machine guns unexpectedly. Therefore, it is not desirable to install them in salients, but to the right or left or in the rear. In the salients only fake emplacements should be constructed...

Laying out boyaus – In order to prevent the enemy from forming an exact idea of our system of defence, the boyaus connecting the different lines may be given an oblique direction. This arrangement gives the appearance of the meshes of a net, in which the boyaus, provided with auxiliary defences and organised as firing trenches, have at the same time defensive functions. [Riegel]...

Everywhere must be kept in mind the necessity of escaping aerial observation.[486]

The British victory at Messines had spurred the Germans to develop a new defensive posture. Allied aerial observation remained a major concern to German military thinking. Newer tractor (engine in front) aircraft such as the RE 8 and longer focal length cameras such as the 'L' Type camera were fielded, allowing aerial observation at higher altitudes with the same quality coverage, particularly with stereo photographs. Intelligence discovered that the Germans were modifying their defences at the front. The Germans' new line of 'strong points' using shell holes further complicated Allied operations. Aerial photographic review of the front sectors soon identified the network and provided in-depth information for subsequent artillery targeting. The shift in fire towards these fortifications both dramatically increased German casualties and caused a rift in their overall defensive network.[487]

During the summer the RFC developed a strategy for air superiority over the Western front from the Lys sector to the coast. RFC pursuit fighter strength was on the increase. They prosecuted the enemy to a depth of five miles from the German front line back to the observation-balloon line. Aerial observation conducted more operations as the battle for aerial supremacy fuelled the largest air battles of the war. Massive dogfights with 30 aircraft on either side commenced. This culminated on the evening of 26 July above the Ypres battleground. An air battle took place involving at least 94 single-seat fighters flying furiously at heights from 5,000 to 17,000 feet. Despite the incredible numbers of fighters engaged, there was no decisive result. In fact, while the dogfight was in progress, four German aerial observation aeroplanes took the opportunity to fly over the British lines and conduct a reconnaissance of the Ypres sector.[488] At 8,000 feet, 30 Albatros Scouts fought seven DH 5s; at 12,000 to 14,000 feet, ten Albatros Scouts fought approximately 30 British single-seater fighters; at about 17,000 feet, ten Albatros Scouts engaged seven British naval Sopwith triplanes. Below the entire dogfight, a few British two-seaters flew at 5,000 feet.[489]

FRIENDS IN HIGH PLACES

By the final year of combat British military aviation pioneers such as Trenchard and the Salmond brothers had proved aerial reconnaissance to be an integral part of the battleground and they now held senior positions of command. These officers created the foundation for aerial operations in that final year, giving the RFC the direction it required to both refine aerial observation and increase pursuit roles necessary to maintain aerial supremacy. Major-General Trenchard assumed the role of the first Chief of the Air Staff (CAS). Field-Marshal Haig protested about the move with a letter to senior British politicians: 'I am convinced that the proper place for Trenchard is in actual command of the Flying Corps in the field in France.'[490] Major-General John Salmond at the incredibly early age of 36 was assigned to the Army Council at the War Office.[491] Major-General Geoff Salmond now managed the RFC in the Middle East. Lieutenant-General Sir David Henderson still played a significant role in shaping the RFC's future. He finally stepped down from the DGMA posting that he had held since 1912 and focussed on establishing a single service that combined RFC and RNAS – the future Royal Air Force (RAF). Brigadier-General Frederick Sykes learned the trade of a senior staff officer. His experiences since returning from the Gallipoli campaign covered a broad range of responsibilities from establishing machine-gun corps for the Royal Army to serving under General Sir Henry Wilson at the Supreme War Council. His skills were being honed for the greater responsibilities that lay ahead.

French aviation also clearly demonstrated innovation and leadership in the ranks, but seniority was limited. Despite the influence of aviation on French military operations for the first three years of combat, the most senior French rank in the Aviation Militaire (chef de l'aéronautique at GQG) at this time was commandant (Major).[492] Despite the absence of senior officers, Allied expertise within the ranks of intelligence and aerial photographic interpreters was impressive. The outcome of three years of intense analytical effort interpreting aerial photographs created a sound resource on which to base the interpretation of German military doctrine and practice. In many respects, the aerial photographic interpreters became the most knowledgeable experts on how the enemy put into practice the art of positional war.

One important feature of Allied aerial reconnaissance at this stage of the war was the introduction of a new generation of state-of-the-art aeroplanes that could both meet the demands of lower-altitude missions against the threat of enemy pursuit as well as operate at high altitude beyond the limits of fighters and anti-aircraft artillery. Allied aerial reconnaissance units now started to acquire Breguet 14 A2, Salmson 2 A2, and DH 4 aeroplanes. One outstanding aeroplane of the era, the Bristol F2B Fighter, was initially designed for fighter reconnaissance. It would subsequently serve the RFC/RAF as one of the best fighter aeroplanes of the war. On one mission in August 1917, a flight of Bristol F2Bs from 48 Squadron flew a photo-reconnaissance mission at an impressive altitude of over 19,000 feet. The flight leader, Captain B.E. Baker, reached an altitude of 22,000 feet during the sortie.[493]

One of the finest aerial reconnaissance aeroplanes of the war, the Breguet 14 A2 arrives at the front on August 1917. This aeroplane was assigned to the first Breguet-equipped escadrille BR 7. (SHD, B78.917)

Lieutenant Weiller's exploits in French aviation achieved legendary status with each assignment. He shot down one German aeroplane flying a Nieuport 17. (Mousseau, *Le siècle de Paul-Louis Weiller 1893–1993*)

CAPITAINE PAUL-LEWIS WEILLER ASSUMES A COMMAND

Nineteen-seventeen crystallised Lieutenant Paul-Lewis Weiller's illustrious military career with an impressive string of accomplishments. On 11 February, while serving in the Chemin des Dames sector, he took a great risk by flying his Caudron G 4 low over German lines and taking aerial photographs to prove to commanders that an artillery barrage meant to prepare a French attack on a German installation had not succeeded. Weiller was wounded in the jaw and lost consciousness, reviving just in time to land the plane safely beyond the French trenches; only to slip back into unconsciousness. He awoke in hospital. Weiller's superb photos convinced Général Mangin and staff of the actual state of German defences resulting in the decision to cancel the attack. Weiller's impressive exploits were personally cited by Général Mangin – his sixth award for valour.[494]

That summer, Weiller was transferred to the Italian front to gain greater experience on aerial operations. Upon return, he served for a few weeks with the British. His courage and accomplishments were so impressive that he was awarded the Military Cross. As a recognised leader of aerial reconnaissance operations, Weiller reviewed how they best served the Allies and came up with the concept that became known as the Weiller Group (*Groupement Weiller*). He recommended a reconnaissance escadrille be established to accomplish strictly strategic missions out to 60km beyond the German lines. The French senior staff at GQG concurred. Weiller was promoted to capitaine and provided with command of Breguet escadrille, BR 224, on 5 September 1917.[495] Weiller could now accomplish what the years of aerial reconnaissance and combat had shown could be done – acquire important information on the enemy well beyond the front lines via aerial photography through dedicated resources.

CONTEMPORARY ALLIED ANALYSIS OF THE WESTERN FRONT CONTINUES – A SUMMARY OF 1917

Allied analysis continued to determine German intentions through in-depth analysis of the changing configurations of the front lines. The standard for German defence at the end of 1916 was based on the Seigfried Stellung model.[496] By the summer of 1917, the Allies were able to determine German positional war doctrine thanks to intelligence acquisition of key orders and supplements. Published Allied manuals described the transition from combat in the first lines to combat within zones of combat. Allied efforts to defeat the Seigfried Stellung prompted German commanders to further define effective order in the front lines. Combat now focussed on fighting zones, not the actual trench lines. Fighting zones comprised the network of trenches to a depth of several kilometres. A defensive zone (called the *Vorfeldzone*) was designed to repulse all surprise attacks. The centre of gravity for combat effectiveness was in the *Grosskampzone*, where the Germans placed the greatest priority for winning the battle. If the terrain was unsuitable for maintaining the position, the *Grosskampzone* was moved back to work from a more defensible position. To the rear of the *Grosskampzone* was the *Ruckwaertige Kampfzone*.

This approach required effective and reliable communication with the rear echelons, good artillery observation and use of reverse slopes to complement artillery observation. Good shelter for infantry and artillery was still a requirement. However, mobility was strongly encouraged to include working from unprepared positions. The defensive doctrine still held sway, involving flanking positions, dugouts, observation stations and posts of command.[497] It was this doctrine that shaped the battles of the last year of conflict, starting with the historic battle at Cambrai.

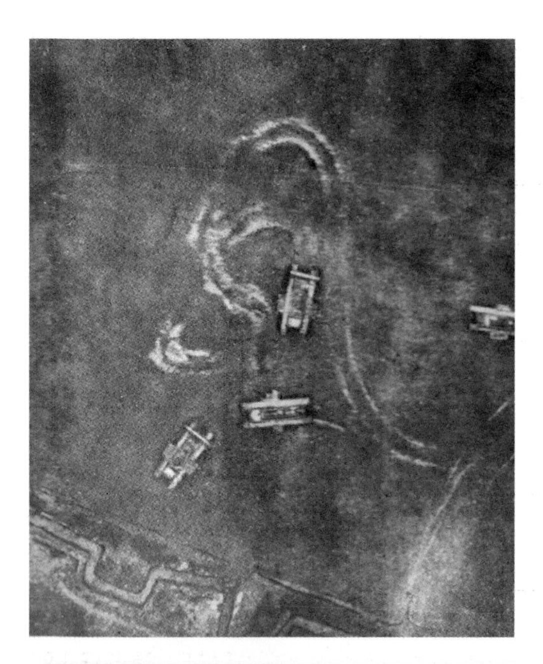

Captured German aerial photograph of British Mark IV tanks at Cambrai. (Edgar Middleton, *The Great War in the Air*)

Aerial photography of First World War tanks represents the first photographic interpretation of mobile armour in history. (Elmer Haslett, *Luck on the Wing: Thirteen Stories of a Sky Spy*)

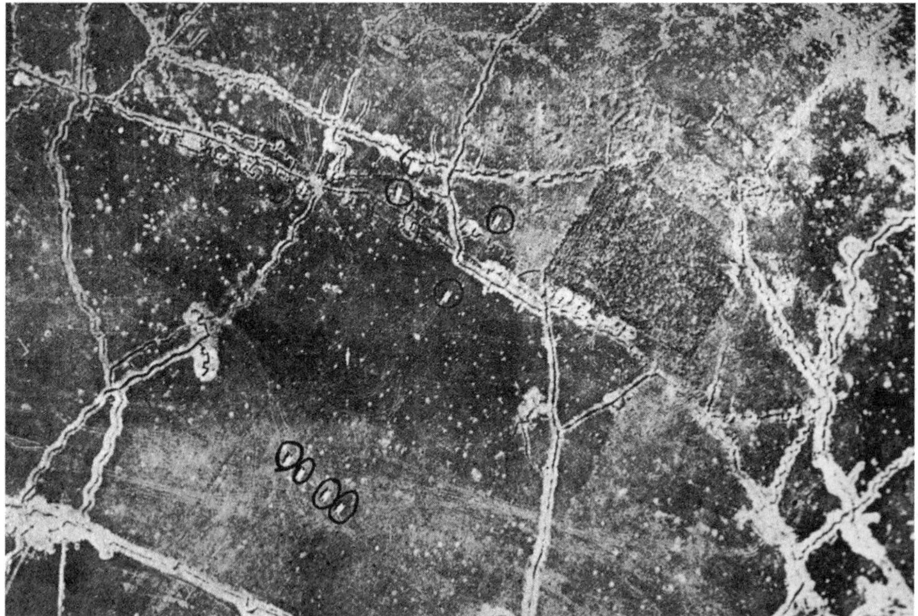

CAMBRAI

(For Cambrai map, see colour section.) The year 1917 concluded with the battle at Cambrai, most noteworthy for the extensive employment of tanks. Previous French experience with a mass armour attack during the Nivelle Offensive had failed mostly because of German three-metre trenches designed to block forward movement. Lessons learned over the year in tank employment revolutionised artillery close support and provided new tactics for the battle. The Cambrai sector was selected as the point of attack because the plain was especially favourable for tank operations. Additionally, the sector had the necessary natural cover in Havrincourt Wood, to mask the assembly of equipment prior to the advance. Surprise was intended to be the key factor in breaching the Siegfried Stellung's extensive defence network. The British eschewed the traditional precursor to the attack, a heavy artillery barrage. The shock effect of the armoured advance was deemed sufficient to neutralise any infantry in the way. Now artillery was assuming a new role. Instead of delivering destruction, it supported mobility by neutralising enemy artillery and any infantry firepower that the tank did not eliminate.[498]

The British paid particular attention to a new technique of 'shooting by the map', by which 'unregistered' or 'predicted' fire range and bearing was calculated in advance for each artillery piece. The lack of a preliminary bombardment and associated analysis gained surprise and eliminated several German batteries at the start of the offensive.[499] RFC operations laid the groundwork for the offensive, providing important reconnaissance information.[500] At dawn on 20 November, screened by artificial smoke clouds but without artillery preparations, a veritable tank army crawled across the rolling plain, smashing its way into the German lines. However, the weather during the attack was not conducive to effective aerial reconnaissance. 'Air observation was impossible at first, dangerous later and always unreliable.'[501] When they did get airborne, the British Army squadrons did conduct aggressive attacks against German airfields, batteries and infantry. However, this time the German aviation reacted quickly with newly arrived squadrons of fighters. In two days of intense struggle the Germans converted RFC air supremacy into 'air parity'. German aviation also effectively integrated with the army battle organisation providing effective aerial reconnaissance and artillery support in the counterattack.

The ineffectual analysis by British Intelligence at GHQ under Brigadier-General Charteris had bothered the British Prime Minister, Lloyd George, for a long time. Despite Charteris' strong influence on Haig, the time had come for his dismissal. Charteris' promulgation of the idea that the Germans were defeated in will and capability was consistently wrong, culminating in the successful German counter-attack at Cambrai. It proved to be the final straw. In December 1917 Lloyd George took action and had Major-General Macdonogh's nominee, Brigadier-General Edgar Cox, relieve Charteris. BEF intelligence remained under the influence of the Ministry's Directorate of Military Intelligence for the remainder of the war.[502] As for Charteris, Haig's loyalty saved him from total humiliation. He was retained as the GHQ's Deputy Director General of Transportation and continued to have the ear of Haig.

Cambrai demonstrated that infantry could now depend on armour to assist in an offensive. Artillery was still vital for all remaining campaigns in the war, but its dominance had to adjust for mobility. The tanks also depended on artillery to assist in their advance. The critical development was that artillery could now concentrate on supporting the breakthrough, leaving armour to breach the front lines. Despite Cambrai's legacy as a relatively successful massive tank attack on the Western Front, the breakthrough did not occur. The siege of the Western Front continued with more massive casualties for another year. The impact on aerial reconnaissance was increasing requirements for aeroplanes that could better support the demand for intelligence from the rear echelon. Positional war was giving way to mobility.

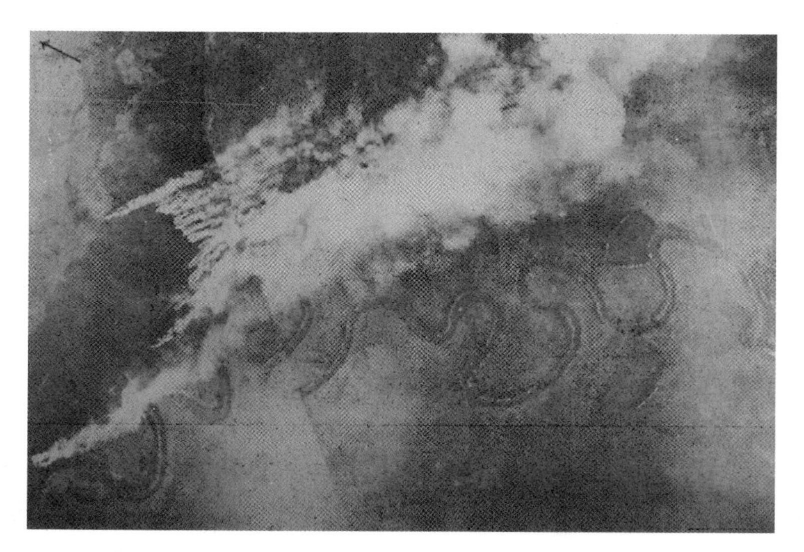

Concealing operations was a requirement for both sides. At Cambrai the weather was misty, allowing tanks to roll forward. When the weather did not help, smoke screens provided cover, as here in the Oise sector. This French photograph was reproduced in a postwar British manual. (AIR 10/1120. 'Notes on the Interpretation of Air Photographs.' February 1925)

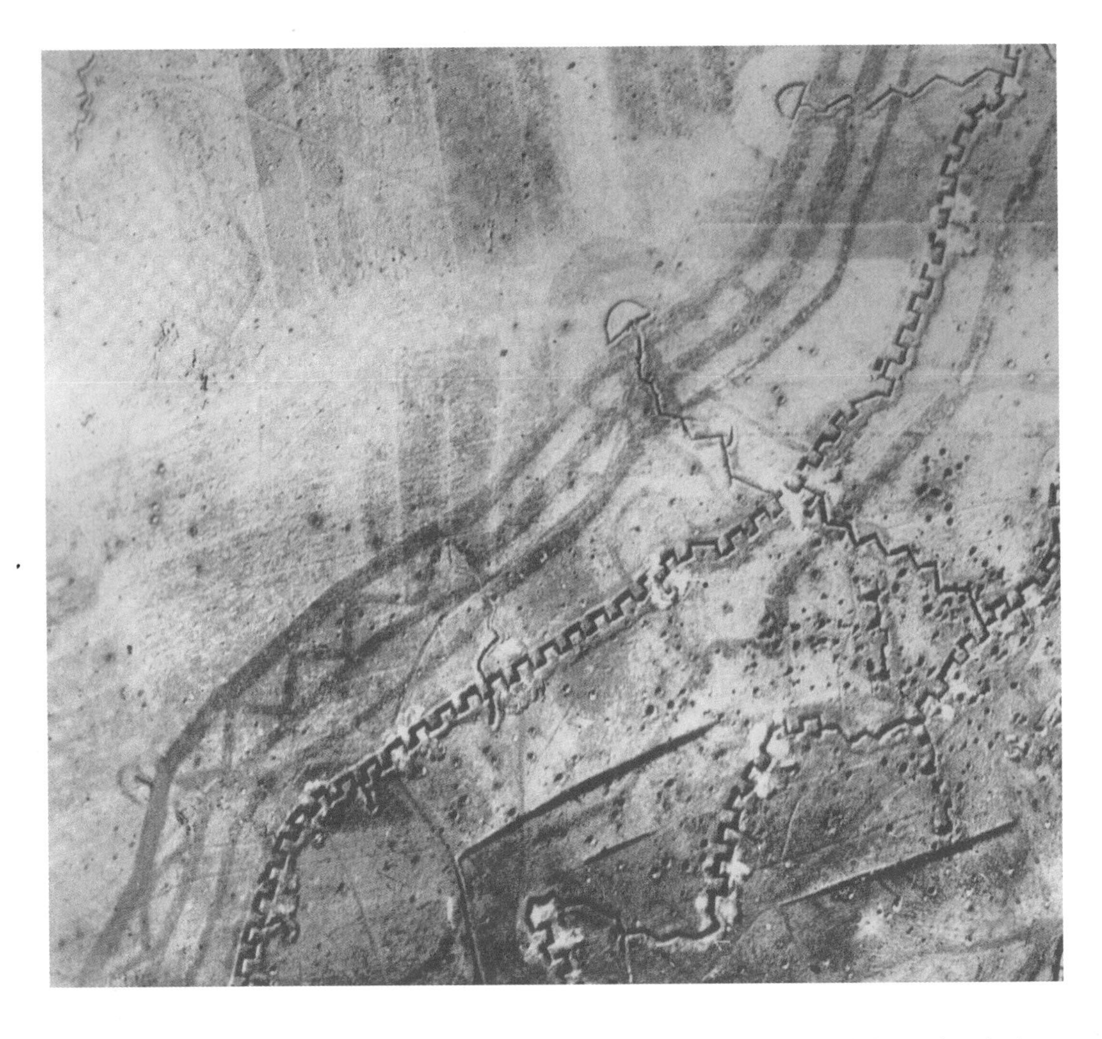

The formidable German lines at Cambrai were photographed and analysed before the assault by British tanks. (Ian Passingham, *All the Kaiser's Men*)

A captured British tank. After the disastrous failure of the tanks at Bullecourt in April 1917, the potential of the tank as a war-winning weapon was demonstrated at Cambrai – though the crucial breakthrough did not come. (USAFA McDermott Library Special Collections (SMS57))

CHAPTER 6

1918: Aerial Photography Refined

PREPARATIONS FOR OPERATION *MICHAEL*

The fall of the Russian Government and the subsequent release of German divisions from the Eastern Front provided the Germans with an opportunity to make a major strategic move in the west, leading to one of the most decisive battles in the war. The Allies were in a quandary, for they knew that German reinforcements were soon to be arriving. The question was where and when the forces would be committed. The status of American reinforcements was also a great concern. Despite the US having been in the war since April 1917, it had too little military presence in theatre to make a difference. The potential was there, but only four divisions had arrived by the end of that year. Of those four, only two divisions had experienced the front lines when the Germans commenced their offensive.

Indications of the impending offensive multiplied throughout February 1918. What intelligence could be acquired by RFC aerial reconnaissance in the winter months focussed on rear echelons. Aerial reconnaissance and photographic missions had observed abnormal amounts of railway traffic, construction of new aerodromes, dumps, railway sidings and hospital camps. In January alone, 14 new German aerodromes had been established just opposite the British Fifth Army sector.[503]

Leadership changes had just taken place in the RFC. Major-General John Salmond replaced Major-General Trenchard as General Officer Commanding (GOC) the RFC in the Field. Trenchard returned to London as the newly established Chief of Staff, Air Council.[504] Salmond directed the RFC headquarters reconnaissance unit, No. 25 Squadron, to concentrate their aerial observation efforts east of the British Fifth Army area. Subsequent reporting provided further indications of increasing German rail and road movements behind the Third and Fifth Army fronts, as well as more sightings of new dumps, aerodromes, and emplacements. The signatures for a buildup left little doubt in the mind of many commanders, including Général Foch and Field-Marshal Haig, that something was going to break loose. Later that month, Salmond assembled his RFC Brigade commanders and gave them instructions as to the importance of obtaining evidence, generating more sorties from available aeroplanes, and increasing the reconnaissance effort to help substantiate evidence of enemy intentions. Air reports and aerial photographs generated at the end of February revealed a marked increase in German road and rail movements and daily extension of the network of light railways. Despite the growing indications, it was intriguing that British aerial observation was not being over tasked to glean any indications of German intentions. A British aerial observer, Lieut. Thomas McKenny Hughes, commented that 'there are not a great number of shoots to do as a rule now and the Corps are quite exemplary in the smallness of their demands for photographs.'[505] However, British BEF GHQ had not come to a definite conclusion as to where and when the attack was to commence.[506]

Four years of positional war had accustomed the Germans to the risks associated with aerial reconnaissance and photographic interpretation. They were always aware of the potential intelligence value of the photograph. Preparations for Operation *Michael* included taking steps against aerial observation. General Ludendorff personally signed off a memorandum entitled 'Protection of Artillery Positions from Air Observation' that expressed senior concern with Allied aerial observation:

> The constant observation of our organisations by the enemy's aeroplanes in trench warfare and the technical improvements introduced in aerial photography compel us to hide our artillery positions from air observation to an increasing degree. Experience has shown that the type of work carried out by our troops up to the present has not led to the desired results. In all camouflage work, greater endeavours than in the past must be made, not only to avoid the effect of light and shade, but also to assimilate the whole emplacement to the colour of the ground. The plates employed for photography from great heights are extraordinarily sensitive to colours and show up

> clearly the slightest difference in shades, even those imperceptible on the ground. Effective camouflage is only possible if, before the emplacement is built, the colouring of the surrounding ground is carefully taken into account. The necessary data can only be furnished by air photographs.[507]

General Ludendorff and the senior OHL staff had at their disposal one of the most important documents on the subject of aerial photography. During the Cambrai counter-offensive of November 1917, the Germans had acquired the British 1917 edition of *Notes on the Interpretation of Aeroplane Photographs* (S.S. 550). With the manual the Germans were able to react to the Allied method of accomplishing aerial photographic interpretation with clarity. Ludendorff's directive continued, 'It is evident that increasing care is taken to conceal emplacements and to defeat the camera. As, however, the Germans usually start to construct camouflage after a battery emplacement has been completed, their attempts are rendered abortive, owing to the fact that the emplacement will probably have been photographed several times during the various stages of construction.'[508]

Ludendorff issued his directive to the forces in final preparation for Operation *Michael*. Steps to be taken during the offensive included:

> 1. Whenever an emplacement is to be built, a specially detailed Divisional Artillery Staff Officer will be responsible for furnishing the battery with air photographs showing all necessary details as regards light effects and colouring.
>
> 2. Photographs will be taken repeatedly during the stages of construction, in order to test the camouflage. The staff officer mentioned above will make the necessary arrangements with the divisional flight concerned.
>
> 3. Once the camouflage is complete, the emplacement should, whenever possible, be again photographed before it is occupied.
>
> 4. The same precautions will be taken in the construction of alternative emplacements, and of positions built in the rear in view of possible retirements.
>
> 5. The staff officer mentioned in para. 1 is responsible for testing by frequent photographs the camouflage of occupied battery positions.[509]

Allied reconnaissance now had to contend with an ever more alert adversary. On the eve of the battle, from 17 to 20 March, early morning reconnaissance reflected the calm before the storm. German aerial activity was minimal with Allied aerial observation activity detecting nothing of interest, a reflection of adherence to the Ludendorff directive. Night reconnaissance was called for. British sorties were made but nothing was detected. German march discipline was at its peak. Whenever the British airmen dropped flares to detect activity, 'all traffic stops, and the troops crouch lifeless under the shelter of trees and houses.'[510] Aerial reconnaissance was not enough to determine German intentions prior to *Michael*. The Allies had been successfully deceived by German operations prior to the offensive. Order of battle estimates showed 190 German divisions in France, supported by those flowing from the Eastern Front to the West. Many of these battle-experienced troops were relocated to the Hirson-Mézières region, over 100km from the Western Front. The Germans had proven from fighting the Russians at Riga (Baltic State of Latvia) that they could stage their forces for an offensive far from the front. French intelligence had conducted a major assessment of that battle and was aware of the strategy. Their armée escadrilles discovered new railway lines leading toward the British sector to reinforce a possible operation. However, the indications did not lead to reinforcement of the Allied line.[511]

British artillery had been active prior to the German attack, firing conventional and gas rounds into the German lines. During one barrage, the British Third Army fired on a target discovered from aerial photography: a number of small white objects in pits east of Bullecourt. The British fired 9.2-inch howitzer rounds, resulting in a massive explosion. The objects had been piles of ammunition boxes. The British quickly followed up with a general bombardment that created over 100 explosions counted, large and small. Similar objects were noticed on aerial photography, found in the hollows and valleys of the German sector opposite the British Fifth Army. The initial thought was that the covered objects were tanks, until the artillery rounds verified they were munitions under wraps.[512]

Keeping to the Ludendorff directives, the Germans moved largely at night and camouflaged their forces by day, thereby diminishing the value of aerial reconnaissance. German attention to Allied strategical reconnaissance capabilities was noteworthy. All aeroplane staging was accomplished far behind the lines to thwart any inadvertent discovery by reconnaissance. Throughout March German aeroplane movement forward was strictly concealed, employing cover from

Opposite: Operation *Michael*, also called the Somme Offensive. (*West Point Atlas*, courtesy Department of History, US Military Academy)

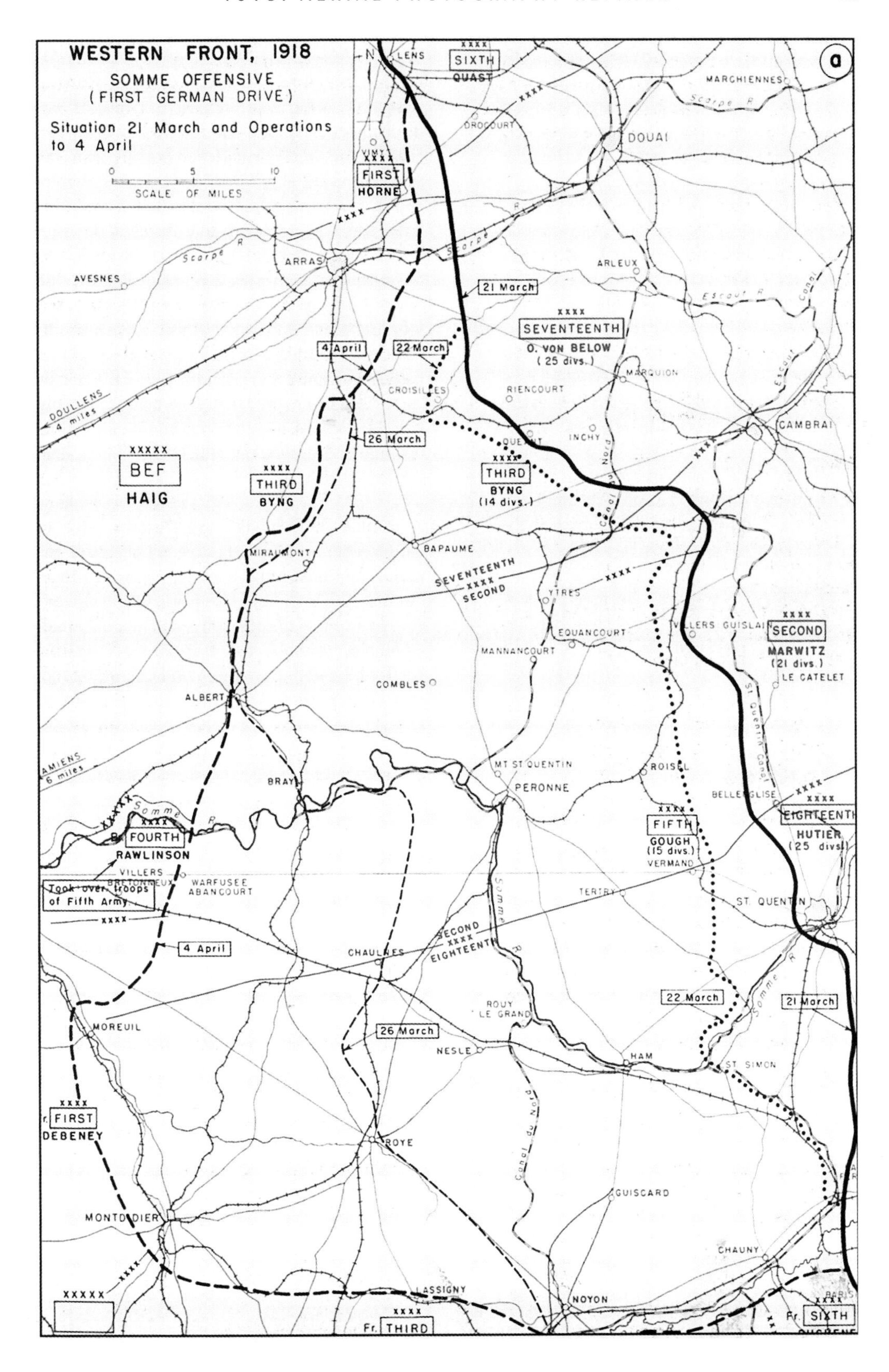
WESTERN FRONT, 1918
SOMME OFFENSIVE
(FIRST GERMAN DRIVE)
Situation 21 March and Operations to 4 April
SCALE OF MILES
SIXTH
QUAST
FIRST
HORNE
SEVENTEENTH
O. VON BELOW
(25 divs.)
THIRD
BYNG
(14 divs.)
BEF
HAIG
SECOND
MARWITZ
(21 divs.)
FIFTH
GOUGH
(15 divs.)
EIGHTEENTH
HUTIER
(25 divs)
FOURTH
RAWLINSON
Took over troops of Fifth Army
Fr. FIRST
DEBENEY
Fr. THIRD
Fr. SIXTH
21 March
22 March
26 March
4 April
LENS
DOUAI
ARRAS
CAMBRAI
BAPAUME
PERONNE
ALBERT
ST. QUENTIN
MONTDIDIER
NOYON
HAM
NESLE
ROYE
CHAULNES
MOREUIL
CHAUNY
GUISCARD
LASSIGNY
DOULLENS 4 miles
AMIENS 6 miles

barns and horse tents. *Drachenballon* (captive balloons) were camouflaged, awaiting orders to be inflated. Any flying activity associated with preparing for *Michael* was accomplished in the rear echelons; 48 hours prior to the first salvos all German ground support units were forward making final preparations, 24 hours later all German aeroplanes were forward and ready.[513]

The weather was also a factor. For four days before the attack, the early morning reconnaissance flights were impossible due to thick cloud and rain. However, challenges from weather were not enough to prevent the British RFC from photographing some of the German preparations.[514] For this campaign, the Germans became masters in masking their radio network. They employed false signals to create a phantom army on the French front, imitating preparations for an offensive against the British in Flanders. By exercising effective radio transmission discipline – limiting communications to short messages – the Germans offered French direction-finding a small signature to work with, concealing the German order of battle. Almost two weeks prior to the offensive, the Germans started to change their call signs daily. Complicating the picture was the frequent German use of ciphered communications, an important radio intercept signature, usually indicating impending operations. German employment of artillery barrages and infantry raids to probe against the French and British lines added to the confusion. It was impossible to tell if these actions were in preparation for the great blow or were simply feints.

Artillery achieved prominence at this stage of the war as a result of bold initiatives by the brilliant German tactician, Colonel Georg Bruchmüller. Destruction was not the best strategy for ensuring victory in positional war. Instead, Bruchmüller advocated 'neutralisation' through demoralising the enemy. The preliminary bombardment was not targeted against obstacles and troops in the trenches. Instead, Bruchmüller advocated that artillery strike the enemy's command and control networks comprising PCs, telephone exchanges and observation posts. A subsequent barrage of artillery fire would be applied to different sectors using a variety of weapons and munitions. The final artillery barrage comprised an intense fire on the front lines for ten minutes followed by a 'creeping barrage' in front of the infantry attack. The Germans also limited artillery registration to add to the element of surprise. German aerial support to registration relied on other sources such as sound-ranging, accurate meteorological data and ammunition consistency.[515]

German artillery units honed their deception tactics prior to the offensive. Bruchmüller was masterful in applying deception. Camouflage was erected before any artillery positions were built. Deception plans included bogus orders, dummy wagon and caisson tracks, and construction of dummy positions along the front. Bruchmüller even misled some subordinate German artillery commanders into thinking their artillery units were to support the main attack – the result was that the deceived were more committed in preparing for the upcoming battle. Bruchmüller used dedicated aerial reconnaissance to check on the preparations.[516]

> Our batteries had half their guns constantly on the move, appearing here, there and everywhere, firing at night a few shots into enemy territory: and when next morning British or French reconnaissance aircraft flew over to take photographs of the new battery positions, there it apparently was, but actually consisting of wheels of farm carts with a wooden beam, all scarcely camouflaged. The British artillery fired fiercely on these dummies and the next night, a few hundred yards away, a new 'battery' popped up. And this deception manoeuvre went on for weeks from north to south, and especially on those parts of the front where no attack was intended to take place.[517]

Haig's former Intelligence chief, Brigadier-General Charteris, even provided his own estimate of the situation. Despite his new assignment working transportation issues, he was still able to access aerial photography and make his own deductions on the growing body of evidence. On 8 March he observed:

> There was a good deal of shelling going on in the front line, and swarms of aeroplanes – mostly our own – out reconnoitring.... Our front there is pretty strong, and should give a good account of itself if it is attacked. The Third Army is rather concerned about some new marks on air photographs in the fields by the roadside of the German back area. The tracks leading to them mean some form of German tank. I think there are too many of them in one place to be tanks. The Fifth Army has something of the same sort on its photographs of areas near St Quentin, and thinks it is ammunition brought up on caterpillar tractors. I am not sure that they are not large handcarts for the supply of small arms ammunition, but if so they must have particularly broad wheels. Whatever they are, they point pretty conclusively to a very early offensive.[518]

OPERATION *MICHAEL*

When the German offensive commenced on 21 March 1918 at 0440 hrs, 6,608 German guns opened an artillery barrage lasting five hours along almost 50 miles of front.[519] The weather favoured the attackers as fog covered the southern Fifth Army sector, prohibiting any aerial reconnaissance of the advance. One sortie, from RFC No. 59 Squadron, did manage to track the German bombardment along the whole front of the IV Corps. Other RFC aircraft were launched to support British artillery counter-battery. By noon that day, six aeroplanes were generating countless radio transmissions to the artillery on a multitude of targets, including artillery and large formations of infantry on the roads heading west. The Germans had neutralised the sector communications. What British Corps aircraft that could fly did not get airborne until noon and what little they could see was not of much consequence. When the fog cleared that afternoon, British aerial reconnaissance saw many targets, but the British artillery had been effectively neutralised.[520] The standard historical interpretation of the first day of Operation *Michael* is that the Royal Artillery failed. The organisation between artillery and aerial observation did not succeed in working through the challenge of the fog on the first day of the campaign. Although 'Contact patrol work and close reconnaissance were maintained by the Royal Flying Corps, and its reports often gave the corps staff the best information which they received as to the movements of the fighting troops and of the enemy's advance.'[521] Despite the fact that the Germans were advancing so rapidly, the British did establish air superiority, allowing their aerial observation to continue.[522]

The impact of the offensive was astounding. Within a week the Germans had driven the Allies back almost 38 miles to Amiens. French intelligence was as stunned as the rest of the Allied command, 'By virtue of my job, I am the best-informed man in France, and at this moment I no longer know where the Germans are. If we're captured in an hour, it wouldn't surprise me,' the head of GQG's 2e Bureau declared.[523] The exemplary gains made by British forces at Messines Ridge the previous fall were quickly overrun in the first days of the offensive. However, the British did generate a significant amount of intelligence at the front. When *Michael* was launched, the Intelligence Summaries of British divisions at the front were credited in post-war assessments as valuable, 'so that an immediate appreciation of all the enemies' intentions was always available.'[524]

The overwhelming German advance in the first days of Operation *Michael* found British resistance to be disjointed. Post-war analysis determined that cooperation between the Royal Artillery and the RFC failed from the start. For long periods, not only in the fog of the first morning but throughout the British retreat, the communication between batteries and aeroplanes was ineffectual at best. British planners had attempted to establish a zone call system to meet the anticipated conditions of open mobile warfare. Major-General Trenchard had emphasised in his memorandum, 'The Employment of the Royal Flying Corps in Defence,' that

> Modern artillery plays such an important part in a defensive action that too much stress cannot be laid on the work of the corps machines. Counter-battery work without air observation is in most cases of little value. The best way, therefore, in which the Flying Corps can assist at this period [i.e., when the attack has begun] is by assuring a continuance in the air of artillery machines. If this object can be

RFC units such as No. 15 Squadron were operating at peak levels during the Michael offensive. The squadron commander, Major H. V. Stammers, reviews fresh reports from the front. (Nesbit, *Eyes of the RAF*)

> attained, it will be a far more effectual help to the infantry and artillery, though invisible to them, than any amount of low-flying or bombing against the enemy front-line troops.[525]

Retreating British forces could not maintain effective communication between artillery batteries and artillery liaison. Artillery telephone lines and radio masts had been disconnected or not set up effectively. As the artillery batteries retreated westward, they were unable to set up and link to the artillery spotting aeroplanes for decisive targeting. Even when the retreating artillery was able to erect their vital radio masts, receiver operators were often unaware of which zones were assigned to the battery. The roar of the artillery created so much noise on the lines as to drown out communications.[526]

Operation *Michael* created a most desperate situation for the Allied commanders. They needed as much information on the Germans' intentions as could be acquired. Intelligence from aerial reconnaissance was deemed 'intelligence that is of the greatest value in present circumstances.' What information could be acquired had to be shared as quickly as possible by whatever means available. On 23 March 1918, Général Foch sent a personal memorandum to Field-Marshal Haig, – a comprehensive survey of the disposition of German forces – and summed up three possible German objectives. First, an offensive in a sector of the front between the sea and the Somme; a continuation of the attacks between Montdidier and Rheims, and a surprise assault on some other part of the front, in Champagne, at Verdun, or in Lorraine.[527] The French staff drew up detailed plans for the reconnaissance escadrilles to conduct both day and night reconnaissance of the railway systems to assess the density of the traffic along the whole German front between Flanders and Alsace. The British were to maintain a watch of the area opposite the British armies. The overall reconnaissance strategy focussed on a systematic aerial photographic reconnaissance of enemy billeting areas. Maps were issued marking the billeting areas and likely manoeuvre areas for German artillery. Aerial observers were also instructed to keep a general look-out for extensive counts of trains and heavy artillery distant from the front.[528]

Three days into the attack against the British sector, Général Pétain alerted the British that there were definite signs that he was about to be attacked in Champagne and mentioned aerial reconnaissance had found German camps, tents, etc. lit up that night.[529] The night reconnaissance from French aerial bombardment sorties in the sector led to the conviction that the Germans' preparations were defensive in nature. The usual signatures associated with an offensive build up did not transpire. The Militaire Aviation did not convince GQG of their suspicions concerning German intentions in Champagne. Even after the German attack commenced at the Somme, the GQG view that Champagne was at risk did not abate. French bombardment aviation remained concentrated against Champagne. German deception also contributed to French confusion. On 23 March they prepared to reinforce the British. However, the next day the order was revoked. A German *Drachenballon* landed on the French lines as if by accident. The gondola contained documents that referenced a German attack on 24 March against French armées in the Champagne sector to the south of the Somme. The deception worked, for the French ordered their bombardment units to operate away from the Somme. On 27 March the bombardment units finally returned to the Somme and commenced operations.[530]

French Général Maxime Weygand, chef d'état-major to Général Foch, followed up a week later in correspondence with the British staffs on maintaining a sense of urgency regarding strategic reconnaissance, 'Reconnaissances to ascertain the direction of the enemy's main movements, and therefore his intentions, should extend at least as far as the line ST. QUENTIN-CAMBRAI-DOUAI, where he may detrain.' Strategic reconnaissance remained critical, 'To ensure that air observation covers every part of the area of approach to the battle zone, it will be necessary for the British Air Service to watch particularly the approach routes leading from Le Catelet to Peronne; Cambrai to Bapaume; Aubigny to Arras; Douai to Lens. The air units with Armies will undertake to follow up any movements discovered in the back areas, as far as the battle front.'[531] Foch's subsequent memorandum on the state of the Allied effort, reminded all participants that whatever information was acquired, it had to be shared.[532]

In the midst of Operation *Michael*, the RAF was born. The hard work of Lieutenant-General Sir David Henderson culminated on 1 April 1918 with the RFC and RNAS merging to create a new independent RAF. The decision did not affect aerial observation and reconnaissance because the new organisation revolved around strategic bombing. The principal motivation behind establishing the RAF was to unite the heavy bombers in an 'Independent Bombing Force' to impose destruction directly upon the industries and population centres of the enemy.[533] Perhaps the most significant change was the departure of those who were instrumental in establishing the RFC legacy from its inception as an aerial reconnaissance arm of the British Army to a well-established air force with both tactical and strategic roles. Major-General Trenchard tendered his resignation as CAS following grave differences of opinion with the Secretary of State for Air, Lord Rothermere. It was finally approved on 13 April. A week later, Major-General Frederick Sykes was appointed CAS.[534] Sykes' former senior, Henderson, then submitted his resignation as Vice-President of the Air Council.[535] The seminal figure of British aerial reconnaissance disappeared from British military aviation for the last time.

The merger of RFC and RNAS did pose issues over standardised equipment for British aerial photography. All RFC and RNAS aeroplanes were assessed for aerial reconnaissance. Major Moore-Brabazon, representing the RFC, and RNAS Lieutenant Charles Gamble, discussed all aerial platforms to determine to what extent aerial photographic missions could be served by the existing aerial fleet. An exchange over the ubiquitous Sopwith 1½ Strutter showed the RNAS preference. Gamble told Moore-Brabazon, 'With respect to XXI (Sopwith 1½ Strutter listed in the memo), we like them for photography, but you do not. I understand the contract is expired and no more will be made. In any case I think they should be fitted for cameras.'[536]

With the reorganisation, concern surfaced in the Royal Navy over meeting the needs of seaplane stations, especially in continuing their ongoing anti-submarine operations. However, the photographic officers of both services were able to work together and resolve their differences. Standards were established on camera fittings. The table of equipment was also revised to suit the requirements of both former RFC and RNAS. All the branches were 'departmentalised' to allow for further expansion as the program required.[537]

When the German's Operation *Michael* started to show indications of winding down, ongoing aerial reconnaissance started to detect a shift in direction by the German forces, to include preparations for another offensive in the Lys salient.[538] Aerial photographs taken in the afternoon of 6 April showed the Germans were preparing an offensive north of Aubers, opposite the Portuguese left. Throughout the morning of 7 April aerial observers reported the main roads full of moving transport immediately opposite the Portuguese sector. Other intelligence sources reported more ammunition being moved into the German support lines. The impression conveyed by the combined air and ground reports was that the German's concentration of forces was nearing completion.[539] Strategical reconnaissance kept headquarters aware of the flow of reinforcements into the battle. Throughout the day aerial observation reported the main roads opposite the Portuguese were full of moving transport. That evening, rail activity was observed increasing behind Lens and eastward from Lille.

On 9 April the second phase of the German's campaign, Operation *Georgette*, commenced against the Portuguese division on the front line. The steps required for preparing the Operation *Michael* attack were repeated for *Georgette*. It began with an intense German artillery bombardment to include the extensive use of gas directed at the Allied artillery, headquarters, and communications. Advancing German storm troops quickly followed. Aerial observation was limited owing to fog in the battle areas. The Germans' drive through the front lines caused Portuguese resistance to fold. The Germans quickly reached the Lys salient. On the evening of 11 April, Field-Marshal Haig summed up the Allies' desperate situation with his 'backs to the wall' Order of the Day, one of the most memorable statements of the entire war.[540]

Two days into Operation *Georgette*, aerial reconnaissance finally caught a break in the weather. Now the Allied air forces were acquiring observations, conducting infantry contact missions and attacking German troops. On 12 April, weather over the battleground was excellent. RAF sorties were able to provide artillery spotting information and low altitude ground support against the German infantry. The advancing German divisions were now subjected to relentless attacks by the RAF squadrons. The effort was memorable. 'More hours were flown, more bombs dropped and more photographs taken than on any day since the War began.'[541] RAF aerial coverage that day produced an impressive 3358 photographs.[542] A daily average of 40 sorties by the RAF provided extensive coverage. When weather got worse toward the end of the month, sorties dropped to an average of 23 per day.[543]

When the offensive came to a halt and the line stabilised at the end of April 1918, practically no work was done along the entire front beyond that which was immediately necessary. The Germans did not attempt to construct a properly organised trench system despite their tremendous gains. In the Lys salient, tractors or tanks were extensively used for the transport of ammunition in the counter-battery area with the tracks of the vehicles becoming very conspicuous on the aerial photographs.[544] Both Operations *Michael* and *Georgette* clearly demonstrated that new tactics could yield spectacular results. Artillery had to adjust to the new phenomena. Instead of massive and prolonged barrages along the front, artillery now resorted to sudden and annihilating concentrations of fire on demand. Excellent reconnaissance and initiative on the part of the artillery commanders was now required. Mobility was reestablishing itself in the thinking of the combatants.

ESTABLISHING AN AMERICAN PRESENCE

America's entry coincided with a flagging of the Allied force, exhausted after almost three years of unceasing combat. The Americans, upon arrival in theatre, were relegated to a novice role vis-à-vis their

Despite the initial difficulties with language, arriving American aviators established excellent relations with their French aviator mentors throughout the remainder of the war. Here the 99th Aero Squadron poses with members of escadrille SAL 1 'Escargot'. (USAFA McDermott Library Special Collections (SMS57))

Allied counterparts in the art and science of positional war. Their own combat experience had primarily been fighting guerilla-type battles in diverse locations like the Philippines, the southwest United States, and Mexico. Positional war of any duration had not been experienced since the last year of the Civil War. American forces had not acquired or integrated the technologies that had been developed by the combatants during the first three years of combat. A major industrial mobilisation was required, but this was possible thanks to America's innate industrial might, which included a vibrant photographic industry. Despite the nation's potential, maintaining an effective reconnaissance capability was a constant struggle right up to the Armistice.

The first missions for the newly created US Air Service involved aerial observation. Aerial units were aligned with the French, who presided over initial training for conducting observation in the sector. The Americans received their first lessons in aerial photographic reconnaissance at Amanty (Meuse) in January 1918. The first class consisted of ten personnel. However, it was ineffective because of the language difficulties. French faculty and American students were unable to understand each other. The Americans were only able to mutter a few words of French, dealing mostly with things to eat. After a few weeks, the French left the school to the on-scene Americans to work with their own available experts on the subject of aerial reconnaissance and photography. The initial student cadre soon became the instructors.[545] Despite this stumbling beginning, the American juggernaut picked up speed and proceeded to build an extensive force of aeroplanes and support operations. By the end of the war, the United States had 45 Aero Squadrons and 21 Photographic Sections assigned in theatre. It was an impressive buildup for a military that had been devoid of any substantive aerial expertise when war was declared.

The US Air Service commenced aerial observation operations in April 1918. The German's Operation *Michael* was concluding in the Somme Sector. The United States established the 1st Corps Observation Group comprising two squadrons (1st and 12th) as well as the contingent of the Photographic Section #1 near the front. The units were assigned the responsibility of covering the American divisions near Chemin des Dames. Both arrived for duty at Ourches flying Spad two-seaters on 5 April 1918.[546] Upon arrival, Major Royce of the 1st Aero Squadron and Captain Griffin, commanding officer of the photographic section, flew over the lines and took the first aerial photographs by Americans in a combat environment. However, they

The French and British provided the arriving Americans with any available aeroplanes in the inventory. An AR 2 was one of the first aeroplanes for the 12th Aero Squadron. (USAFA McDermott Library Special Collections (SMS57))

An example of cooperation between the French and Americans is shown with a Spad XI from escadrille SPA 42 flown by a French pilot and an American artillery officer observer in May 1918. (Gorrell Report, NARA)

did not have the luxury of an established photographic laboratory infrastructure to work with. The first American Expeditionary Force (AEF) aerial photographs were developed, washed, and prints made in a photo car parked along the side of the road. They had to make do while the photographic laboratory buildings and support barracks were being constructed.[547]

US Air Service squadrons began active operations over Toul, a quiet sector of the front occupied by the American's 26th (Yankee) Division. The 1st Corps Observation Group was the primary US aerial unit in the region with two Aero Squadrons assigned. The 1st Aero Squadron conducted aerial photographic missions, while their counterpart squadron, the 12th Aero Squadron, was assigned artillery spotting missions.[548] The 1st Corps Observation Group eventually consisted of the 9th Aero Squadron, 24th Aero Squadron and 91st Aero Squadron. The 91st Aero Squadron in time became the most famous of the US Air Service aerial reconnaissance units, supporting the AEF with long-range reconnaissance.[549]

The initial US aerial photographic effort was abysmal. Almost all of the sorties attempted were either cancelled by weather or suffered aeroplane motor problems. What sorties were generated did not yield any coverage of value. Only 10 plates were acquired in a period of two weeks and were of no value in alerting the American forces to German intentions. When the 26th Division fought the first major engagement by US forces against the Germans at Seicheprey on 20 April 1918, US aerial reconnaissance was ineffective.[550] Despite the lacklustre start, the US Air Service honed its skills with each campaign, ultimately supplying important information on the activities of the enemy by means of both visual and photographic reconnaissance.[551] It is noteworthy that the work of the 1st Corps Observation Group was seldom hampered by the presence of enemy pursuit aircraft. Thus, practically no experience in combat was gained. However, the enemy anti-aircraft fire in the Toul Sector was exceedingly dense, active and accurate; resulting in the pilots becoming adept at evading that threat after a month of operations.[552]

The 1st Corps Observation Group learned the aerial reconnaissance trade through extensive interaction with the French. They established an operations room at Group Headquarters to serve as an information and intelligence center. Large-scale maps of the Toul Sector and of the enemy's defensive organisation, charts, diagrams, and tabulations of all available tactical and technical information collected by the operations and intelligence officers were displayed in this room.[553] American aerial reconnaissance support for British units in the sector included a few squadrons assisting battle planning. Prior to the combined Australian-American attack on 4 July 1918 at the battle of Hamel, US Air Service units provided 1,200 copies of aerial photographs. The first American operationally deployed photographic unit, Photographic Section #2, rounded out the 1st Corps Group, providing superior support throughout. The only threat to ground operations came from an occasional German aerial bombardment.[554] Edward Steichen later had high praise for the Photographic Section #2 organisation. 'I do not believe that there was a better or more efficient outfit for the producing of aerial photographs in any of the combatant armies.'[555]

THE SECOND MARNE CAMPAIGN

The increasing arrival of fresh American forces to the front heightened German concerns and prompted new campaigns. The lessons from Operations *Michael* and *Georgette* showed the Allies still were vulnerable to a well-planned German operation. *Michael* had depleted the British Army to the point that the French Armée had to reposition their reserves to the northern sector of the Western Front. Ludendorff used this opportunity to generate a major offensive against the French VIème Armée, positioned on the Chemin des Dames ridge north of the river Aisne. The preparation for battle included 3700 guns being massed on a 38-mile front opposite the French. Aerial reconnaissance assets from both France and Britain were committed to the sector. The British sent their Ninth Corps south to the French-controlled Aisne sector, supported by RAF No. 52 Squadron. The squadron's RE 8 observation aeroplanes were successful in reporting the buildup, especially when they detected large dust clouds from forces and weapons on the move toward the front lines. Despite the indication of the German's movement toward the sector, the French Armées' 2e Bureau was slow to respond. Early on the morning of 27 May, the Germans unleashed one of the fiercest bombardments of the war, followed by a rapid advance across the Aisne. Three days later the Germans captured Soissons and reached the Marne at Château-Thierry. Fierce resistance by two American divisions (2nd and 3rd) and French reserves managed to stop the German advance. Throughout the drive, German pursuit activity had been aggressive, keeping the Allied aerial reconnaissance to a minimum.

American exposure to the utility of aerial photography came quickly as the Germans advanced on Château-Thierry. The rapid German advance prompted fears with the GQG that it could be the final breakthrough to Paris. Artillery targeting in this sector was hampered by lack of any large-scale maps. French and American topographic personnel were rushed to the sector and through the use of photography enlarged a smaller-scale map in a desperate attempt to provide rear echelon artillery with sufficient data. At the same time, rapidly acquired aerial photography created sufficient mosaics to create a more accurate larger scale map to prioritise artillery targeting. It demonstrated to both Allies the value of the resource against the German thrust.[556]

On 9 June, the Germans conducted one of their last offensives against French forces in the Compiègne sector. This time, the offensive lacked surprise. French intelligence analysis had successfully determined the Germans' intention. An intercepted message sent by German headquarters to the staff of the German *18. Korps* occupying the sector read, 'Rush munitions Stop Even by day if not seen.' The implication of the message was that the Germans were massing for an offensive from the town of Remaugies. The intelligence analysis estimated that Ludendorff was trying to salvage his salient by attacking southward on a line between Montdidier and Noyon towards Compiègne. Subsequent aerial reconnaissance and prisoner interrogations confirmed the intended manoeuvre. Fifteen German divisions engaged in the offensive advanced six miles before a spirited counterattack led by five French divisions under Général Charles Mangin stood their ground, forcing the Germans to retreat. The French IVème Armée commander, Général Henri Joseph Eugène Gouraud, practised a detailed analysis of the aerial photographs in his sector. He countered the German offensive in his sector with a remarkable defensive arrangement comprising two lines. The first line was a trip wire, lightly held and evacuated at the first signs of activity. Gouraud's forces then proceeded to a more powerful network on the second line, which was out of range of German artillery.[557] It proved to be Ludendorff's last chance to achieve decisive strategic results with his armies on the Western Front. It was now the Allies' turn to establish the final offensive.

GQG correctly ascertained German intentions concerning the July offensive, thanks to intelligence capabilities. Two weeks prior, French aerial photographic reconnaissance systematically covered the entire Marne salient, to include Rheims, Château-Thierry, and

Soissons. The analysis led to the GQG's forecasting the German drive to within a 48-hour window, determining the exact axis of the main drive. German intentions were revealed through standard signatures of an offensive buildup, including new second-line defences, supply depots and ammunition dumps and the enormous increase of traffic on the highways. Also, additional ammunition dumps were observed near the front. There was a 'sudden great increase in the activities on the roads immediately behind the German front lines; the construction of ramps and new roadways leading from battery positions and trenches to facilitate the egress of troops; and one particularly damning indication of attack, the fortunate location by French aerial observers of a German pontoon train, located not far from the Marne and just north of Dormans.'[558] French ballon observateur activity complemented the ongoing observation. The sector had reinforced the observation posts (the French term for this function was *service des renseignements de l'observation du terrain* (SROT)) with additional telephone lines. When the Germans finally attacked, the French PC was immediately informed of the advancing forces and artillery activity.[559]

THE CHANGING DYNAMIC OF AERIAL RECONNAISSANCE

As the war reached its final months, aerial tactics changed to meet the threat posed by pursuit aeroplanes. 'Reconnaissance planes may have to give battle' assessed a French post-war Aeronautics manual. A merger of the two missions became the focus of aéroplane development in the final years of the conflict. 'Aviation was to seek out the destruction of enemy aeronautics, assure use of aerial observation and impede enemy aerial observation. Aeronautical units were at the disposal of the army and placed in appropriate echelons.'[560] This included A3 reconnaissance aéroplanes of long range subordinate to the armée and armée group and reconnaissance aéroplanes of short range attached to the troops. However, the French discovered that the larger A3s (Letord, Morane-Saulnier T) complicated their management of assigned escadrilles. Aviation Militaire made the decision to disperse the long range A3 category to individual reconnaissance and in some cases, pursuit escadrilles, to provide each unit with a reconnaissance function to better support armées or CA as necessary.[561]

The final year of the conflict saw a concerted effort by all of the Allies to establish a strong pursuit presence in support of aerial reconnaissance. British pursuit squadrons assigned greater numbers of aircraft to escort aerial observation, to ensure continuation of the observation mission.[562] Aerial reconnaissance mission planning followed a routine that included a rendezvous with the protecting pursuit squadron.[563] As the war neared its conclusion, the Americans eventually determined that their best observation formation was several aerial photographic aircraft flying together. State-of-the art aeroplanes such as the Salmson 2 A2 and DH 4 made this configuration defendable. The lead aeroplanes accomplished the photo mission; the remaining aeroplanes conducted defence. However, there was no denying that additional Allied pursuit planes in the area also benefited the mission.[564] A post-war French assessment conveyed the prevailing attitude: 'Combat planes may sometimes bring back valuable information to the Command.'[565]

German aerial defence was pushed to the limit in the final months. There were not enough aeroplanes to engage the burgeoning Allied offensive. The lack of air superiority also had an impact on their infantry:

> All regiments have repeatedly complained that low-flying enemy aeroplanes are not interfered with by our fighter machines. In reply to these complaints, the Group points out that its resources in aircraft are so low, and those available are so busily engaged on reconnaissance work, that they cannot also be expected to engage enemy machines carrying out low reconnaissance over our lines. The infantry must arrange for its own defence against enemy aircraft to a greater extent than heretofore.[566]

Greater emphasis was placed on masking infantry operations. The 2nd Guards Reserve Division issued the following directive on 1 August 1918: 'The camp, viewed from the air, should produce the impression of being deserted. Tracks will be definitely marked out; all tracks leading into a camp will be continued through it and will lead into other tracks or roads. Camouflaged netting is to be put up over tracks at frequent intervals.'[567]

Allied offensives now had purpose. The last months of the war demonstrated aviation's dynamic role as the conflict transitioned from the stalemate of positional war to manoeuvrability in the final weeks. All Allied combatants pushed their respective air forces to be fully engaged in combat. For the French, the Aviation Militaire committed their escadrilles to an aggressive campaign of long-range bombardment, night and day aerial reconnaissance, infantry contact and fighter patrols in group (*groupe de chasse*). The French possessed more aéroplanes than any other aerial combatant and it was clearly demonstrated in the final weeks of conflict.[568] An in-depth experience base of aviators accompanied the state-of-the-art aéroplanes now in use.

France's aerial reconnaissance leader, Capitaine Paul-Louis Weiller, led the advance in new strategical reconnaissance roles with the Breguet 14 A2. His vision was put to the test in the ten months that he commanded escadrille BR 224. He discovered that systematically flying over the German rear echelon provided more intelligence on German intentions in a given sector. The best formation, Weiller reasoned, was a group of large-scale reconnaissance flights committed to photograph the rear echelons to a depth of 100km as frequently as conditions permitted. The Breguet excelled in this strategy. Under *Maréchal* Foch's personal direction, two flights of Breguet 14 A2s were dedicated to acquire rear echelon coverage. It later comprised two escadrilles, BR 45 and BR 220, under the unit title, the Weiller Group.[569] Just prior to the armistice, the Weiller Group added Caudron R XI aéroplanes from escadrille R XI 246.[570] Weiller applied high altitude aerial reconnaissance tactics to his strategical sorties. His Breguet 14 A2s were flown without an observer and all extra weight removed to reach a ceiling of 6000 metres. This also expanded the coverage area for the aerial cameras.

The Weiller Group's success was making an impression on the Supreme Allied Commander, *Maréchal* Foch, with in-depth strategical reconnaissance combined with superb analysis based on aerial photography. A thorough understanding of the rear echelons gave Foch the advantage in deciding the Allies' next move. Weiller Group predictions on enemy intentions were made as far in advance as one month of the actual operation. The critical data were provided on a regular basis to Foch, sometimes personally by Weiller at all hours of the day or night. His strategy reflected what Weiller Group maps revealed – directing the maximum effort toward sectors being abandoned by the Germans. It put greater pressure where it counted – the Germans were not able to counter Allied manoeuvres with such precision.[571]

Capitaine Paul-Lewis Weiller led development of French strategical aerial reconnaissance. By March 1918, Weiller was decorated with the legion of honour and the Croix de Guerre with several palms. (Mousseau, *Le siècle de Paul-Louis Weiller, 1893–1993*)

THE SUMMER OFFENSIVES

The reassignment of American forces from the Toul sector to Château-Thierry was the beginning of large-scale American military operations against the German lines. The German assault on Paris in June culminated in their forces being stopped at Belleau Wood and Château-Thierry. The Americans followed up with their first offensive at Aisne-Marne in mid-July. This involved five American divisions working in concert with three French divisions. Aisne-Marne represented the first of many strategic offensives, countering the last efforts by the Germans to establish an offensive strategy and set the stage for the conclusion. More resources and better weather aided aerial photographic reconnaissance. The initial dismal performance by the US Air Service was quickly forgotten as aero squadrons increased their sortie rate and acquired more relevant coverage for ground operations. More effort was also placed on distributing photographs in a timely manner. The First Army Corps placed photographs into categories of 'First Needs' that delivered by aeroplane drop the packet of prints to the PCs for Corps and Divisions. Recipients included Intelligence (G-2), operations of Corps HQ, and Intelligence offices supporting key divisions in the Corps (42nd, 77th, 82nd, 90th). 'Second Needs' distribution went to echelons below division.[572]

American forces commenced planning to regain Soissons and drive the Germans from the Aisne salient. On 18 July, the Allies launched a counter-offensive between Soissons and Château-Thierry. It was one of the first shifts from positional to mobile war. The resulting mobile campaign complicated the well-established ground and air liaison. Infantry contact patrols now became the more important aerial mission, with aerial photography assuming a less critical role. It also had an impact on the traditional artillery spotting role with batteries manoeuvring on a more frequent basis. Continued success prompted the Allies

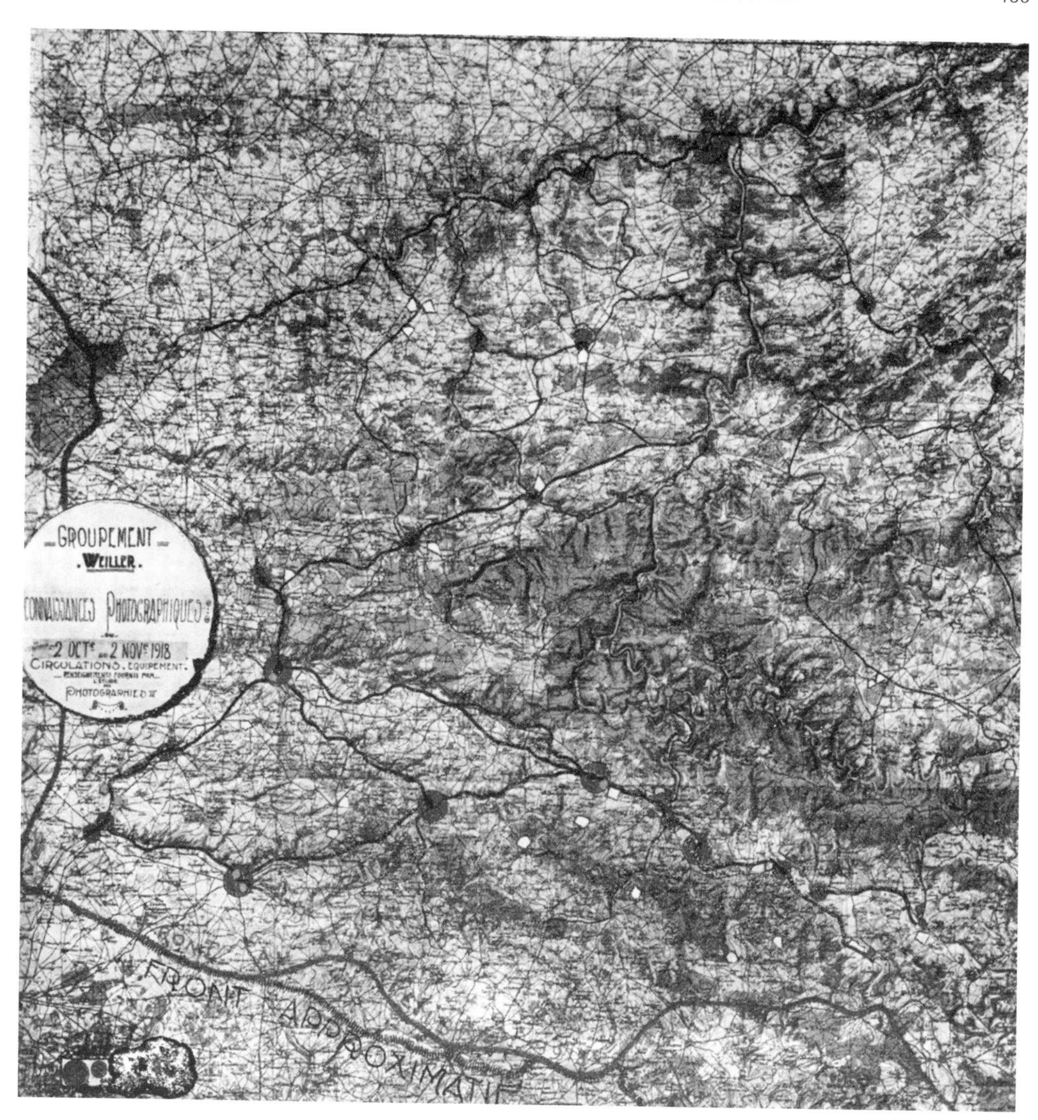

The Weiller Group made strategical reconnaissance a successful tool for determining enemy intentions. The map shows key German targets in the rear echelon. (de Brunoff, ed., *L'Aeronautique*)

Escadrille BR 220 comprised one of three escadrilles for the Weiller Group. This Breguet 14 A2 from BR 220 provided coverage of the rear echelons. (SHD, B 83/1783)

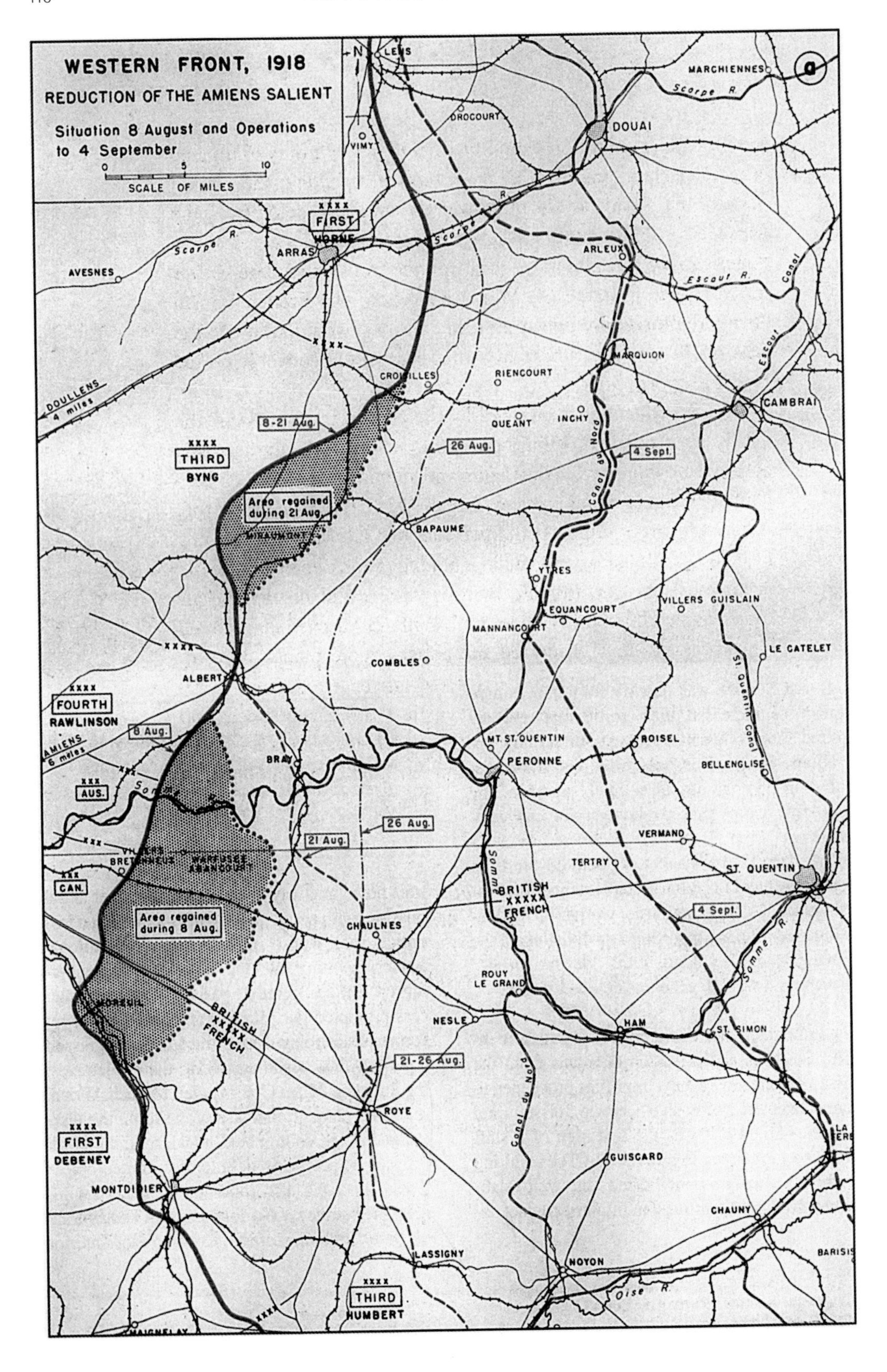
WESTERN FRONT, 1918
REDUCTION OF THE AMIENS SALIENT
Situation 8 August and Operations to 4 September
SCALE OF MILES
LENS
MARCHIENNES
DROCOURT
DOUAI
VIMY
FIRST
HORNE
ARRAS
AVESNES
ARLEUX
MARQUION
CROISILLES
RIENCOURT
QUEANT
INCHY
CAMBRAI
DOULLENS 4 miles
8-21 Aug.
26 Aug.
4 Sept.
THIRD
BYNG
Area regained during 21 Aug.
MIRAUMONT
BAPAUME
YTRES
EQUANCOURT
VILLERS GUISLAIN
MANNANCOURT
LE CATELET
COMBLES
ALBERT
FOURTH
RAWLINSON
8 Aug.
AMIENS 6 miles
MT. ST. QUENTIN
PERONNE
ROISEL
BELLENGLISE
BRAY
AUS.
26 Aug.
21 Aug.
VERMAND
VILLERS BRETONNEUX
WARFUSEE ABANCOURT
TERTRY
ST. QUENTIN
CAN.
BRITISH FRENCH
Area regained during 8 Aug.
CHAULNES
4 Sept.
ROUY LE GRAND
MOREUIL
NESLE
HAM
ST. SIMON
21-26 Aug.
ROYE
FIRST
DEBENEY
GUISCARD
MONTDIDIER
CHAUNY
LASSIGNY
NOYON
BARISIS
THIRD
HUMBERT
MAIGNELAY

to reconsider aerial reconnaissance roles. As the 2nd Marne achieved its objectives, Maréchal Foch reviewed plans to increase pressure on the Germans through coordinated, local ground and air offensives achieving definite but limited aims. The RAF also modified their aerial reconnaissance strategy. No longer did they concentrate on static billeting areas. Coverage of interior railway networks became the priority to help gauge German efforts to sustain the fight.[573] The US Air Service now emphasised aerial reconnaissance as an integral part of the American battle planning. Within the 1st Observation Group, the 1st Aero Squadron accomplished long-distance photographic missions, adjustment of divisional heavy artillery fire, and long-distance visual reconnaissance. The 12th Aero Squadron and the 88th Aero Squadron conducted short-range visual reconnaissance, short-range photographic missions, adjustment of light artillery fire, and infantry contact patrols.[574]

The British commenced their counter-offensive in August in the one area that posed the greatest threat to their logistical network – Amiens. Amiens was such a critical risk that the British Fourth Army commander, General Sir Henry Rawlinson, cautioned just prior to the German *Michael* Offensive, 'There can be no question that the Amiens area is the only one in which the enemy can hope to gain such a success as to force the Allies to discuss terms of peace.'[575] The subsequent battle that summer proved a major success for the British Army. Major-General John Frederick Charles (J.F.C.) Fuller, one of the most famous British military strategic thinkers of the twentieth century, cited Amiens as '....the greatest mechanised battle fought in the last war'.[576] The British honed their applications of force with a more effective mix of armour, artillery and infantry. Employing successful tactics proven at Cambrai, surprise was key. Batteries were deployed at night, ammunition dumps concealed from aerial observation, and more accurate surveys of the rear echelons ensured greater accuracy of fire. Aerial photographs showed an ageing defence network of trenches, mainly dug in 1915. The real strength of the German defence lay in a diversified array of concealed machine-gun locations placed throughout the battlefield to stem the British advance.[577] However, the Allied armoured offensive proved overwhelming and the German sector collapsed. General Ludendorff's memories of this moment would be forever engraved in history. 'August 8 was the black day of the German Army in the history of the war.'[578]

Opposite: Amiens. (*West Point Atlas*, courtesy Department of History, US Military Academy)

Artillery staffs now focussed on striking deeper targets. The concept of strategic surprise gained respect as they proceeded to annihilate rear echelons.[579] The next major battle was the breakthrough of the Siegfried Stellung at Drocourt (east of Vimy Ridge) in September. Additionally, Cambrai was finally breached without a German counterattack. The sudden access to mobile operations had an impact on Allied Intelligence. Intelligence no longer followed a well-established routine, as it had in collecting against a fixed geographical point on the battlefield.[580] The last months of the conflict strained French and British aircrews. The near fatal effect of Operation *Michael* with its long-term critical engagement had a severe impact on the Allies' ability to generate aeroplane sorties. The shift from positional mode of operations to a much more flexible mobile war concept, reminiscent of the first months of 1914, placed additional burdens on routine reconnaissance missions. In 1914, aerial operations covering a mobile battlefield were limited in scope, reflecting the limited available resources. The concluding military operations were much more massive in scale, requiring extensive coordination to ensure the best use of an overstrained resource. Accommodating the requirements of the last months of the British offensive on the northern front with aerial reconnaissance was described as 'being very difficult at times'.[581]

THE AMERICAN AIR OFFENSIVE AT ST MIHIEL

Ongoing battle experience over the summer helped solidify AEF appreciation of the value of aerial reconnaissance. In early August, the American 3rd Corps participated in a minor offensive with the French VIème Armée along the River Vesle. Aerial photography was acquired by the 88th Aero Squadron assigned to 3rd Corps.[582] US Air Service aerial reconnaissance successfully accomplished vertical and oblique coverage of the entire River Vesle sector to a depth of seven and a half miles. American field commanders were very impressed with oblique coverage, resulting in more requests for that product to aid their battle planning.[583]

The Americans had been at war for almost 17 months before they were ready to conduct their own campaign against the Germans. Now the AEF had an opportunity to clearly demonstrate their value to the Allies by planning a major operation against the St Mihiel salient, held by the Germans since 1914. St Mihiel had been a battleground many times, with the strategic city of Metz as the ultimate objective, as it served as the logistical lifeline for German forces. However, AEF planning focussed on the first objective of liberating St Mihiel and the immediate area. Recapturing the strategic city of Metz with its vital railway hub would come later. The intelligence base comprised French Plan Directeur and other army maps. The US Air Service was immediately tasked to acquire the necessary aerial photographic coverage.[584] St Mihiel stirred the imagination of the American planners, both infantry and aviation. For the latter, Brigadier General William 'Billy' Mitchell envisioned a dynamic role employing around 1,500 aeroplanes to engage the enemy not only with bombardment and infantry contact missions, but also in an extensive aerial reconnaissance role. During the early part of September, a large amount of aerial photographic work was accomplished throughout all hours of daylight.[585] The AEF G-2-C Topography unit alone generated and distributed 2,320 oblique prints of the battleground prior to the offensive.[586] There was extensive preparation of maps, charts, assemblages, and clay reliefs of the territory over which the AEF advanced. The effort was impressive. In one day, a single American photographic production facility (Photographic Section #4) processed 250 plates and produced 11,212 prints.[587]

By September 1918, the AEF Air Service was made up of one army observation aero squadron, seven corps observation aero squadrons, a day bombardment aero squadron and fourteen pursuit aero squadrons. In anticipation of the St Mihiel offensive, the French put a large number of pursuit and day bombardment escadrilles at the disposal of the Americans and under Mitchell's command. It became the largest aggregation of aeroplanes assembled for a single operation on the Western Front during the entire war. A total of 1,481 aeroplanes comprising 701 pursuit, 366 observation, 323 day bombardment, and 91 night bombardment supported the offensive.[588] The Americans had finally arrived. For the first time they fielded an aerial force larger than the Germans.

Army reconnaissance sorties monitored targets up to the city of Metz, covering aerodromes, hospitals, ammunition and supply dumps, road, railroads and rear area works of all kinds.[589] One aerial observer, Lt. Lawrence L. Smart of the 135th Aero Squadron, remembered flying every day that the weather allowed, accomplishing aerial photography, observation and *avion de réglage*. Both Corps and Army Headquarters wanted every inch of the sector photographed daily to enable them to tell if the Germans were anticipating an attack, or if they were digging any new trenches, barbed wire or gun emplacements, or preparing troop movements. Smart recalled, 'By comparing daily photographs of a sector, new construction or movements of large groups of men could be detected with accuracy.'[590] They photographed the objectives

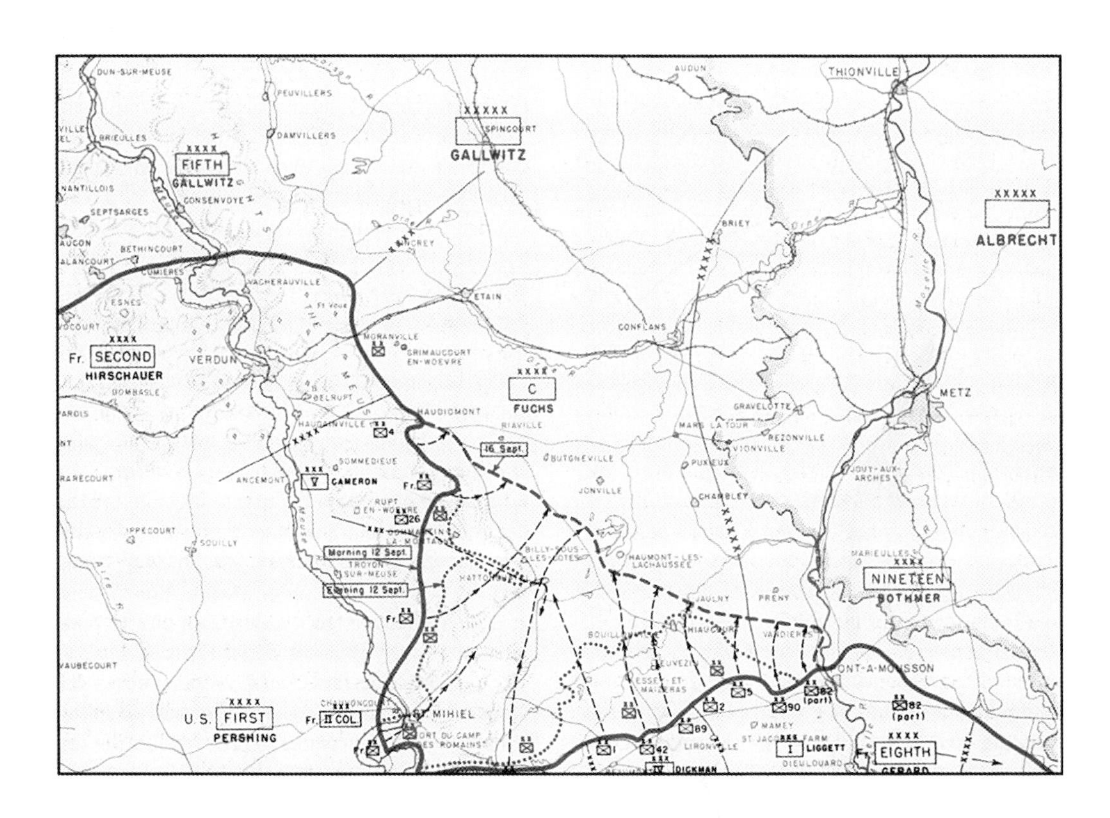

St Mihiel. (*West Point Atlas*, courtesy Department of History, US Military Academy)

many times. 'Intelligence was very busy studying photographs taken constantly by our observers, and studying them for machine-gun emplacements, battery positions, and anything of unusual significance, such as new defences and troop concentrations.'[591]

Another aerial observer, 2nd Lt. G. T. Lindstrom, recalled his role with photographic reconnaissance preparations in support of one of the American divisions. His aeroplane flew at an altitude of 2,000–2,500m about 15km to the rear of the German lines. When the St Mihiel attack commenced, Lindstrom flew low-altitude missions at 350m taking oblique photographs of the German's front-line positions. A formation of other aeroplanes from his aero squadron as well as pursuit fighters from other aero squadrons rounded out the mission, providing protection. Lindstrom recalled that aerial photographic reconnaissance required concentration on four variables: the course of the plane, where the enemy pursuit planes were flying, location of enemy anti-aircraft fire, and the whereabouts of the protecting aeroplanes.[592]

Bad weather curtailed coverage during the initial drive. However, other intelligence sources come into play. American radio intercept stations (goniometric service) were able to locate the position of all enemy radio stations. These stations were usually located close to the enemy's headquarters, providing the intelligence analysts not only with the approximate location of the command centre, but also an estimate of the numbers of enemy regiments and battalions in the sector. All the combatants waged a different kind of war when it came to goniometric operations. Deception techniques were common, for example generating intense radio traffic in quiet sectors and curtailing radio activity in the area where a surprise attack was planned. Just before the attack at St Mihiel, intelligence suspected German forces had withdrawn. However, goniometric analysis determined that German radio stations were still active in the sector. Armed with that evidence, the AEF Commander, General John Pershing, directed that the attack follow a heavy artillery bombardment.[593] The second day of the offensive allowed some sorties to acquire low-level oblique coverage of key towns in the area. However, the overall lack of aerial reconnaissance stifled analysis of enemy locations and intentions. During the final hours of the St Mihiel campaign, the Americans were not able to confirm the withdrawal of the Germans from the sector. This posed problems about where to commit forces as well as how to determine the extent of the German retreat.[594]

Allied aerial reconnaissance aeroplanes accomplished an enormous amount of aerial photographic coverage of the forward area. Between 21 and 24 September, AEF staffs sorted and distributed approximately 8,500 oblique and 20,000 vertical photos. French photographic interpreter support was a crucial part of the overall effectiveness of the operation. They established a special distribution of photos in preparation for the offensive. This was planned and carried out under the direct supervision of France's leading photographic interpreter, Capitaine Pépin. Photographic packages were furnished to French chars d'assaut headquarters in sufficient number to give each French and American tank oblique photographs of their assigned sector as well as a set of about 30 vertical shots.[595] When the offensive commenced photo sections compensated for the bad weather and produced extensive, low-level, oblique coverage for the rapid advance of the Allies to cut off the Germans at the salient.[596] In spite of the impressive statistics, post-war reflections on the lessons learned by America's aviation effort at St Mihiel suggested disappointment with the lack of accomplishment by aerial photography: 'In general our back-area photographic information was unsatisfactory, owing to the rapidity of our advance and the bad weather, which combined to keep our Squadrons photographing the areas just ahead.'[597]

St Mihiel proved the AEF's ability to conduct a major offensive against the Germans. The performance of US Air Service aerial reconnaissance reinforced Allied confidence in the ability of American forces to proceed to more demanding campaigns. After St Mihiel, Maréchal Foch had his staff receive and examine aerial photographs of the sector to confirm the nature and strength of the new Allied line established by the victorious Americans. The evidence was conclusive and the verdict was in Pershing's favour. Foch approved the next offensive at the Meuse-Argonne.[598]

MEUSE-ARGONNE

(For map see colour section.) The Meuse-Argonne campaign was the last major Allied offensive of the war. The movement of men and material was made entirely under cover of darkness, all activity suspended and men concealed during daylight hours. Consequently, the roads leading into the area were scenes of activity at night as troops and artillery, ammunition, and supplies moved forward. French soldiers within the sector remained in the outpost positions until the last minute in an effort to hide Allied intentions. Battlefield preparation was extensive. Photographic mosaics were prepared for the AEF's 1st, 3rd, and 5th Corps, covering the front of each. A mosaic of the Meuse River was made for the 3rd Corps. Staff packages containing oblique coverage along the front lines were prepared for the commanders and staff.[599] The Meuse-Argonne became one of the largest battles ever fought by Americans, involving over 600,000 soldiers.

Planning for Meuse-Argonne required extensive analysis of several high-interest German targets in the region, such as parts of the Siegfried Stellung known as the *Kriemhilde Stellung* and *Freya Stellung*. As in the St Mihiel operation, the weather for Meuse-Argonne was critical leading up to and during the offensive. There were occasional ideal flying days, but atmospheric conditions ranged from poor to impossible and proved a great handicap in carrying out extensive aerial operations. Haze and fog obscured the ground to a great extent in the early morning and late afternoon during the last four weeks of the offensive. Aerial photography from 26 September to 11 November 1918 was in almost every case confined to taking photographs of well-defined areas of particular interest to the Intelligence (G-2) section of the corps and Army staffs, rather than searching for targets of opportunity. All corps air services adjusted to the weather challenge, readying their aeroplanes for immediate launch, given a break in the conditions.[600]

Throughout the campaign, the AEF G-2 continually asked the question 'Does the enemy intend a withdrawal, and if so, to what position?'[601] Initially, interpretation of aerial photography acquired during the battle convinced French and AEF interpreters that the Germans were making practically no progress in strengthening their Kriemhilde Stellung defence. However, at the same time, the photographic interpreters were unable to determine German intentions for staying put or withdrawing.[602] As the Meuse-Argonne battle brought the war to an end, tasking aerial observation and photographic coverage became an exercise in keeping pace with the rapidly advancing front line. The 1st Army assigned corps area-of-interest information to the Corps Squadrons and Corps Intelligence. The Meuse-Argonne offensive prompted the army office to redraw the new front line frequently, a phenomenon not seen since 1914.[603] Oblique coverage supported the advance of the infantry. Vertical coverage was viewed as more appropriate for cartographic work.[604]

An impressive count of 56,000 prints was produced and distributed in four days by the American Photographic Sections.[605] Headquarters planning staffs had envisioned a maximum daily outlay of prints from a single production facility to be 10,000. In spite of the challenges posed by limitations to coverage due to the weather, a huge number of plates were acquired and prints developed. Photographic Section # 5 supporting the 3rd Observation Group produced 11,500 prints.[606] The AEF Tables of Organization initially had contemplated having one photographic section assigned to each observation group. The volume of prints meant a rethink on the manning necessary for the photographic section to sustain the rate of production. In the closing days, the AEF assigned two complete photographic sections to the 1st Army Observation Group to handle the volume.[607]

The effort by the American aerial reconnaissance elements was impressive. There was competition both within the squadrons and throughout the group. The aero squadrons were in a race to keep coverage coming in that would be of value to the staffs.[608] During the Meuse-Argonne advance, in four days alone, the American observers took 100,000 pictures of the enemy line.[609] By the end of the war, one aeroplane out of every four flying on the Western Front was there to provide aerial photographic coverage.[610] The final mark of success was the impression that aerial photography made on the ultimate user, the infantry in battle. Steichen recalled two cases where the opinion was highly favourable. One American infantry colonel referred to the aerial photographs he received from the US Air Service as the most vital assistance he had ever been given. The details available from studying the photographs had contributed mightily to the success of his unit's action and as an important determinant in reducing casualties. Another post-battle critique came from a young lieutenant in speaking of his battle experiences near Vaux. He commented that the only data he had there that were worth a 'tinker's damn' came from three aerial photographs some one had 'policed up' for him.[611] High praise for American aerial photography's contribution came from General Pershing: 'No army ever went out with the information as to what was in front of it as the American army did at St Mihiel and the Argonne.'[612]

THE OFFENSIVE TO THE FINISH

A resumption of mobility in warfare in the final days of the war prompted rethinking of aerial reconnaissance practices. As the Allies advanced at a rapid rate, established procedures were shown to be ineffective. The AEF's 3rd Army Corps noted on 5 November 1918, in its Operational Report, 'Photography was not attempted as the rapidly changing situation did not warrant such missions.'[613] Most of the aerial photographic missions from 26 September to 11 November focussed on well-defined areas specified by the intelligence sections of the corps staffs. The weather during this period determined target priorities with sorties during breaks in cloud coverage.[614] Oblique photographs continued to be a favourite of the advancing infantry, providing excellent views of the objective. German centres of resistance were the ongoing reconnaissance objectives assigned to elite flying units that were thoroughly familiar with the exact locality to be photographed, reducing the possibility of failure.[615]

Caudron G.4 aerial reconnaissance flown over the treacherous mountains of the Austrian South Tyrol on 24 April 1917 by Lieutenant Palli and observer De Claricini. (Gen. Basilio Di Martino)

Lieutenant Natale Palli, the greatest Italian aerial reconnaissance pilot of the First World War. (Gen. Basilio Di Martino)

Lamperti aerial camera – the mainstay of *Servizio Aeronautico* aerial reconnaissance. (Ives, *Airplane Photography*)

SVA 5 from 87a Squadriglia returning from one of the most famous aerial missions in the war – eight aeroplanes flew to Vienna, dropped propaganda leaflets and took aerial photographs. (Gregory Alegi, *Andaldo SVA 5*)

In the final days of the conflict, the Allies employed advance guards of all combat arms to follow the German forces. This advance team determined enemy positions of resistance so that the main body of follow-on forces avoided becoming entrapped. Weather remained a challenge to the end. To the north, British aerial observation met with little success in providing tactical coverage to support the advance team. A few aeroplanes were employed to report to the corps headquarters the whereabouts of the leading troops and the position of the front line hour by hour. In general, the final days of the war saw the RAF engaged in minor bombing sorties of railway junctions, cantonments, reserves and transport. Aerial observation played a role in determining enemy intentions, especially through observed massing of enemy reserves for counter-attack, artillery spotting when possible, and spotting infantry contact on the roads.[616] British aerial photographic coverage was now attained at higher altitudes with more powerful cameras, acquiring greater detail of a broader area more rapidly. Moore-Brabazon touted the possibility of photographing the entire battle line in the short space of 2½ hours.[617]

As the German armies withdrew in the final days of the war, concern arose over a possible 'last stand'. Aerial photography was called on to determine possible lines of natural defence that could aid ongoing preparations for mobile warfare.[618] However, aerial photography's primary purpose in the final days was to prepare the commander for rapid advance. Lines of communication networks were evaluated in detail, prepping the mobile battlefield for impending obstructions and points of resistance. Roads, railways, rivers, and key bridge crossings were subjected to analysis. Any networks that had been blown by mines were now the subject of interpretation, to include damage repair estimates. This cut down the time allotted to repair the network and ensure that momentum was not impaired. 'Without such information the advance would have taken considerably longer than was the case.'[619]

Following the Armistice in November, the Allies moved into the German Rhineland and prepared to advance further should the peace negotiations at Versailles in the Spring end in failure. Intelligence operations in the occupied Rhineland sought out information on Bolshevism creeping into the region. Aerial operations continued over the Rhine, providing an assessment of any signs of German forces near the border of the occupation sector and helping to maintain force readiness. Aerial photographs were acquired of the scenic views along the Rhine and Moselle Rivers; they were often used to grace the squadron walls. In the six months leading up to the Treaty of Versailles personnel were more focussed on returning home and to civilian life than honing their skills in aerial photography. Occupation forces along the Rhine tried to maintain consistency, reviewing what had been published and retaining data for operations.

Such was the case with the Pépin study on aerial photographic interpretation. It remained filed away at the American headquarters at Koblenz along with other documents generated after the conclusion of the fighting. Other campaigns still active in the world, such as in Mesopotamia, would keep British aerial photography active. However, by the early 1920s the military seniors wondered if it was necessary to keep training in an art that had had its day – to many, positional warfare did not seem to have a future.

Members of the 99th Aero Squadron pose in front of their Salmson 2 A2 two weeks after the Armistice. (USAFA McDermott Library Special Collections (SMS57))

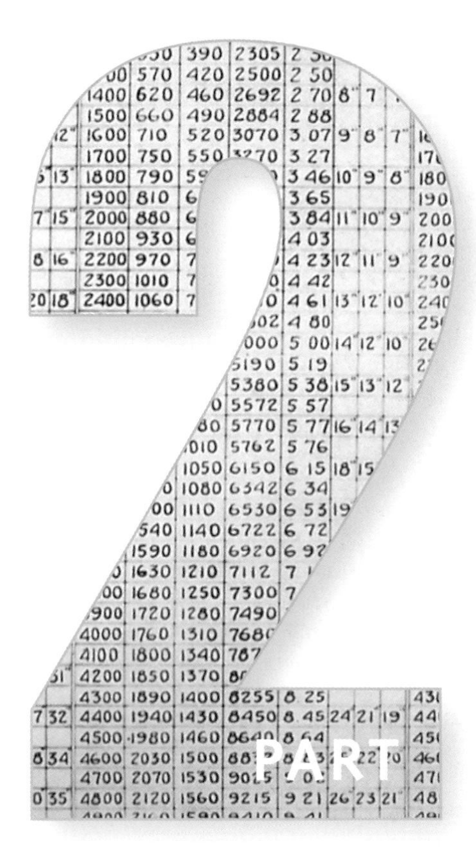

The Architecture of Allied Aerial Reconnaissance and Photographic Interpretation

CHAPTER 7

The Operation of Interpretation

Aerial photography in the First World War engendered three distinct operations: 1) location of enemy works on the print or glass plate (interpretation); 2) application of aerial photographic data to map construction (restitution); and, 3) exploitation of intelligence for subsequent operations against the enemy (such as adjusting artillery fire).[620] Allied dependence on these processes increased substantially as the aerial photograph became an accepted form of intelligence data. Interpretation established the science of photography as a legitimate source of intelligence, restitution served as the framework for cartography, and exploitation assumed the ultimate role of validation of the many sources of information that served the combatant.

THE BASIC SCIENCE OF PHOTOGRAPHY

Aerial photography created a landscape of black and white images oriented to the points of the compass and a corresponding map. The first step behind interpreting a photograph required understanding sunlight direction so that the relationship of image and shadow cast could be correctly established. Lessons in aerial photography introduced this requirement:

> A photograph reduces the earth to a pattern of tones ranging from dark to light. This ground pattern is produced primarily by colour, which in turn is enormously modified by light and shade. Embankments and hollow places produce shadow. In connection with light and shade, it should be understood that objects need not be of dark colour, or in shadow, to photograph dark. They may be light coloured and in full sunlight and yet the quality of their surface may be such that they contain enough shadow to photograph dark. In short the most important factor in determining shadow, excepting contours and cavities, is texture [of the perceived image]. Any irregularity or roughness of surface produces shadow, and any flattening down or smoothing of surface reduces it.[621]

Understanding shadow was critical to the discipline. Without it an interpreter could easily confuse a hole with a mound.[622] One technique employed by the interpreter was to adjust the light source during the examination to simulate sunlight. This gave value to the shadows, enabling the reliefs of the hollows to be differentiated, and the height or the depth of the obstacles to be judged in comparison with the length

Analysing shadows was critical to the aerial photographic interpreter's discipline. (*Étude et Exploitation des Photographies Aériennes*)

of the shadows of known objects.[623] One post-war description further embellished the discipline:

> The interpreter must know shadows not as an artist knows them, but rather as a scientist knows them. The shadow is his best friend. By its length and shape he not only pictures to himself the object which cast it, but he can also judge the size and shape of other objects close by, whose own shadows are perhaps indistinct or misshaped. Without the shadow, he could hardly tell an embankment from a trench; and he certainly couldn't hope to distinguish between a gun emplacement and a haystack, unless somebody told him that one of them was shooting.[624]

Interpretation was rarely left to a beginner, even in wartime.[625] The interpreter had to know photographic reality. Light and shade did not directly relate to normal vision. The photographic image represented the amount of light reflected, rather than the degree of 'actinic' (photochemically active radiation such as that produced from sunlight) effect of the objects reproduced.[626] Thick vegetation appeared dark while a smooth road surface showed as a white line on a print. Light reflection and absorption created the final image.[627]

PLANNING THE AERIAL PHOTOGRAPHIC MISSION

Extensive planning was required prior to conducting the mission. Every aspect of photography had to be considered before the aeroplane took off on the sortie. Preparation included selection of camera, determining the flight path (ground trace) of the aeroplane, altitude flown, and considerations for defence. The camera's configuration depended upon the aerial photographic mission and the type of coverage required. Aerial photography was a flexible medium thanks to a variety of focal lengths and plate sizes, providing the planner with accessibility to a broad range of photo scale to ground coverage.[628] Prior to each aerial reconnaissance sortie, the aircrew and photographic interpreter would consult photograph charts to determine the altitude, camera lens, photographic plate size and speed of the aeroplane required to successfully cover the terrain to be photographed.[629]

Mission planning aerial photography was an exact science. The AEF photogrammetry chart captures the key ingredients for all required data, to include camera lens, scale, desired overlap, and aeroplane altitude. (Gorrell Report, NARA)

SCALE AND GROUND AREA COVERED BY LENSES OF GIVEN FOCAL LENGTH ON GIVEN SIZE PLATES

SIZE OF PLATE	13 X 18		18 X 24		18 X 24		18 X 24	
F.L. OF LENS	26		26		50		120	
AVAILABLE PLATE AREA	12 X 17		17 X 23		17 X 23		17 X 23	
ALTITUDE IN METERS	METERS COVERED	SCALE	METERS COVERED	SCALE	METERS COVERED	SCALE	METERS COVERED	SCALE
1000	461 X 654	1-3840	654 X 885	1-3860	340 X 460	1-2000	142 X 192	1-835
1250	575 X 817	1-4790	817 X 1105	1-4820	425 X 575	1-2500	177 X 240	1-1040
1500	691 X 981	1-5760	981 X 1328	1-5780	510 X 690	1-3000	213 X 288	1-1210
1750	805 X 1145	1-6710	1145 X 1550	1-6740	595 X 805	1-3500	248 X 336	1-1460
2000	922 X 1308	1-7680	1308 X 1770	1-7670	680 X 920	1-4000	284 X 384	1-1670
2250	1036 X 1470	1-8630	1470 X 1990	1-8650	765 X 1035	1-4500	319 X 432	1-1880
2500	1152 X 1635	1-9620	1635 X 2240	1-9620	850 X 1150	1-5000	356 X 480	1-2100
2750	1268 X 1798	1-10560	1798 X 2430	1-10580	935 X 1265	1-5500	391 X 528	1-2300
3000	1382 X 1960	1-11520	1960 X 2660	1-11540	1020 X 1380	1-6000	426 X 576	1-2505
3250	1498 X 2123	1-12480	2123 X 2875	1-12530	1105 X 1495	1-6500	461 X 624	1-2715
3500	1615 X 2209	1-13460	2209 X 3100	1-13020	1190 X 1610	1-7000	497 X 672	1-2920
3750	1690 X 2450	1-14090	2450 X 3320	1-14450	1275 X 1725	1-7500	533 X 720	1-3140
4008	1845 X 2620	1-15380	2620 X 3540	1-15550	1360 X 1840	1-8000	568 X 768	1-3340
4250	1960 X 2780	1-16320	2780 X 3760	1-16390	1445 X 1955	1-8500	604 X 815	1-3560
4500	2075 X 2942	1-17230	2942 X 3980	1-17360	1530 X 2070	1-9000	638 X 864	1-3750
4750	2190 X 3110	1-18250	3110 X 4210	1-18320	1615 X 2185	1-9500	675 X 912	1-3975
5000	2305 X 3278	1-19200	3278 X 4435	1-19300	1700 X 2300	1-10,000	710 X 960	1-4170

CHART № 101

PHOTO SECTION-AIR SERVICE-A.E.F.

AERIAL PHOTOGRAPHIC PRODUCTS

Aerial photographic plates produced an array of products by which the photographic interpreter could conduct analysis and dissemination of material for staffs and infantry combatants. Production included negatives, positives, photographic proofs, enlargements, mosaics and stereoscopic views. Mission planning also specified the type of photograph that was acquired. The three formats – panoramic, vertical and oblique – became the most recognisable signature of the aerial photograph. Camera focal length was the key to mission planning. The shorter the focal length, the larger the

TABLE OF INFORMATION
FOR TAKING AERIAL PHOTOGRAPHS

0.26 CAMERA

ALT.	GROUND COVERED		SCALE	1m/m ON PHOTO	TIME BETWEEN EXPOSURES		
	LENGTH	WIDTH			140 k	160 k	180 k
1000m	650m	460m	3847	3m84	9"	7"	6"
1100	710	500	4231	4 23			
1200	780	550	4616	4 60	11"	9"	8"
1300	850	600	5000	5 00			
1400	910	640	5385	5 38	13"	11"	10"
1500	980	690	5770	5 75			
1600	1040	730	6154	6 15	15"	13"	12"
1700	1110	780	6539	6 56			
1800	1170	830	6924	6 90	17"	15"	13"
1900	1240	880	7308	7 30			
2000	1310	920	7693	7 70	19"	17"	15"
2100	1380	970	8077	8 07			
2200	1440	1010	8461	8 46	21"	18"	16"
2300	1510	1060	8846	8 84			
2400	1570	1100	9230	9 23	22"	20"	18"
2500	1640	1150	9615	9 61			
2600	1700	1200	10.000	10 00	24"	21"	19"
2700	1770	1250	10 384	10 38			
2800	1830	1290	10 769	10 76	26"	23"	20"
2900	1900	1340	11 151	11 15			
3000	1960	1380	11 538	11 53	28"	25"	22"
3100	2030	1430	11.925	11 92			
3200	2090	1470	12 307	12 30	30"	26"	23"
3300	2160	1520	12 691	12 69			
3400	2220	1570	13 078	13.07	32"	28"	25"
3500	2290	1620	13 460	13 46			
3600	2360	1660	13 845	13 84	34"	30"	26"
3700	2430	1720	14 229	14 22			
3800	2490	1760	14 612	14 61	36"	32"	28"
3900	2550	1800	14 998	14 99			
4000	2610	1840	15 383	15 38	38"	33"	29"
4100	2680	1890	15 767	15 76			
4200	2740	1930	16 152	16 15	40"	35"	31"
4300	2810	1980	16 536	16 53			
4400	2870	2030	16 921	16 92	42"	37"	32"
4500	2940	2070	17 305	17 30			
4600	3010	2120	17 690	17 69	44"	38"	34"
4700	3070	2160	18 074	18 07			
4800	3130	2210	18 459	18 45	46"	40"	35"
4900	3200	2260	18 843	18 84			
5000	3260	2310	19 228	19 22	48"	41"	37"
5100	3320	2345	19 613	19 61			
5200	3380	2390	20 000	20 00	50"	43"	38"
5300	3450	2435	20 385	20 38			
5400	3510	2480	20 770	20 77	52"	45"	40"
5500	3580	2530	21 155	21 15			
5600	3640	2570	21 540	21 54	53"	46"	41"
5700	3700	2620	21 923	21 92			
5800	3770	2660	22 310	22 31	54"	48"	43"
5900	3830	2710	22 695	22 69			
6000	3900	2760	23 080	23 08	56"	50"	44"

0.50 CAMERA

ALT.	GROUND COVERED		SCALE	1m/m ON PHOTO	TIME BETWEEN EXPOSURES		
	LENGTH	WIDTH			140 k	160 k	180 k
1000m	440m	330m	1923	1m 90	6"	5"	5"
1100	480	360	2115	2 10			
1200	530	390	2305	2 30	7"	6	6"
1300	570	420	2500	2 50			
1400	620	460	2692	2 70	8"	7	7"
1500	660	490	2884	2 88			
1600	710	520	3070	3 07	9"	8"	7"
1700	750	550	3270	3 27			
1800	790	590	3460	3 46	10"	9"	8"
1900	810	620	3650	3 65			
2000	880	660	3840	3 84	11"	10"	9"
2100	930	690	4032	4 03			
2200	970	720	4230	4 23	12"	11"	9"
2300	1010	750	4420	4 42			
2400	1060	790	4610	4 61	13"	12"	10"
2500	1100	820	4802	4 80			
2600	1150	850	5000	5 00	14"	12"	10"
2700	1190	880	5190	5 19			
2800	1230	920	5380	5 38	15"	13"	12"
2900	1270	950	5572	5 57			
3000	1320	980	5770	5 77	16"	14"	13"
3100	1370	1010	5762	5 76			
3200	1410	1050	6150	6 15	18"	15"	14"
3300	1450	1080	6342	6 34			
3400	1500	1110	6530	6 53	19"	16"	14"
3500	1540	1140	6722	6 72			
3600	1590	1180	6920	6 92	20"	17"	15"
3700	1630	1210	7112	7 11			
3800	1680	1250	7300	7 30	21"	18"	16"
3900	1720	1280	7490	7 49			
4000	1760	1310	7680	7 68	22"	19"	18"
4100	1800	1340	7870	7 87			
4200	1850	1370	8065	8 06	23"	20"	19"
4300	1890	1400	8255	8 25			
4400	1940	1430	8450	8.45	24"	21"	19"
4500	1980	1460	8640	8 64			
4600	2030	1500	8832	8 83	25"	22"	20"
4700	2070	1530	9025	9 02			
4800	2120	1560	9215	9 21	26"	23"	21"
4900	2160	1590	9410	9 41			
5000	2210	1630	9600	9 60	27"	24"	22"
5100	2260	1660	9790	9 79			
5200	2300	1690	10.000	10 00	28"	25"	23"
5300	2340	1720	10 195	10 19			
5400	2390	1750	10 385	10 38	29"	26"	24"
5500	2440	1790	10.580	10 58			
5600	2490	1820	10.770	10 77	30"	27"	25"
5700	2540	1850	10.960	10 96			
5800	2580	1880	11.155	11 15	31"	28"	26"
5900	2630	1910	11 345	11 34			
6000	2680	1940	11.540	11 54	32"	29"	27"

1.20 CAMERA

ALT.	GROUND COVERED		SCALE	1m/m ON PHOTO	TIME BETWEEN EXPOSURES		
	LENGTH	WIDTH			140 k	160 k	180 k
1000m	190m	140m	833	0m83			
1100	210	150	915	0 91			
1200	230	170	1000	1 00			
1300	250	180	1084	1 08			
1400	270	200	1166	1.16			
1500	290	210	1245	1 25			
1600	300	230	1330	1 33	4"	3"5	3"
1700	320	240	1415	1 40			
1800	340	250	1500	1 50	4"5	4"	3"5
1900	365	270	1585	1 60			
2000	390	290	1670	1 67	5"	4"5	4"
2100	410	300	1755	1 75			
2200	420	310	1840	1 84	5"5	5"	4"5
2300	440	325	1915	1 91			
2400	460	340	2000	2 00	6"	5"	4"5
2500	480	355	2085	2 08			
2600	500	370	2170	2 17	6"	5"5	5"
2700	520	385	2255	2 25			
2800	540	400	2340	2 34	6"5	6"	5"
2900	560	415	2425	2 42			
3000	580	430	2510	2 51	7"	6"5	5"5
3100	600	445	2595	2 59			
3200	620	460	2680	2 68	7"5	7"	6"
3300	635	470	2765	2 76			
3400	650	480	2830	2.83	8"	7"5	6"5
3500	670	495	2915	2.91			
3600	690	510	3000	3.00	8"5	8"	7"
3700	710	525	3085	3 08			
3800	730	540	3170	3 17	9"	8"5	7"5
3900	750	555	3255	3 25			
4000	770	570	3340	3 34	9"5	9"	8"
4100	780	580	3425	3 42			
4200	800	590	3510	3 51	10"	9"5	8"5
4300	820	600	3595	3 59			
4400	840	620	3680	3 68	10"5	9"5	9"
4500	860	630	3765	3 76			
4600	880	650	3850	3 85	11"	10"	9"5
4700	900	660	3935	3 93			
4800	920	670	4020	4 00	11"5	10"5	10"
4900	940	680	4105	4 10			
5000	960	700	4190	4 19	12"	11"	10"
5100	980	710	4275	4 27			
5200	1000	720	4360	4 36	12"5	11"5	10"5
5300	1020	730	4445	4 44			
5400	1040	740	4530	4 53	13"	12"	11"
5500	1060	750	4615	4 61			
5600	1080	760	4700	4 70	13"5	12"5	11"5
5700	1100	770	4785	4 78			
5800	1120	780	4870	4 87	14"	13"	12"
5900	1140	790	4955	4 95			
6000	1160	800	5040	5 04	14"5	13"5	12"5

Coverage to support restitution was planned according to the data portrayed on this chart. (Gorrell Report, NARA)

area that could be photographed. The aerial photograph was further divided into three groups corresponding to the collection method. Continuous photographs were taken so that each succeeding photograph had enough overlap to show continuity of coverage. Separate pinpoint photographs were taken to acquire a specific target within the photographic plate. Finally, stereo coverage required the maximum amount of overlap in order to accomplish the necessary stereo effect.[630] Of the three, stereo coverage was the most useful format for detailed interpretation.[631]

When the Allied forces were introduced to a new sector, the best and most effective way of introducing the terrain and configuration was through the panoramic photograph. The first recorded panoramic was acquired by the French on 20 October 1914.[632] Panoramic photographs provided the necessary study of a sector ensemble up to the horizon. Broad details such as the general form of the terrain, trench lines, study of routes, and the like were acquired from the photo angle.

The camera focal length available to the French and Americans acquired three areas of coverage such as displayed on the chart. (Gorrell Report, NARA)

Using the standard camera (French 50cm focal length, for example), a panoramic view of the sector provided a relief impression of the ground showing terrain features of valleys, hills, and natural lines of communication as well as locations and present state of woods and villages. The aérostat/ballon captif aerial platform was the best source of panoramic photographs, providing a precise view of the opposing sector. The area beyond the coverage of the balloon could be marked off and acquired by aeroplane reconnaissance.[633] A panoramic view or photo mosaic of the sector soon decorated the squadron operations room.

Oblique photographs differed from panoramic views in that they were acquired at a lower attitude and covered less terrain than those cameras possessing a shorter focal length lens. It was the favourite of the line officer infantryman prior to an offensive, for it provided a view angle that approximated the sector to be attacked. The officer could now plan the itinerary for the attack. Oblique photographs were usually acquired by an observer either holding or mounting a camera on the machine-gun turret rail and shooting laterally away from the airframe of the aeroplane. This was a favorite technique used by French observers. The coverage provided the unskilled student with a better understanding of slopes, form of the terrain

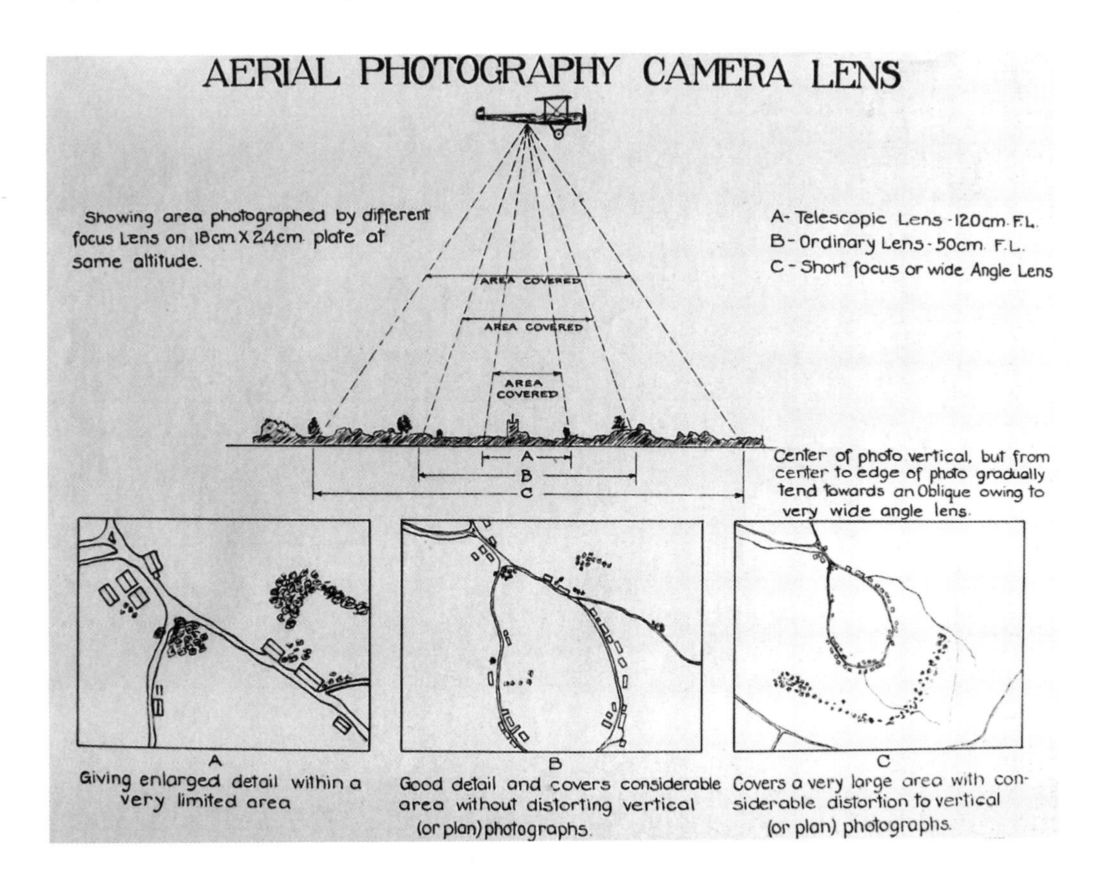

Panoramic photographs provided a broad view of the battle terrain, such as this trench network in the Vosges Mountains. (*Étude et Exploitation des Photographies Aériennes*)

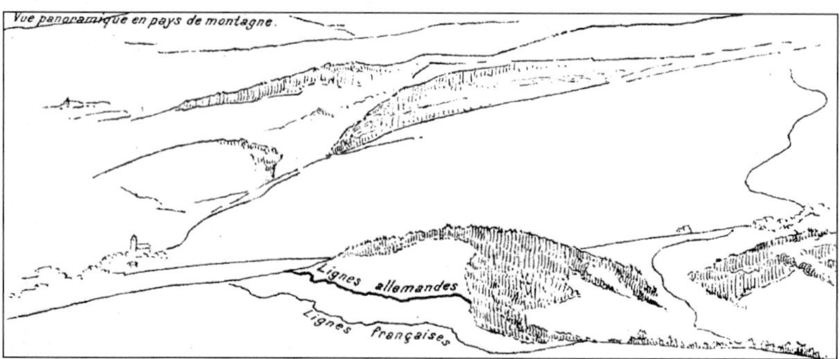

The draughtsman's line drawing highlighted appropriate details for cartographic and intelligence review. (*Étude et Exploitation des Photographies Aériennes*)

behind the trenches, depth of sunken roads, loop holes, entrances to shelters and dugouts and those works which were not apparent from vertical views. Oblique views were made at angles from 12 to 35 degrees from the horizon, aiding analysis of the details of shelters, gas installations, and trenchworks behind an abrupt slope not observable from the ground.[634] Oblique coverage was also an invaluable aid towards further detailing the features on a map.[635] Oblique photography required sensitivity to the parameters of flight. An oblique photograph required a high shutter speed to eliminate the effect of rapid movement in the foreground.[636] Flying at a few hundred feet, leaning from the cockpit with a camera held so that its line of sight was at the desired angle also contributed to the challenge. It was a hazardous assignment, but one that was greatly appreciated when the infantry received the coverage. The following is a typical example of oblique reporting: 'Photographic missions took 20 oblique exposures from Eply to Les Menils; 17 photographs from Moselle to Bois de l Voivrotte; 18 photographs in vicinity of Remenauville, Regnieville and Bois du Four; 12 exposures N. of Remenauville; 20 oblique exposures of front lines W. of Moselle.[637]

The vertical photograph was the most popular aerial photographic format used in the war, providing the most invaluable tool to intelligence for determining exact locations of key targets such as machine guns, minenwerfer, shelters, and frequented tracks. The vertical view was the primary source of photographs used in the restitution process.[638] Acquiring the pinpoint vertical coverage required a sufficient focal length to secure an adequate size image, a shutter speed to counterbalance the speed of the plane, and a sufficiently wide lens aperture to give adequate exposure with the required shutter speed.[639] Photos taken from the perpendicular were the most desirable for several reasons. As one post-war assessment stated: 'All aerial photographs should be taken vertically; otherwise we would have distortion which makes the reading difficult and inexact.'[640] Vertical photos were the document 'most rich in every nature' and were used extensively to form mosaics, the photographic map at a desired scale.[641] It was important that the entire enemy area opposite each division be

Opposite top: Mission planning for oblique coverage determined the altitude to be flown and the area required. (Gorrell Report, NARA)

Opposite bottom: A superb oblique shot of the German trench network with interpretation details. (*Illustrations*, S.S. 550a)

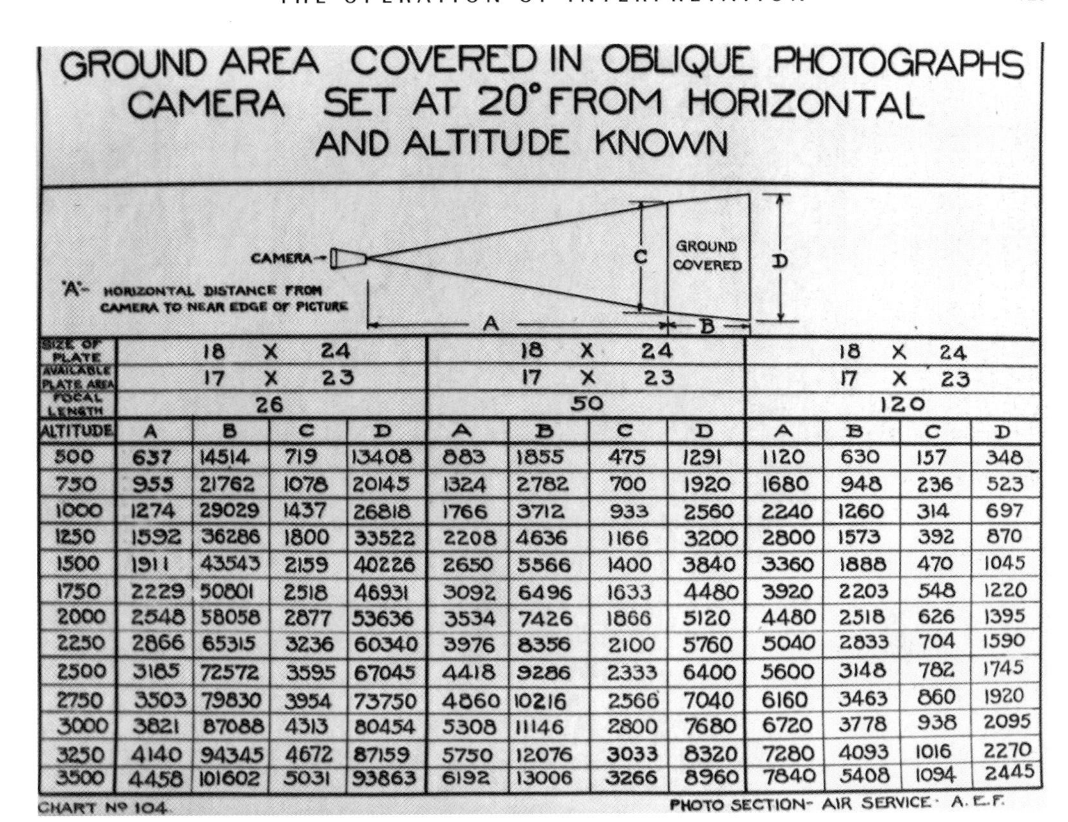

GROUND AREA COVERED IN OBLIQUE PHOTOGRAPHS
CAMERA SET AT 20° FROM HORIZONTAL
AND ALTITUDE KNOWN

SIZE OF PLATE	18 X 24				18 X 24				18 X 24			
AVAILABLE PLATE AREA	17 X 23				17 X 23				17 X 23			
FOCAL LENGTH	26				50				120			
ALTITUDE	A	B	C	D	A	B	C	D	A	B	C	D
500	637	14514	719	13408	883	1855	475	1291	1120	630	157	348
750	955	21762	1078	20145	1324	2782	700	1920	1680	948	236	523
1000	1274	29029	1437	26818	1766	3712	933	2560	2240	1260	314	697
1250	1592	36286	1800	33522	2208	4636	1166	3200	2800	1573	392	870
1500	1911	43543	2159	40226	2650	5566	1400	3840	3360	1888	470	1045
1750	2229	50801	2518	46931	3092	6496	1633	4480	3920	2203	548	1220
2000	2548	58058	2877	53636	3534	7426	1866	5120	4480	2518	626	1395
2250	2866	65315	3236	60340	3976	8356	2100	5760	5040	2833	704	1590
2500	3185	72572	3595	67045	4418	9286	2333	6400	5600	3148	782	1745
2750	3503	79830	3954	73750	4860	10216	2566	7040	6160	3463	860	1920
3000	3821	87088	4313	80454	5308	11146	2800	7680	6720	3778	938	2095
3250	4140	94345	4672	87159	5750	12076	3033	8320	7280	4093	1016	2270
3500	4458	101602	5031	93863	6192	13006	3266	8960	7840	5408	1094	2445

CHART Nº 104. PHOTO SECTION- AIR SERVICE- A.E.F.

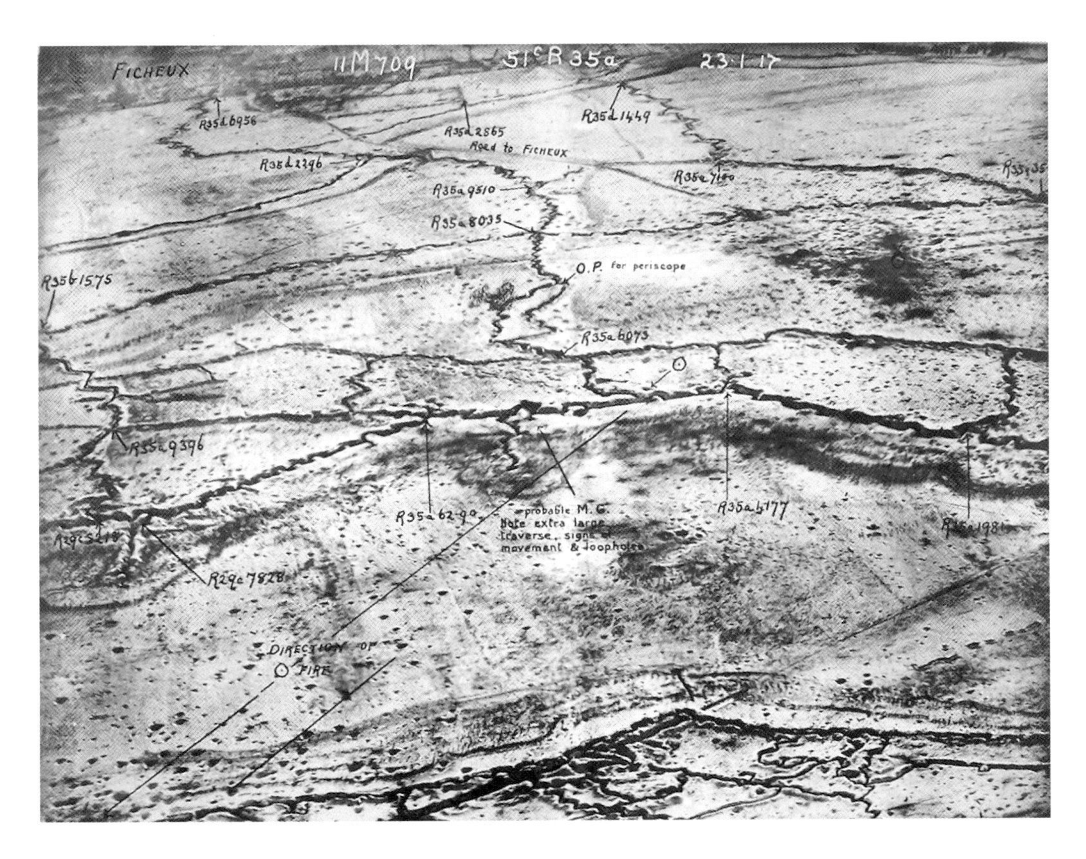

mapped to a considerable depth. Missions were assigned in such a manner that the photographs would overlap and form a continuous mosaic map when pasted together.[642] An accurate mosaic required photos with an equivalent scale. To achieve this required a consistent photograph altitude and vertical alignment. Another concern was differences in the time of the photograph, which could create shadow variance. Barring those variables, the mosaic provided the interpreter with a comprehensive view of an enemy trench network as well as an automatic layout for a map.[643] With vertical coverage and mosaics, the military decision maker now possessed a more precise medium to appreciate both the natural obstacles and hostile defences.[644] As early as 1915, the mosaic became an important tool for planning. A British liaison to the French Aviation Militaire noted that 'photographic maps are continually being made of the whole line, and when any important change is found a re-issue is made.'[645]

Stereoscopic photographs or stereographs were the most valuable aid in detecting terrain contours, relief, and an estimate of object height and depth. The view from stereo coverage exaggerated the third dimension. The normal tool for interpretation was a magnifying glass. The stereoscope provided two overlapping photographs taken at the same altitude, at the same hour covering the same points. Stereo coverage employed either vertical or oblique photos. The stereoscopic oblique photograph added value to mission planning by showing undulations of the ground which portrayed features of tactical importance that were not obvious from the map itself. Instructions for stereoscopic coverage required placing one photograph alongside a companion photograph matching the coinciding features. As the photographs were drawn apart to the point that the eyes could directly focus on the views, a process aided by the simple stereoscope, objects stood up. The photographs had of course to be placed in the order in which they were taken. For example, the left-hand photo was positioned on the left-hand side of the viewing lens. Overlooking this simple requirement resulted in the features appearing reversed; houses appeared as cavities and cuttings became embankments.[646] The stereoscope was a well-established fixture of the photographic world long before the conflict, providing the public with a

Photographic planning for vertical coverage showed the ideal coverage area to support the restitution process. (Gorrell Report, NARA)

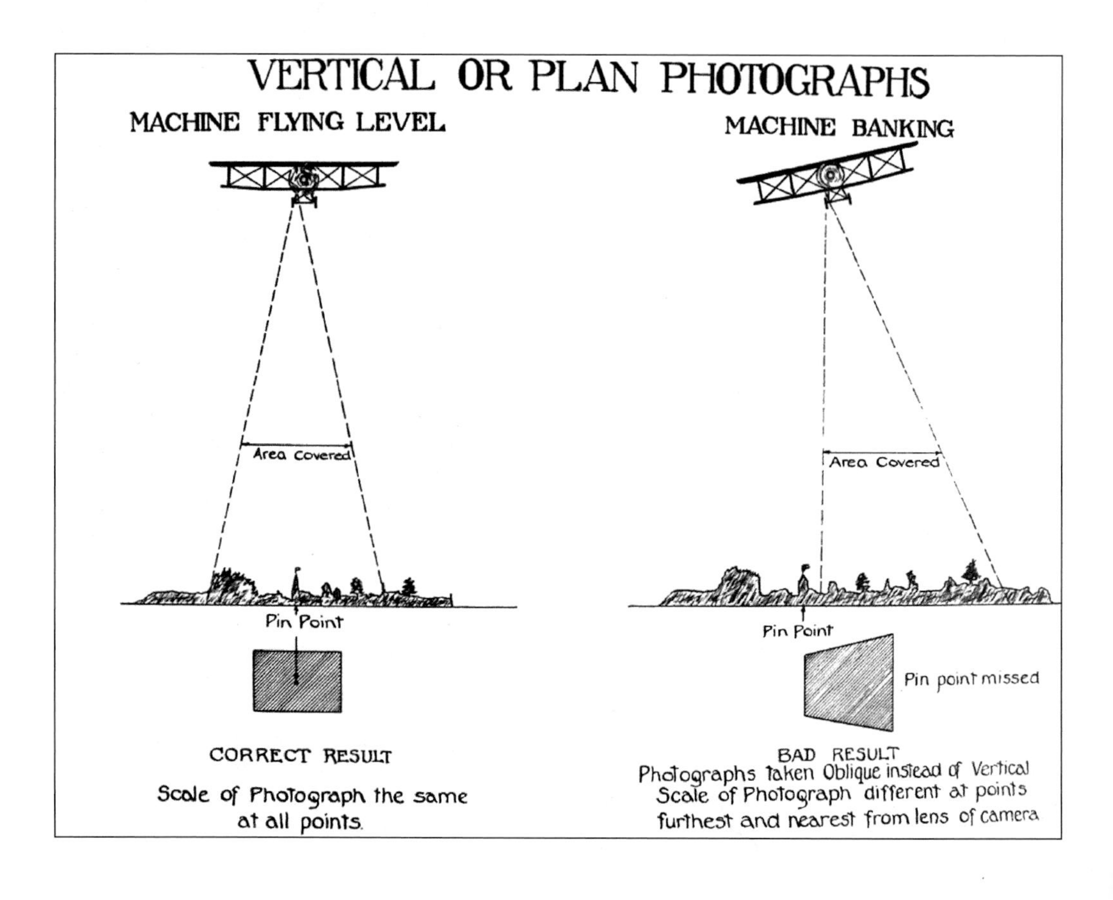

Above left: A draughtsman would assemble the photographs and align them so that the assembly would form a continuous mosaic map. The British aerial photograph format was 4 x 5 inches, requiring a higher number of photos to cover a given sector than larger plates. (Imperial War Museum) **Above right**: The stereoscope provided the photographic interpreter with the tool to identify and explore important features from overlapping aerial photographs. (Gorrell Report, NARA)

divertissement and experts with a new way of recording the world.[647] By the mid-1800s, the stereoscope phenomenon became a topic of intense scrutiny. A leading nineteenth-century theorist of vision, Hermann Ludwig Ferdinand von Helmholtz, wrote at that time, 'These stereoscopic photographs are so true to nature and so lifelike in their portrayal of material things, that after viewing such a picture and recognising in it some object like a house, for instance, we get the impression, when we actually do see the object, that we have already seen it before and are more or less familiar with it.'[648] About 70 years later, the stereoscope helped analysts discern battlefield reality.

The demands of war promoted the application of physics and the science of colour to further enhance the stereoscopic view. Moore-Brabazon recalled using slide projection (lantern-slide) images tinted with red and green to highlight the stereoscopic effect of the terrain features. The photographic interpreter audience viewed the coloured projections with red and green spectacles that produced a stereoscopic effect from the photographic pairs.[649] This interpreter technique continues to the present day.

ANNOTATING AERIAL PHOTOGRAPHS

It became apparent from the beginning that aerial photography required a standard identification process to portray essential information describing the moment that the photograph was acquired. The British made reference to this as early as December 1914. Their direction specified a standard for each print: 'Every negative and every print shall have the following marked upon it, viz., Date, hour, and place of exposure. A line indicating the approximate direction of true North. Height at which flying. Condition and atmosphere and light. Without accurate information on the above points, a photograph no matter how good in itself, is likely to be useless for military purposes.'[650] Since aerial photography was proving to be an invaluable information tool, clarifying information directly applied on the plate and print accentuated the value of the intelligence. Sortie information was compiled and processed as soon as possible, to include: Pilot name; Observer name; Date of Photograph; Time of Photograph; Altitude of the Aerial Platform; Mission; Camera Type; Camera Focal Length; x and y coordinates of the target; Number of Plates exposed; Number of Magazines used; Planned Itinerary; and significant observations.[651] Once the photo plates were developed, annotations were transcribed on the photographic plate and became the identification moniker for all subsequent prints. For French aerial photographs the data included: name of escadrille airfield; escadrille; principal target (trench, wood, village); coordinates of the photograph; focal length of the camera; date; hour of the photograph; altitude that the photo was taken; and north arrow.[652] The following is an example of French photographic annotation:

> La Cheppe [location]. F22 [Escadrille Farman 22]. 454. Nap-Py [principal target]. 536/725 [coordinates of photograph]. F=50 – [50cm focal length] 22.8.16 – [date] 10H [1,000 hours] NR 3,000m [altitude].[653]

British photograph annotations were aligned to 1/40,000 artillery maps. The maps were divided into 6000- yard squares identified by capital letters A to X. The areas were further subdivided into 36 numbered 1000 yard squares, which were further subdivided into 4 lettered (lower-case) 500-yard squares (abcd). Points within the 500-yard square were pinpointed by 4 figure coordinates, eastings (left to right) then northings (from bottom to top). A full-grid reference annotated on the photograph read as a series of alphanumeric symbols. The references were identical for larger scale 1/10,000 and 1/20,000 maps derived from the 1/40,000 map.[654] An example annotation provides a breakdown of the aerial photograph's identity:

5AE 25 37	Negative series number, local serial of the print
7.8.18 7am F 8 1/4	date, time, focal length of camera
62D V25bd 26ac	62 D [map sheet number]
V	[identification of the 6000-yard square]
25bd, 26ac	[identification of 500-yard squares covered by aerial photographic print] 655

Some British annotations did not apply to the artillery map standard and reflected less data. The Squadron number, Wing (initial letters), serial number of the photographic plate and date of photograph were provided.[656] For example, the 680th photograph taken by No. 30 Squadron, 31st Wing, 18 April 1918 was simply displayed on the photograph as: '30 C.A. 680, 18-4-118.'[657]

The US Air Service mostly adopted the French standard for annotating aerial photographs. The top margin of the plate/photograph would be marked with the following information: Corps number; Aero Squadron number; location of photograph; coordinates of the photo center; date; hour; altitude of photograph; and focal length. The location of the photograph included the village or farm in the vicinity. When appropriate a map number was assigned. Guidance suggested making interpretation marks on the photo, to include nature of the terrain (compared with the Plan Directeur), highlighting belts of wire and machine-gun locations.[658] The photographic section sometimes included the serial number of the plate to further add detail to the annotation.[659] Here is an example of an American serial number photographic annotation: 'US 1AC (1st Army Corps) 1 Sq (1st Aero Squadron) Bg13 Eply (target location) 3860.236.7 (coordinates of the centre of the photograph) 29.8.18 (date of photograph) 16h, (hour that photograph was taken) 500 (altitude in metres photo was taken) 32 (focus).'[660]

HISTORY OF COVERAGE

One of the foundations of aerial photographic interpretation was the ability to exploit the history of coverage of areas and targets. This technique, now known as negation and research, with negation being the ability to go back over time to see what in an image had changed, and research being the analysis of what happened, had its origins with aerial photography of The First World War. The ability to analyse a target through a collection of photographs acquired over time became the mainstay for analysis. The French recognised this early on in the war and established a complete photographic and tactical dossier history of the sectors.[661] Target analysis was refined with each aerial photograph acquired and compared. The frequency of coverage was based on sortie availability and the sector. French policy regarding the cycle of coverage was discussed in post-war analysis: 'Such work necessitates a constant supervision on the part of the aviation and the methodical taking of photographs of all the zone to be surveyed. In an Army Corps it may be accepted that the first position is to be photographed on an average once a week and the rear area once a month.'[662] A post-war discussion described

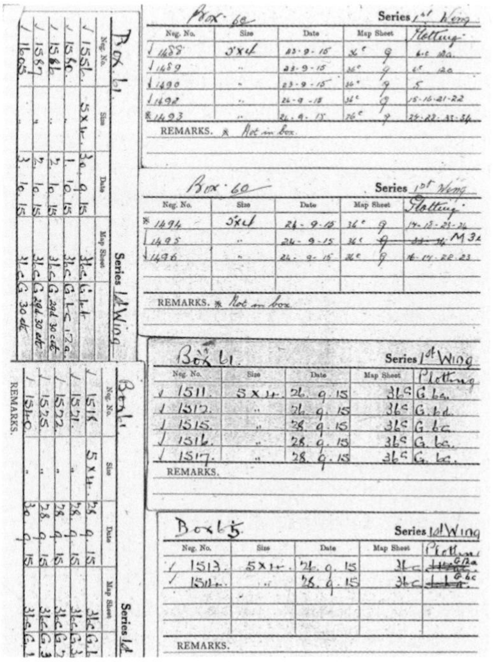

The British detailed all appropriate data for filing aerial photographs according to the negative number, date of coverage, and corresponding map sheet. (Courtesy Nicholas Watkis)

the added value that history of coverage provided the interpretation discipline:

> In order to obtain good results from a study of a photographic reconnaissance, one must not be content only to study the photographs of this reconnaissance but must complete them with photographs taken previously, which will allow one to see the small changes which are often of great importance. It is necessary to bring to this study a great deal of method and minute exactness. The whole surface of the photograph must be explored in a logical order; one must distrust his imagination and control it as often as possible.[663]

The following is an example of analysis from a history of coverage: 'Region of Vilcey. No. 943 – 14/8/18 – 695–460. A comparison of this photo with one taken June 30, shows no marked increase of activity between the two dates. The collection of buildings at point 369.5–246 shows indications of being an important distributing point and might be perhaps a Pioneer headquarters.'[664]

FILING AND DISTRIBUTION

Experience showed that all information acquired through aerial photographs required careful archiving to make it accessible and quickly retrievable.[665] However, cataloging and filing were not without problems. Filing was accomplished at all echelons without a standard established. The French advocated that information be kept in a sector dossier. The dossier contained all vertical or oblique photographs concerning the sector covering both French and German areas.[666] British squadrons pasted photographs organised by serial number into a photo album-type book. They also maintained another file based on coverage of a given five-km square. Despite the attempt to create order as the quantity of photographs increased, both systems were not without their critics. A review of French and British practices led the AEF Assistant Chief of Staff for Intelligence to comment in a memorandum: 'One of the most serious problems of the office has been the photo filing system.'[667]

The British established a separate filing and distribution process within each army. Branch Intelligence Section personnel assigned to the British Second Army prepared a daily chart showing the location of area coverage. Filing of photographs occurred after the decision was made as to whhich aerial photographs were to be retained. A daily summary along with a chart displaying the photos taken were forwarded to distribution centres. All negatives were compared with a baseline map recording the coverage (a record map). Plates that were deemed duplications were then thrown out. The plates that were retained were given serial numbers according to the covered sector. As soon as the photograph was accepted, the area covered was blocked off on the daily chart and the photo number entered in the space. Photos were filed according to area and indexed under the category of area and serial number. Distribution was based on the requirements of the unit serving in the sector.[668] The British Fifth Army policy was to have all photographs on a specific target logged in and target records kept in alphabetical order. Photographs were filed in numerical order by squadron. Maps were annotated with the area covered by the photograph and the photo was filed according to the map section. A cross-index book was kept to provide ready reference to photos of particular objects (such as dumps, etc.) that were not displayed on the maps.[669]

Over time, the British adjusted their distribution to meet both the requirements of the combatants and the realities of the front line. The initial large quantity of photographs for the front line units was reduced in favour of intelligence being transmitted via message or communicated verbally. Whenever a sufficient number of changes warranted attention, the unit received a photograph mosaic and an accompanying document with explanations. The end result for British units was that they received more manageable amounts of aerial photographs with supporting information.[670] The standard British aerial photograph processing included the 'rush copy', filing copy, and a copy for the dossier. The dossier copy was circulated and all relevant information was culled from the photograph.[671] The remainder of the prints, disseminated according to the prescribed squadron distribution, quickly followed the rough print. When tasking ordered a large number of prints beyond normal distribution, the squadron prepared and forwarded a transparency to Army or Corps photo centres.[672] Photograph distribution was arranged by squadron, each forwarding their material in separate envelopes.[673]

US Air Service Corps aero squadrons were instructed to send an advance copy of aerial photographs for prompt study. However, the procedure was discontinued as 'every squadron either discontinued the practice or carried it out so tardily as to make it worthless.' Photographic production and interpretation attempted to support all known functions that benefited from the analysis. The distribution differed with the assigned mission. 'Each Corps Squadron sent us thirteen finished copies of every photo. Two were for our files, one for artillery information service, four for artillery observation group, c/c First Army Air Service, and six for Photographic Aérienne, French Second Army.'[674] Even at the lower echelon squadron level, the distribution list was fairly extensive, including 1st Army, A.E.F.;

Headquarters 1st Brigade Air Service; 96th Bombing Squadron; G.H.Q., A.E.F.; G-2, G.H.Q., A.E.F.; Observers and Pilots and the Branch Intelligence Officer file.[675]

US Air Service aero squadrons emphasised order and ready access. The AEF Assistant Chief of Staff for Intelligence praised the American x and y coordinates (measured to one tenth of a km) filing system employed at the aero squadrons. A master photo file was established to retain the original copy and extra copies as required. Photos of enemy aerodromes and a key map was established to monitor the sector. A review of the process by the AEF G-2 was a reminder of what worked best behind a successful filing system: 'Perhaps the best solution is a thoroughly competent and interested clerk in charge of this very important file.'[676]

Post-war direction from Brigadier General Mitchell to US Air Service aero squadrons reminded all that photographs required x and y coordinate filing in conjunction with the Plan Directeur process. In practice, a card catalogue was established listing more information on the size of the camera, coordinates, index number, altitude, date, and a qualitative description of the photograph. A map was used to illustrate areas covered by photographic missions. Plate coverage was outlined, the index number and date attached, and the area covered by the mission coloured. The maps illustrated the sorties accomplished during the week. Once a month a second map was prepared displaying the entire area covered by aerial photographs.[677] The classification and filing of all aerial negatives became a post-war priority to archive the battleground. The objective was to have a complete set of photographic negatives to portray a different view of the war. Ground photographs of all the important points pictured in aerial photography were taken to clarify the aerial targets. Additional photos were acquired from French and British files.[678]

EVOLUTION OF THE SCIENCE AND ART OF PHOTOGRAPHIC INTERPRETATION

The earliest writings on aerial photographic reconnaissance clearly acknowledged that aeroplane photographs formed a most useful source of information for planning a military operation. Once the interpreter accomplished the review of the photograph, he had at his disposal a visual framework for understanding the enemy's defensive organisation, its topographical location, and a comprehensive reference to the evolving military formation. By the end of the conflict, those well experienced in the interpretation discipline considered the aerial photograph to be an 'inexhaustible mine of information' contributing to an understanding not based on theory but purely empirical data.[679] Three phases for interpreter analysis were developed. The preparatory period involved studying the target area and applying the research to the plan of attack. The destructions period determined the state of progress of the Allies' impact on enemy works. Finally, the execution period comprised analysis of the attack to determine the extent of Allied occupation.[680]

According to the earliest aerial photographic interpretation manuals, written in 1915, the overhead image of the battleground was 'now visible, except machine guns and guns placed in trenches and covered over.'[681] The interpreter's analysis went beyond the visual cues and developed a perception of reality through the exercise of deductive logic. Data that were not relevant in determining the threat were to be eliminated from the mental process. Upon completion of this phase of deliberation, the interpreter had to pass the intelligence to the staff in a timely manner so that operation orders for both artillery and infantry could be generated.[682]

Establishing an aerial photographic reconnaissance-aware military culture remained a challenge throughout the war. The experiences of the British and French showed that it took several years of aggressive promotion on the part of a few advocates to gain support from command, staff, and eventually the military forces being supported. By 1916, aerial photography was recognised as essential to the campaign. However, when American forces started arriving in Europe in late 1917, the same challenges resurfaced. Major Steichen ruminated on the issue:

> Photographic reconnaissance is a most difficult and ungrateful part of aerial observation. It gets more recognition and credit from the enemy than any one else. Captured enemy documents indicate that their General Staff was very much worried about Allied aerial observation, and particularly the planes with the all-recording camera soaring miles up in the air overhead.[683]

The key to the conversion of sceptics to aerial photography and interpretation was education. The French led the way in determining the purpose of aerial photography from an academic standpoint. French photographic experts related it to prevailing military thinking. Leading the effort was Capitaine Jean de Bissy. His initial pamphlet, *Note concernant l'interprétation méthodique de Photographies Aériennes*, 19 November 1915, helped establish the precedent for what was to become aerial photographic interpretation training.[684] A team of French photographic interpreters (Couderc, Pépin, Rousselet, and Jaulin et Paulon) subsequently produced the first known photographic manual. Published in 7 January 1916, the *Bulletin du Service de la Photographie Aérienne aux Armées* provided a brief introduction to the experiences on the Western Front as conveyed through aerial photography. The US version was published in 1918, and entitled the *Bulletin of the Aerial Photography Department in the Field 1918*. It was an intelligence document, compiled from various captured German documents that directed standards for military structures, and from line drawings of German trench networks found on the body of a dead German soldier. The manual provided a glimpse of the German policy that governed their operations at the front. The seminal work would serve as the intelligence foundation for subsequent published manuals.[685] In 1916 Capitaine de Bissy produced the first comprehensive manual that correlated text with photographic templates of relevant targets.

The baseline French aerial photographic interpreter manual for the last two years of the war was *Étude et Exploitation des Photographies Aériennes*. This work became a critical source for explaining to the intelligence community the role and science of the aerial photograph. Authorship was not specifically credited in the text, but the preponderance of evidence from a selection of writings shows the main author to be Lieutenant Eugène Pépin. This brilliant academic provided a textual foundation for the philosophy behind aerial photographic analysis, combined with descriptions of the entire battlefield. Pépin was recognised later for galvanising the American photographic interpreter effort during the final months of the war.

Brigadier-General Charteris spearheaded British aerial photographic interpretation training support based on exchanges with the French during their ongoing effort to create doctrinal manuals. Charteris took their insights and commenced producing a similar training manual for British forces. The British lead for aerial photographic interpretation, Moore-Brabazon, considered that the Intelligence staffs were not getting enough information from the photographs. He proceeded to acquire the services of the British portrait painter, Sir Oswald Birley, to lead the assembly of the manual of instruction.[686] The Army Printing (AP) and Stationary Services (SS) produced the manuals. AP and SS were well known by BEF staff on account of the rapidity with which they turned out printing, including illustrations. The AP and SS also reproduced aerial photographs in large quantities.[687] The effort culminated in the publication of the first British manual for Photographic Interpretation, S.S. 445, *Notes on the Interpretation of Aeroplane Photographs*, and the companion S.S. 445a, *Illustrations to Accompany Notes on the Interpretation of Aeroplane Photographs* in November 1916. Moore-Brabazon gave his intelligence counterparts credit for the final publication. 'They produced during the war one of the finest works on the interpretation of aerial photographs I have ever seen. It was a great show.'[688]

The second British photographic manual, S.S. 550 and companion S.S. 550a, with the sam titles, were published in March 1917. The distribution was extensive, down to the British battalion, machine-gun company, and trench-mortar battery. The War Office in London alone received 3,100 copies.[689] *Notes on the Interpretation of Aeroplane Photographs* was also given to the Allies. American recipients included the American Day Bombardment School at Clermont-Ferrand.[690] The British published the third and final version, S.S. 631, in February 1918. Later that summer the Americans published additional copies of the S.S. 631 text through the Government Printing Office. It was a sensitive document showing the quality of the coverage, the ability to pick out key German targets in the trenches, and details provided by draughtsmen. The reader was warned: 'This document is the property of the United States Government, and the information in it is not to be communicated, directly or indirectly, to the press or any person not holding an official position in the Government service.'[691]

When the Americans arrived there was an urgent need to educate them about the purpose of aerial photography, to introduce staff officers and senior leaders to its potential in battle planning and, most importantly, to integrate them quickly into Allied operations.

At the end of 1917, two intelligence conferences attended by Allied aerial photographic experts and staffs recognised that need. The French provided the AEF with its *Étude et Exploitation des Photographies Aériennes*. With this document in hand, the pioneer American photographic interpreter team of Major James Barnes and 1st Lieutenant Edward Steichen became the prime movers in standing up the first American aerial photographic interpretation teams, using this manual and whatever other manuals and handbooks they could acquire. Their efforts resulted in two American translations of the French document. A translation by members of the AEF staff, *Study and Utilisation of Aerial Photographs*, appeared as a field manual in July 1918.[692] A companion version was produced by the Army's Division of Military Aeronautics and published through the Government Printing Office entitled *Study and Exploitation of Aerial Photographs*.[693] A final, more literal translation, titled *Study and Use of Aerial Photographs*, was produced after the war by the American Mission with the Commanding General Allied Forces of Occupation, Mainz.[694]

The initial American lack of analytical expertise on positional warfare was clearly discernible. In retrospect, shortly after the Armistice, the Assistant Chief of Staff G-2, 1st Army, AEF, made a recommendation: 'The officers charged with the duty of interpretation of photos found themselves handicapped by their ignorance of actual conditions. Of many of the objects which they learned to recognise on photos they had never seen an example on the ground. It is thought that if a photo intelligence officer is without actual experience in the combat areas, he should be sent forward on a tour of observation. A few trips in an airplane would also assist in giving him the point of view of the air.'[695] Steichen echoed the sentiment. Access to the training manuals was not enough. The art of interpretation required an intimate knowledge of the battlefield:

> The mistakes that are [made] by a student in interpretation whose knowledge has been gained entirely from the study of books and pictures are flagrant; a few hours spent on the battlefield studying the ground and organisations in connection with aerial photographs previously, are of more value than any amount of classroom work. The interpretation of photographs is in no sense an exact science, but is largely a matter of astute deductions, and men picked for this work must have qualifications that will make it possible for them to develop along such lines.[696]

At war's end, aerial photography had an infrastructure that assured timely acquisition of required information and dissemination to the combatants. Mission planning, annotating photographs, distribution, and production of manuals became integral to the manner in which the photograph was employed. The interpretation operation achieved a permanence that would support military operations for the remainder of the century. The cartographic process evolved with aerial photography's ascendance.

A stereoscopic pair of the historic city of Ypres showing extensive damage. (IWM, courtesy Nicholas Watkis)

CHAPTER 8

The Framework of the Aerial Photograph

CARTOGRAPHY AT THE BEGINNING OF THE CONFLICT

When the fast-paced mobile campaign of the first Battle of the Marne shifted to the trenches on the Aisne, it set the stage for aerial photography to become the standard for planning and conducting the battle. Integral to the entire planning process of the war was the map, providing a range of purposes from firing charts to geologic surveys.[697] Prior to the Marne battle, the only way a combatant could acquire a remote understanding of the terrain was through ground observations transferred to a map. Details of the earth's surface used to define a map varied with whoever conducted the survey. When the BEF arrived on the continent in August 1914, the maps of the impending battlefield were based on Napoleonic era French maps and Belgian surveys. Both were found to be inadequate and unreliable for supporting accurate artillery fire.[698] Aerial photography created an information breakthrough when its results were applied to the map. It was the ideal medium to illuminate intelligence based on visual observation. 'The map is a guide for travelling over the terrain.'[699] The map provided all combatants the means to prepare for battle by studying the terrain from a perspective that couldn't be achieved from traditional viewing angles. Thanks to the aerial photograph, 'the other side of the hill' was no longer a mystery.

Effective targeting for artillery demanded cartographic accuracy. Before 1914, British coastal defences through Ordnance Survey established a high standard for target surveys using instruments that accurately fixed on a target and measured the location. German prewar targeting included rudimentary flash-spotting triangulation and measurements. The French established the *canevas de tir* or artillery board to enhance their targeting. The canevas de tir was a rigid, gridded board with the location of artillery, SROTs, and enemy targets plotted with range and bearings applied. The aerial photograph would soon revolutionise the entire process.

Cartographic accuracy was complicated by the fact that none of the Allies shared a cartographic standard before the war started. A consistent map scale did not exist. In 1914, the British standard for maps of the impending battleground were at a much smaller 1/250,000 scale compared to the French standard of 1/80,000.[700] The 1/80,000 *Carte d'État Major* was the universal topographical map representing the whole of France and based on the Cadastral Plans (large scale land registration plans). The entire map consisted of 274 sheets, each sheet having a name and number, the name being that of the most important town on the sheet. The names of the four adjacent maps were printed in the centre of each side of the border. The 1/80,000 scale was ill adapted to calculations.[701] At the end of 1914, the French *Service Géographique* enlarged their 1/80,000 map to 1/20,000 scale. However, it proved a useless endeavour for the details were inaccurate and out of position.[702] Prior to the war, a comprehensive Belgian national survey provided a cartographic foundation for operations at Ypres and the surrounding area. The maps were fairly detailed for the time. As befitted the Low Countries' flatter terrain, vertical contours on the map were measured in single metres, rather than the normal 5-metre interval. The Belgian 1/10,000 scale drawings had been evacuated from Antwerp in 1914 and were subsequently reproduced for Allied use.[703] Man-made changes to the landscape totally transformed cartography in the Great War. The destruction virtually eliminated topographical landmarks in the battle area. In some sectors forests disappeared. Cartography was challenged by the obliteration of man-made features such as buildings and roads. Traditional surveying became seriously challenged by the elimination of common benchmarks such as church steeples. With each barrage, maps of the region became obsolete, a factor that added to the inevitable 'fog of war'. Aerial coverage became the sought-after solution to maintain a common cartographic standard with each daily round of devastation.

THE DISCIPLINE OF RESTITUTION

Restitution, the method of transferring information from the photograph to a map, became an integral part of the cartographic process in support of daily operations and offensive planning. It was a demanding process, requiring pinpoint accuracy to meet artillery targeting requirements. Maps updated by this process were critical to intelligence analysis. In a post-war analysis, Brigadier General Dennis E. Nolan, AEF G-2, wrote, 'There is no part of a system of military intelligence more important than is the furnishing of maps and topographic information to the army and the nation.'[704] It served all combatants throughout the conflict. The prolonged stasis on the battleground characteristic of positional war amplifieded the demand for detail on the enemy. The science of restitution served both map construction and aerial photographic interpretation. The aerial photograph became the primary template for the map. In turn, the map, augmented through the restitution process, guided the aerial photographic interpreter throughout his exploration of the photograph. The traditional sources for developing the map, such as surveys, did not compare with aerial photography in both accuracy and timeliness. If restitution was to be effective, it required expertise in determining the appearance of objects as viewed from above. Prior to the examination of a photograph, intense study of the best available map was required to gain knowledge of the ground configuration and prominent features in the area. Scale was determined by taking a known distance between two points on the map and applying it to the same two points on the photo. The intimate relationship between aerial photograph and the corresponding map became an important feature of the evolving reconnaissance process, each lent authority to the other.

An understanding of relief, height, slopes and identifying both natural and artificial objects on the surface became key to conducting the positional war campaign. The battleground required a responsive map development process to portray what the aerial photograph revealed. Positions of strength, flanks, protection from enfilade fire, points of command, and the like, became a dynamic part of the cartographic ensemble. Every opportunity had to be exploited to study actual ground objects identified on a photograph. Restitution also included the interpretation of man-made terrain within villages, sunken roads, slopes, earth prints, earthworks, breastworks, intersections of routes and railways. Restitution converted captured hostile trenches and positions into classrooms for the aerial photographic interpreter. Restitution's dominance remained a chief influence on French decisions on how to portray intelligence information, as the ability to 'put [information] in the form of a map or a sketch ... is preferable to any others.'[705]

MAP SCALE AND THE PORTRAYAL OF THE BATTLEGROUND

The level of detail described on a map correlated to the purpose and the scale. Years of experience on the same battleground institutionalised battlefield map development. Both French and British maps were produced at a range of scales to accommodate the respective operational mission. Smaller-scale maps provided a broad overview of terrain. Larger-scale maps provided the reader with intimate details. Thanks to aerial photography, a more accurate standard was attained. French maps of the battleground gained a reputation for being so accurate that advancing infantry consulted them to determine the exact tactic to be employed against key German defence points such as machine-gun nests. Artillery planning used maps to know in advance where they were to target the barrages. Reserve forces also employed maps to organise their operations in support of the offensive.[706]

Core Allied maps comprised the French Plan Directeur and the British Firing Map (also known as the Trench Map). Both served the same purpose, providing the commander of the battlefield with a situation update. The maps ranged in scale from 1/5,000 to 1/50,000, together creating an overall map of operations throughout the battlefields of France, Belgium and even into Germany.[707] The Plan Directeur became the focal point for French battle planning. Général Pétain reinforced the role of the Plan Directeur in developing his priorities for daily operations. He stipulated that the level of detail be commensurate with the map scale.[708] This meant that no one map scale sufficed for the conduct of the campaign. Large scale 1/5,000 maps were designed as infantry maps for use in attacks within a given sector, showing the German trench organisation in fine detail. The 1/5,000 version was derived from aerial photographs, aerial and ground observation and prisoner interrogations.[709] The large scale provided a schematic of the enemy disposition from the defensive point of view, whereby the probable distribution of the troops in the various trenches and shelters was emphasised, to include the halting points, the emplacements of the reserves and the probable directions of counterattacks. With each annotation, an explanatory notice was attached to each study or printed directly on the bottom of the card. The photographic information was applied by hand drawing the perceived information, and then kept up-to-date. Distribution was limited, so that only a select few such as the battlefield commander received timely copies.[710]

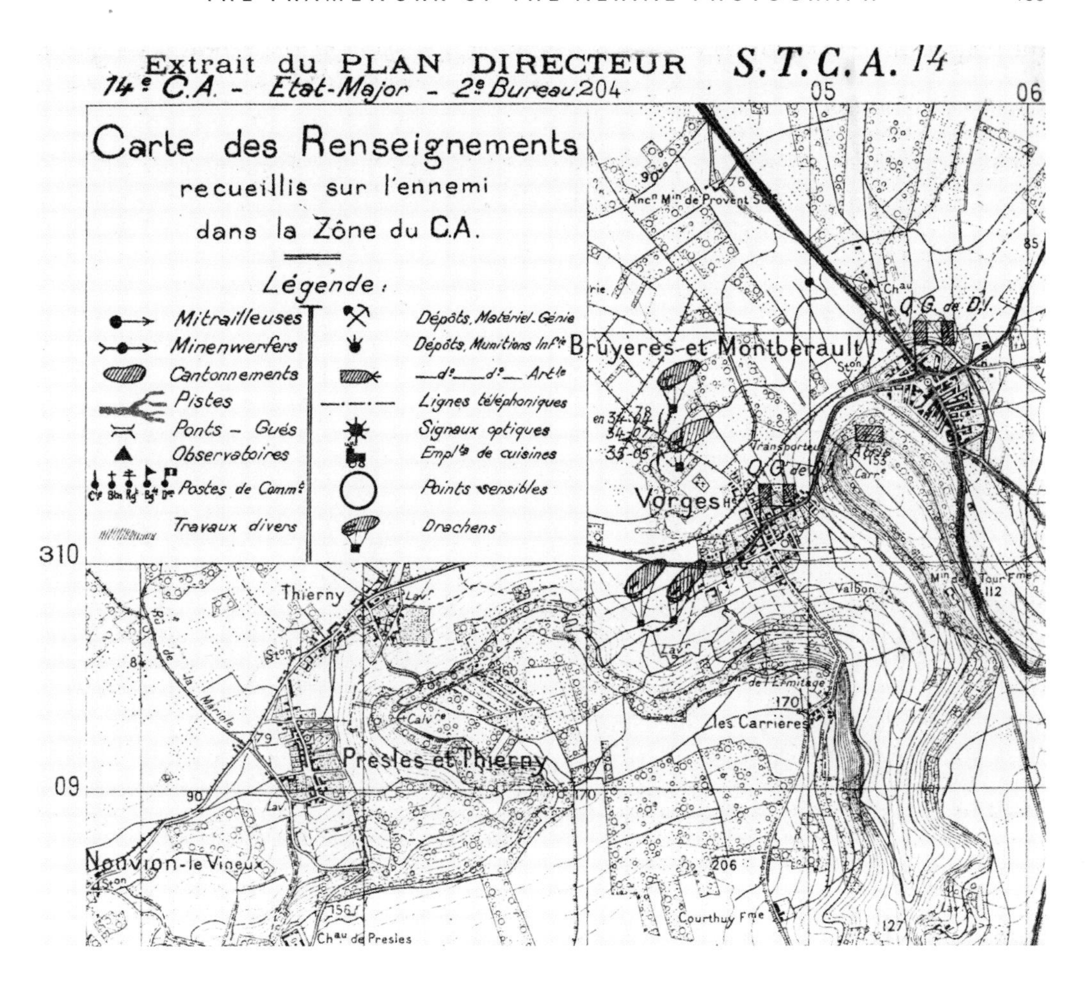

The *carte des renseignements* portrayed the intelligence picture on the 1/10,000 Plan Directeur. (SHD)

The first British maps were devoid of any depiction of the British lines, through fear of maps at the front falling into the hands of the Germans. Confusion set in as troops in the trenches tried to figure out their own sector. The British then resorted to field-generated maps that not only portrayed the trench, but also included other additional sensitive information. The situation was out of hand when the troops depended on captured German maps to show their own lines. In September 1915, while preparing for the offensive at Loos, the BEF established a limited edition map of their own trenches. Distribution was limited only to the Brigade Headquarters and above. The impact of aerial photographic reconnaissance and restitution proved that maps had to include both sides if they were to be of value. By the end of 1916, the British produced such maps for general distribution.[711] Special planning revolved around the 1/10,000 French Plan Directeur and British Firing Map. Pétain himself labelled the 1/10,000 as the 'complete document which enables the [assault unit commanders] to issue detailed orders with all desirable precision.'[712] Both maps had similar purposes, meeting the needs of the infantry at the front. Theg map provided forces with a large-scale plan of villages and important strong points. Some maps annotated the front-line trenches and communication trenches according to their relative importance. The 1/10,000 Plan Directeur map included additional detail, covering a range of significant data on the enemy forces to include vital organisations or points, shelters, flanking organisations, CP, ammunition depots, optic and telephone stations.[713] The 1/10,000 map was also used by artillery and staff to show details for accomplishing accurate fire against wire, obstacles, dugouts, trails and other important components of the trench network organisation. They were dated. Without the date, these

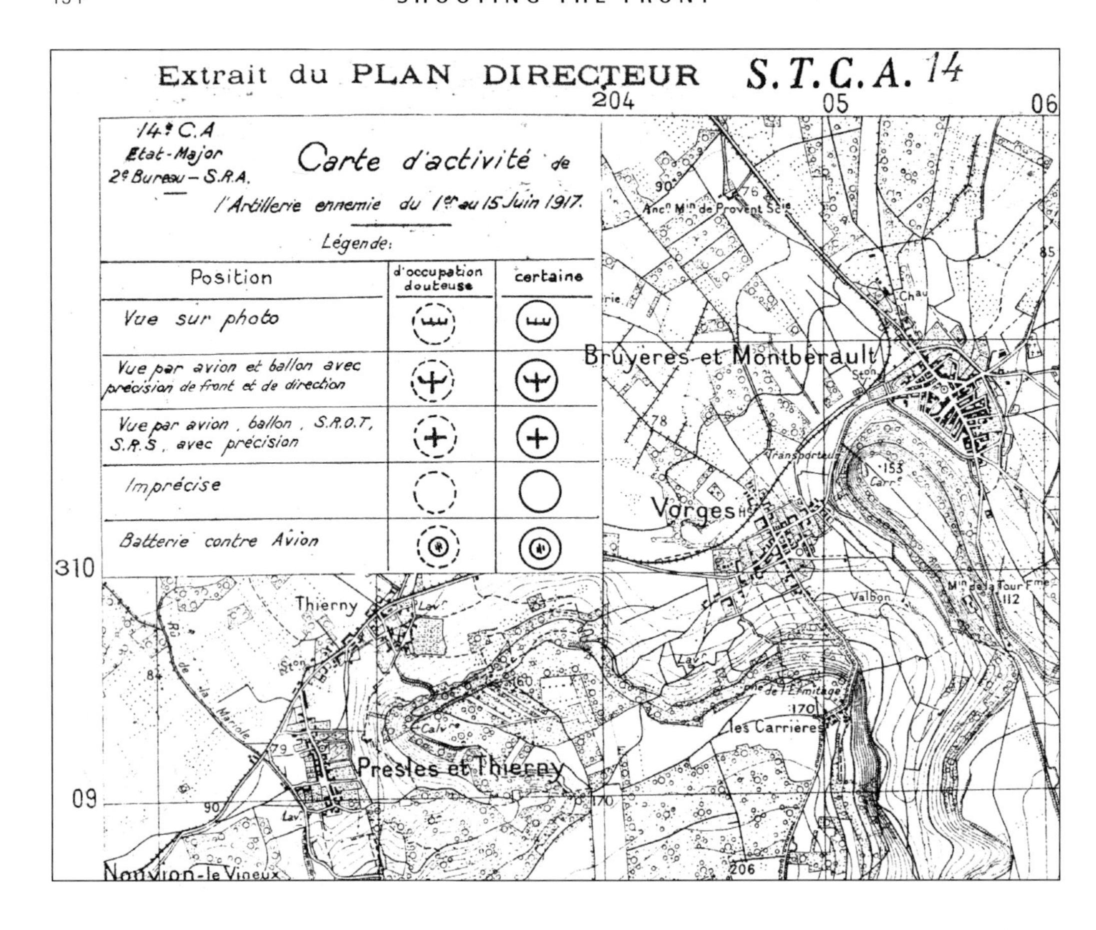

The *extrait du Plan Directeur* illustrated the intelligence source, verified by aerial photography. (SHD)

Opposite: British Firing Maps portrayed extensive detail to support artillery operations. (Peter Chasseaud, *Artillery's Astrologers: A History of British Survey and Mapping on the Western Front 1914–1918*)

maps were of little or no value.[714] By 1916, the 1/10,000 series would serve as the most accurate targeting map for the artillery. Sound-rangers and flash-spotters used it to plot their data to determine locations of German artillery.[715] The Allies also used the 1/10,000 to show their offensive organisation.

The Lambert method of projection, using Lambert kilometric coordinates, was used for Firing Maps. The Lambert method was one of three survey processes, which included the Polyhedric Projection and Bonne's (*Projection du Depot de la Guerre*) method.[716] When Foch assumed Supreme Allied Commander, he directed that mapping become uniform among the Allies. 1/20,000 maps were produced based on the Lambert projection, to include a grid overlay that became the standard map reference. The changes were late in coming; actual implementation took place after the Armistice.[717] The vital role of targeting artillery was done with the 1/20,000 map. The French Plan Directeur 1/20,000 served as the primary map scale for planning artillery objectives as it showed both battery location with calibres, and used different colours to reflect levels of activity. The map included an area of at least 5km inside Allied lines and the first 10km of German-occupied territory. Artillery planning used the modified scale map with pasted-on sheets marked with specific updated details.[718] Likewise, French tracking of German activity in the rear echelons included the process of updating sections through pasted notes. The rear echelons were marked on the 1/20,000 scale, with updates acquired from aerial photography on key installations such as munitions dumps, railroad activity, and any observed change in German defensive organisations. The French Armée headquarters used the 1/20,000 as their principal scale to evaluate the battle.[719] The French had produced a 1/20,000 Plan Directeur as early as the 1914 Battle of the Aisne made from visual aerial reconnaissance and

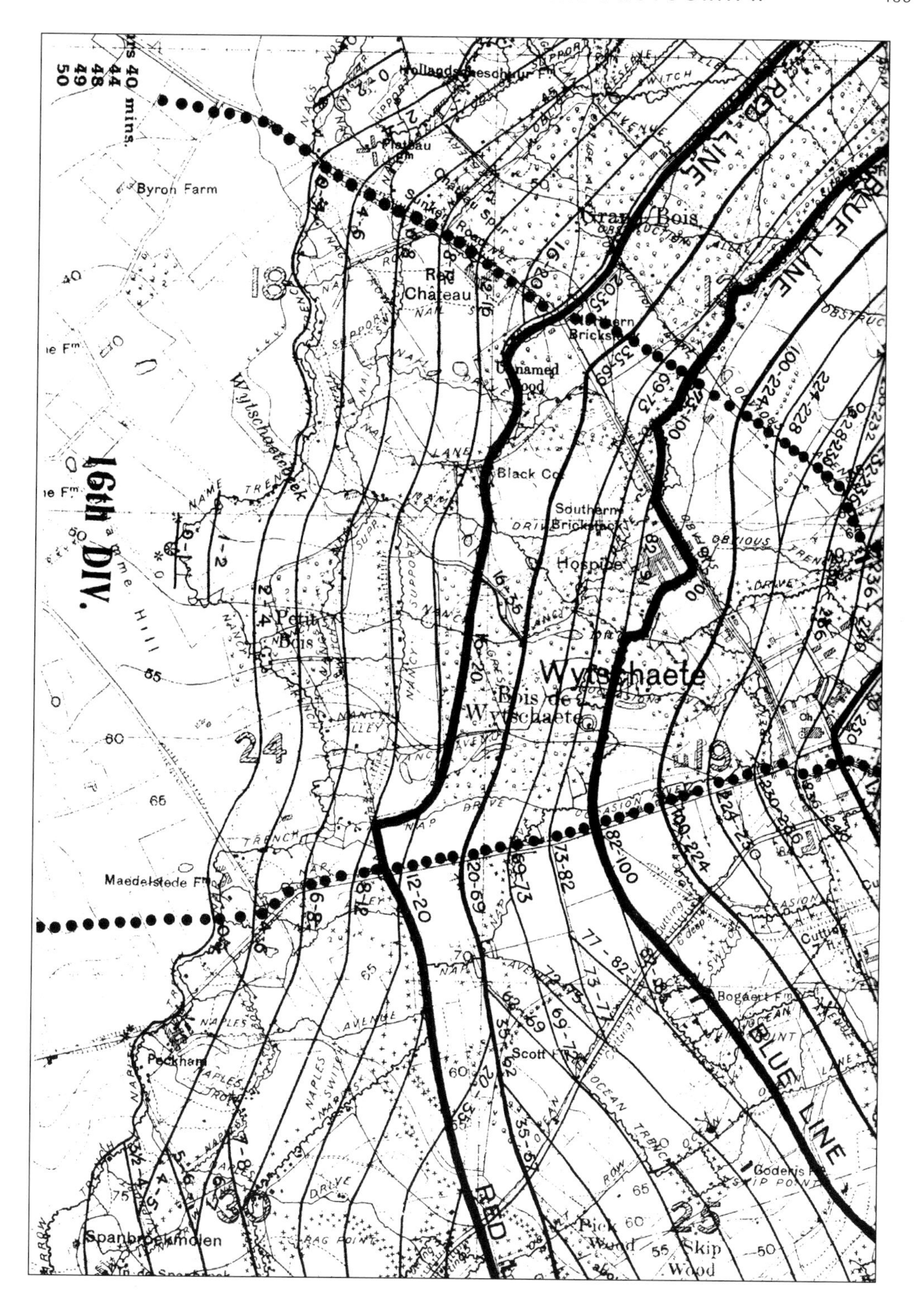

aerial photographs. The British had lithographed copies made for use by the artillery and the RFC. Targeting data marked on the maps included German trench and battery positions.[720]

British efforts to chart all enemy artillery naturally reflected the information sources at their disposal. In 1916 at the Somme, a 'positions' map was created showing all enemy battery positions located from all collection sources, to include aerial photography, flash-spotting and sound-ranging. Each gun emplacement was marked on the map for directing counter-battery fire. Later a Hostile Shelling Map was created for tracking the extent of daily enemy artillery fire. Unlike most information acquired from aerial photography, this map was primarily compiled at the division level. The 1/20,000 map helped portray where artillery was being directed within a particular sector of the front. Reports were forwarded by units in the line through the brigade headquarters showing the numbers of rounds fired, the calibre of the shell and the area shelled.[721] The Hostile Shelling Map was kept for one week. The weekly updates were compared, providing an accurate estimate of the distribution and volume of enemy fire. If the Hostile Shelling Map identified a certain number of rounds being fired for several days in succession against a small section of trench or wire, it suggested that wire-cutting prior to a projected enemy raid was in progress. The unit in the line was warned accordingly. When German artillery fire covered a certain divisional sector or was noticed being directed against trench junctions and communications, it suggested either normal harassing activity or a possible registration of new guns. Any increase in the general volume of fire directed against the divisional front as a whole as well as any marked change in distribution of fire became apparent from the map.

The French and British employed different scale maps as the area coverage increased and the scale decreased. The British maintained a daily map of the Army Area (1/40,000) on which all photos taken in one day were plotted, and within each square covered by one photo is marked the photo number. There was a standing order that the entire army area, extending from 4,000 to 18,000 yards behind the enemy front line, was to be photographed once each fortnight. As soon as negatives were developed they were compared with the record map. Many plates were thrown out on account of duplication. Only those accepted were given serial numbers according to the sector. As soon as the photograph was accepted the area covered was blocked off on the daily chart and the photo number entered in the space.[722] British long-range railway guns used the 1/40,000 scale map for their operations.[723] The British also used the 1/40,000 to track German artillery positions opposite the British front. It was updated fortnightly.[724]

The French GQG used the 1/50,000 to show the strength of the infantry and artillery to be used in the battle. A similar scale map portrayed the details of the enemy organisation.[725] Sketch maps showed active batteries as well as previously occupied or possible emplacements. The French considered the 1/50,000 scale the best source for portraying defensive organisations and their logistical networks.[726]

American processes paralleled the French. Information flowed to the Army Headquarters in the same manner as the French Plan Directeur. The AEF Army Headquarters photo office was responsible for all information on enemy works, regardless of the source (including captured enemy photographs). Their primary source was interpreting vertical photographs to find new activity or changes with the existing works. All changes were noted on the works chart, a tracing created over a 1/50,000 map. A master Plan Directeur was kept in triplicate and updated at all times.[727]

Aerial bombardment relied on the 1/80,000 ensemble map with all objectives in the target area sketched out. When planning bombardment, 1/10,000 tracings from aerial photographs were drafted onto the map to provide the aircrews with more detailed descriptions of the targets. The map and annexed tracings were produced by 2e Bureau and coordinated with the Service Aéronautique and the *Groupe de Canevas de Tir*.[728]

CARTOGRAPHIC UPDATES

Plotting daily coverage became a well established process for all echelons. For operations in the front sector, maps provided the most effective way to track the changing situation. 'Enemy Work' reflected enemy defences portrayed on a map covered by a sheet of tracing paper marked in squares with a letter to identify the location. Once information on the enemy was confirmed by various sources, the data were marked on the tracing paper. New trenches under construction, trench improvement, construction of new switch lines or organised shell hole defences were traced by the draughtsman and applied to the map. It was reviewed weekly. The tracings of the preceding weeks were reviewed, providing an indication of any observed activity being offensive or defensive in nature, as well as additional targets for artillery. All information showing the various features of enemy activity in a given sector were confirmed by the aeroplane photograph. The French annotated new enemy works on tracing paper covering a small-scale map known as *Croquis de reconnaissance*. Prints were then produced on bromide paper, and forwarded to headquarters for staff awareness of ongoing changes and further distribution to the army sector of interest.[729] When a new division came into the sector, one of the first actions was to transfer that information to an up-to-date map of the battleground.[730]

The daily restitution method of the British Photo Sections was to use tracing paper that had lines corresponding to the 1/20,000 scale map of the area. The tracing paper was marked with the number and description of the map used, the scale of the map, the

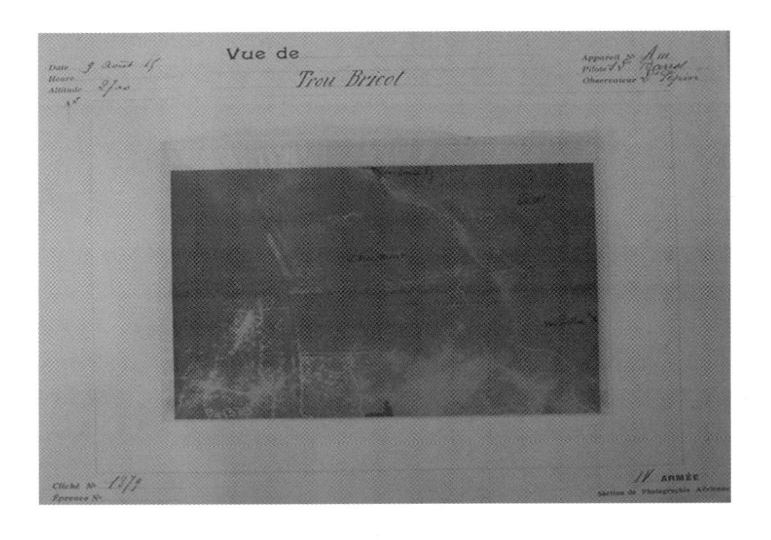

The French target file called the *croquis de reconnaissance* (sketch recognition) provided target name, print, and annotations of enemy works on tracing paper. (Eugène Pépin Papers, (MA&E)

Each aerial reconnaissance squadron produced a key map showing the photographic coverage acquired. 99th Aero Squadron's key map covers 30 October 1918. (USAFA McDermott Library Special Collections (SMS57))

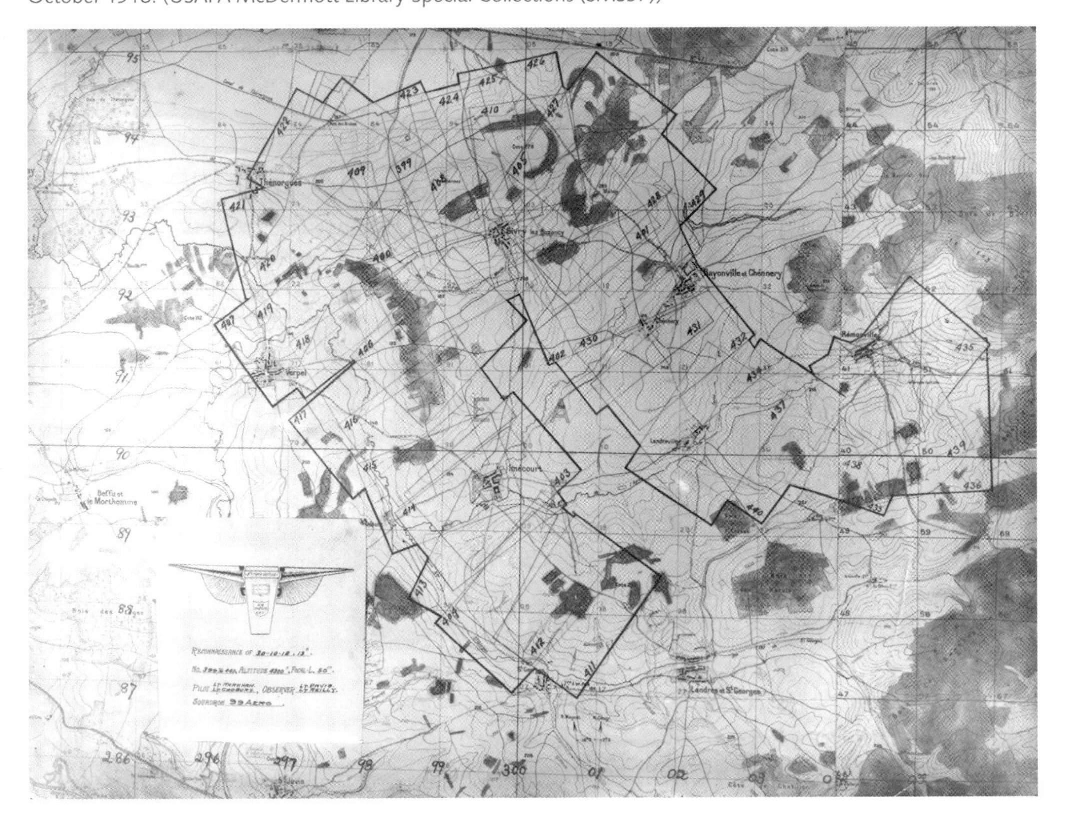

date the photograph was taken and an identifying large 'square' letter that corresponded to a similar square on the photo being examined. Further lines were drawn on the tracing paper to equate to the number of the small squares found on the map area covered by the aerial photograph. These steps allowed the photograph to be 'placed' on the map. The tracing paper was then attached to the photograph and forwarded on to the party of interest with a clear understanding of correlation between print and map.[731] The British also plotted on a 1/40,000 map all relevant information gained from the photographic analysis. In addition to the 1/40,000 map, there was a smaller format 1/100,000 map covering the same area. The 1/100,000 map was used to produce tracings for publication.[732] The Americans had each aero squadron send a key map along with the photographs to show where the coverage had been acquired. After a few days, the key map tracings underwent a colour change to show the interval since the last date of coverage. This process assisted the squadron in prioritising the coverage requirements for each sortie.[733]

The maps showed all enemy batteries that had been seen in action, or were known to have been in action, during the previous 24 hours. All escadrilles and ballon observateur companies received their copy of the Plan Directeur map every morning. In addition to the Plan Directeur, objective maps were also made, indicating in different colours every foot of enemy entrenchment, every weapon of defence, every item of military value. The objective maps were the reults of extensive preparation and were constantly revised, re-edited and reissued to the combatants. Every feature of the German trenches, to include paths, approaches, dugouts, barbed-wire entanglements, machine guns, trench mortars, and PCs was repeatedly examined and updated. The information was recorded on countless photographs under a variety of lighting conditions, altitudes and timeframes.

The photographic role was to acquire vertical coverage of a general view of the trench lines (small scale); details of enemy organisation (large scale); and oblique photographs of the rear areas. Visual reconnaissance and aérostat observations were to complement the photographic analysis with traffic and movement reports on roads and enemy artillery.[734]

COLOUR SCHEMES FOR MILITARY CARTOGRAPHY

An extensive cartographic colour scheme portrayed every major facet of the terrain and the threat. From the initial stages of the conflict, the maps became an array of hues that portrayed a clearly defined network of trenches labelled with key symbols of significant weapon systems or centres of resistance. Everything displayed on a map had to be visually clear and obvious and not overwhelm the reader with detail. Colour schemes applied to all maps, from the large scale Infantry Maps to the small scale staff maps remained consistent to allow quick identification of the targets of interest. The maps that evolved during the war imparted significance and value to each element of terrain and enemy work. Various techniques were applied to include adding bright colours applied with a copying brush or hand drawn on specific targeting maps.

The basic colour schemes to show map features varied among the Allies. The French used blue for German (enemy) trench networks and fire lines; red for Allied trench networks and traffic routes; green for shelters and bivouacs; yellow to illustrate wire entanglements; brown for railways and red to highlight key threat locations such as a machine-gun emplacement. The British showed the German trench network in red and the Allied lines in blue. In 1918, they changed to conform with the French standard.[735] Smaller scale maps had different colour schemes. The colour scheme for the French 1/200,000 series had roads printed red, railroads black, rivers blue and contours brown.[736] Map symbols were coloured to illuminate high-value targets such as machine-gun locations, mines, points of resistance, PCs and ground observation points. A brown-shaded circle highlighted artillery targets such as junctions of fire trenches and principal communication trenches. Those targets that combined features were identified with multiple colours. A communication trench adapted for fire was described by a red line followed by a blue line on the side of the firing parapet. A blue line coupled with a green line represented a trench with deep dugouts.[737]

Some centres traced their maps in a different colour scheme so that the reader understood that the information depicted on a given location was dated.[738] French interpreters used shading with red ink to display the effect of artillery fire. Those areas shaded heavily by hachure or lines drawn close together portrayed completely bombarded and destroyed areas. When artillery was less effective, the area was slightly shaded or a circle drawn in red ink around the area. The finished map was forwarded to the artillery units to help them gauge the effectiveness of fire.[739]

The colour scheme historically traced the impacts of incoming artillery rounds in a given sector. Different colours represented the different calibres of enemy shells; particularly the three smallest guns, the 77 mm, the 4.1 howitzer, and the 5.9 howitzer. One dot of the colour indicated the calibre of the round and was marked on the map at the place where the shell exploded.

CARTOGRAPHY'S EVOLUTION FOR PORTRAYING FUTURE WAR

Cartographical support of positional war adopted a routine. As soon as the GQG staff had committed to go ahead with an offensive, the ground in front of the army was carefully studied, mapped and photographed. This was not just the area covering the front sector, but extending far back over the lines.

Future war would never see the likes of Passchendaele. Cartography had to decipher the topography of destruction. (IWM)

Passchendaele, 16th of June 1917

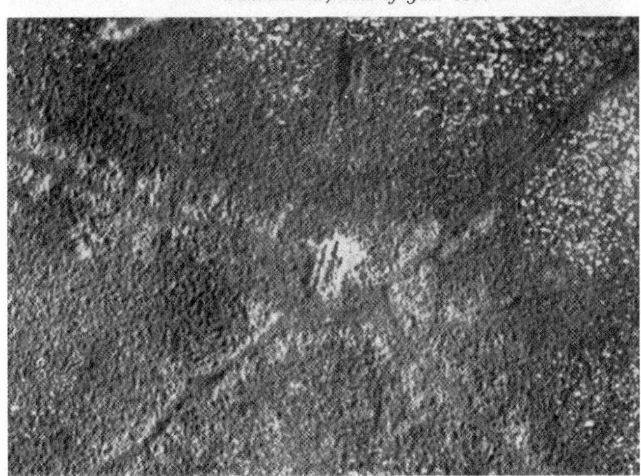

Passchendaele, 5th of December 1917

British Firing Maps made good use of aerial photography to help establish a reference baseline that was of value, even when a rapid advance or retreat caused a major shift in the front line.[740]

Advances in the use of tanks, followed by the breakthrough of the German lines in the last months of the war, forced the cartographic process to change. Aerial photography provided a foundation for determining the topological base; however, the sub-surface geological makeup of the battlefield now had to be considered. The first studies focussed on analysis of water sources to determine the impact of the existing water table on the planned trench network. When positional war started to give way to mobile war, additional consideration was paid to determining 'passability' limits for heavy pieces of equipment such as the tank in the countryside.[741] When the British fought the Battle of Passchendaele in late 1917, the conditions of the terrain were abysmal. The mud made mobility impossible for both human and machine. However, the British Tank Corps introduced the first 'going' map, which portrayed the hazardous areas in the sector that were most likely to be flooded or boggy, and to be avoided by tanks at all costs. In some cases, even the best surveys did little good. Tanks, infantry, and artillery not only bogged down in the horror that was the mud of Passchendaele; some even disappeared without a trace. In the last year of the war, the French were the first to produce a 'passability' map to support military operations. Several prominent French and Belgian geologists were attached to the *Service Géographique* to prepare 'tank maps' that delineated the physical conditions produced by the surface and near-surface soils and materials of future battle grounds to prepare forces for manoeuvres.[742] When mobile warfare commenced in the final months of 1918, the French employed 1/20,000 maps to provide the necessary detail to support faster tempo operations.[743] Another challenge to mobile operations was that the larger-scale 1/10,000 map did not sufficiently cover the rapid advance of the forces. After the Battle of Amiens, the British dropped the 1/10,000 scale and the 1/20,000 on occasion, since their forward momentum made them redundant. They had to resort to a scale of 1/40,000 or smaller.[744] With the transition to more mobile operations, the cartographer added to the confusion of the moment. Now the colour scheme varied, with both German and Allied organisations being portrayed in red and blue.[745] Fortunately, the confusion was relieved by the Armistice.

The cartographic institutions lasted well beyond the Armistice. Knowledge of terrain, topology and the geology that defined the battlefield for this war guided the approach for future conflicts. For France, it encouraged the extension of the battle line within the confines of front fortresses. The self-fulfilling prophecy for future war expressed by map development culminated in an apparently formidable defence strategy that spurred the construction of the Maginot Line of the 1930s.

CHAPTER 9

The Exploitation Operation

American lectures to AEF personnel on the role of intelligence at the front expressed the prevailing thinking among the Allies as to the value of aerial photographs.

> The maximum efficiency in gathering intelligence can be attained only by a balanced and thorough distribution of the different missions of information; so that a complete coordination of work may bring about a complete surveillance of the enemy, and so that eventually all information may come to a single head where it may be checked for error. Photographs are one of these sources of information and because of the great mass of detail presented with a maximum of accuracy they are to be considered the most important.[746]

THE INTELLIGENCE ROLE

Interpretation and restitution operations were indispensable, but were not enough to extract meaning from the intelligence source. The measure of effective exploitation went beyond the recording in these initial steps. Effective exploitation went to the heart of the matter by seeking a greater depth of understanding of the situation, nature of the events in progress, the value of the intended operation and the details of the enemy forces committed to the operation. Determining the reason why was the constant theme for analysis. Discussion of the art of photographic interpretation now concentrated on ascertaining the enemy plan of defence or attack by taking into account the German methods of combat. How this process was followed varied for each Allied nation.

Through consistent and persistent analysis, intelligence could determine what constituted the status quo on the front lines, thereby allowing the defenders to 'permit the plans of the enemy to be discovered in time'. Searching for signatures associated with an offensive operation was critical. Telling signatures included construction of roads and railroads, increased presence of supply depots, sightings of new artillery battery positions, and new trench work into No-Man's Land. Exploitation showed that offensive preparations required a considerable amount of time,

French photographic interpreters engaged in analysis. Their ultimate objective was to legitimise all military activity against the enemy. (SHD, from Christienne and Lissarague, *A History of French Military Aviation*)

and proper surveillance and intelligence ensured that the enemy could not 'pass by unperceived'.[747]

Determining enemy intentions went beyond 'role playing' the adversary. For commanders, it also meant extensive interaction with colleagues in the intelligence discipline.[748] The experience of positional war showed the participants that extensive knowledge of defensive tactics was essential. The French pioneer interpreter Capitaine De Bissy succinctly discussed the methodology: 'In order to determine accurately the existing defensive organisation, it is ... necessary to exercise one's discretion and to eliminate.'[749] French analysis promulgated the view that the German adversary was methodical and closely followed senior policies. Conferences brought together intelligence and photographic specialists to share ideas and techniques for intelligence collection and analysis. The more senior the rank of the participants, the broader the scope of information that was shared. A valuable outcome of the meetings was that a closer cooperation was established between the infantry, artillery and aviation.[750]

SOURCES OF INTELLIGENCE

Prior to the outbreak of the Great War, intelligence was primarily derived from human sources, compiled from espionage, military attachés, and repatriated persons. By 1918, intelligence was an expanding discipline covering traditional methods and a host of new technologies. Sector surveillance of the battlefield had become the established bedrock for intelligence exploitation. Aerial photography exploitation was viewed as the most significant of the intelligence disciplines, for it governed eleven primary sources, to include prisoner interrogations; information from repatriated men; spies; patrol reports; aerial observation from aeroplanes and balloons; ground observation reporting from infantry and artillery; sound-ranging; flash-spotting; captured documents; and radio intercept/direction finding (radiogoniometry).[751] The following description suggests the value of this fusion of sources:

> The intelligence service obtains much information of value to the interpreter, through prisoners, deserters, sounding boards, sounding telephones and documents. Prisoners and deserters give first hand the location of machine guns and works of the front lines in general, and often of rearward positions. Sound boards [sensors that triangulated sound waves from artillery fire] tell us where to look for batteries, for a matter of fact it is not the duty of the interpreter to discover batteries but to locate them exactly in a given area. Orders are tapped from the enemy telephones through sounding telephones. Finally we have the documents captured from the enemy; complete rules regarding the construction of field positions have been obtained in this way.[752]

Daily updates for determining enemy intentions and operations revolved around human intelligence, acquired primarily from prisoner and deserter interrogations. Both French and British relied on this source to substantiate ongoing analysis. Charteris described the British process for prisoner interrogation: 'Most of the information which a prisoner has is information in detail regarding the enemy defensive works on his own immediate front. To extract this information from him requires time. It is sometimes necessary to take the prisoner back in the front line trenches, or to Observation Posts and almost always necessary to examine him with the assistance of aeroplane photographs.'[753] An American G-2 staff member recounted the French prisoner interrogation process:

> The Chief of the Bureau calls his interpreters (who are all officers) into conference and carefully explains to them the nature of the operation which resulted in the capture of the prisoners; shows them on Plan Directeurs and relief models where the action was carried out, and makes clear just what information he is particularly anxious to get and to get first. He then charges one interpreter with the examination of minenwerfer, another with machine gunners, signal men, telephone listeners or riflemen, and so forth. This done, each interpreter goes to his office, where with map and aerophotograph at hand, he proceeds to examine the prisoners who have, in the manner indicated above, been selected as qualified to give the particular kind of information he is after. These interrogations consume a great deal of time, usually from fifteen to forty-five minutes for each prisoner, and in case the interpreter gets into a friendly visit or an argument with a German, it may last even longer.[754]

Prisoner interrogation used the mosaic to portray the sector in explicit detail. Armed with these data, the interrogator could trace with the source (prisoner or deserter) the itinerary taken from the rear to the front-line trench network, confirming the interrogation with specific details from the photo such as an isolated tree, house, or any other visible feature. The prisoner's cross examination could describe myriad targets such as railroad stations, platforms, switches, stores; itinerary routes, tracks; rest camps, ration depots, distribution depots; narrow tracks; as well as the makeup of the enemy's PC (captain through colonel ranks). Other significant interrogations led to location

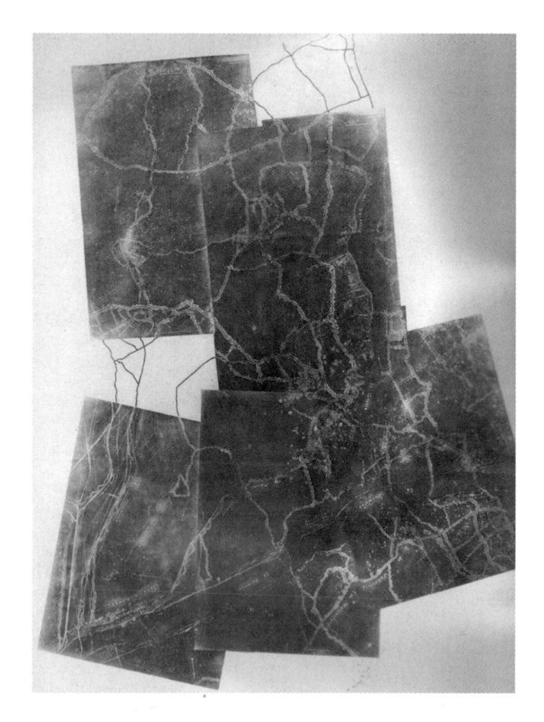

Aerial photographic mosaic of a German regimental sector. The interpreter would construct the intentions of the adversary once the sector was described. (*Étude et Exploitation des Photographies Aériennes*)

of machine guns; limits of the sector; distribution of the units; orders in case of alert, the latter enabling the supporting/strong points (*Stutzpunkte*) to be located and further analysed. Further interrogation focussed on those prisoners associated with the artillery to acquire more information on battery emplacement, observation post locations, and the associated echelons of command. The limits of aerial photographic validation of interrogations came when 'invisible' targets such as 'organisations under wood,' Stollen, tunnels, and mine works were mentioned. This meant interrogation was the sole source of information for many critical details.[755]

Prisoners and raids on enemy territory often provided personal letters and correspondence. Both the British and French kept thorough lists of enemy postal details to include addresses, postal sectors and other units mentioned in material acquired from prisoners.[756] An additional source of human intelligence were repatriated civilians as they made it to Allied lines from German-held territory. Interviews were normally conducted by intelligence personnel at the sector. One example of an interrogation of repatriated civilians:

> The two French civilians who came into our lines near Pont-a-Mousson on Sept 2, have described in detail a water supply system by which water from the Fontaine du Soiron, 66.8–48.3, is pumped to reservoirs at 3 points, each about 1 km distant (N.W., S.W and S.) whence it is to be piped to points on the Hindenburg line from Dommartin, S.E. to Mont Plaisir Farm, for use in concrete work. This system is partly visible on photos, where it has the appearance of buried cable trenches, and it has been so represented on maps. The civilian's statement is entirely consistent with the photographic evidence and indicates that the further strengthening of the Hindenburg line is to be looked for at the points indicated.[757]

Intelligence gathering took place throughout the front at all echelons. Each Ally had its own patrol procedures to collect and disseminate information. When information was necessary for refining last-minute operations, a unit was sent on a trench raid either to capture prisoners for interrogation, gather available evidence from the enemy trenches, or both. Information from observation posts, patrols, sentries on the line, and artillery liaison officers was forwarded to higher headquarters for further analysis and dissemination. Aerial photographic reconnaissance was called in to verify significant reporting.[758] French SROT support to either artillery or command forwarded information to a central office (2e Bureau) and the appropriate commander. Sectors were monitored and activity tracked via the Plan Directeur. Prior to an offensive operation intended to penetrate deeply into the German lines, SROT were located on the heights having a visual command, 'not only in front of the lines but also of the intermediate and rear zones.'[759]

Target acquisition of enemy artillery evolved as a science in the First World War using principles of sound and light. The process became so refined on both sides that by the time of the Armistice once an enemy artillery battery commenced fire it was quickly registered to a precisely known location and became a target for counter-battery fire. Sound-ranging (the French term was *section de repérage par le son* (SRS)) used microphones – usually six microphones set up along a 9,000-yard sector – to record the sounds of artillery rounds. With the rounds travelling at 1,100 feet per second, they created measurable arcs that were plotted on sector maps. The signals from the microphones were tracked and superimposed on a regional map. The data were synthesised using a mechanical device called a 'computer'. The resulting information was sent to Allied artillery units by telephone. Sound-ranging equipment was also used to track and correct friendly fire.[760]

Flash-spotting applied optical measurements to determine the location of enemy artillery. The essential

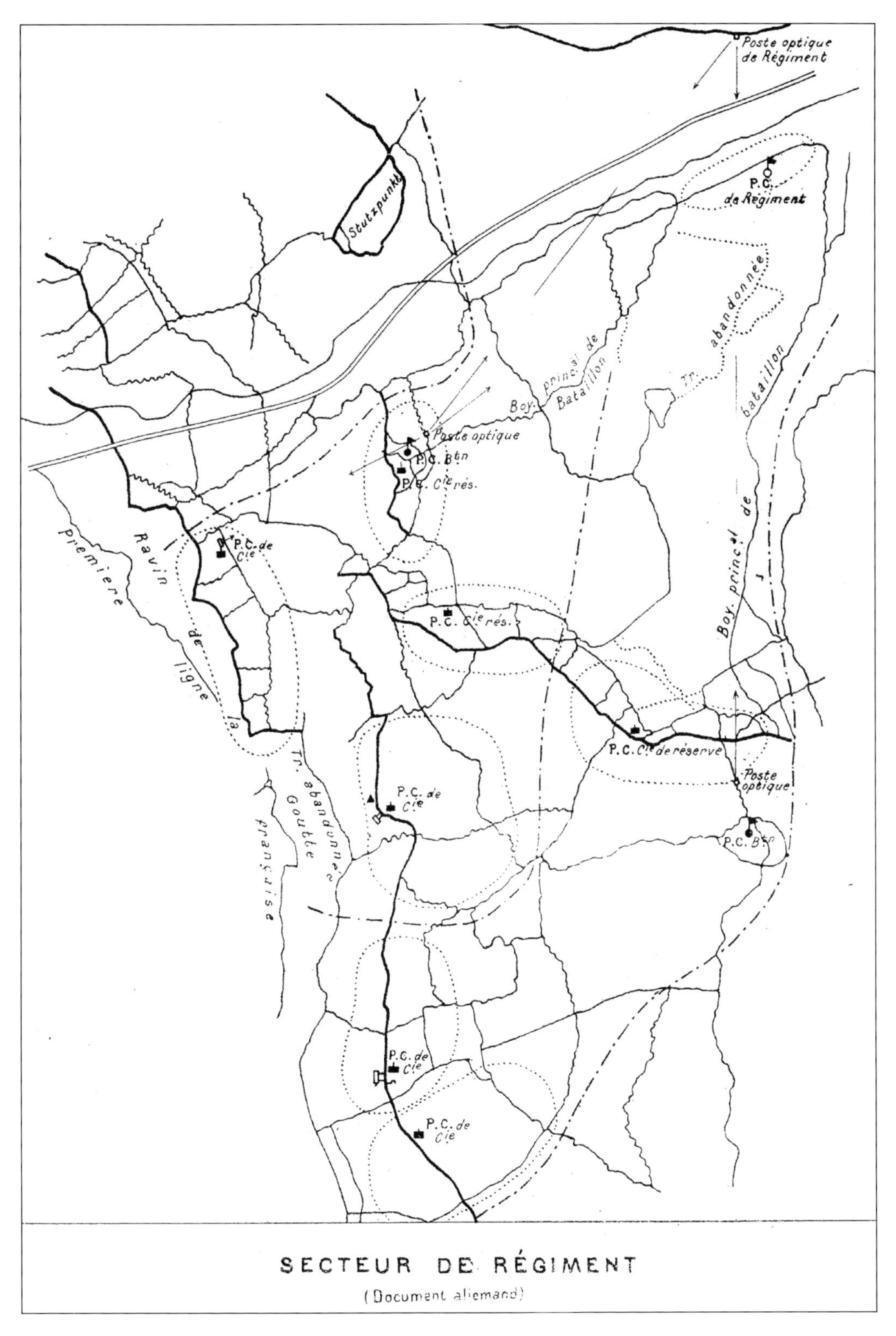

The companion line drawing of the German regimental sector. (*Étude et Exploitation des Photographies Aériennes*)

equipment for flash-spotting control at headquarters was the flash and buzzer board and the plotting board synchronised the observations. A telephone switchboard connected to the posts. The plotting board employed a 1/10,000 map that traced each response. All calculations on identified artillery batteries were forwarded to artillery for counter-battery operations. In some cases, flash-spotting supplemented the work of the ballon observateur.[761] The combination of sound-ranging and flash-spotting became a vital part of the intelligence network in place to support targeting. Aerial reconnaissance was an integral player in confirming the locations indicated by the optical and acoustical sensors.

Captured documents were considered the ultimate tool for 'perfect interpretation', particularly if aerial photographs were available showing coverage on the date of the document. Sketches contained within the document often included notes regarding machine guns, minenwerfer, dummy complexes and other projects of interest. By correlating the aerial photograph with the captured document, the analyst could identify obscure features. The interpreter's art required a knowledge of the details of trench organisation, as found in the published German regulations. Captured documents were forwarded to the photographic interpreter for in-depth examination. The interpreter's appreciation opened the door to understanding enemy intentions, followed by a broader strategic analysis accomplished by the staff. Both sources of analysis depended on various offices for translating documents. The German-French translation was the responsibility of both the GQG's 2e Bureau and the Technical Section of the Engineers. The interpreter and 2e Bureau colleagues 'role played' to better understand enemy decision-making. As the French manuals reminded their readers, the commander's responsibilities included providing for every possible phase of attack, and for successive lines of defence and all enemy lines of approach. This prepared the French defence to counterattack and eject the enemy from any captured territory.

The successes at Tannenberg and the Marne clearly illustrated the value of radio intercept and radiogoniometric (direction finding). A Radio Intelligence Subsection served as the unit designed to accomplish both radio interception and goniometric functions. A wire-tapping section also intercepted German ground telegraph lines (the French term was *télégraphie par le sol* [TPS]). Radiogoniometric units determined enemy order of battle by measuring the direction from which radio signals emanated. Traffic analysis focussed on identifying enemy radio procedures and call signs. This analysis was a highly favoured method for confirming enemy order of battle and determining the depth of echelons in a given sector. Enemy movement was also revealed by radio intercept. It allowed Allied forces to position their own forces to provide effective counter against an enemy attack.[762]

THE ULTIMATE ARBITER OF INTELLIGENCE

The early French writings on aerial photographic analysis provided descriptions of the intelligence exploitation role. Everything on a photograph was of intelligence interest. The snapshot would 'succeed in penetrating the thought of the German commander, to live the life of the men holding the trenches, and to foresee the action that they will instinctively take in case we attack.'[763] Those who collected and utilised intelligence data were obligated to read and interpret aerial photography. In case of conflicting information or data, the photograph was acknowledged by the French as the source that settled the discrepancy.[764] The medium had critical value because it showed collective priorities as well as target details. Whenever an enemy photographic reconnaissance aeroplane was recovered on the Allied side, intelligence personnel acquired the plates or film and processed it for analytical review.[765]

When the Americans arrived and tried to establish an effective intelligence architecture, the AEF G-2 staff made extensive visits to both French and British headquarters. The subsequent reports echoed the prevailing attitude toward aerial photography's leading role. An AEF Intelligence staff officer (Major Riggs) described the sentiment while visiting the headquarters of the lème Armée in February 1918:

> It was continually impressed upon me by all the officers of the 2e Bureau (G-2) that information obtained from photographs was the most important information obtained and constituted the greater part of the total information used by the Bureau. In fact, I got the impression that information of the battlefront from other sources was considered unreliable unless it could be verified by photographs. The study of photographs was considered all important, not only for information concerning defensive works, but also for information of the enemy's intentions, as deduced from indications of routes of circulation, new camps, etc., in the rear.[766]

Opposite top: Example of French escadrille target folder with photograph and *croquis de reconnaissance* line drawing. (Eugène Pépin Papers, MA&E)

Opposite bottom: High resolution aerial photographs were included in the target dossier. (Library of Congress Brigadier General William Mitchell Papers)

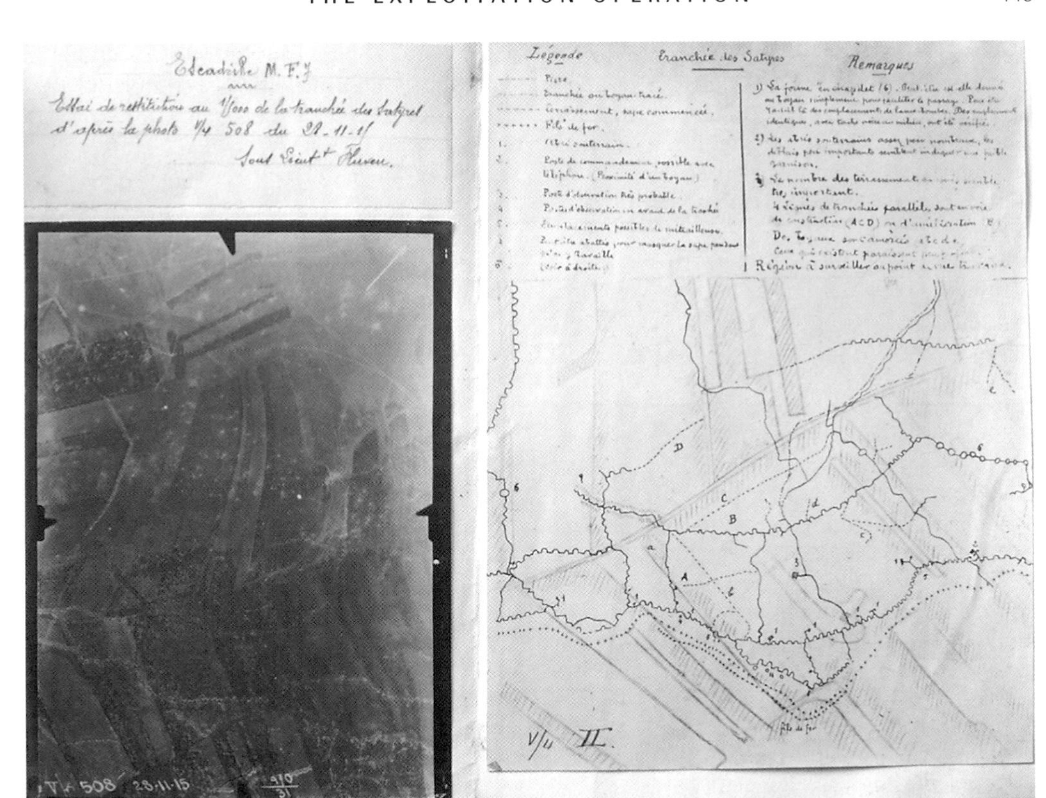
Escadrille M.F.7
Légende
Tranchée des Satyres
Remarques
VA 508 28.11.15
V/4 II.

TOURNES
770-360-F.50
BELVAL
772-360 F.50

The final arbiter for any debate was the GQG's 2e Bureau, acknowledged as the most definitive interpreter of aerial photographs for intelligence purposes.[767] The photographic medium was important, but it was clearly understood by those interpreting the photograph that full exploitation required staff coordination with other intelligence acquired by 2e Bureau.[768] The process was termed 'a drawn summary of a reconnaissance'.[769]

Aerial photographic interpretation was a team effort through three phases of exploitation. The Intelligence Officer usually reviewed one set of photographs, while a draughtsman compared a duplicate set of the photographs with the history of coverage (earlier photographs of the same area) to detect any new works or defences created in the meantime.[770] The exploitation process at both these phases was accomplished by attaching a piece of tracing paper to the photograph and highlighting objects that required further attention. When the first prints were produced in the photograph laboratory, the draughtsman created sketches of key objects on the tracing paper. The sketches were delivered along with the prints. The draughtsman's work was accomplished in coordination with the aerial observer who flew the mission.[771] These 'short notes' attached to the maps also included impressions of the enemy's organisation gained from the study of photographs and of the ground.[772] The process of generating intelligence products, or the third phase of exploitation, employed a broad array of specialists. The main product for the French (and US staffs working within the French sector) was the visual array created from aerial photography. Pépin outlined the list of products in his lectures to the AEF staff, as shown in the table below.

STRATEGIC ANALYSIS

French strategic analysis served as the model for aerial reconnaissance. Weiller Group aerial photographs covering the rear areas deep behind the front lines were printed by a laboratory staffed with approximately 20 photographers. In turn, the prints were interpreted by a team of 30 specialists, trained by the École des Beaux-Arts. Each interpreter was the expert on a certain sector. History of coverage comparison was made with each photograph, noting changes in equipment observed on the ground, shifts of depots and bivouacs, and hospitals being set up in rear areas.[773] The average daily collection was about 1500km^2. Aerial photographs from morning sorties were processed and analysed during the day and reports disseminated in the evening to appropriate commands. Intelligence from the afternoon collection was delivered prior to 0700 hrs the next morning.[774] French strategical analysis became so effective that reports were able to predict with confidence approximately one month in advance the place where the Germans were preparing to attack or abandon a sector. This clearly left behind the traditional reliance on capturing and interrogating prisoners, an intelligence process that usually provided only a 24-hour alert to enemy intentions.[775]

INTELLIGENCE IN RETROSPECT

It became obvious by 1917 that aerial reconnaissance and photographic interpretation had become a critical information source. However, what plagued intelligence analysis was the inordinate weight given to determining German order of battle as the basis for gauging enemy intentions. British operational intelligence emphasised order of battle to determine the German fighting configuration in a given sector. A senior British analyst explained 'As everyone knows, the basis [of intelligence work] is the building up of the enemy's order of battle, for when this has been done the identification of one unit is prima facie evidence of the presence of the division to which it belongs and possibly also of the corps or even army.'[776] This flawed inferential method led to continued frustration for the Allied forces. It took a breakout from the positional war to reevaluate what constituted good analysis of enemy intentions.

Shortly after the war, the French remained the strongest advocates for aerial photography exploitation.

> The data that aerial reconnaissance can furnish and which are the most useful for the study of the intentions of the enemy are those concerning the organisation of the terrain, the batteries, the depots, the aviation fields, the camps, the vehicle parks, the circulation on roads, the intensity of movement on the railroads, the activity in the railroad stations, etc. It is readily seen that the possibility of a minute examination of the photograph of the region flown over permits a much more detailed examination of the ground than the one made in the course of a visual reconnaissance. In addition, the photographic documents give to the information an unquestionable authenticity and great accuracy.[777]

The effort made to acquire and disseminate exploitable information was wasted unless intelligence reasoning was brought to bear upon it. However, such reasoning was imperfect. Allied Intelligence throughout The First World War had mixed results. Detection of critical manoeuvres through aerial observation at the Marne, aerial photographic coverage for Neuve Chapelle, responsive actions at Fort Douaumont following timely analysis of important photographs, and the reaction to aerial photographs that countered German manoeuvres in the final months of combat in Champagne contributed to the final victory. Yet the failures were equally significant. The inability to detect Stollen at the onslaught of the Verdun campaign in 1916 as well as the dramatic, unpredicted breakthrough of Operation *Michael* of 1918 clearly demonstrated that enemy intentions were not effectively determined and acted upon.

With every new dimension of information collection came an effective counter by the enemy through changes in operational procedure or new techniques in camouflage and deception. The environment for intelligence from aerial reconnaissance remained dynamic throughout the war up to the Armistice.

FRENCH PHOTOGRAPHIC INTERPRETATION PRODUCTS[778]

Studies	Results
1. A study of the general organisation of the German Forces.	A coloured schematic map showing sectors, headquarters, supply centres and lines of communication; plus a collection of photographs annotated for each sector.
2. A study of the enemy defensive(perhaps offensive) system.	A map showing the lines of defence, sheltered zones or arms caches and the possible directions of counter-attacks.
3. Investigation of new works.	Annotated sketches and photographs.
4. A detailed study of Infantry organisations.	A large-scale map showing machine-guns, bomb throwers, etc. and annotated photographs.
5. Investigation of battery emplacements.	Description or sketches of batteries.
6. A study of the German Artillery (with its observation posts, telephone centres, etc.)	A map of the artillery showing the principal groupings, etc.
7. A study of sensitive points, of possible targets for our artillery.	Description of targets. A map of the distribution of calibres according to the destruction to be effected.
In times of great activity:	
8. A verification of all destructions.	Rough sketches.
9. Reports of Infantry advance (lines of march, etc).	Sketches.

CHAPTER 10

French Intelligence and Photographic Interpretation

Of the nations fighting on the Western Front, the French had the greatest stake in aerial reconnaissance and photographic interpretation; it was after all their soil that had been invaded and devastated. The number of reconnaissance aircraft produced and operated throughout the war reflected French intention to keep their eyes firmly on the Germans. Further testament to French commitment to the art of aerial observation and photographic interpretation came from the priority placed on reconnaissance following the war. French aeronautical doctrine made it clear that the primary role of the aeroplane was to 'assist the Infantry to conquer and hold terrain. Like the Cavalry, it fulfills its role by fighting and by furnishing information.' The air arm assured security, 'both near at hand and at a distance of the great Units, by furnishing them with detailed information regarding the positions of the enemy, of the arrangement of artillery, of the occupation of camps bivouacs, and aviation terrains, and by informing the chiefs in the course of the battle of the general progress of the battle, the attitude and movements of the enemy, and preparation for counter-attacks. These two types of missions necessitate reconnaissance at more or less long range, at sight, and by photographs.'[779]

For French aerial reconnaissance, command and control commenced with the GQG, commanded by a senior général such as Général Joffre, Général Pétain, and Maréchal Foch. GQG directorates included a 2e Bureau and 3e Bureau (Operations). The commanding général was supported by the Aviation Militaire chef d'état-major for all issues concerning aviation. The Aviation Militaire chef d'état-major commanded all pursuit and bombardment aviation and a certain number of reconnaissance ballon observateur and escadrilles. In addition, he was responsible for logistics supporting aeronautical missions.[780] The Aviation Militaire played an important role in managing the French Intelligence process especially when collection involved aeronautical resources. The Directorate controlled photographic and intelligence services to both 'clarify and facilitate the work' of the reconnaissance escadrilles. The Aviation Militaire photoreconnaissance staffs worked on aerial routes in support of bombardment as well as the assembly of aerial maps of entire sectors. Since the Aviation Militaire was involved in acquiring intelligence, the branch also was responsible for processing and disseminating that information. The primary Aviation Militaire office responsible for managing the

Général Maurice Duval served as *le Chef du Service Aéronautique au GQG* under Pétain and Foch. He led the use of air power thinking in the final year of the war. (SHD, from Christienne and Lissarague, *A History of French Military Aviation*)

aeronautical resources assigned to the reconnaissance mission was the Aeronautical Intelligence Service (*Service de Renseignements de l'Aéronautique* (*S. R. Aéronautique*)). S. R. Aéronautique served as the French authority on intelligence involving aviation, to include German aviation activity, aeronautical establishments, aviation tactics, organisation and material and destruction of organisations. The office also addressed a broad spectrum of intelligence issues such as German troop movements, artillery activity, and determination of extent of destruction. S. R. Aéronautique acquired intelligence analysis from the 2e Bureau to manage the aeronautical resources and to provide refined information to the GQG's 2e Bureau and 3e Bureau's ongoing battle planning.[781] The underlying idea of this structure was that the Aviation Militaire chef d'état-major was the final authority on determining the extent that information was acquired and disseminated from aeronautical sources, to include photography. No document could be printed and distributed without chef d'état-major approval.[782] Cooperation was key to success at Aviation Militaire.[783]

The Aviation Militaire chef d'état-major was the most senior pilot-aviator, chosen to assist the French commander. Général Joffre selected Capitaine Barès as the first chef de l'aeronautique at GQG in 1914, putting 'new blood into an organisation whose principal qualities would be deliberate audacity and intense activity.'[784] His staff included a number of capitaines who were cited as 'a bold young crew with inquiring minds, which constitutes a veritable intellectual centre oriented toward the practical problems posed every day by the air war and the unexpected developments of the ground operations.'[785] Barès was followed by Commandant Paul-Fernand du Peuty as chef d'état-major. De Peuty was an aviator full of enthusiasm for fighting, but lacked the necessary staff expertise. When the Nivelle offensive failed, Général Pétain assumed command and replaced du Peuty with the most dynamic person to occupy the position for the entire conflict, Colonel Maurice Duval. Despite being a certified pilot-aviator, du Peuty left aviation to assume a command in the infantry. He would die at the end of the year leading his men in an assault.

Colonel Duval did not have actual flying experience. Prior to this move, Duval served as commander of a light infantry brigade. Colonel Duval applied an intellectual rigour to the allocation of aerial resources. In his daily routine he made a point of acquiring the viewpoints of a broad range of experts. Duval envisioned that battlefield mobility would return along with a mobile aerial front. Air supremacy would be his constant aim. He directed his organisation with an unwavering will but working towards compromise through persuasion and with a flexible attitude. He would tirelessly examine arguments and demand as many facts as could be found. A series of conferences within GQG included the senior staff. They would air their views and Duval would respond with what he had planned. The meetings were acknowledged as being both very informative and productive.[786] Duval's stature would increase as he continued in his position. He would ultimately be promoted to *aide-major général* (assistant major general).

THE CULTURE OF THE DAILY CONFERENCE

The daily flow of intelligence for the French revolved around key senior intelligence personnel, who discussed every evening key events and the information acquired.[787] The meeting was labelled the 'daily reunion of the 2e Bureau'.[788] The chef, 2e Bureau normally presided over a conference of intelligence officers from all the staffs, to include artillery, aeronautical staffs, radio-telegraphical (*radiotélégraphie*) units, corps radio officers, and sometimes the chef, 2e Bureau. The daily discussion revolved around the update of a variety of Plan Directeur maps that portrayed enemy order of battle and dossiers that described German defensive organisation, artillery, aviation, and mine locations. As the battle became more intense, the frequency of discussion increased between the staffs.[789] A similar pattern took place at all French sectors at the front. The analytical process became so well established within the French ranks that the British became aware and adopted some of the more productive processes. British commentary concerning the manner by which the French established their intelligence role by 1916 referred to an extensive and continuous coordination effort, 'to a certain extent automatically'[790] between the various intelligence branches and throughout the various headquarter echelons. Cooperation was a prominent feature of the ongoing dialogue, thanks in part to the culture that developed from the close proximity of the various offices. It also facilitated rapid tasking and collection of aerial reconnaissance as the situation demanded. A daily dialogue on ongoing operations produced a plan of investigation (a collection plan in today's parlance). The participants worked towards a consensus and the plan was subsequently carried out.

By 1918, American G-2 staff had observed the same pattern. The French VIIIème Armée held daily conferences between the 3e Bureau and 2e Bureau, with the chef, 2e Bureau leading the discussion. The atmosphere was described by an American visitor as enthusiastic. This intimate cooperation was further cemented by the daily schedule of three meals attended by Intelligence and Operations personnel, when the day's work was discussed.[791] Daily conferences with the photographic interpreter also helped develop priorities and an awareness of campaign progress:

> Daily conferences at 2 p.m. are the means by which the Chief of the Bureau keeps in touch with the interpreters as a group and keeps things well in hand and moving. Here each interpreter makes an oral report on what he has done and learned. Many discussions between the Chief and some interpreter or between a couple of interpreters take place. Thus matters of detail are thrashed out, but the Chief with his ever-recurring admonition of 'Hurry Up!' (*activez vous*) kept business moving as expeditiously as possible. When the interpreters are done with their work, they shift to other subjects.[792]

French military intelligence excelled in the daily preparation of the Plan Directeur map. Its well-established process gained the favour of the leading generals. Général Pétain commented in 1917 while reviewing the map production process that instructions regulating Plans Directeur 'have always given full satisfaction; it seems therefore there is no reason to bring any modification to them.'[793] It was a well organised team of experts that produced the French targeting plan. 2e Bureau, S. R. Aéronautique, artillery and other staffs met, providing a definitive restitution on maps and sketches that reflected the most current and agreed intelligence. The intelligence source was also indicated on the Plan Directeur. The whole system depicted successive enemy lines by their respective strength, taking into account any intelligence at hand. Intelligence indications not confirmed by photographs or other precise documents were highlighted by special signs.[794] Aerial photographic interpreters not only helped in the daily production of the Plan Directeur, but also generated written intelligence reports every evening.[795] The Plan Directeur's list of artillery targets was drafted every evening. The process also served as tasking for the escadrilles to accomplish subsequent observation. When conflicts arose over photographic interpretation of a given target area, the chef d'état-major was called in to decide on the final position prior to approving the target list.[796] Draughtsmen were employed to inscribe the latest changes based on the latest intelligence on the master Plan Directeur. At midnight, the completed, up-to-date master was sent to Paris where the Service Géographique completed the printing by the early morning. By 0400 hours, the first thousand copies were delivered to the respective units involved in the operation.[797]

British assessments of the process showed a respect for the way the French attended to detail in the various intelligence elements. In late 1916, maps produced for French operations at the Somme represented a compilation of intelligence gleaned over time. The information was extremely comprehensive, providing the artillery with detailed targeting for the initial offensive. Targets included the most vital points of the enemy's defences (battle headquarters, observation posts, telephone exchanges, supply centres, routes used for bringing up reinforcements). The British commented that the accuracy and efficacy of French artillery 'surprised our opponents'.[798]

STAFF RESPONSIBILITIES

The merging of intelligence was the dominant feature of Plan Directeur map production. At the armée echelon, all 2e Bureau information relative to the German's defensive organisation was forwarded to the *chef d'Cartographie* who was the acknowledged French expert on the sector. An American G-2 staff officer visiting the French headquarters described the chef d'Cartographie as one who clearly understood the battleground, possessing a very intimate acquaintance with the terrain and its organisation along the entire armée sector. The officer personally made visits 'going not only into the foremost observation posts but even into No-Man's Land'.[799] The chef d'Cartographie reviewed all available aerial photographs, prisoner interrogations, daily reports and summaries appropriate for the given sector. Since the crux of the French intelligence effort was aerial reconnaissance and photographic interpretation, knowledge of aerial photography gave the chef d'Cartographie a special stature amongst intelligence counterparts from the other disciplines. The chef d'Cartographie possessed great skill in the interpretation of photographs, scanning large numbers very quickly in search of evidence of new works, various enemy activities and enemy intentions. He directed the photographic interpretation effort within the respective sector. When aerial photography discovered a new battery, railway, or other targets of significance, the chef d'Cartographie personally annotated the master Plan Directeur. Other intelligence sources such as prisoner interrogations were validated by him through other interpreters or through personal interrogation of the more useful prisoners. Again, intelligence gleaned from these sources was annotated on the Plan Directeur. Daily reports from the array of functions at the front, intelligence summaries and miscellaneous sources of all sorts were transferred onto the Plan Directeur. Once the chef d'Cartographie approved, the Plan Directeur was then turned over to the draughtsmen along with final verbal guidance to ensure the final map portrayed the exact intelligence overview.[800]

One of the offices involved in the development of the final Plan Directeur was the Army Topographical

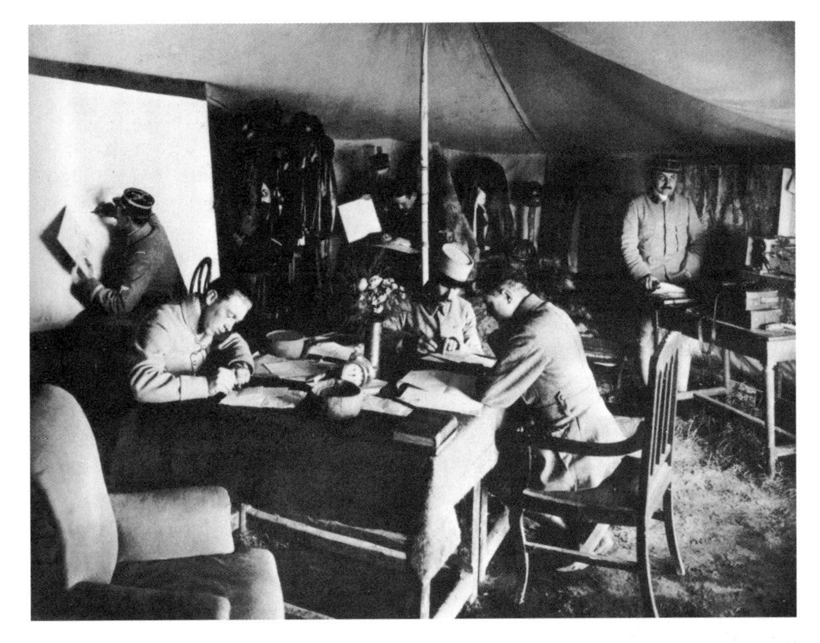

French 2e Bureau analysis and reporting underway. Note the Plan Directeur map is being updated. (BDIC, from Ezra Bowen, *Knights of the Air*)

Groupe de Canevas de Tir des Armées staff at work near the Verdun battleground. (BDIC)

Section, known as the *Groupe de Canevas de Tir des Armées* (GCTA).[801] The GCTA received the results of the deliberations amongst intelligence seniors and conducted precise and final restitution of the aerial photographs being applied to the Plan Directeur. A cell of officer photographic interpreters provided quality control for work furnished by the 2e Bureau of the subordinate CA concerning targets deep in the rear zone of the enemy's front, to include determining withdrawal positions, communication nodes, targets for both artillery and air bombardment, military establishments, and other areas of interest. This information led to the final restitution of the information onto the artillery-firing map. Pétain made it clear in his 1917 instruction that the GCTA 'should be entrusted with the drawing of all maps for which a certain amount of accuracy and technical knowledge are required.'[802]

Assisting GCTA in restitution topographical work accomplished at echelons below the GQG and French armée were the offices of the *Service Topographique Corps d' Armée* (STCA) and Divisional Topographical Section. The French intelligence architecture was repeated at the CA level. The CA Intelligence Service comprised the 2e Bureau, Artillery Intelligence Service (*Service de Renseignements de l'Artillerie* (SRA) French designation; AIS, British designation), S. R. Aéronautique and the Radio-Telegraphical Intelligence Service. Each CA element was responsible for their respective sources of intelligence. They worked together on synthesising key elements of enemy activity, to include the order of battle, troop movements, situation of enemy artillery, defensive organisation, destruction, aeronautical activity, etc. CA information resources such as reconnaissance escadrilles forwarded their information to the 2e Bureau.[803]

The CA chef d'état-major assigned primary responsibility for fusion of the respective intelligence sources to the chef, 2e Bureau des CA, who in turn directed the research and ensured coordination among the members. The officer in charge of STCA worked under the immediate supervision of the chef, 2e Bureau des CA. The STCA served as an intelligence hub in its own right, acquiring information from headquarters, Aviation Militaire and the artillery staffs. Within the respective CA sector, aerial photographs relevant to Corps' objectives were processed by the STCA for restitution.[804] This office also applied topographic details. The STCA sometimes produced more timely maps to reflect the current battle situation. By 1916, cartographic information portraying enemy defences was being produced on an hourly basis.[805] It remained a very dynamic process as the STCA interpreted aerial photographs, revised artillery plans and drafted special plans as ordered by the chef d'état-major. The STCA was usually co-located with the 2e Bureau offices near the landing ground of the assigned escadrilles, providing a possible model to the British for the Branch Intelligence Officer (BIO) function at the Squadron. The British saw great value in the STCA being close to the front where the escadrilles operated. This avoided crowded offices at headquarters and facilitated the aerial photographic interpretation process. One British visitor commented on the STCA's operation in glowing terms. 'At Verbery the arrangements are for general photographic work in connection with distribution mapping, and analysis of photographs for the Army Corps. Here the mapping was splendid and carried out with the greatest care for detail. The work was very strong in stereo photography.'[806]

INTELLIGENCE FOR ARTILLERY

The French SRA centralised all intelligence concerning artillery targets. In practice, the officers of this branch dealt almost exclusively with the enemy's batteries and observation posts. In several CA, the SRA functioned like the 2e Bureau, operating near the landing grounds of aerial observation escadrilles. The SRA further studied batteries and associated target sets such as cantonments, bridges, stations, depots of munitions, etc. SRA officers were responsible for working out their own means of obtaining intelligence or using means placed at their disposal such as SROTs, artillery spotting, sound-ranging and observation sections. The ongoing interpretation of photographs stopped only after approval of the chief of the 2e Bureau. Additionally, the SRA assisted the French batteries in determining accuracy of fire.[807] Sound-ranging and observation sections were armée-level responsibilities, but they were authorised to support the CA with intelligence directly emanating from the assigned sector. The SRA personally ensured the rapid exchange of intelligence between the various artillery sections.

Restitution of aerial photographs noting enemy defences in the sector was a daily event.[808] The SRA attended the daily 2e Bureau des CA conferences to assist in the daily compilation of battery target lists. Once the targets were selected, they were forwarded to higher headquarters offices for validation. Despite

Service de renseignements de l'artillerie (SRA) provided intelligence support to the artillery. The aerial photograph was used to further determine German artillery locations and associated target locations. (Eugène Pépin Papers, MA&E)

their intimate involvement with the targeting process, the CA staffs were considered incapable of calculating target coordinates with sufficient accuracy. The armée staffs replied as soon as possible with final coordinates to allow the CA to construct their final battery list for their own artillery units. In turn the armée forwarded the definitive target locations for publication with the armée's SRA Summary. This helped avoid duplicate target assignments.[809]

At the end of every day, especially during active periods, the process for producing the Plan Directeur helped both observation and artillery to prioritise the extent of destruction to be accomplished in the next barrage. Artillery planning also took into account, from aerial photography, what shots had struck outside of the planned destruction zone.[810] Artillery targeting information was extensive. CA SRA received various reports from intelligence sources throughout the sector, to include sound-ranging and observation Sections. A daily battery list was prepared that provided coordinates of batteries (in action) and confirmed locations of enemy observation posts. Diagrams were then constructed showing the general disposition of the enemy's artillery.

The common cartographic baseline also served as the foundation for planning an offensive operation. French assault unit commanders used the Plan Directeur document to issue detailed orders 'with all desirable precision'. The Plan Directeur met the map requirements for 'artillery objectives which appear to be specially signalled on the Division front'. The Plan Directeur also enabled the various Artillery Commands to set up 'action plans'.[811] In the period of preparation for an offensive, rough sketches accomplished by the aerial unit (escadrille or aérostat) were drawn from the aerial photographs and provided expeditiously to the 2e Bureau offices. After examining the photographs, the CA compiled reports of artillery and aerial observation, and sketches of the new battleground. They were edited by the STCA and distributed the very same night to artillery staffs down to and including Group Commanders.[812]

French 2e Bureau des Armée staff interaction with the CA 2e Bureau was carried out through orders, instructions and personal liaison between the officers of both echelons. Staff visits between the two headquarters were frequent. When a CA was assigned to a certain armée, the chef, 2e Bureau des CA was personally attached for two or three days to the 2e Bureau des Armée's to help personalise the staffs' relationship. When the time came for the CA to proceed to the front line, the CA staffs were familiar with the armée's headquarters methods and procedures.[813]

The French intelligence architecture maintained a similar structure to support divisional operations. The division's 2e Bureau officer worked under the direction of the 2e Bureau des CA. The pattern of interaction repeated itself. French division staff coordinated with several offices to concentrate on key issues concerning the division's sector. Coordination was accomplished with the SRA, the division's own Topographic Service (*Section Topographique de Division d'Infantrie* (STDI)), STCA, and S. R. Aéronautique. At the division, the photographic interpreter examined front-line enemy organisations and tabulated all information supplied by the troops in sector and maintained frequent liaison with the unit intelligence officers at the front. Reports of patrols, prisoner interrogations, verbal reports from listening posts, repatriate information, and aerial photographs taken by the CA and armée escadrilles flowed down to the division as appropriate. The French division also employed a photographic interpreter to work for the 2e Bureau and Topographical Service. Staff instructions to the division included directions to ensure all divisional elements were provided with maps, sketches, photographs and Plans Directeur. The French emphasised that no information furnished by any source was in itself complete. All information received was to be compared with the information of the same nature already at hand and with the general situation already known.[814]

THE NETWORK OF FRENCH AERIAL RECONNAISSANCE

French aeronautical doctrine emphasised the primacy of timely reconnaissance and analysis. At the moment of the attack, information required immediate transmission and circulation to the infantry commanders. Aérostats, aeronautic liaison and observation aéroplanes were three main resources to accomplish this. Order of battle resources for the two phases included an infantry aéroplane (*avion d'accompanement de l'infanterie*) assigned to each infantry division and an observation aéroplane assigned to each AL group. Additional aéroplanes were assigned to provide relief and more coverage during observation or fire missions. Relief sorties reinforced ongoing fire or observation missions as well as substituting for those aéroplanes already committed. Reconnaissance in these two phases required definitive coverage. 'When beyond the zone of fortified positions, the Army Corps aeroplanes endeavour to see and to photograph the position of the supporting troops and they observe movement in the rear'.[815] Any photographs taken by the escadrilles were interpreted by the various services at the landing ground site.

CA escadrilles accomplished aerial and photographic reconnaissance for the CA as well as registration for any special batteries such as *batteries de position* (batteries left in quiet sectors without motor transport or escadrille support) or batteries placed at the disposal of the CA. The escadrille supporting the divisions provided *avion de réglage* support for divisional artillery, contact patrol and reconnaissance and photography. SAL escadrilles working with heavy artillery batteries accomplished registration-only missions.[816] The size of a French escadrille in the war varied, but an ideal complement was ten aéroplanes.[817] In 1918, RAF reporting on French Corps Aviation showed at least 45 aéroplanes assigned throughout a typical CA echelon. Within the CA, a nominal count of 10 aéroplanes comprised a CA-subordinate escadrille with the specific responsibility of acquiring either aerial observation or photographic reconnaissance. Additionally, the escadrille would provide registration for any special batteries assigned to the CA. Finally, the escadrille would support its division with additional reconnaissance. For every two divisions that were organic to the CA, an additional escadrille with a nominal count of 10 aéroplanes provided immediate support. When additional divisions were added to the CA, each acquired its own escadrille complement of 10 aéroplanes. An artillery-spotting escadrille with 15 aéroplanes was also assigned for every AL regiment. This unit would only provide registration for the batteries.[818]

The French employed armée escadrilles for aerial observation and photographic reconnaissance, air raids and pursuit. Orders for reconnaissance sorties and air raids were drafted by the 2e Bureau des Armée and signed off by the armée chef d'état-major. For operations in support of the CA, the CA chef d'état-major approved reconnaissance missions generated by the 2e Bureau.[819]

On the ground, telephones connected the CA Service aéronautique to the various headquarters in the sector. If telephones failed, the operation had access to TSF and aerial message bags dropped from available aeroplanes. The French kept horses at advanced landing grounds so that the observer could immediately ride off and make his report at Headquarters.[820] The French maintained a procedure for tasking escadrilles for aerial reconnaissance and photographic missions. The 2e Bureau determined from the 3e Bureau, SRA, S. R. Aéronautique and other units what information was required to meet their objectives. The 2e Bureau forwarded their reconnaissance plan to the respective commander of the armée or CA service aéronautique. Mission priorities were then determined and photographic assignment took place. The French photographic observer assigning the mission was responsible for knowing the exact targets and region requiring aerial photographic coverage. The observer had to know the special purpose that the photograph served so that the mission planning best served the scale and final layout of the master Plan Directeur. Furnished with this information, the French observer carefully prepared the reconnaissance, and coordinated with the chef, SPAé, to ensure the correct camera with the right focal length and magazine was installed and operating. The observer then worked with the pilot on flight patterns for the mission.[821]

Since the strategic intelligence obtained by the armée service aéronautique was considered less perishable than that supplied by the CA Service aéronautique, the French did not always commit to having an intelligence office at the landing ground. An officer of the armée service aéronautique staff, whenever possible the 'Operations' officer, was assigned the duty to check intelligence obtained by reconnaissance and to forward it to the 2e Bureau. In case of time-sensitive situations requiring immediate information, the officer was considered the appropriate person to get the information to those responsible quickly. Photographs taken on reconnaissances were immediately developed and proofs distributed as soon as possible. The armée service aéronautique not only sent a preliminary report (or print) of intelligence obtained but also thoroughly examined the sources of information and produced a Daily Summary of Intelligence illustrated by sketches giving deductions drawn from the information gathered during the day. The armée service aéronautique also assisted SRA with aerial photographic interpretation to update battery index cards.[822]

The 2e Bureau of the armée service aéronautique sent a copy of all photographs acquired from the reconnaissance sorties to the respective armée or 2e Bureau des CA, SRA and to the group assembling the Plan Directeur. The interpretation conducted at the escadrille included the analysis of the aerial observer. Each photograph was reviewed and used to modify and annotate the current interpretation as required. The chief of the 2e Bureau tactical assistant and officer in charge of photographic interpretation made the final interpretation. High interest items found from the photographs were circled in red ink.[823] At the CA, drawings were made of changes and improvements in the trench system and reproduced for issue to the staff and to various units in the CA. Copies of all photographs and drawings were sent to the higher echelon armée staffs for further verification and study in the 2e Bureau and also for use by the Canevas de Tir for keeping Plan Directeur maps up to date.[824] The French also employed the *Bureau De Cartographie* to add all special information on the battle maps. Special information included detailed and confidential information relative to the location of German batteries, minenwerfer, machine guns, ammunition and supply dumps, means of transportation and communication, dugouts, rest camps, railheads, hospitals, and dressing stations.[825]

The entire intelligence process made a marked impression on those who witnessed it first hand. One French aviator recalled, 'The reading and interpretation of the photographs has to be done immediately, for it is of the greatest importance that the information given by the photographs be obtained quickly.'[826] The work routine embodied a philosophical perspective on the fighting life in general. 'All the work in a photographic reconnaissance should be done on the same day the reconnaissance is made. The officer should take no rest until everything is finished and the prints sent off, for example, by motorcyclists. It is not uncommon for photographic officers to go several nights without sleeping. What difference does it make? They know the importance of their work and should perform their duties with devotion. When rain comes (and it comes often in France) they will rest.'[827] At the CA echelon, French escadrille photographic work was centralised by the SPAé Photographic Officer. He was assisted by observers and was responsible for the development, printing, interpretation and distribution. During a battle, the CA service aéronautique commander attached one of his staff officers to the headquarters of the respective CA to assist in providing the information in a timely manner.[828] Escadrille members were encouraged to participate at daily meetings conducted at the landing ground. It fell to the 2e Bureau to coordinate the inputs from all in attendance. The professional photographic staffs from attending British and American forces had the highest regard for the French process of plotting, translating and interpreting photographs. Steichen's earliest impressions were forwarded to the Signal Corps Photographic Division: 'It is a most excellent system and you may remember it was referred to by Major Campbell as being very much better than the English system.'[829] The Americans inherited the French system when they took over a sector at the front.

THE FRENCH SPAÉ

French intelligence expanded its presence throughout the force by the continued formation SPAé to support aerial photographic requirements. The originator of France's SPAé, Capitaine Grout, helped expand the awareness of the forces as to the potential of aerial photography by encouraging technical photographic expertise in the commands and units. The continued demand for aerial photography prompted the acquisition of more photographers and draughtsmen. A vigorous recruitment drive for such skills took place, seeking those trained in the art amongst graduates of photographic or drawing schools, and aspiring amateurs. The services rendered by the SPAé des armée were considered so significant that the French commanders made the decision to equip CA escadrilles, escadrilles assigned to divisions, and Heavy, Day Bombardment escadrilles with SPAé.[830] French SPAé became integral to the reconnaissance escadrilles assigned throughout the French armée. In the last year of the war, the French further institutionalised photographic support by assigning photographic officers to both armée and CA headquarters. The photographic officer at the armée echelon coordinated SPAé functions pertinent to that echelon. At CA, the photographic officer directly supervised the SPAé's assigned both directly to the CA and to the infantry divisions that were CA-attached during the course of operations.[831] Key SPAé personnel included the Officer Tactical Photographer who preplanned the mission with the aircrew prior to take off. The Technical Officer was the senior enlisted who ensured the laboratory and other functions were accomplished successfully. This warrant officer was considered the heart of the SPAé, providing the wisdom of years to the younger members of the operation. Other members included draftsmen and specialists in photographic maintenance.[832]

Members of a French photographic laboratory looking at a recently processed photograph. (SHD)

SPAé laboratories were completely equipped to handle a range of tasks, to include loading and downloading photo camera magazines and plate holders, developing and drying plates and prints, recovering various chemical products and enlarging and projecting prints. The photographic laboratory drawing room was also responsible for marking of photographs after identification; constructing the map of the areas covered; constructing sketches of the reconnaissance; assembling photographs to make a mosaic and mounting of stereoscopic views.[833] Post-mission processing of information from the aircrew and cameras was impressive. 'Two hours after the Observer photographer has returned to the aviation camp all the pictures he has taken should be sent in all directions, to all departments, regiments, batteries, and the Staffs that need them. In the French Army they make about 80 copies of each picture, and each Observer always brings in 20 to 30 pictures ... It is not uncommon for a squadron photographic section to make 1,000 prints a day; to accomplish this rapid development and printing, each French squadron has a photographic section composed of specifically trained men. Automatic machines develop and print the photographs.'[834] Once aerial photographs were processed and interpreted, the French took the photographic proofs and inscribed on the left side of the negative or on a thin sheet of gelatin fixed on the negative a summary of their interpretation.[835] Negatives were filed for ready access. A British visitor noted that 'The negatives are all carefully catalogued, being kept in the original plate boxes.'[836] In addition to volume, the speed of production and timely reporting was noteworthy. 'In some cases, for example with contact patrols, we have to get out our photographic intelligence with the greatest speed. The French photographic service uses a process of developing, printing, and drying which produces one or two prints on paper, completely dried, ten minutes after the photographic aeroplane has come down.'[837] The SPAé also forwarded the necessary photographs to the STCA and the SRA and assisted in the final development of the Plan Directeur.[838] As part of their role, the SPAé maintained the necessary cameras for the aeronautical units, the standard inventory being 26cm, 50cm and 120cm cameras.[839]

Intelligence distribution was well established by 1916. Photograph distribution was equally prolific. A representative example was the French VIème Armée distribution. CA escadrille photographs were selected by the commander and distributed by the Service aéronautique. Distribution from CA to armée included ten non-interpreted prints to the armée's key staff to include, 2e Bureau (1 print), 3e Bureau (1 print), Canevas de Tir (2 prints), Artillery (1 print), Heavy Long range Artillery (1 print), Engineers (1 print), commander of the service aéronautique (1 print), *Groupe des Armées du Nord (GAN)* (Group of Northern Armies) (1 print). A follow-up distribution of more detailed interpretations was repeated to those offices that required specific attention to detail, to include 2e Bureau (1 print), Canevas de Tir (1 print) and service aéronautique (1 print). Aerial photographs acquired by armée escadrilles had 12 prints generated for the staff and to the Groupe des Armées du Nord. At the CA level, distribution included prints for the staffs of the 2e Bureau, CA, Artillery and the respective division.[840]

The French armée air park provided rear-echelon logistic support to SPAé. The facility possessed a precision mechanic and a cabinet maker to repair material and inspect new material. The storehouse was supplied with all photographic material such as plates and paper. SPAé commanders were supplied directly from the armée air park directly under the control of the Assistant Photographic Officer of the Armée.[841]

Photograph distribution included a broad range of military organisations and schools. The *Service Entrepôt Special d'Aéronautique* No. 2 at Nanterre assembled and classified the respective photographs to facilitate the range of training requirements for the French military. When the US entered the war, aerial photographs were forwarded to the schools at Cornell University and Rochester, NY. In the immediate post-war era, the photographs were used extensively by the various commissions assigned to conduct the peace treaty and to determine the extent of damage in the liberated regions.[842]

The French led all combatants in the First World War in developing an effective reconnaissance mission. In the following years, they emphasised reconnaissance as an equal to the evolving roles of pursuit and bombardment. By the late 1930s the French Air Service was called 'Information Aviation', with two primary classes, 'Observation Aviation and Reconnaissance Aviation'.[843] The intelligence process thus maintained coherence with aviation's capability. Pursuit continued to ensure freedom of action for observation missions. French aviation continued to think defensively, based on the prevailing influence of positional war and their existing fleet of aéroplanes, which were limited in range and strike power.[844] It was the continuation of the defensive mindset of the inter-war years that hindered the understanding of enemy intentions. Innovation in reconnaissance did little to overcome shortfalls in French military strategy in the inter-war period. Even when capable of acquiring information on the enemy, the French military did not meet the challenge of new concepts such as rapid mobility with an appropriate counter. It resulted in the country's capitulation under the onslaught of the Germans in 1940.

THE ANALYST: CAPITAINE EUGÈNE MARIE EDMOND PÉPIN

Capitaine Eugène Marie Edmond Pépin (27 June 1887–28 April 1988) was one of the greatest pioneers of aerial photographic interpretation in the First World War. His service to France and her Allies made aerial photographic interpretation an institution for military operations in the twentieth century. He was a man of great diversity and vision, an expert in architectural art, but best remembered for breaking new ground through aerial photography and in the application of law to space travel and exploration. Pépin came from an academic background and applied his brilliance to development of the most important intelligence source of the war. His ability to weave German positional war doctrine and military policy into the interpretation of the aerial photograph advanced the role of aerial photography as a fundamental tool of intelligence.

Pépin was exposed to the leading edge of intelligence and academe from the start of his military career. While serving compulsory military duty in 1907, he was introduced to new sources of intelligence, including the rapidly evolving science of intercepting wireless telegraphy transmissions. Following this initial period of military service, Pépin pursued a graduate degree at the Sorbonne in legal history and international law. His academic credentials enabled him to become a leading figure in the new field of international law as applied to aviation. In 1910 he served as secretary to the first international conference on aerial navigation. During this period, Pépin met James Brown Scott, a legal scholar who established the foundation for modern international law. Pépin's network of contacts in this evolving arena reflect his intellectual acumen and understanding of the complexities associated with the phenomenon of aviation.[845]

The French aerial photographic interpreter giants of World War 1, Capitaine Jean de Bissy and Lieutenant Eugène Pépin, prior to a sortie in an AR 1. (SHD, 76/1168)

Pépin Portrait. (Pépin Papers, MA&E)

When war broke out, Pépin was commissioned and assigned to aviation, where his intellect was quickly applied to supporting the expanding role of aerial reconnaissance. The influential Lieutenant Maurice Grout acquired Pépin and had him assigned as Chief of the *IVème Armée* SPAé. As the war progressed, he soon became responsible for eight French armée SPAés.

Pépin applied his academic brilliance to refining aerial photographic interpretation and moving it from the realm of novelty to becoming a definitive science. His efforts helped establish French aerial photographic analysis as the intelligence standard among the Allies. His knowledge soon drew the attention and approval of Général Pétain, who assigned him to his staff to provide aerial photographic interpretation expertise for campaign planning and to train French officers in aerial photography. He was instrumental in introducing intelligence into mainstream military thinking. Amongst his achievements, he assisted in the development of French policy relating to camouflage. In August 1917, he served as the GQG 2e Bureau's representative to the *Commission Centrale de Camouflage*, where he personally developed the primary instruction manual for French camouflage and aerial photography.[846] He also led the development of the French Armée's aerial photographic interpretation manuals, including *Étude et Exploitation des Photographies Aériennes* and a translation of the US manual, *Study and Use of Aerial Photographs*. Both served as the standard by which aerial photographic interpretation demonstrated its utility as an intelligence medium. His written description provided a comprehensive assessment of the visual configuration of the German battleground network. It was one of the most extensive and unique first-hand descriptions of warfare at that time.[847]

When the Americans arrived in 1917, Capitaine Pépin was personally assigned by Général Pétain to the newly established French Mission that supported senior AEF staffs with in-depth knowledge on how to conduct modern positional warfare. In September 1917, Pépin became the personal advisor to General Pershing on aerial photographic interpretation. He continued to support Allied planning throughout 1918. During the lead up to the Meuse-Argonne offensive, he was assigned to the Enemy Works Subdivision of the Headquarters First Army, providing comprehensive interpretation of the battlefield.[848] When the Meuse-Argonne campaign commenced, he was a leading analyst supporting American headquarters assessments of threat locations.[849]

At the Armistice, Pépin's expertise was sought to help draft the peace treaty with Germany and its allies. He was reunited with James Brown Scott in the drafting delegation that defined the borders of several new nations. His work in the aeronautical arena continued as well, through leading several forums that addressed the evolving role of aerial navigation.

In the inter-war years, Pépin wrote several books describing the culture of the France he knew and loved, including an in-depth review of cathedrals in the Loire River valley. These were a direct reflection of his mastery of aerial photography, covering the terrain in detail, while expounding on the cultural and historical significance of the subject matter. In his work on the cathedral at Chinôn, he not only wrote the book, but also took the photographs.[850]

Pépin was also a leading figure in French international law during the turbulent inter-war period, serving in the French foreign ministry throughout the 1920s. Through his position in the ministry, he became actively involved in Latin American issues. Later, in the 1930s, he extended his interests to Asia, where he served as a legal councillor for the Japanese foreign ministry (*Gaimusha*). In that capacity, he served on the Lytton Commission investigating the disagreement between Japan and China over Manchuria. When the Second World War started, Pépin again served with the French Army as the senior officer in charge of aerial photography. After Germany defeated France, he was retained in the government as a director of the Economy Section, evaluating the consequences of the German occupation on the French economy.[851]

Following the Second World War, Pépin became a renowned leader in the field of international civil aviation and aviation law. He became general counsel for the French government's aerial transport office and served as General Secretary for several international law conferences addressing the role of modern aviation.

In his later years, he served as Director of the Institute of International Air Law at McGill University in Canada. In this capacity, he took on yet another pioneering role in jurisprudence relating to the evolving world of aerospace. His judicial formulation was seminal in determining the extent of national boundaries. His analysis uncovered gaps that existed in international regulations governing air traffic. In the age of Sputnik when legal experts in aviation and aerospace law had to come to terms with the technical advances that advanced the exploration of space, Pépin's analysis addressed the new role of satellites revolving around the earth. As a result of Pépin's leadership, an international body of law recognised the sovereignty of a nation as extending only to the atmospheric space above its territory.[852]

Pépin in later years. (Pépin Papers, MA&E)

CHAPTER 11

British Intelligence and Photographic Interpretation

ESTABLISHING AN INTELLIGENCE LEGACY

One of the greatest legacies of the First World War was the ability of the French and British to develop parallel, complementary intelligence organisations. The years of conflict helped erode many old differences as dependency increased in order to ensure each Allied partner held up its side of the front line. Both nations were expending their human capital at a terrible rate in a long series of offensives with marginal gains. The seemingly endless campaigns prompted the two Allies to consider less costly ways to conduct the war. Greater attention to intelligence not only aided the commander in the conduct of the campaign, it also saved resources. For Great Britain and her Commonwealth Allies, the organised intelligence network that arose from these years of conflict created the primary infrastructure that served their military for the remainder of the century.

British intelligence was a well organised and disciplined formation designed to serve the BEF commander and operations. Like the French, the British viewed aerial photography as an invaluable source to validate information acquired through other means. What differences existed between the two Allies mainly concerned the final product. French attention to cartography was loosely copied by British forces in the production of their own Firing Map. However, the British paid greater attention to Intelligence Summaries to describe what the aerial photograph revealed. Lack of total agreement on the best means to accomplish the intelligence mission could be summed up in the basic differences in the culture of the two Allies. They worked together as well as could be expected, but each had their own views. One British senior, AVM Brooke-Popham, remarked on this in his post-war notes:

> I think one sees the difference between these methods in the two services even today. The French individually more brilliant but relying on a few star pilots to win the war and constantly failing for want of discipline and organisation; the British standing on a firm foundation unmoved by temporary set backs or by drastic reduction, able to face with calmness an indefinite expansion.[853]

THE GENESIS OF MODERN BRITISH MILITARY INTELLIGENCE

The most notable legacy of British military intelligence was the role established by Lieutenant-General Sir David Henderson in making the discipline an integral part of the modern British military.[854] Testament to the effectiveness of Henderson's aerial reconnaissance process came from American observations. Early in 1915, the American attaché to London, Colonel George Squier, reported:

> The presence with the British Army of an aeroplane squadron assigned to each Corps, and in addition a squadron and an aircraft park at GHQ, places in the hands of the Intelligence Service one of the greatest aids to obtaining information, which as yet has been developed. Aeroplanes are now regularly used, not only for obtaining information concerning the enemy, and his dispatches, but also in cases of a large Army on the move, [where they] are a very efficient and ready means of furnishing the Commanding General with the tactical positions of his own troops as well. All reconnaissance reports from aeroplanes are sent immediately to the Intelligence Officer of the different Corps, and generally the aviator reports in person with his report, in order to be interrogated by the Intelligence officer. These reports are forwarded from the different Corps without delay to the Director of the I.D. at GHQ.[855]

By 1918, Henderson's intelligence process was fully integrated into the framework of military operations. US Army liaisons sent to the various British staffs to review their intelligence process showed the depth of analysis well established throughout the front. Major

David M. Henry, AEF G-2 staff, reported on the network of information collection and analysis:

> Before my visit to the British Front I had heard often, as everyone has, many statements of the amount of Intelligence obtained by reconnaissance and aerial photography, and I accepted those statements, as they were made as a matter of course. To what extent Intelligence is thus obtained I do not believe anyone can understand fully who has not seen the actual work of collecting it as it is done by the Squadrons day by day, and the extent to which those whom it concerns depend upon it for the information necessary to enable them to carry on operations. There seems to be no development in the Enemy's system of defences that is not both quickly observed and photographed. From the front line trenches back through the counter-battery area there are thus made known the location and formation of all trenches, machine-gun emplacements, fortified shell holes, tank traps, dumps, trench mortar positions, anti-aircraft defences, artillery pits, wire, dug out positions, roads, paths, everything in the nature of defensive construction or preparation ... As this Intelligence is collected day by day changes and new developments in connection with all of these things are quickly shown. In no other way would it be possible to obtain a large amount of intelligence, and probably none of it could otherwise be so accurately and so thoroughly obtained.[856]

Henderson's assignment as an acknowledged expert in military intelligence to lead the RFC proved fortuitous. His impact on determining what constituted successful reconnaissance in war laid the groundwork for RFC achievements in the opening salvos and remained a noteworthy influence for the rest of the war.

THE INTELLIGENCE HIERARCHY

British intelligence was active both in London and with the GHQ in France. The most senior British intelligence offices were in London at the War Office. A distinguished colleague of Henderson was Lieutenant-General Macdonogh. Macdonogh served as DMI following his term as Field-Marshal French's senior intelligence person. French had the highest regard for Macdonogh despite the altercation at the Battle of First Ypres:

> Our intelligence service had been admirably organised, and was working most effectively under the able direction of Brigadier-General Macdonogh. I cannot speak too highly of the skill and ability displayed by this distinguished officer throughout the whole time during which we served together. His service was invaluable; his ingenuity and resource in obtaining and collecting information, his indefatigable brain, and the unfailing versatility and insight with which he sifted every statement and circumstance were beyond all praise. He trained an excellent staff, who valued his leadership, for he had an extraordinary power of getting the most and best work out of every one. His information as to the enemy's movements was remarkably accurate, and placed me throughout in the best position to intercept the enemy's probable intentions.[857]

Lieutenant-General Macdonogh proved to be one of the most brilliant British generals in the war, conducting an array of intelligence operations against the Germans that gleaned many key strategic intentions. Unfortunately, Macdonogh met resistance from the British General Staff, particularly Field-Marshal Haig. Macdonogh's key intelligence arm, known as the Secret Service, had determined with certitude the date, time and location of the 1918 *Michael* offensive. Despite the information, the BEF did not reinforce those sectors that bore the brunt of the attack. The Channel ports to the north remained the primary concern. Additional troops were being withheld by the British Prime Minister, Lloyd George, who feared that Haig would initiate another campaign of attrition, such as had occurred at Passchendaele in the Ypres sector the previous autumn.[858] On a personal level, Macdonogh had to bear the brunt of the senior British military staff's bias as it concerned his religious affiliation. Macdonogh was a recent convert to Catholicism. Key antagonists included Haig and Charteris.[859]

General Haig's intelligence chief, Brigadier-General Charteris, managed intelligence from 1915 to December 1917. Charteris did succeed in institutionalising aerial photography within the BEF, but overall, he was less than exemplary in his management of British intelligence at the front. General James Marshall-Cornwall, who served on Haig's staff, wrote 'I soon discovered that the views held by Charteris, and reported by him to Sir Douglas Haig ... differed widely from the estimates made by the Director of Military Intelligence at the War Office.' It was felt that Charteris deliberately misled Haig. According to Marshall-Cornwall, Charteris 'honestly thought that by suppressing all pessimistic evidence about enemy potential, moral and physical, he was strengthening his chief's determination to win the war.'[860] Lloyd George shared Marshall Cornwall's view: 'General Charteris, who was an embodiment of the Military Intelligence which he directed, glowed with victory. For him the news was all good.'[861]

The leading British intelligence officer of the First World War was Lieutenant-General Sir George Macdonogh. Macdonogh is seen in a post-war photograph with Maréchal Foch. (IWM)

Despite the problems associated with Charteris' tenure with the BEF, he formalised the intelligence structure, installing an Intelligence Corps that led to new areas of military operations, to include tank units and expanding use of radio intercept.[862] Charteris also established the British infrastructure for aerial photography. His initial staff comprised two regular army officers and three wartime officers; all of whom developed skills in the arena of photographic interpretation. One of these wartime officers, Captain Carrol Romer, a barrister and surveyor, devised the first rules for scaling air photographs. He is regarded as the first British photographic interpreter.[863]

THE BRITISH INTELLIGENCE ARCHITECTURE

GHQ's intelligence was disseminated through two intelligence products: the Daily Intelligence Summary and the Daily Summary of Information. Both were geared towards providing relevant intelligence to the Commander-in-Chief. GHQ used these reports to support ongoing tactical and strategic planning. Aerial reconnaissance and photographic interpretation reports received at GHQ were also forwarded to General Haig.[864] At the front, special functions were established to support intelligence processing. For example, the British Commonwealth (Canadian) staff was supported by an aeroplane photo man (a photograph interpreter or photographer) to assist in the generation of their Intelligence Summary.[865] The GHQ also established a special RFC unit to acquire intelligence. Known as the 9th Wing, the organisation operated independently of the squadrons in the RFC Brigades. The 9th photographed and reported on key strategic German targets such as aerodromes, railway movements and made 'special observations'.[866]

At GHQ, Section I-a-3 was established to monitor the zone immediately in the rear of the German army, generating a report on enemy intentions for the Chief of Intelligence. The section collected statements of repatriates, spies, prisoners, army observation reports and aerial photographs. Aerial photographs of the towns and adjacent areas were compiled and made

Captain Charles Romer worked on Brig-Gen Charteris' staff and is remembered as the first British photographic interpreter. (IWM)

available immediately for review. Train movement was monitored extensively to determine front sustainability.[867] British intelligence also created a network of prisoner interrogation centres. A Prisoner-Examining Officer was appointed at Corps to work with prisoners contained in 'cages'. Further examination of key prisoners was accomplished in those cages. The average interrogation time was ten minutes. If the situation warranted, interrogation was extended. Information from prisoners was subsequently collated and disseminated to all parties concerned. A Document Officer was appointed to manage all captured material. Extensive intelligence was acquired from letters, postcards, and printed matter such as orders confiscated during prisoner interrogation.[868] In 1917, the RFC adopted the Brigade formation to better support the British Army with aviation resources, to include increased aerial photographic coverage. The RFC Brigade comprised an Army and Corps Wing to provide aerial support.[869] The RFC Brigade in 1918 showed the expanding influence of aerial photography. Five out of the six squadrons supporting the Corps Wing included photographic personnel.

Within the Brigade the RFC appointed a Brigade Photographic Officer to overcome the lack of coordination resulting from the variety of policies that proliferated throughout the RFC wings. An additional photographic officer was attached to each wing. Orders for aerial photography were issued by the Brigade to the Army Wing, and a squadron would be tasked to fly the mission. Sometimes special large-scale photographs and long-focus obliques were asked for.[870] Brigade Headquarters was also responsible for photographic printing. Once prints arrived at Army or Corps headquarters, the RFC responsibility ceased. The subordinate headquarters continued distribution to the front-line trenches.

British Army Wing Squadrons covered the strategic area behind the counter-battery zone and operated as far back of the lines as was practical.[871] Corps Wing Squadrons were limited to the counter-battery zone, extending back a distance of about 2,500 yards from the enemy front-line trenches. The photography covered up to 8,000 yards behind enemy lines.[872] Photo coverage of villages, towns, bomb targets dumps, billets, and enemy headquarters required two copies for each addressee on the distribution list.[873] The Branch Intelligence Sections attached to Corps Wing Squadrons generated one copy of all mosaics, one copy of any photograph of particular interest – photographs taken with new cameras or showing new forms of enemy defensive construction, using concrete for example, new fortifications, shell holes, or any other new feature deemed significant. Any interesting stereoscopic or oblique coverage was also included in the distribution.[874] Corps and Army commanders placed greatest value on intelligence acquired through the tactically focussed Corps Wing Squadrons. It was indicative of the British belief in the value of aerial photographic reconnaissance that when the sorties were overdue, backup aeroplanes were tasked to complete the work.[875]

Intelligence dissemination to the line became well-orchestrated, especially for the Canadian Commonwealth units that pushed for increased intelligence awareness to better support their personnel. During trench warfare the first duty upon entering a new sector was an immediate reconnaissance in detail of the new defence, to fully understand the sector topography, and make oneself fully cognisant of the general topography of the Brigade front, the disposition of the Brigade front, and the disposition of the units within the Brigade.[876] To assist intelligence awareness within the Brigade, intelligence staffing included one staff captain, twelve observers, one aeroplane photo clerk, two draughtsmen and one clerk. They received a mosaic photograph showing the wiring and other features of the German system. They also received photographs of the map showing the German trench system as well as an oblique photograph of the landscape. The mosaic, the photograph of the map and the landscape photograph was packaged in a way that provided easy handling.[877]

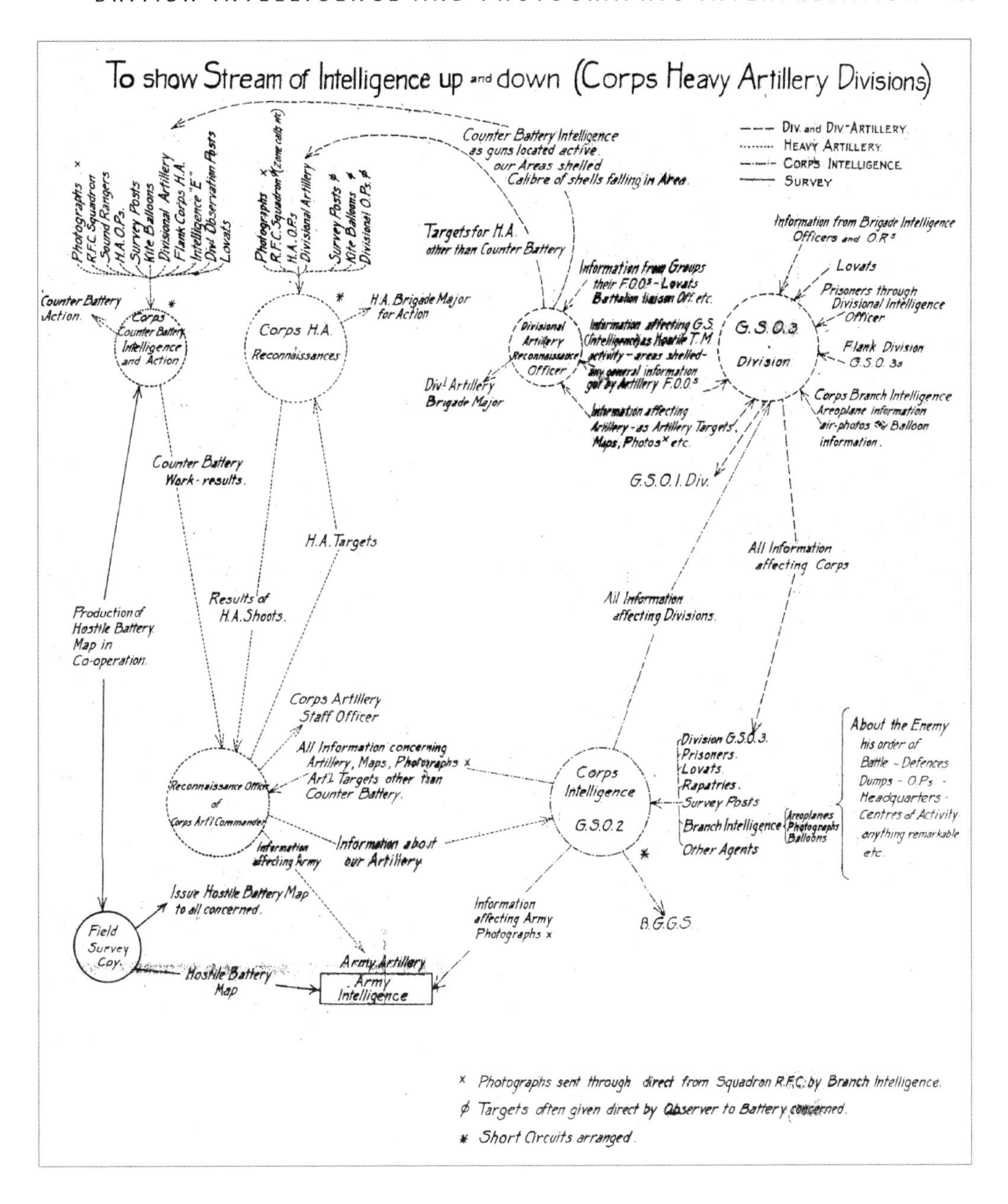

British Commonwealth units established very effective intelligence networks. In the Canadian Corps the intelligence officer served as the directing officer and maintained close liaison with the operations officer; with daily interaction. Their desks at headquarters even faced each other.[878]

KEY INTELLIGENCE POSITIONS

The British intelligence network was well established at the lower echelons. Under the staff title of General Staff Officer (GSO) Intelligence, the organisation was configured in numerical order. The senior Intelligence staff member was GSO 1. The GSO 1 was the British army equivalent to the acknowledged French expert on the sector, the chef d'Cartographie. The British GSO 1 understudy was the GSO 3, responsible for writing the Army's daily summary on artillery activity. GSO 1 addressed enemy losses and related subjects; enemy intentions (prisoners' statements); and captured documents. Analysis included the effect of British air operations and artillery fire, the liaison between enemy infantry and artillery and other similar miscellaneous subjects. Two British Armies, First and Second, were deployed in the Ypres sector, site of some of the worst fighting along the entire Western Front. The hotly contested area required as much intelligence as could be acquired to maintain force readiness. The British Second Army was on photograph distribution from their three subordinate army squadrons, four corps squadrons, flank army squadrons and GHQ squadrons. All incoming information was filed according to squadron. The results were posted on the daily map.[879] The British First Army produced an Intelligence Summary and Summary of Information that included intelligence from aerial photographs.[880]

At the Somme sector, the British Fifth Army kept a similar structure. The GSO 1 was normally an assigned Lieutenant-Colonel, responsible for maintaining enemy order of battle, maps and diagrams showing disposition of enemy forces, and captured documents. GSO 2 Intelligence Officers were responsible for interpretation of aerial photographs and enemy rear echelon organisations. One GSO 2 worked enemy defences; the other provided augmentation on the enemy defence function and rear work organisations. The subordinate Corps BIO personnel comprised an Intelligence Corps Officer, NCO (clerk), orderly and two draughtsmen. The BIO received and consolidated orders for photos; interviewed pilots and observers; prepared a consolidated report; indicated map co-ordinates on photos; kept a record map showing new work indicated by photos; and distributed aerial photos.

Like the French chef d'Cartographie, the Air Photo Officer served as the senior authority on aerial photography. The Air Photo Officer personally examined all aerial photos relevant to the army sector (roughly 500 per day). This officer's reputation was such that he could locate any photo in the sector without reference to the coordinates. He remained liable at any time to be called on by the GSO 1 for an opinion of the enemy's intentions or to provide a summary of new works. The Air Photo Officer produced a summary of the German defence line, providing detailed descriptions and strength estimates. Select topics of aerial photographic analysis included indications of enemy raids, enemy movement, enemy air activity and fires and explosions. The Air Photo Officer included his draughtsmen in the discussion to prepare the map or target folder.[881] The British First Army employed its Air Photo Officer function to acquire all Army and Corps aerial photographs covering the front. Individual photographs were applied to construct a mosaic for a common reference point. An American visitor in early 1918 observed that they had 'thousands of individual photographs, and now they try to paste these photographs together and by rephotographing reduce the many photographs to one.'[882]

The British intelligence process was efficient. Aerial photographic reconnaissance exemplified the efficient network that was established. Squadrons serving the Corps strove to have their initial prints readied half an hour after landing and a secondary issue was despatched six hours later. The division had an aeroplane photo man review the photos, record new observations covering German activity and alert the intelligence staff of developments. Comparisons of old and new coverage was a standard process to detect new defences under construction.[883] One British infantry officer recalled the aeroplane map was a key reference for summarising patrol work in No-Man's Land. His commanding officer had the patrol retrace their steps on a photograph.[884] They also examined photographs to corroborate information from other sources for the divisional intelligence summary. GSO's validated the summary before dissemination. The aeroplane photo man also served as distribution manager for aerial photos relevant to the division. During interpretation of photos, any items of interest were fed into the intelligence channels through the intelligence officer on-scene.

THE BRANCH INTELLIGENCE OFFICER

Within the RFC, the intelligence function evolved quickly. Early on in the war, the central figure for squadron-level intelligence collection and dissemination was the squadron commander. As its role increased, aerial photographic interpretation became the responsibility of the squadron's senior observer. The lack of an intelligence network in place quickly became apparent. The inability to correlate aerial sources of information with other intelligence sources (prisoner interrogations, radio direction) limited the potency of intelligence.[885]

After the Somme experience, the RFC reevaluated their information infrastructure and decided on a new formation. In January 1917, Branch Intelligence Sections were formed comprising an Intelligence Corps Officer, two Field Survey Company draughtsmen, a clerk and an orderly. The Intelligence Corps Officer was now at the squadron headquarters, establishing a stronger acquaintance between intelligence and the aircrew. Branch Intelligence Sections were integral to the Corps, with information generated at the squadron being immediately disseminated to Corps for action. The BIO function was established from the Branch Intelligence Section. In early 1918, an American visitor from the AEF G-2 to the British squadrons succinctly summarised BIO responsibilities:

> The first essential for this Officer is that he should be an expert in the reading and interpretation of aeroplane photographs. Unless this is so, he cannot carry on. He must know to whom to send the appropriate items of intelligence and must be able to make up his mind which of several items coming in at the same time should be communicated on first. It is imperative that he should keep in close touch with the staffs of all the formations in his Corps and with the Intelligence Staff at Army H.Q. He should get to know these officers personally – also the staffs at Wing and Brigade H.Q. RFC. He should make a point of being hand in glove with the C.C. (squadron commander) of the Squadron to which he is attached – also with all the pilots and observers of that squadron. If he does not do this he will miss a large amount of information which he would be able to collect otherwise. He should endeavor as far as possible to make his office a place where the pilots and observers drop in for a chat about their work (as well as other things); this will prove of great assistance to his efficiency.[886]

The BIO was the ultimate broker for British aerial photography. Despite working at squadron level, the BIO was the primary conduit for aerial photographic intelligence to Corps and Army commanders. American observers highlighted the BIO's responsibility as the ultimate validation point for other intelligence sources.[887] Aerial photograph requests were worked through the BIO in conference with the Squadron Commander. Army units at battalion and higher sent a request to the BIO, who in turn highlighted the area of concern through pin flags on a large-scale map. If the contemplated photography was not excessive, the BIO sent the squadron's flight commander a memo outlining the target request. The squadron commander followed up and detailed additional aeroplanes if the request proved excessive.[888] Following the request for aerial photography, the BIO and squadron commander assigned observers and pilots their missions. The BIO further explained the Corps' intelligence requirements and provided all relevant information such as maps, sketches, and photographs.[889] In addition, the BIO maintained daily contact with Corps Intelligence and the Corps Heavy Artillery intelligence liaison. Frequent interaction reinforced awareness of every requirement, allowing for greater initiative for new observations as well confirming standing requirements. Upon completion of their flight, both pilot and observer reported to the squadron headquarters and the BIO for post-flight interrogation and to deliver the exposed photographic plates. The plates were turned over to the laboratory; in turn, prints and aerial photographic interpretation reports were quickly generated. Extra photograph copies were always maintained. The squadron sent time-sensitive, unwashed and unmarked photos of important targets to the army headquarters as soon as the plates were developed. The early morning shoots arrived at headquarters in the afternoon. Evening shoots arrived later that night.[890]

While the laboratory was producing the prints and photographic interpretation was underway, the aircrew filled out a blank form showing what had been accomplished, with significant observations noted. In the British First Corps, the crew personally entered their reports in a blank book with the following format headings: Date; Pilot & Observer; Machine; Report. Second line: Fires; Explosions; Trains; Balloons; Hostile Artillery; Floods; Hostile Aircraft; Miscellaneous. At the end of the day, the information in the book was consolidated and forwarded to the Corps Intelligence Officer. The report of train movements was phoned to Army Branch Intelligence.[891] After the aircrew debrief, the BIO telephoned observation and photographic interpreter reports on the average two to three times a day. Aerial photographic prints were disseminated according to established procedures. Rapid dissemination was accomplished by messenger. Less time-sensitive target photographs were distributed later in the day. Major Henry commented on the communications set up for British dissemination. 'One factor that seemed of special importance in the easy and quick distribution of Intelligence and in the working of the whole Intelligence organisation was the very extensive and complete telephone system behind the British lines. From any Squadron or from any Headquarters any other place that was wanted could be gotten practically immediately.'[892]

The BIOs generated operational reports to their counterparts at the RFC Headquarters Intelligence Section three times a day. The 'Situation Wires' were telephoned between eight and nine o'clock in the morning, at one o'clock in the afternoon and after the day's flying. Information to the RFC/RAF included weather conditions, number of artillery adjustments, bombs dropped, photographic plates exposed, enemy aircraft activity, casualties, and other matters affecting the general situation in the Brigade operating area. RFC Headquarters also received copies of all photographs taken during the day. The number of copies forwarded depended upon the nature of the photograph. During active operations, a special messenger delivered high priority photographs as soon as they were printed.[893] In the evening the Squadron Commander generated a typewritten summary called the Squadron Record Book. The BIO also prepared for Headquarters distribution and in the evening a typed summary, 'Branch Intelligence Section, – Army. Daily Summary'. This summary was a compilation of all information in the reports given to him during the day by observers and pilots. The final BIO routine for the day was to keep the situation map updated and the area photographed that day outlined. Finally, a summary of significant intelligence was attached to the map.[894]

Within the British RFC and RAF, the Photo Section worked under the direction of the BIO.[895] Two proofs were prepared from each new plate, date and map co-ordinates were entered on the back of each in pencil, after which the corresponding notations were entered on the plates. As proofs were checked with the record map to determine coordinates, the proofs were examined for new work, notations indicated on slips clipped to proofs. Such new work was entered on the record map as soon as all proofs had been coordinated. Proofs were distributed to Army, Corps, and other headquarters as required. Requests for additional copies were expected by noon the day following distribution of proofs. The BIO sent off additional copies as he determined the need. The photos were then filed by map sections. A cross-index book was kept to provide ready reference to photos of particular objects (such as dumps, etc.) that were not shown on the maps.[896]

Intelligence manning throughout British forces expanded to meet demand. To support the reconnaissance, the British assembled the necessary staff to work both reporting and interpreting aerial photography. Two draughtsmen and one clerk were assigned to each unit.[897] The British used their draughtsmen to complete the aerial photographic interpretation. In the last year of the conflict a draughtsmen was assigned targets based on expertise and he kept up-to-date records and maps of his target. Every squadron in the Corps Wing and two Squadrons in each Army Wing were equipped with a photo section consisting of one non-commissioned officer and 20 men. Estimates at the end of the war showed personnel strength assigned to the Photo Section at 250 officers and 3,000 of other ranks.[898]

The experience at the front line shaped the intelligence requirements for the infantry. Initially, the British assigned a small cadre of personnel to accomplish the intelligence for the sector. A British battalion had one officer assigned responsible for all related functions. By 1918, manning increased to handle the multitude of intelligence requirements. The intelligence staff of a single Canadian Infantry Battalion included an intelligence officer, one scout officer, eight observers, eight snipers, one aeroplane photo interpreter and draughtsman.[899] During duty at the front line, the Intelligence Officer kept battalion intelligence maps updated, adjusting as required from intelligence reports. Flank activity was constant and required both monitoring and coordination with counterparts. Knowledge of the enemy at battalion level was frequently coordinated with artillery liaisons, resulting in ongoing target assignments against wire and other obstacles.[900] Observation reporting from elements throughout the battalion sector flowed through the battalion Intelligence officer. Every square inch of terrain required observation post coverage. Detailed, large-scale maps and aerial vertical and oblique photographs were employed extensively. Aerial oblique photographs proved invaluable, especially when the shot was taken above the observation post location. This gave the same general view and together with the map enabled a very comprehensive study to be made of the ground. The battalion aeroplane photograph man and draughtsman worked directly under the intelligence officer. The two personnel assisted the fusion of intelligence by integrating aerial photograph information with the other reports. The division occasionally hosted two-week aerial photograph training by the Corps' aerial photography expert with the objective that those in attendance be able to accomplish simple aerial photographic interpretation to serve battalion operations.[901]

At the outbreak of the war the RFC had only one non-commissioned officer and four men assigned for general photographic duties. Aerial photography requirements were not to be established until the following year, resulting in the initial cadre being reassigned for other, more relevant work. An exponential increase in manning occurred as aerial photography gained a foothold and became an integral part of British intelligence. Photograph development and printing was initially carried out in squadrons by non-commissioned personnel. The next phase in photographic manning involved establishing Wing Photographic Officer's positions. These personnel were drawn from squadrons and assigned photographic functions. The first photographic staff role was primarily experimental in nature. The organisation was 'elastic and hard-and-fast establishments and systems to a large extent avoided'.[902] The rapid demand for photographic experience soon over-tasked squadrons. Those units not facing increased photographic sorties loaned their photographic personnel to the squadrons experiencing a greater load. The mission was to be served regardless of whether the augmentation was at the expense of non-participating squadrons.[903]

When the RFC established the Brigade echelon, intelligence officers were assigned to the subordinate Army and Corps Wing. However, the absence of a senior photographic officer at the Brigade Headquarters was soon felt.[904] The value of the Photographic Officer function was quickly recognised as more requirements for aerial reconnaissance were established. Soon the function multiplied at all levels, particularly after the RAF was established in 1918. Had the war continued, the transformation of aerial reconnaissance positions underway would have resulted in a Photographic Officer being assigned to each RAF squadron in France.[905]

An AEF visitor's portrayal of the RFC aerial photograph dissemination process. (Conger Papers, USAMHI)

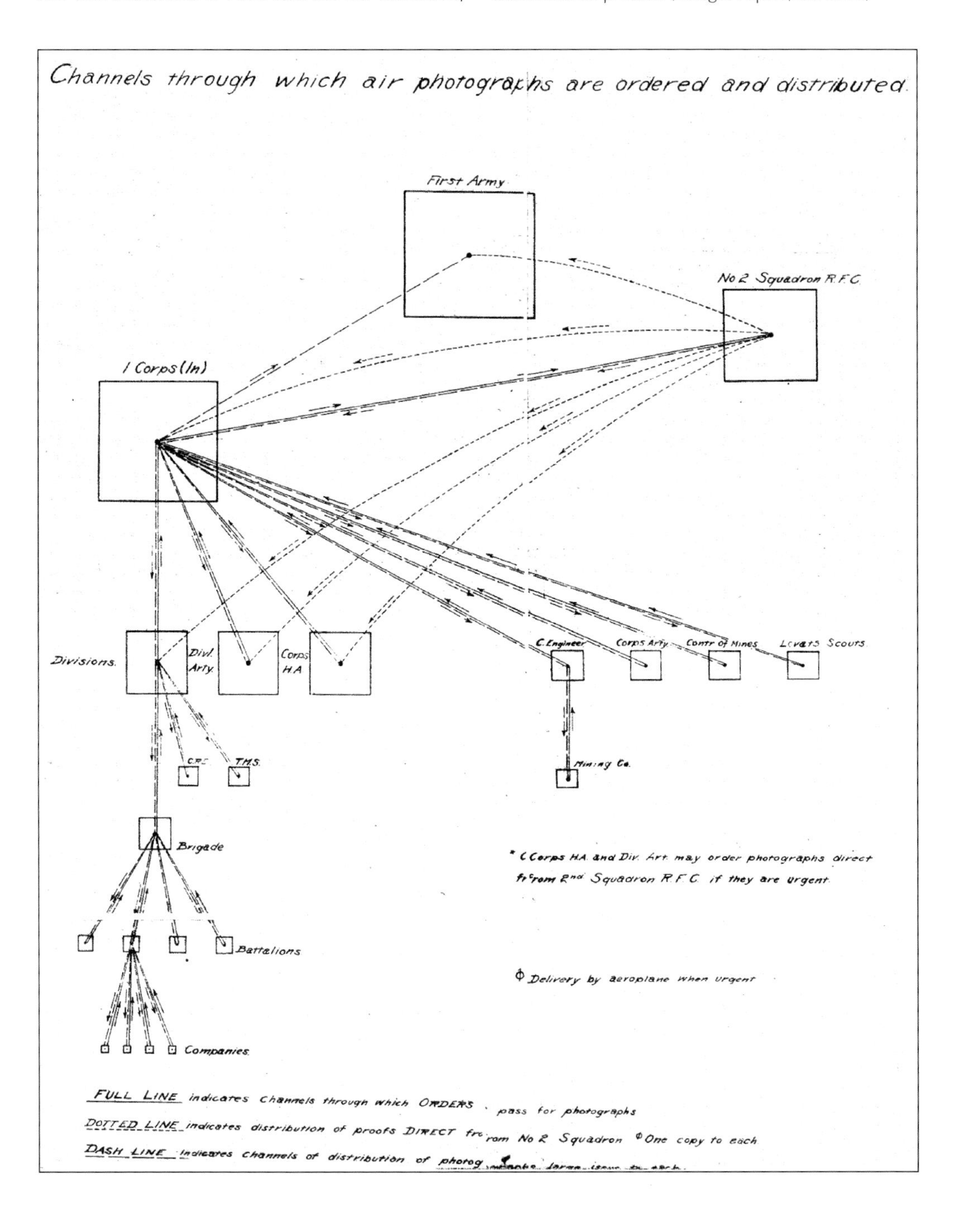

British Corps intelligence was tactically focussed, concerned with front-line activity and recording rear sector activity. They established a systematic process for obtaining aerial photographs of a given area. Initially, their channel of communication only addressed corps special requirements. However, liaisons with other collection sources such as the Counter-Battery Office (responsible for flash spotting) soon led to photographic verification of the ground observations. Corps squadrons began to work with other collection sources, resulting in more refined targeting through aerial observation. The culmination of the effort was a greater acceptance of aerial photography for validation of other sources.[906] The German artillery activity was continually monitored through ground networks and aerial photography. The British kept situation maps at all echelons labelled with intelligence inputs and source. The frequency of interaction between the staffs was dependent on the tempo of operations. Daily conferences were held during active operations to keep all offices appraised of the ongoing situation. Attendees included Field Survey Company, observation groups and sound ranging section, representatives from the Balloon Company (working artillery spotting functions recording flashes and movement), Intelligence Officers from Corps Squadrons, Divisional Artillery and Counter-Battery Offices from adjacent Corps.[907] Timely reporting to high priority tasking was accomplished by British squadrons in a variety of ways. The first rough print from a negative (the print was called 'rush copy;' one that was not annotated or washed) was immediately forwarded by despatch rider from the squadron to the BIO at the Army or Corps Headquarters. Standard time for 'rush copy' was five hours.[908]

Distribution of the initial intelligence product was the responsibility of the aerial unit that collected the intelligence.[909] Once the British Corps received the copy from the squadron, an 'advanced copy' of key photographs was printed. The Corps forwarded these to the subordinate division's GSO Intelligence for further examination. The division subsequently provided the division's General Officer Commanding (GOC) relevant photographs containing appropriate remarks. After review by GOC and GSO Intelligence, the photos were properly marked and filed by the Corps Intelligence Office. The division also received secondary distribution to inform the remaining staffs and units. Standard division distribution for a single photograph was approximately 35 copies. The British Second Army also maintained an Air Photo Section to analyse enemy defences and rear organisations. Two officers, the Air Photo Officer and 3rd Intelligence Officer, worked on intelligence from aerial photographs. Seven draughtsmen assisted the officers in the interpretation of coverage. Three draughtsmen examined photos over the whole front. The remaining four examined aerial photographs for key targets in an allotted sector of the front. Each of the seven draughtsmen examined photos covering the entire army front, each paying attention to a special subject, such as dumps, airodromes, roads, railways, bridges and hospitals, trench area defences, and photograph filing. At the British Second Army, BIO personnel were the focus of intelligence-gathering from photographs. They interrogated pilots and observers of photographic squadrons and prepared a chart showing the location of the area covered by the photographs. Once the decision was made as to what prints were to be retained, the photos were filed. A daily summary and the chart of photos taken were distributed to the respective centres. Intelligence concentrated on finding new hangars, sheds, or airodromes. All reports on train movements were forwarded to Army Branch Intelligence and consolidated in the daily summary. Requests for special photos were sent from the Army Intelligence Office through RFC Brigade Headquarters to Wing and Squadron commanders. There was a standing order that the entire army area, extending from 4,000 to 18,000 yards behind the enemy front line, was to be photographed once each fortnight.[910]

BRITISH PHOTO SECTIONS

The Photo Section configuration was based on demand when the first unit was established in 1915. Manning consisted of one officer, two NCOs and six men.[911] By 1916, the high demand for photographs overwhelmed the centrally managed wing photo sections, resulting in critical prints and information arriving late. This was especially true for artillery, which required constant updating of photographs to supplement their Firing Maps. Under the April 1916 reorganisation, producing and disseminating aerial photography was at squadron level. The army corps staff gave orders directly to the commander of the attached squadron for all photographs within five miles of their corps front. Beyond this area, photography was the responsibility of the Army Wing. To speed up the delivery of prints a small photo section in each corps squadron and in each army reconnaissance squadron was in place.[912] Demands for aerial photographs continued to increase. In February 1917, 40 cameras were issued to the RFC. Plate production was estimated at 2,400 dozen. Supply started to meet the demand nine months later, when monthly camera production increased over six-fold to 340 cameras, and plate production was up five-fold to 9,600 dozen.[913]

In the last year of the war, logistical support for the Photo Sections was managed by the British Air Ministry in London. Since the Air Ministry focussed on strategic issues, wings and squadrons were left with most of the analysis. The Air Ministry's Equipment and Personnel Branch did not engage in policy development as it concerned aerial photography. This was, frankly, through ignorance about the subject. One British memoir recalled the absence of senior deliberation on aerial photography in a rather sharp tone: 'It is strange how the general staff officers fall short in imagination with regard to Air Photography which [was] left without sound and technical advice.'[914]

The exponential increase in demand for aerial photographic product during the battle of the Somme overwhelmed the photographic equipment resident at the squadrons. One attempt to increase production through half-tone printing failed owing to slow production and unsatisfactory results. The demand for large photographs of trench areas required a more sophisticated laboratory. The British AP and SS then offered to assume responsibility for the demand. In October 1916 they established an experimental section at Amiens, equipped with a mechanical photo-printing plant, to produce up to 5,000 whole-plate bromide prints a day. The volume of production was massive. The photographs – including those for Intelligence and Police purposes – taken or reproduced rose in number from 25,000 in 1916 to 2½ million (whole plate equivalent) during the first ten months of 1918. The results were so successful that a similar production facility was authorised for each Army. Photographic work extended to panoramas, mosaics, maps and stereoscopic photographs, as well as reproduction of Graves Registration photographs and police photographs of deserters, escaped prisoners, and spies of both sexes. Also, general outdoor photography for operations, training and record purposes was accomplished. Based on the demands for electricity demonstrated by the Amiens prototype, each subsequent laboratory was equipped with its own mobile generating plant. Logistic needs were served by newly established 'technical stores.'[915]

British enthusiasm for aerial photographic technology modified their intelligence to prioritise the visual over the remaining intelligence sources. Increased photographic production translated into commitment of resources. With it came more personnel skilled in the discipline of interpretation. Throughout the conflict, aerial photography was integral to every British military operation. The British were leaders in developing a strategic role for reconnaissance, supporting their bombardment operations against the enemy. It became the rootstock for the professional intelligence service that exists today.

THE RENAISSANCE MAN: LIEUTENANT COLONEL JOHN THEODORE CUTHBERT MOORE-BRABAZON

J.T.C. Moore-Brabazon (8 February 1884–17 May 1964) was a true Renaissance man. His aristocratic roots allowed access to opportunities beyond the reach of the average British citisen. Moore-Brabazon took on whatever challenge piqued his interest. He took to motor racing with a flourish, winning one of the first races in the history of the sport, the Circuit des Ardennes, in 1907.[916] He indulged his appetite for high-

J.T.C. Moore-Brabazon made British history as the first citizen to acquire an aero license. (J.T.C. Moore-Brabazon, *The Brabazon Story*)

Left: The Aero Journal *Flight* featured Moore-Brabazon as one of the great aviation pioneers of 1909. (*Flight*, 6 November 1909)

Above: Moore-Brabazon with 'the first pig to fly'. (Moore-Brabazon, *The Brabazon Story*)

risk adventure to the new phenomenon of aviation making British aviation history as the first to acquire a British pilot's licence.[917] Moore-Brabazon went on to become the first Englishman to fly an aeroplane (a Voisin) in England. His passion for competition led to victory in the first all-English aeroplane race sponsored by the *Daily Mail*. The event spurred on the establishment of the British aeroplane industry. Moore-Brabazon became a British spokesman for the age when he reflected, 'I think we were all a little mad, we were all suffering from dreams of such a wonderful future.'[918]

Moore-Brabazon possessed an incredible personality and drive. Known as 'Brab' to his friends, he 'would always be one to tell a story or two'.[919] Yet he was passionate in the extreme to achieve the best through his efforts. Shortly after the outbreak of the war, he enlisted in the Royal Flying Corps and was commissioned a second lieutenant in the Special Reserve.[920] Moore-Brabazon's reputation for taking on new challenges was recognised. His early endeavours with wireless transmissions led to an assignment on the technical side, exemplified by the RFC's No. 9 Squadron. He would confess later to being initially petrified of the ways of the military, feigning laryngitis while in command of troops during a march. Thanks to the discreet assistance of the unit's Sergeant-Major, Moore-Brabazon soon learned the basics of command at close quarters. However, his aristocratic background required adjustment to the realities of military decorum. 'You will obey your superior officers,' the No. 9 Squadron commander once remarked early on. Moore-Brabazon replied, 'Superior officer? – senior, if you please, sir.'[921]

Moore-Brabazon found his niche in the rapidly evolving world of aerial reconnaissance. 'It suddenly dawned on me that photography from the air was bound to come, and that few could know much about it.'[922] The RFC Chief of Staff, Colonel Frederick Sykes, tasked Moore-Brabazon to lead the RFC's development of aerial photography. To gain an appreciation of the technology, he first met with members of the French *Section Technique*, the leading research office in aerial photography on the Allied side.[923] He was soon joined by two other brilliant British experts on photography, C.D.M. Campbell and Victor Laws. The three of them together constituted an impressive well of expertise. However, when they arrived at the RFC's No.1 Wing, Moore-Brabazon remembered, 'We were about as welcome as measles.'[924] They persevered and created the prototype 'A' Type Camera. 'Six box type cameras were evolved, this being the real R.F.C. camera, and from that moment photography improved a great deal.'[925]

Throughout the war, Moore-Brabazon was the driving force behind British aerial photography's evolution. He led the development of the entire suite of British aerial cameras employed. His scientific

Major Moore-Brabazon. (Nesbit, *Eyes of the RAF*)

approach to all challenges led to the resolution of several key issues limiting aerial photography, such as designing camera mounts to reduce vibration and using photographic flares for night photography.[926] By the end of the war, he spearheaded the Allies' role in many aerial photographic endeavours, including acquiring optics for cameras, assisting US aerial photography staffs in training and organisation and supporting the ongoing reorganisation of aerial photographic interpretation roles at squadron level and wings. Reflecting later in life on his accomplishments, Moore-Brabazon wrote, 'As to my own job, I always thought it wonderful that there was never a rounder peg in a rounder hole ... It was an extremely busy and complicated job, but I say without fear of contradiction that at one time I knew more about aerial photography than anyone else in the world. That is a bold claim to make about any subject, but I am perfectly convinced that I am right.'[927]

After the war, Moore-Brabazon continued to add to his remarkable curriculum vitae. He led the development of the British aviation community. At the same time, he pursued a successful career in politics. In December 1918, Moore-Brabazon was elected to Parliament for the Chatham Division of Rochester, and retained his seat during the Bernard Law and Stanley Baldwin era. In 1924, he was appointed Parliamentary Secretary, Ministry of Transport.[928] Moore-Brabazon served as Minister of Transport and later as Minister of Aircraft Production during the Second World War. He was forced to resign in 1942 following the opposition Labour party's censure over a remark he made in a private forum that the Germans and Russians should exterminate each other. After the war, Moore-Brabazon became an elder statesman, serving in the House of Lords and remained very active in a broad range of activities.[929]

Lt-Col. Moore-Brabazon at a 1918 photo conference. (Gorrell Report, NARA)

CHAPTER 12

American Expeditionary Force (AEF) Intelligence and Photographic Organisation

Notwithstanding the horrors of the war, the Americans could still look upon it as the 'Game'. There was a spirit of competition invoked by youthful exuberance. As 2nd Lieutenant Prentiss M. Terry, 91st Aero Squadron BIO, concluded in his manual on intelligence, 'The Position [of intelligence] affords you the Opportunity to play a Positive Part in the Game which will help toward Beating the Boche and that is what you are fighting for.'[930] Victory meant teamwork and the arriving American forces were the rookies in a well-established game. The US Air Service had to come to terms with obvious shortcomings that came with the nation's overall lack of preparedness for total war. It opted to follow both French and British approaches and develop eclectic methods. The French Plan Directeur and the British BIO served as fundamental building blocks for the Americans' intelligence and aerial photographic interpretation while accommodating the AEF's own needs. The American military culture soon developed an awareness and appreciation of aerial photography's role as an intelligence source supporting military operations. The ultimate benefactor for the remainder of the century from this experience was military intelligence itself.

The Allies voiced serious concerns over America's inadequate preparation six months after entry into the war. One British staff officer's Memorandum to the Air Board conveyed the feeling: 'American thoughts and methods do not appear to be at all well understood in this country. At the same time there is so much to be gained from ensuring harmonious action between the two services that it appeared well worthwhile to ensure if it be possible.'[931] The lifeblood of the British and French was ebbing quickly as the war entered the fourth year. The Americans had to quickly come to an understanding with their Allies about what was required to fight a well-established positional war. 'When America declared war on Germany the striking feature was their total unpreparedness in every branch and this was added to in the case of the whole subject of military aeronautics by an absolute lack of knowledge in every department of this subject. This lack of knowledge existed not only among the general public but even with their aviation experts, both Naval and Military.'[932]

Lieutenant Colonel George O. Squier served as military attaché to London during the first year of the war. He later assumed command of the US Army Signal Corps. (Maurer, ed., *The US Air Service in World War I*)

US military intelligence was also totally incapable of handling the challenges of modern warfare. Most of the state-of-the-art intelligence processes, including radio intercept, flash- and sound-ranging and most important, aerial photography, did not exist within the American army. What little was known did not carry any weight with the army hierarchy.[933] Prior to arriving overseas, American military staffs decided on a tentative intelligence organisation based on the British model. The father of US Army Intelligence, Colonel Ralph Van Deman, employed the Military Intelligence (MI) paradigm of both a 'positive' and 'negative' approach to Intelligence. Positive referred to useful information about the enemy, to include intelligence from aerial photographs. Negative intelligence referred to enemy undercover agents to acquire and determine information (today's word is counterintelligence).[934] MI-7 became the element within the War Department that handled cartography and to a limited extent aerial photography.[935] The other key figure in American military intelligence was Major General George O. Squier, who established the Aviation Section of the US Army Signal Corps. Squier had been military attaché to London for four years prior to his assignment to the Aviation Section in May 1916. His reporting helped establish his awareness of evolving military operations and intelligence processes employed by the British and spurred American military interest in developing similar programs prior to entry into the war. He was later promoted and made Chief Signal Officer in February 1917; he continued in charge of the Aviation Section for most of that year.[936]

As the US mobilised for war, the US Army Signal Corps became the first branch to create an American aerial photographic program. They quickly realised that if the program were to be successfully implemented, they had to defer to the Allies for military aerial photography technology. At first, they commenced a buying spree, picking up any available camera equipment to rectify their shortfall. The US Army Signal Corps purchased large quantities of motion-picture cameras, hand-held cameras and view cameras. However, these acquisitions did not meet the requirements for aerial photography. The Americans were quickly coming to terms with the realisation that they did not possess their Allies' three years of wartime experience in aerial photographic reconnaissance. In 1917, the Signal Corps deferred to their Allied counterparts for guidance as to the standard photographic process and equipment.[937] Likewise, the Americans were forced to acquire maps through any means possible, 'begging, borrowing, and perhaps stealing maps of everything from Dieppe to Damascus ... We needed maps for every possible contingency that might confront General Pershing.'[938]

Reorganising the American military to meet the initial demands of mobilisation resulted in the Signal Corps becoming foster parent to the US Air Service. The British and French were quick to recommend that a separate independent air service similar to their air ministries be established. The recommendations were partially accepted and changes were implemented.

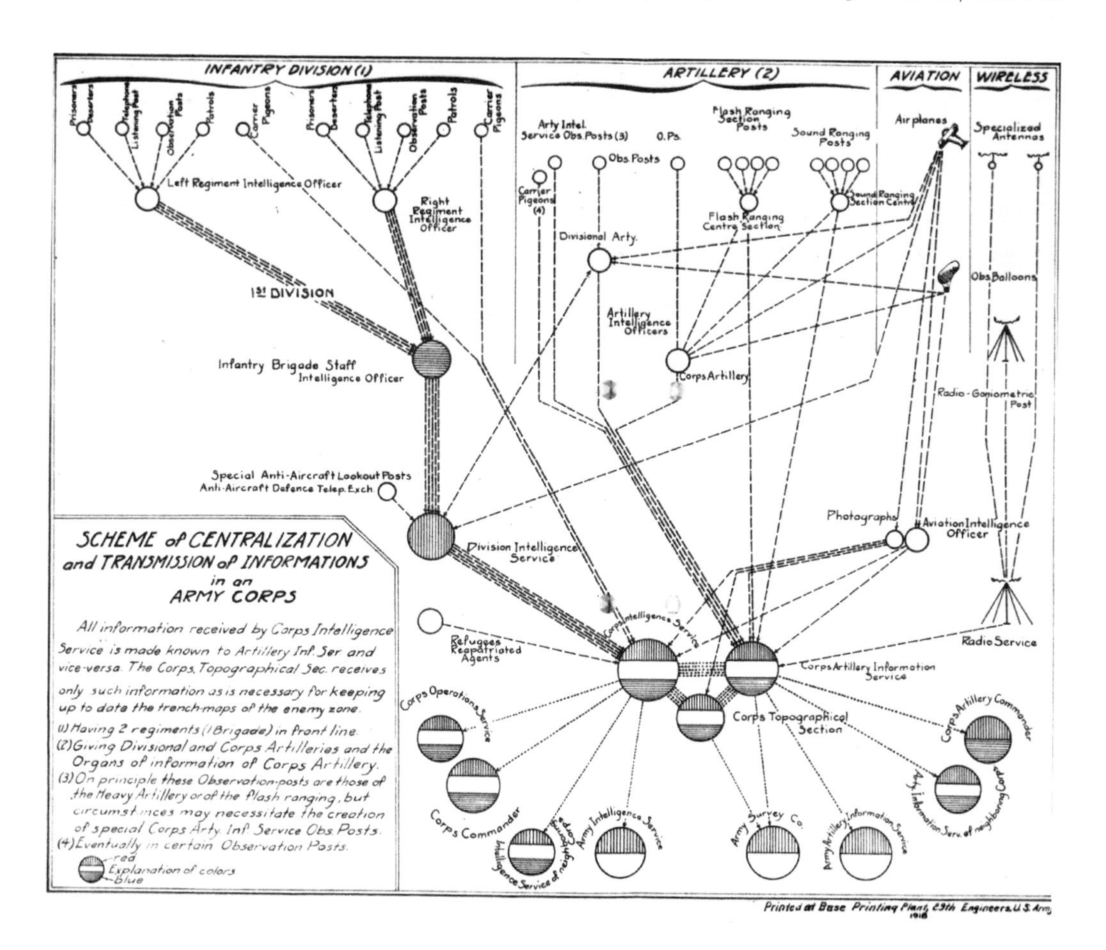

The complexity of AEF intelligence roles in 1918 is described in the contemporary schematic. (NARA, RG 120)

Squier continued in the capacity of Chief Signals Officer. A Division of Military Aeronautics was established, with Brigadier General William L. Kenly assigned as Director. Military Aeronautics served as the administrating directorate for aeronautical training, to include developing aerial reconnaissance.[939]

Despite the US War Department's initial efforts to institutionalise aerial photography, it was the AEF that became the primary proponent for aerial photography. Upon arrival in France the initial AEF cadre took important steps with the French and British Intelligence authorities, acquiring significant amounts of data based on three years of in-depth analysis and study. The staff commenced careful study of their Allied counterparts' intelligence architecture, establishing observation facilities to monitor ongoing operations. Shortly after arriving in theatre, staffs commenced visits to the GQG's 2e Bureau, IIIème Armée, 11ème CA, and 35ème Division d'infanterie. The Americans then proceeded to the British headquarters and witnessed Ypres operations with the British Second Army.[940] The Americans were able to follow up on the July 1917 British offensive at Messines Ridge, a battle which was considered by the British to have seen the most successful use of intelligence in the war up to that time.[941] Concern with the best way to manage military intelligence also had an impact on the US Air Service. Discussion evolved over the months into a debate about who was responsible for the intelligence resources. By the spring of 1918, American staff officers were sent out to examine what best suited air intelligence. Major D. M. Henry and Major C.F. Thompson of the US Air Service reviewed both Allies' operations. The decision to work intelligence through the BIO became the prevailing opinion for the staffs.[942]

Two of the greatest advocates of the US aerial reconnaissance role in the First World War were Colonel William Mitchell (right) and Lieutenant Colonel Lewis Brereton. (Maurer, ed., *The US Air Service in World War I*)

Both French and British intelligence programs were worth emulating. Their adherence to known intelligence theory and practice had built-in benefits. Active liaison between the French and British intelligence staffs illustrated the value of progressive ideas. In recalling the relations between the newly arriving Americans and the well experienced French and British staffs, Brigadier General Nolan recalled that the British were 'very anxious that they would make us understand how serious all these things were. The French had the same solicitude but they were less tactful in expressing it ... Their attitude was more "Well, this is the way it is; you can take it or leave it." The British attitude was "this is the way it is: for God's sake take it, take our word for it".'[943]

The Americans decided that the key factor in speeding intelligence was geopolitical and ethnic identity. Like the Americans, British forces had to organise intelligence to serve an expeditionary force operating in a foreign country. Also, there was of course language. The American military assessment that both British and French intelligence systems accomplished the same results in very much the same manner naturally favoured emulation of the more adaptable British system. American military seniors had their own expeditionary challenges. The end result was that US military policy and regulations on intelligence mirrored those of the British. This approach was suggested to General Pershing and approved.[944] However, intelligence resources such as maps required adaptation to whoever had the supply. The Americans borrowed the French system of designating accurately a given point by numbered coordinates in grid lines. Such maps proved invaluable when Americans were subsequently assigned to French sectors.[945]

In addition to establishing an intelligence framework relevant to operations, the American aviation community had to contend with a flurry of rotations among the senior commanders for aviation. American adjustment to the method of aerial reconnaissance as perfected by French and British experiences required a pronounced degree of self-initiative. As it evolved, the US Air Service in theatre was always on the move and transforming itself. The first Chief of the US Air Service, Brigadier General Foulois, accompanied the AEF headquarters when it moved from Paris to Chaumont. A few months later, the entire Air Service staff separated from the AEF headquarters and established itself at Toul. Along with the geographic separation, ongoing reorganisations caused the American senior aviators to be reassigned. In January 1918, the 1st Corps of the AEF became the central administrative command of all the American forces on the French front. Commanded by Major General Hunter Liggett, the 1st Army Corps appointed Colonel Mitchell as their first Chief of Air Service. His aerial organisation served two masters, the French VIIIème Armée (which had tactical control) and Major General Liggett (administrative control). When the first American squadrons arrived in April 1918, they were initially sent to the quiet Toul sector in eastern France. Mitchell moved his headquarters to Toul to maintain oversight and control.[946] A month later, Pershing appointed Brigadier General Mason M. Patrick as the Chief of the US Air Service, relieving General Foulois, who was appointed Assistant Chief of Air Service, Zone of Advance (ZOA). Mitchell was placed under Foulois and provided an opportunity for aviation command as the Brigade Commander, 1st Brigade, Air Service.[947] These changes translated into a greater dependency on Allied experience and policy for the remainder of the war.

US forces had no doctrinal basis for air operations when they arrived in France in 1917; however, by the end of the first year, they had managed to establish a base of understanding of how aerial observation and photographic interpretation worked to support the demands of the forces. Their emphasis remained tactical. Strategic policy evolved in the post-war era.[948] The Americans subscribed to common intelligence practices employed by both French and British. The AEF G-2 staff produced on a daily basis their own version of the Plan Directeur, distributing three copies for headquarters.[949] When the American 1st Division arrived in the Toul Sector in January 1918, they were under the control of the French CA, and were required to conform to the practice of a French division within the sector. This included adding a topographic officer and draughtsman to the division to 'make hasty drawings, sketches, and diagrams for the General Staff of the Division, study aeroplane photographs of enemy territory, to keep up to date the sector maps showing both Intelligence and Operations Information, to distribute maps, and to collect and forward to the Corps at prescribed intervals the corrections in trenches and other military features for incorporation in new editions of the battle map at Army Headquarters'.[950]

The Americans duplicated French techniques for employing army reconnaissance assets for photographic validation of the other intelligence sources. Intelligence manning soon included aerial photographic interpreters to assist in the production of intelligence summaries.[951] As it was for their Allied counterparts, aerial photography was emphasised as a primary source of intelligence. 'The making of aerial photographs of enemy territory is considered one of the most vital functions of the Air Service'.[952] At the higher echelons, senior officers relied on the G-2 for advice on aerial photographic reconnaissance. The G-2 subdivision prepared the First Army Summary of Intelligence, providing substantive information on 'enemy works, based on the traditional intelligence sources such as prisoner interrogations, validated by aerial photography'.[953] In addition, GHQ, AEF published every midnight a Battle Order map that portrayed the known location of enemy troops in the sector. This was distributed to aero squadrons for review by the aircrew before each sortie.[954] It was testament to the self-starting First Army's G-2 office work that the airplane photograph office was originally set up 'with no clearly defined sphere of operations'. By the summer of 1918, the function expanded to assume the responsibility for all information on enemy works, whatever the source.[955]

At lower echelons, the American intelligence process simulated the British. The hub of American intelligence was the function of the BIO. The BIO was assigned to the photographic group by the AEF General Staff G-2. BIO had responsibility for collection, compilation, and distribution of all information on the enemy pertaining, directly or indirectly, to aerial operations. As it pertained to aerial photographic interpretation, the BIO had responsibility for assembly, study, interpretation, and distribution. Aerial observer responsibilities included support to the BIO in photographic interpretation.[956] Once the sortie results were explained, the BIO was the responsible office for updating everyone on new developments. Key actions such as enemy troop movements, transport activity, railroads, dumps and aerodromes and other targets of interest were included in the mission report. The 'did not observe' comment was just as important as the significant sighting. 'The observer must at all time report every thing that he sees, regardless of how unimportant it may seem to him'.[957]

In the summer of 1918, the 91st Aero Squadron, the lead US Air Service strategical reconnaissance squadron, published an extensive manual on intelligence for both aircrews and intelligence personnel. It represented an impressive effort by the squadron's BIO, 2nd Lt. Prentiss Terry. The manual covered every known facet of the squadron's role in acquiring information, to include procedures for reviewing available intelligence, maps to be reviewed and a series of self-generated forms designed within the squadron to include the 'Observer's Report' and 'Summary of Information Obtained from Photographs'. This comprehensive effort provided squadron members with the most detailed understanding of the intelligence process at the squadron level. It provided a detailed overview of an American observation squadron, to include review, analysis and dissemination of aerial photography. Terry's manual included the only known illustration of a standardised format for aerial photographic interpretation. The header to the form was a basic description of the function: 'That the best results of

Photographic study may be obtained some form should be used whereby the information derived from that study may be fully written down and a record kept as to the Photographs upon which the information is found and the map area which covers it.'[958] Such efforts did not escape the attention of Capitaine Pépin. He personally reviewed the manuscript to ensure the document captured the essence of intelligence collection and processing[959]

To directly support the American divisions assigned to the front, the AEF acquired topographic expertise from its own 29th Engineers. The personnel were subjected to training in interpretation and restitution of aerial photographs and provided with whatever supplies could be acquired. As more AEF divisions arrived, the supply of engineering expertise from the 29th Engineers was depleted. The solution was to create a mobile training team (today's parlance) and instruct the division staff on topographic skills prior to proceeding to training in the line under the French.[960] US Air Service photographic reconnaissance tasking came from Army or Corps intelligence offices to the commanding officer of the Observation Group. The Group's operations officer assigned the mission to a squadron and designated the size of the camera to be used. The squadron operations officer then assigned the mission to the observer and the Group's photographic officer. The BIO then commenced preparing the intelligence effort for the mission that started upon receipt of the photographic order.[961] Upon completion of the aerial photographic reconnaissance sortie, the camera was downloaded, plates developed and prints produced. A photo laboratory technician recalled the process:

> When dry, the prints were removed from the rack and sorted in sets for each negative and rushed to the Intelligence Department. Here our work ended. The Intelligence Department and General Headquarters worked out the strategy to be pursued and the prints to be distributed to the artillery, infantry, and other departments so operations could be coordinated.[962]

Timely reporting was an indispensable goal. Army squadrons were to produce photographs for the consumer within six hours of the reconnaissance aeroplane's landing.[963] The British baseline for 'rush copy' prints remained at five hours. Communicating the results of the observation sorties began at the landing site. The 1st Corps Observation Group had telephone lines set up to link the Group and Squadron headquarters to the pilot and observer tent on the field. Long-distance lines were maintained with tactical command posts and higher Air Service headquarters. Radio liaison between the Group and all points in the area was established from the radio station. In case of communication failure, the US Air Service advance units at the front also employed pigeons. The Group maintained two pigeon lofts.[964] The American 1st Army G-2 maintained a Radio Intelligence Section not only to monitor enemy radio stations, but to conduct 'the policing of our own telephone lines near the front for dangerous conversation which would be overheard by the enemy.' Those that violated security were reported to responsible authorities.[965]

America's most notable contribution toward developing aerial reconnaissance and photographic interpretation lay in establishing the entire network for aerial photographic support. At the time that war was declared in April 1917, the War Department and the United States Signal Corps were charged with organising the nation's military for war, to include organisation, supply of personnel and materiel, and training. The Signal Corps' major priorities included establishing an air service. Aerial photography was not worthy of staff consideration. However, thanks to persistent efforts by a core element of photographic experts, the US Signal Corps established the Photographic Division (2 August 1917). Under the lead of the pre-war explorer Major James Barnes, the Signal Corps' Photographic Division commenced a review of existing literature on the subject, to include British intelligence reports and French aviation photographs. Probably the most important figure in the seminal period of American military photography was Barnes' deputy, Edward Steichen. Barnes acknowledged that Steichen was an extremely credible figure in the US military's first efforts with aerial photography. Without Steichen's 'aid and advice, the Photographic Division might at any time have gone on the rocks.'[966]

For the Signal Corps central office everything had to be created from scratch. A list of approximately 100 applicants with known photographic experience was compiled and solicitations forwarded for recruitment.[967] Military training camps, photographic laboratories and a logistic network for photographic supplies had to be established and personnel trained. Supplies and equipment have to be generated from within the United States, but the Allies still had to allocate already scarce resources. Grooming a new generation of photographic personnel required establishing a curriculum and instructors possessing the immediate skills required for aerial photography.[968] All of this occurred during intense competition over priorities in a growing list of urgent requirements for the mobilising army. Considering the total lack of familiarity amonst the senior ranks with aerial photography, the accomplishments of this core group were amazing. They stuck to their message throughout, and wisely so. 'Nearly nine-tenths of the

Major Barnes (right) and Captain Steichen established the Photographic Division of the US Army Signal Corps. (James Barnes, *From Then Till Now: Anecdotal Portraits and Transcript Pages from Memories' Tablets*)

information for the offensive or defensive' operations in the war were being produced by 'the work of cameras under trained direction in the air.'[969] In November 1917, Barnes and Steichen arrived in theatre to establish the Photographic Division for the AEF. They came across an AEF staff already overwhelmed with trying to establish a credible military to aid the Allies. The two officers were left to their own discretion and they faced major challenges in establishing the program. Assistance was limited both from within theatre and the home offices. No photographic material arrived from the US until June 1918.[970]

The American cadre established a strong network with the Allied aerial photographic pioneers, including the British photographic team of Lieut.-Col. Moore-Brabazon, Major Toye and Major Campbell; and the French Capitaine Pépin and Lieutenant Labussiere.[971] They participated in key conferences regarding photographic standards and training in December 1917. Both British and American photo experts attended a British conference at Blendecques that addressed aerial reconnaissance and photographic interpretation roles with Army and Corps Intelligence. British emphasis on the subject was reflected in the spectrum of attendees: 45 officers represented Aerial Observation, Army and Corps Intelligence, Heavy Artillery, Counter Batteries, Field Survey and Topographical Service attended.[972] A week later, both Barnes and Steichen met with the French at Châlons and adopted the French manual, *Étude et Exploitation des Photographies Aériennes* for AEF aerial photographic training. Despite this critical interaction with the Allies, the AEF senior staff was too overwhelmed by other pressing challenges to pay attention to Barnes and Steichen. Photographic reports were ignored. Despite their frustration, Barnes and Steichen played a significant role serving as the American representatives to several international forums on photography, where discussions addressed standardising lens and plate size.[973]

The efforts by the Barnes and Steichen advance team proved very successful in the end. They established training facilities and curriculum, represented the American position on aerial camera standards, acquired

Thanks to the leadership of Barnes and Steichen, Photographic Sections became an integral part of the US Observation Group Aero Squadrons. The 5th Photographic Section supported the 99th Aero Squadron. The unit commander, Lieutenant Sileo, is with a colleague in front of the photographic van. (USAFA McDermott Library Special Collections (SMS57))

the vanguard of a significant aerial photography team, and initiated the logistic infrastructure necessary. Their culminating effort was the delineation of responsibilities between the forward deployed ZOA and the rear echelon support Services of Supply (SOS). It was agreed that the best personnel were to work ZOA operational missions. Their efforts helped prepare the way for establishing the American's first Photographic Section, supporting the operations of the newly arrived 1st Aero Squadron in April 1918.[974] The Herculean effort by the photographic vanguard established Photographic Sections both in the US and in theatre with the AEF. The demand for aerial photographic skills continued to increase while resources remained extremely limited. One Steichen proposal requested photographic elements at ten separate organisations, to include Headquarters, Air Service, headquarters of an army, photographic laboratories with observation and bombardment squadrons, Air Depots, and the Photographic School at 2nd Aviation Instruction Center (AIC). However, the proposal never took shape owing to lack of proper equipment even when borrowing what could be spared from the Allies. Only two laboratories (No. 1 with the First Corps Observation Group at Ourches, and No. 2 with the 91st Aero Squadron at Gondreville) were partially functioning by April 1918.[975]

The Barnes and Steichen photographic team separated in February 1918 to work for different offices. Barnes was eventually recalled to Washington DC to further assist War Department efforts to standardise the Photographic Section. Upon arrival, Barnes was reassigned to Rochester, NY to assume command of the photographic school. Steichen remained in Europe past the Armistice and became the most knowledgeable US expert on aerial photography.[976] Thanks to the tireless efforts of Barnes and Steichen, the first two US Photographic Sections arrived in April 1918 – but without any usable equipment. Extensive efforts by the AEF's Photographic Division were meant to ensure that equipment for the required photographic laboratory would arrive unscathed. The marked crates labelled as photographic equipment showed up in England, and inside were 24 long-handled shovels. Nonetheless, the esprit of America's initial Photographic Section cadre was not to be dampened. The personnel proceeded on to France and marched with shovels at their shoulder. The incident was subsequently described as an effort 'to camouflage this important agrandisement [sic] of the forces of the A.E.F.'[977]

LESSONS LEARNED – THE AMERICAN WAY

The volume of coverage soon exceeded the capability on the ground to process and disseminate the information. At the end of the war, reviews showed the infrastructure supporting observation was inadequate. Division of responsibility for photography between the Air Service and Intelligence Section was one area that needed attention.[978] Another was effective rear echelon coverage. 'In general our back area photographic information was unsatisfactory, owing to the rapidity of our advance and the bad weather, which combined to keep our Squadrons photographing the areas just ahead.'[979]

After the Armistice Brigadier General Mitchell issued a 'Provisional Manual of Operations of Air Service Units' on 23 December 1918. He was working out of Coblenz, Germany in the Army of Occupation. His vision for structuring aviation to be more dynamic included a strong reconnaissance arm. His reconnaissance target priorities were 'all enemy works, such as dumps, lines of defence in rear of the front lines, concentration points, aviation fields, hospitals, and railroad yards.' He foresaw that aerial reconnaissance had to focus on the emerging mobile targets that used roads and railroads to accomplish their objective. Mosaics of lines of communication were necessary to provide the mobile battlefield commander with a definitive look at the overall force in motion. Mitchell recognised that the Army had to be intimately involved throughout the reconnaissance process. Priorities were to be levied 'as far in advance as possible, [and] the objectives to be photographed, together with the order of their importance'.[980] His guidance reemphasised the need for timely and effective aerial reconnaissance. 'During an attack, especially, speed in distribution of the photographs is of the utmost importance, and delivery of the first urgency prints should be made by airplane.'

Mitchell's vision was further refined by Lieutenant Colonel William C. Sherman, Chief of Staff of the First Army Air Service. Sherman authored a 'Tentative Manual for the Employment of the Air Service' in early 1919, which was subsequently published in April 1919 under the title 'Notes on Recent Operations'. Sherman covered aerial reconnaissance and the supporting infrastructure of the photographic laboratory and intelligence. His manual emphasised the BIO function and described the basic functions which had become institutionalised during the war.[981] It also illustrated the normal sortie altitudes required for effective reconnaissance – 3,000 metres was mentioned as the usual altitude for aerial photography. A solo aerial photographic plane was capable of operating at higher altitudes (5,000 to 5,500 metres) to accomplish the mission.[982] Perhaps what was most significant in Sherman's manual were statements that addressed an aerial photographic reconnaissance in transition. Like Mitchell, Sherman addressed a more mobile battlefield that required a more flexible reconnaissance effort. 'During stable trench warfare tactical maps are based on data furnished by Corps and Army aerial photographs of the enemy's territory. Photographs during war of movement are of little value in studying the enemy organisation but serve to acquaint the Command with details of terrain ... During open warfare photographic missions are only occasional and are requested to clear up map obscurities or most other specific demands for information.'[983] In his view, remaining with the traditional aerial reconnaissance process would not effectively serve the changing dynamic of combat.

The most significant US assessment came from the leader of America's aerial photographic program in theatre, Major Steichen. On 26 December 1918, Steichen addressed the entire photographic interpretation issue with his manifesto. (See Appendix E for the Steichen Report.) The necessary depth of understanding of photography, he asserted, could not come from the typical military introductory course. The art of the photograph required a complete understanding of the photographic process. Frequently throughout the war, the mistakes made at the squadron level in photographic interpretation were due to ignorance of the process of photography.[984] His declaration was an amazingly astute assessment of the role of photographic intelligence, both in the sense of lessons learned in the war and with respect to its potential in future conflicts. However, his recommendations were to be ignored by AEF seniors. What transpired in the inter-war years was accommodation for observation, but not for photographic intelligence. The latter mission was relegated to the National Guard. Thanks to the National Guard, the aerial photographic experience in the inter-war period was successfully retained. One Illinois Guard unit in an exercise was able to photograph the front lines and deliver to the battlefield commander the wet print in eight minutes.[985] When the United States entered into the Second World War, the experience base was limited at best, lacking the wisdom embodied in Steichen's vision.

The American role in intelligence within the First World War demonstrated rapid adaptation to the mission once the commitment was made. The transition to a more effective program was greatly aided by the assistance of British and French staffs. What best exemplifies the marshalling of American resources to the task at hand was the creation, training and eventual mobilisation of Photographic Sections to support aerial reconnaissance. When the Armistice arrived, not only did the Americans have 21 Photographic Sections in theatre, but more were on the way.

THE ARTIST – MAJOR EDWARD JEAN STEICHEN

If anyone exemplified the best in aerial photography in the First World War, it was the photographic master Edward Jean Steichen (27 March 1879–25 March 1973). Passion, intensity and drive were his trademarks. His daughter described him as an artist who 'liked things to look the way they look. Show things as they are.'[986] He applied his genius for photography to military applications and became the most articulate spokesman for the medium throughout America's involvement in the conflict.[987]

The Steichen family emigrated from Luxembourg to the United States when Steichen was eighteen years old. Three years later, Steichen returned to Europe to study painting in Paris. While there, he met the great photographer Alfred Stieglitz, who was in the forefront of the movement to have photography recognised as art. Steiglitz recognised Steichen's talent and potential and he strongly encouraged his pursuit of photography.[988] When the war commenced, Steichen was living in northern France. As the Germans advanced on the Marne River, Steichen cabled a contact in New York City to determine the best course of action at that precarious moment. 'Advise strategic retreat' was the prompt reply. He departed just before the German Army arrived.[989]

Edward Steichen holds a German-made aerial camera while posing in front of a Breguet 14 B2 aerial bombardment aeroplane. (US Signal Corps Photograph, NARA)

When the United States entered the wa, Steichen – aged 38 – was eight years over the age limit set by the US Air Service. However, due to his reputation as a photographer, he was commissioned a first lieutenant and assigned to the branch responsible for photography, the Signal Corps.[990] His proud mother commented, 'The United States government has one of the great creative artists of the world now training young men to go over the top with cameras instead of guns.'[991]

Aerial photography's foundation was well established long before Steichen entered the ranks. In the summer of 1917, he met the RFC photographic pioneer Major C.D.M. Campbell, while Campbell was campaigning for support of aerial photography in Washington DC. Campbell quickly filled Steichen with enthusiasm for aerial photography's potential.[992] Steichen soon teamed up with the big game hunter photographer, James Barnes, to initiate an aerial photography program for the US Army. He said of Barnes, 'All he knew about photography was whatever he had learned on a hunting safari when he made a motion picture of wild animals.'[993] However, outranking Steichen, Barnes became the voice of aerial photography for the Signal Corps. Nonetheless, Barnes recognised that Steichen was the true expert in the field of photography, who served as the lynchpin for achieving success in any role that the AEF had in store for aerial photography.'[994]

Both men left for France in October 1917 with Major Barnes continuing his leadership of the Photographic Section and Steichen serving as technical advisor for aerial photography. They set up an office in Paris, trying to make a difference in the rigid thinking about aerial reconnaissance, while reminding themselves that 'the whole photographic situation for the overseas Air Service' was an entirely new phenomenon.[995] Over the following months, the Allied staffs quickly recognised Steichen's expertise. Among aerial photographic veterans such as Campbell, Steichen's reputation advanced. He wrote to Barnes that Steichen was 'the finest master of photograph technique in the world'.[996]

Major Victor Laws and Captain Steichen together—two aerial photographic giants of the First World War. (Gorrell Report, NARA)

When Barnes transferred back to the United States, Steichen assumed overall responsibility for the AEF's photographic operation. It was challenging work, but Steichen excelled, showing strong managerial ability, skill in coordination and the capability to handle personnel and materiel.[997] Additionally, he excelled as a photographer. He learned to appreciate the camera's ability to render great detail. His photographic art became sharp and clear, providing the image quality that the war demanded in aerial photographs.[998] This quality became a hallmark of his art in later years. On occasion during the war, Steichen had joked that anyone turning in a fuzzy photograph would be court-martialled.[999]

In late July 1918, Steichen met one of the greatest influences of his military career, Brigadier General Billy Mitchell. Both men were vigorous, energetic and driven in pursuit of improving aviation's role in warfare. In addition, Steichen personified the kind of aviator that Mitchell admired. In Mitchell, Steichen found a senior military mentor, while Mitchell had in Steichen his aerial photographic expert. Their relationship continued beyond the Armistice.[1000]

Steichen summarised his approach to his work by recalling on the night of the Armistice, 'I had never been conscious of anything but the job we had to do: photograph enemy territory and enemy actions, record enemy movements and gun emplacements, pinpoint targets for our own artillery.' Steichen knew what the job as intended to achieve. 'We had had to improvise all along the way with inadequate equipment and materials and inadequately trained personnel. But the photographs we made provided information that, conveyed to our artillery, enabled them to destroy their targets and kill.'[1001]

In the postwar era, Lieutenant Colonel Steichen supported military aerial photography until his Reserve retirement in 1924. He returned to his photographic art and became a noted celebrity. His photography increasingly reflected attention to detail, reflecting the demands imposed on aerial photography. The Second World War brought him back to action. In the fall of 1941, at the age of 62, Steichen attempted to enlist in the Army Air Corps, but he was turned down because of his age: 'When I arrived, the officer in charge took down my name and address politely, but when he came to the year of my birth, 1879, he put down his pen with an air of finality and told me he was sorry, but I was beyond the age limit for induction into active service.'[1002] After Pearl Harbor, he was offered a commission in the Navy Reserves as a lieutenant commander. 'I almost crawled through the telephone wire with eagerness,' he later wrote. He was assigned to organise a team of photographers to record naval air activities. He was aboard the USS *Lexington* aircraft carrier in 1943 and at the battle of Iwo Jima in 1945. At war's end, Steichen concluded his illustrious military career as a Navy Captain in charge of all naval combat photography.[1003] Towards the end of his life, he mused on his distinguished military career, saying, 'I've spent over six years of my life in uniform. That's quite an indictment of civilization ... War is a monstrous piece of human stupidity. And I can't look at it any other way.'[1004]

CHAPTER 13

An Interpretation Methodology of the Front Lines

'OBSTACLES TO OUR PROGRESS'

A summary perspective was provided for arriving American aviators in 1918 prior to their exposure to the operational realities of flying over the front:

> It has been the plan of the German staff, when it is their intention to put up a prolonged and stubborn resistance, to organise at least two positions, one in the rear of the other, and in the neighborhood of one or two kilometres apart, and under certain circumstances, these distances may be considerably increased. Positions in the front line are as it were suspended on a succession of organised strong points, these strong points constituting a stabilising influence on the whole line. Between the first and second line there are shelters for the reserves of the sectors, centers of distribution of supplies, and a large number of field or short range guns.[1005]

The breadth and depth of aerial coverage of enemy territory created a wealth of material to ascertain enemy intentions. The entire landscape was scrutinised for any changes in configuration. The photographic interpreter became as sensitive to the policy of the Germans as much as the actual German combatant. Any distinctive signature was carefully noted, quickly followed by indications of intention disseminated to the commander. The descriptions that came forth from interpreting the photographic print established a new language, one that helped shape the culture of modern warfare.

TRENCHES

When the 1914 war of manoeuvre stalemated at the river Aisne and Ypres, the military science of positional war commenced. No-Man's Land experienced an intensity of destruction never witnessed up to that time. Capitaine Jean De Bissy remarked in his first assessment of aerial observation a year after that first trench was dug, 'Trenches are obstacles to our progress.'[1006] The initial trench configuration was based on terrain. Spadework produced 'broken lines ... adapted to the ground as well as possible'.[1007] Each defensive point represented a 'small salient', carefully constructed to ensure a continual defensive perimeter reinforced by flanking detachments.

'Obstinate defence' became a standard term to describe German intentions. 'When the enemy intends to make an obstinate defence he organises at least two positions, one behind the other.'[1008] Positional war's initial defensive configuration comprised a basic design of two principal lines of resistance, lines that linked the North Sea to the Swiss border. Throughout the German trench network, it appeared they wanted permanence and expected to stay in one place for a long time. The French had no intention of establishing a permanent defensive network to keep the Germans eternally at bay. Their conflict required retaking their sovereign territory and driving the Germans out. On the Allied side, the underlying assumption of all construction was that they expected to move forward at any moment.[1009] The German front lines of 1915 that faced French and British forces comprised several lines of trenches with the closest trench facing No-Man's Land serving as the real line of resistance. The prototypical trench network that first year of positional warfare comprised a front-line firing trench reinforced by a second trench at a distance of 100m to as much as 2km separation.[1010] The initial German defensive works comprised two or three successive lines, spaced to limit the lethality of the heavy artillery barrage. The trench configuration was also designed to complicate enemy artillery firing patterns by forcing the gun crew to readjust when they fired at another target. The second trench line paid particular attention to the terrain, frequently employing the rear slope for additional cover.[1011] The number of trenches increased to accommodate the flow of infantry between the first trench and the second and third trenches of the front-line position. Since

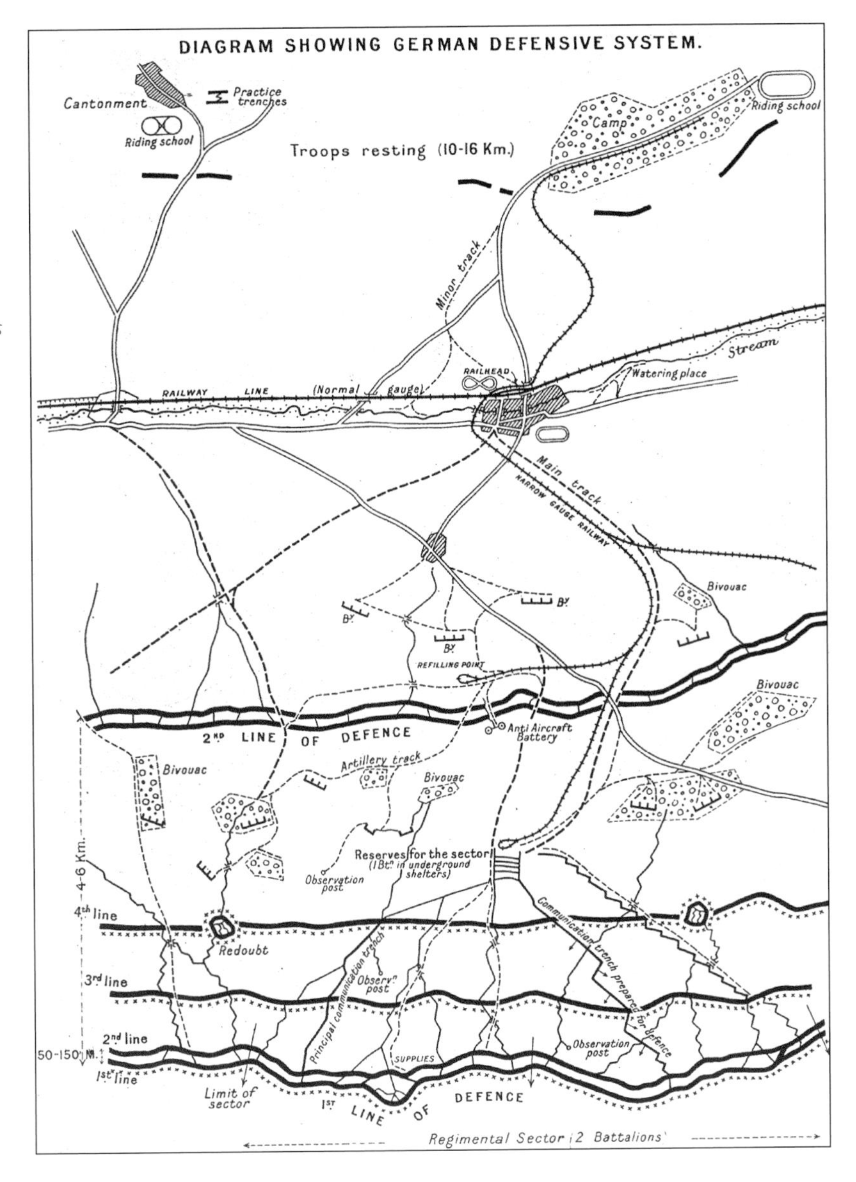

Capitaine Jean De Bissy's map was one of the first line drawing renditions, based on aerial photographic interpretation, to describe the German forward area in detail. (J. de Lannoy De Bissy, *Illustrations to Accompany Captain De Bissy's Notes Regarding the Interpretation of Aeroplane Photographs*)

trenches were configured to the terrain, they appeared irregular in shape and form on the aerial photograph. Such irregularity also ensured protection from flanking enfilade fire.[1012]

The aerial photograph usually portrayed the trench as a black line, further identified by turnouts in the line called traverses and fire-bays. The traverse was designed to limit the threat posed by enfilade fire to the trench passageway. Both reflected an appearance of dark and light lines that varied with the direction of light. Identifying trench design assumed a variety of elaborate and intriguing descriptions, to include 'Greek', the broken line, the indented line and the sinuous line.[1013] On slopes facing the Allied lines, communication trenches resembled a serpentine.[1014] Sometimes loops in the trench were established at intervals to afford free passage of troops coming in and going out.[1015] The primary trench design analysed from photography showed the main thoroughfare employing a large traverse every 30 metres. Traverses in sectors with reduced manning were usually spaced 15 to 20 metres apart. Photographic analysis determined that typical traverse measure 2 to 4 metres wide at their top, with a length of 8 to 10 metres. In very exposed positions, traverse length was reduced to 4 to 5 metres. In places exposed to flanking fire, special recesses for sharp shooters and machine gunners were included. Trench formation was defined by shadows from the

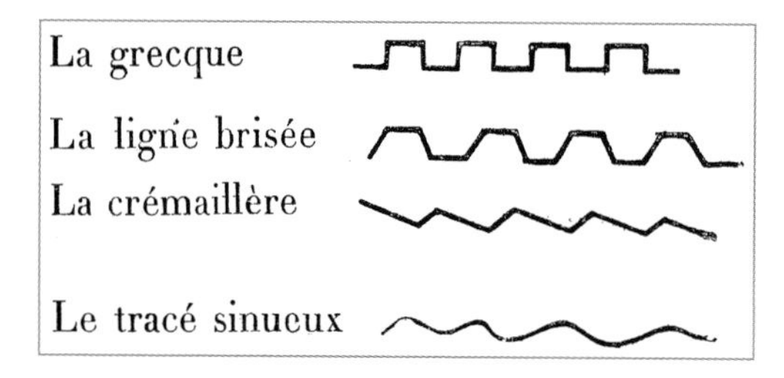

French designation for trench designs: 'Greek,' the broken line, the indented line and the sinuous line. (*Étude et Exploitation des Photographies Aériennes*)

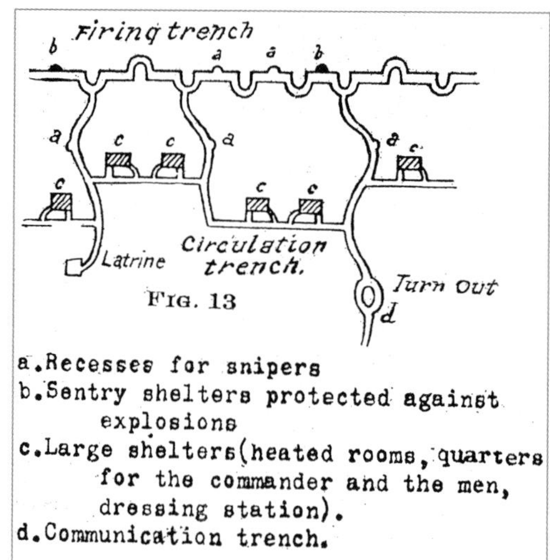

The first trench facing No-Man's Land was referred to as the firing trench. The schematic illustrates the details of the trench network to include the communication and circulation trench. (*Bulletin du Service de la Photographie Aérienne aux Armées*)

2- to 4-metre-wide piles of earth and sandbags piled on facing both front and rear.[1016] The other feature of the trench were 'dependencies,' otherwise known as trench commodities to promote sanitary living conditions. The signature for cesspools and latrines was a small rectangular hole at the end of the small trench (*boyau*).[1017]

Determining the centre of resistance and most frequented traffic routes was a priority for aerial reconnaissance. The key avenue in the trenches, commonly known as the communication trench (boyau de communication) had to be found. This main thoroughfare linked front-line fire trenches to the rear sectors. This configuration let the infantry manoeuvre safely within the trench labyrinth to defend against attack and provided an expeditious route for reinforcement. The communication trench design was dictated by the limitations of trenchworks. The forward firing trench was dug deep and narrow to provide the best protection. However, the combatants soon learned that this design hampered the infantry's ability to circulate. To avoid having to enlarge the trench and improve force mobility from within, a series of smaller circulation trenches were organised along the main firing trench.[1018] The communication trench length varied with the terrain. It averaged two metres in depth and extended over 1.5km in distance. Traffic density played a major role in the design. The communication trench was generally broader and straighter to accommodate the urgency of response in reinforcing the firing trench against attack.[1019]

As the war continued, the front line transitioned from a trench containing the main infantry elements to a select cadre of sentinels, scouts and observers. Shell holes and holes dug out by sharpshooters became the priority for analysis. Activity from patrols and secondary defence activity became detectable. Trench improvements provided visual clues that the enemy was intending to reinforce the present location rather than try to gain new ground.[1020] Lack of activity also helped gauge the value of the terrain. Sometimes deep recesses were found in old trenches that had broken down during prolonged periods of bad weather, suggesting the trench had declined in defensive priority. An additional signature was any absence of shadows in the trench line, which suggested the sector contained either dummy or incomplete trenches.[1021]

When taking up a new position, the Germans designed trenches based on natural strong points. The network prepared for defence had been first backed up with a series of fire trenches to support additional operations. These 'centres of resistance' comprised several fortified areas. At least two but more often three successive positions or systems of defence were prepared. Trench lines linked strong points such as villages, farms or woods in the forward area.[1022] Villages were especially attractive when the houses had cellars. These obstacles helped the defenders hold against an attack long enough to allow reinforcements to arrive. Machine guns and minenwerfer were added. Labels helped identify the trench configuration. The very front line was termed fire position or observation position, followed by the cover position. Support and reserve positions covered the rear sector.[1023]

The trench firing pattern included firing notches and sniper's posts. To the photographic interpreter, these appeared as T-shaped appendages to the main trench. Firing trenches were distinguished from all other trenches by the presence of fire steps and loopholes for the gunners. The aerial photograph portrayed fire

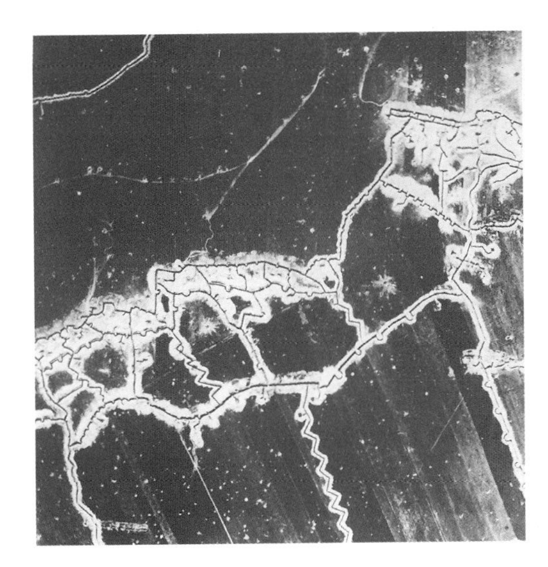

By 1915, each sector was intensely scrutinised to determine which sections of the trench showed any signs of activity. Analysis focussed on the important centres of resistance against Allied operations. (*Étude et Exploitation des Photographies Aériennes*)

The companion line drawing showed each section of the trench and associated activity. (*Étude et Exploitation des Photographies Aériennes*)

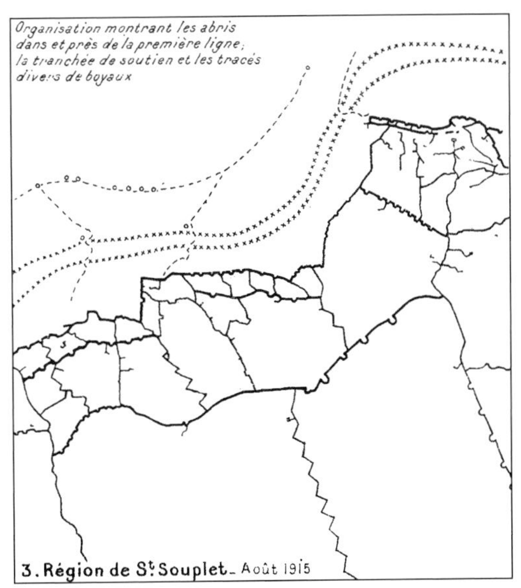

steps as tiny black notches having the appearance of sawteeth. Lookout holes in the trench network were usually located in a smaller adjacent covered trench. A splinter-proof shelter was usually observed nearby.[1024] In the rear trenches, the Germans configured the arrangement to ensure effective cross-fire and flanking fire towards the front sector or between the prepared elements of the trench itself.[1025]

Divisions divided into brigades and regiments at predetermined points in the line. At the lowest echelon of company and platoon, each unit proceeded to their own predetermined quarters in the trenches. Sectors were linked by several communication trenches, which decreased in number as the distance from the front line increased.[1026] Analysis of sector and unit echelon alignment commenced once a network of communication trenches was determined. Deployment activity, reinforcement routes, and centres of resistance were the subject of investigation. To help rationalise the sector layout, the interpreter assumed the role of the enemy sector commander. The most important lines of communication were described by level of activity, purpose and possible unit assignment.[1027] The approximate sector boundary was determined by where tracks terminated. Natural divisions between sectors were sometimes created from topographical features, such as canals, rivers, main roads or railway lines. It was generally assumed that three battalion sectors represented a regimental boundary.[1028]

Life within this environment settled into a well-established routine. Soldiers assumed a nocturnal existence to avoid detection from the enemy front-line observation and overhead aerial reconnaissance. Sentries manned the parapet using a periscope to observe signs of activity and alert snipers of possible targets. The remainder of the infantry stayed under cover. The three lines, firing, support and reserve, appeared deserted during the daylight hours. Observable activity commenced in the evening. Firing trenches served as the springboard for work parties. The communication trench became the major lifeline of the trench network and replenished all the adjoining sectors. When conflict resumed, the setting was transformed. An alarm sounded and the ground shook with a constellation of artillery fire, very lights (flares) and signal rockets. At sunrise, the garrison stood to arms and repeated the cycle.[1029]

Reporting trench activity from photographs followed a standard format. The most detailed descriptions were included in Intelligence Summaries published at the higher headquarters, such as the following:

> Photos of October 1st show that little work has been done on the enemy withdrawal position around CHAMPIGNEULLES. The prisoner's statement appearing in our Summary of October 1, that new trenches had been built filling in the gap southeast

Trenches in 1917 showed greater construction activity along the front line. (*Étude et Exploitation des Photographies Aériennes*))

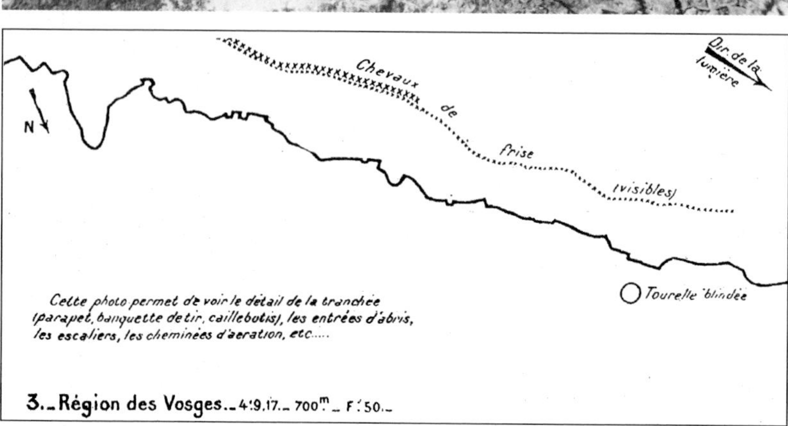

The companion line drawing details a sector in the Vosges region. An armoured turret (Tourelle blindée) is identified toward the right of the photograph. (*Étude et Exploitation des Photographies Aériennes*)

of CHAMPIGNEULLES, between 295.4–1287.1 and 297.6–286.5, is disproved by these photos, as are also his other statements regarding new trenches. A new trench about four hundred metres long follows the counter slope of a hill at 296.5–283.1, running along the north edge of some woods one km southeast of MARQ. Other photographs of October 1st serve to strengthen the belief that the enemy is intending to make a strong resistance on the position which he at present occupies along the ridges which run north of and parallel to the EXERMONT stream between GESNES and EXERMONT. Photographs were taken on reconnaissance crossing this line at two points – in the region of the town of EXERMONT and at a point between the RUISSEAU du GOUFFRE and the Bois de la MORINE. In the first region a new trench is seen running from the point 300.4–1281.1 due east to the point 301.6–281.1, the latter point being about 300 metres directly north of the town of EXERMONT. Elements of a communicating trench approach this line from the north and give further evidence of the intention of the enemy to hold and consolidate this position.[1030]

A French Vème Army Bulletin of 9 July 1918 simply reported: 'Aeroplane Photographs, July 7, 1918, 7th Army Corps: The photographs of this reconnaissance show the line of defence bordering the dirt road Licy-Hutevesnes is occupied.'[1031]

SAPS (SAPES D'ATTAQUE)

One primary indication that an attack was being readied was the discovery of a sap (derived from the French term, *sappe*). Saps were below-grade excavations or tunnels constructed to provide the offensive with jump off trenches immediately preceding the charge. Their presence signalled to the aerial photographic interpreter that an attack was possible within one to two days. Saps offered another benefit to the attacker by providing

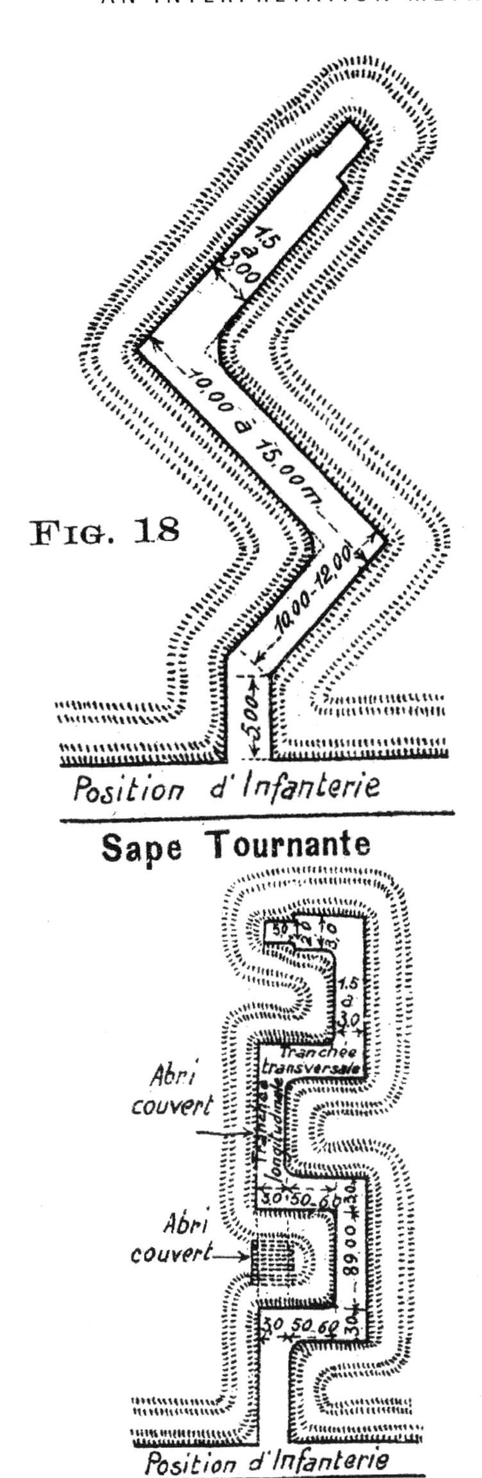

French manual shows the standard dimensions of a sap structure. (*Bulletin du Service de la Photographie Aérienne aux Armées*)

an advance line to configure a new firing trench. Saps were designed to rapidly link together and form a new front line. The interpreter checked for sap construction by regularly examining thin, threadlike paths from the front line trench to listening posts. Earthen spoil was removed to the rear areas as covertly as possible to hide evidence of ongoing sap construction. Sap design had several descriptive labels to include curved (most common), serpentine, and indented.[1032]

FORTIFIED SHELL HOLES

Fortified shell holes evolved from the devastation on the battleground. Like the sap, shell holes supported front-line operations while at the same time providing additional protection. The Germans found it economical to align shell holes to produce a new front line. General Haig commented on the new defensive configuration in his 28 August 1917 diary:

> Trenchard reported on the work of the Flying Corps. Our photographs now show distinctly the 'shell-holes' which the enemy has formed into 'strong points'. The paths made by men walking in the rear of those occupied first got our attention. After a more careful examination of the photo, it was seen that the system of defence was exactly on the lines directed in General Fritz von Arnim's pamphlet on 'The Construction of Defensive Positions'.[1033]

The aerial photographic interpreter searched for evidence of an occupied shell hole by examining the adjacent areas for footpaths. The fortified shell hole was normally singled out from the rest of the craters in the area since it had a darker appearance. Any excess earth spoil was usually thrown into an adjacent unused shell hole. A fortified shell hole was selected for its proximity to a well-drained area and passage to a shelter. Camouflage was added to provide additional cover.[1034] An American 1st Army Corps intelligence summary during the 8 July 1918 battle near Belleau Wood illustrated the significance of potential enemy resistance from fortified shell holes. 'July 6 to July 7, 1918. Oblique photos from French squadron taken July 5, 1918, show position of fragmentary 1st line German position from Bois de la Brigade de Marine to Vaux. Co-ord. 76.5–62.5 to 80.2–59.2 Occupied pits, pits probably occupied, paths and shallow trenches.'[1035]

SHELTERS

From the beginning of the conflict, harsh conditions dictated an immediate need for shelter (French:

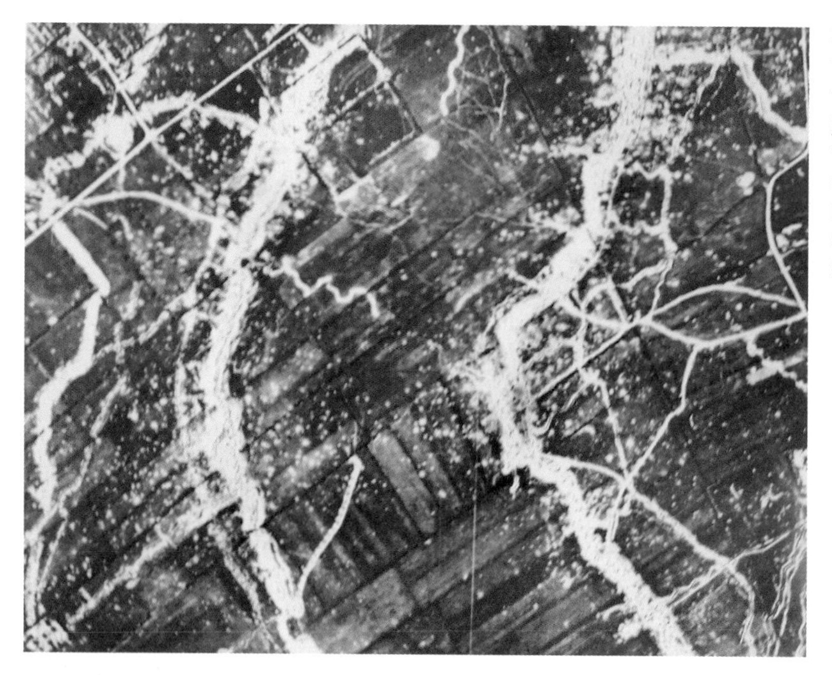

Saps were constructed from the main trench to serve as a jumping-off point for troops. Here the sap clearly extends into No-Man's Land. (*Illustrations*, S.S.550a)

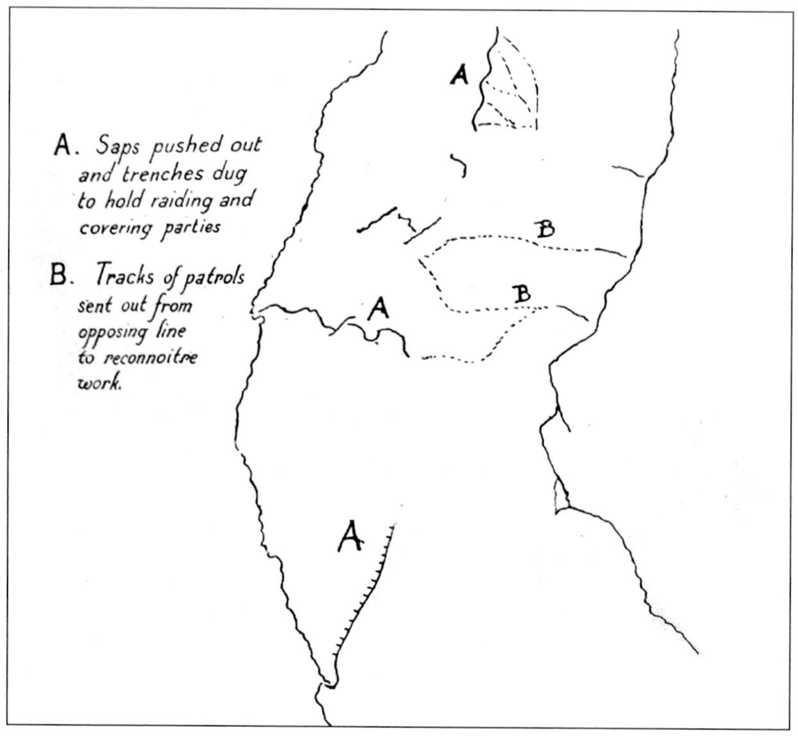

Line dawing of sector indicating the progress of sap construction. (*Illustrations*, S.S. 550a)

abris; British, American: dugout) to provide some measure of cover from munitions as well as allow refuge for the infantry during their deployment. Allied shelter construction remained rudimentary as a way of reminding themselves that positional war on French and Belgian territory was temporary. German construction was more refined, suggesting permanence for the occupying forces.[1036] The routine for the first year of positional war had infantry units responsible for the defence of the trenches conduct their normal routine during the day in shelter, either in the actual first line (those troops immediately responsible for defence) or farther in the rear (reserves) at a distance from the trenches varying between 800–2,000 yards (1–2km).[1037] Whereas, in the latter half of the war, the most important shelters were found in the second and third line trenches.[1038] In the first year of the war, the Germans constructed a variety of protective shelters: splinter-proof, surface, underground shelters, and covered trenches. Splinter-proof shelters were dispersed among the trenches to limit the risk from direct hits. The excessive moisture of the front line soil forced the inhabitants to build their shelter to a standard that included a double layer of roofing beams with at least a metre of earth thrown on top. These shelters did not extend above the parapet to any noticeable degree so as to provide protection against small projectiles. Splinter-proof shelters proved very difficult to locate with aerial photography. The search would focus on an area 12–15 metres to the rear of the firing trench or communication trench being served. Sometimes an abnormal amount of rubbish in front of the firing trench provided an indication that a shelter was nearby. Deep underground shelters were built from the many hard limestone caves that existed in northern France. These could withstand the pounding from heavy artillery. As the network approached the rear sector, the normal shelter consisted of covered trenches.[1039]

Shelter dimensions normally equated to the size necessary to house garrison strength for the area. A standard estimate of personnel assigned to a shelter was two groups of nine men (one non-commissioned officer and eight men each). Any observed construction activity helped lead to the discovery of shelter entrances. One logical place for an entrance was the corner of a traverse to help ensure maximum protection from incoming shell fire. Whenever a shelter was discovered from aerial photography, the interpreter would further examine the area for any signs of tunnelling. Ventilation shafts, small tunnels and rectangular openings were sought to gauge the shelter's total accessibility to the main trench. Communications in the vicinity led to the PC that controlled the forces in the area. Oblique coverage contributed immensely to understanding the potential dimensions.[1040] Those shelters to the rear were more dispersed along existing transportation networks.

The most famous shelters in the war were the Stollen, constructed by the Germans for safely holding troops and weapons and used to great effect before the commencement of the Verdun campaign in 1916. The walls of the average Stollen shelter were estimated to be 1.5 metres by 2 metres, with the construction descending to a depth of 15 to 20 metres beneath the surface. French intelligence identified four Stollen shelters in the Verdun complex: *Hangstollen*, *Fabienstollen*, *Gluckaufstollen*, and *Kuchenstollen*. The Verdun experience reminded the aerial photographic interpreter that underground facilities such as Stollen could have disastrous consequences on operations and required constant surveillance and analysis.[1041]

Shelters were a prime target for artillery fire. Besides discovering the actual shelter location, the interpreter would also determine the shelter's ability to withstand bombardment. Once the calculations were complete, the information was relayed to the artillery planner for consideration. A scale of four degrees of resistance, from a simple shelter to cavernous dwellings, was then posted on the Plan Directeur to alert the artillery to the degree of difficulty for achieving destruction.[1042]

BARBED WIRE

Barbed wire was applied row upon row in a broad swath. As an obstacle, barbed wire helped define the traverse of No-Man's Land. It was considered an auxiliary defence (*défenses accessoires*), designed to reinforce the trench network. Barbed wire also strengthened defences for lookout posts, bases of resistance and battery emplacements. Later, fortified shell holes, villages, hedges and ditches were observed with barbed wire belts. Barbed wire was not arrayed in parallel to the fire trenches. Instead, it reinforced the trench flanks against a broad attack line of advancing infantry. German trench design after 1916 complicated the layout. They established a network of barbed wire compartments. From an aerial photograph, the compartment assumed a triangle-shaped geometric form, 6–10 metres thick and connected by cross entanglements. The design was spaced throughout to limit destruction from artillery and other weapons. Since most barbed wire belts were constructed at night, the following day aerial coverage would be examined for signs of foot tracks along either side of the belt.

Planning operations in No-Man's Land required a precise knowledge of barbed wire location and design. This shaped the strategy and tactics of forces that were advancing in an offensive or conducting a trench raid. The battalion at the front was responsible for accomplishing required analysis of the front sector. Their observation routine determined any change in status, describing the condition of the wire in the sector while at the same time updating any reference map. Most intelligence on barbed wire was obtained through prisoner interrogations. Aerial reconnaissance and photographic interpretation validated the reports and reinforced final planning. Once the plan was set, artillery was then called upon to clear what wire obstacles lay in front of the battalion. 20,000 artillery

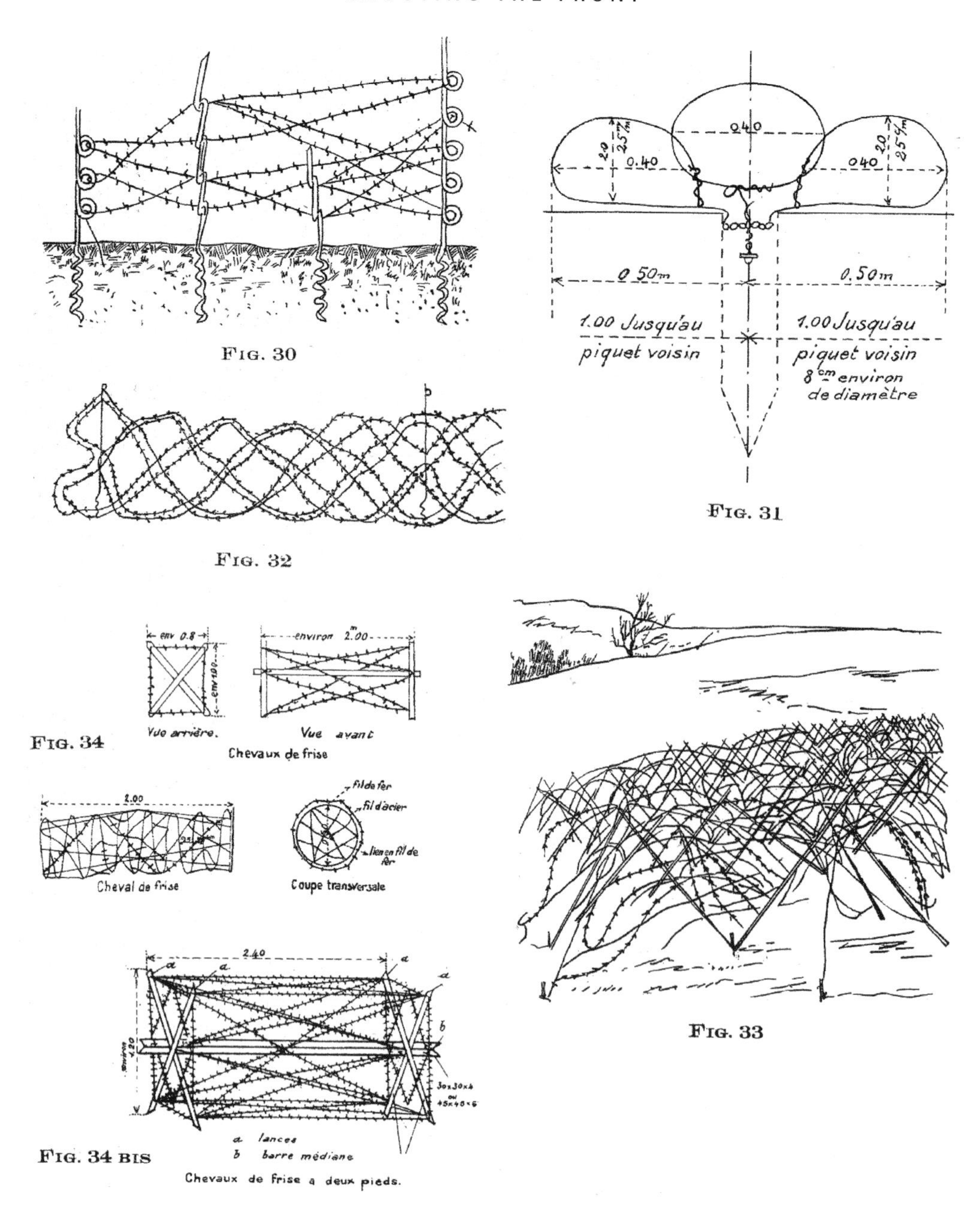

French manual shows the various barbed wire configurations. (*Bulletin du Service de la Photographie Aérienne aux Armées*)

rounds fired on a 50-mile front was the standard that artillery used to reduce the wire obstacle. An incomplete barrage heightened the lethality for Allied troops on the offensive. Enemy machine guns could rapidly deploy to locations to cover gaps in the wire created by the barrage. When the Germans finalised the Siegfried Stellung, wire covered the area in front to an average width of 10 metres and over one metre in height.[1043] Barbed wire bands at the front were not continuous. Operational planning required openings to allow movement of sentinels out into No-Man's Land. A wire entanglement called a *chevaux de frise* or knife rest (a wooden or metal frame on which wire is fastened) was constructed to obstruct any passage. Additional

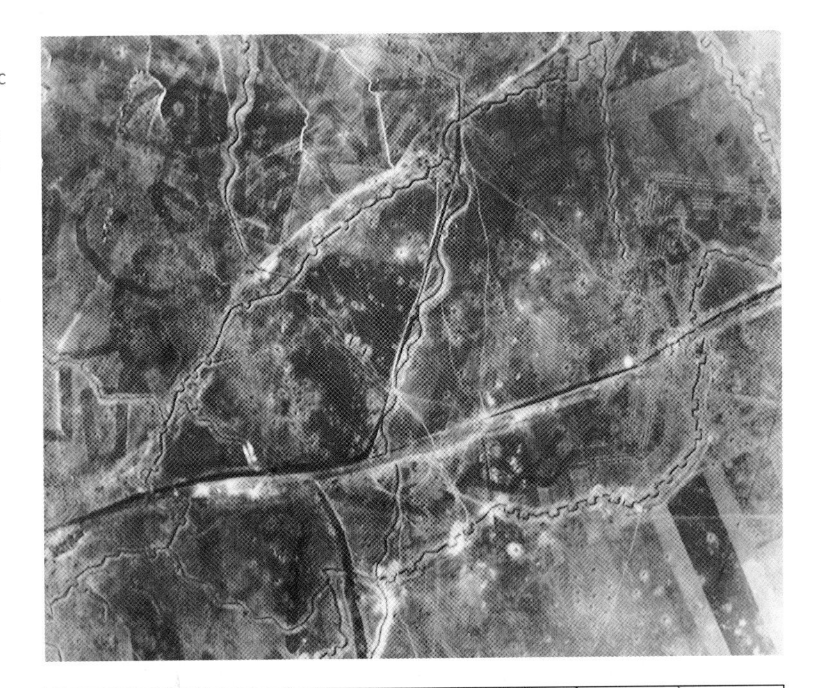

The vast array of barbed wire produced a geometric design on the battlefield. (*Étude et Exploitation des Photographies Aériennes*)

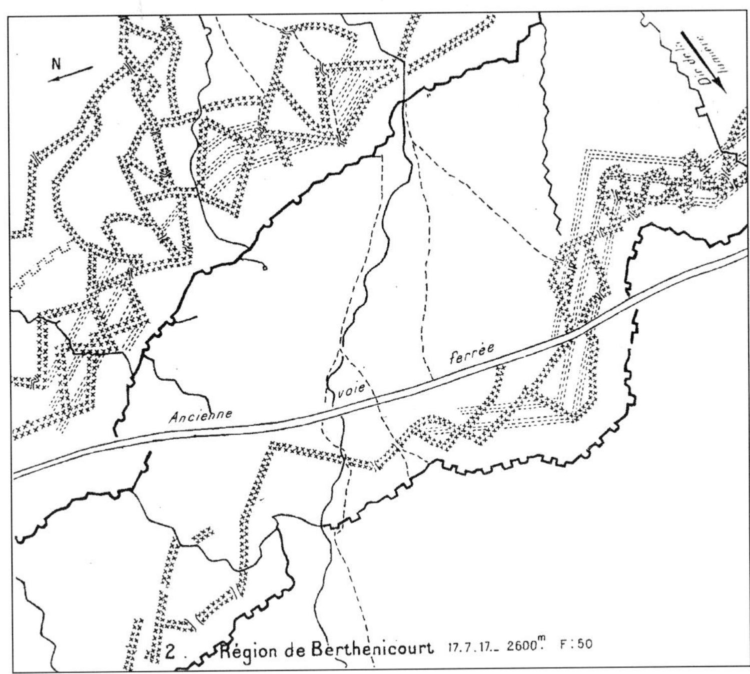

Companion line drawing details the elaborate configuration. (*Étude et Exploitation des Photographies Aériennes*)

obstacles included *abatis* (felled trees) branches, and military pits (*trous de loup)*. When sentinels and patrols ventured out, they followed a zig-zag route through the defences. Such routes were easily detected on an aerial photograph.

Visually detecting barbed wire on the front line was not simple because the wire presented a variety of signatures. Aerial photographs portrayed the bands as a dark ribbon, light grey to almost black depending on the newness of the wire and the contrast of the colour with the texture of the ground. When the wire was laid down it provided a visible sheen that was easily detected. Over time barbed wire tarnished and blended in with the terrain. This called for the interpreter to analyse the

Mineshafts were ideal for providing safe haven for troops in the front sector. Geologic formations provided an additional barrier against artillery fire. (*Étude et Exploitation des Photographies Aériennes*)

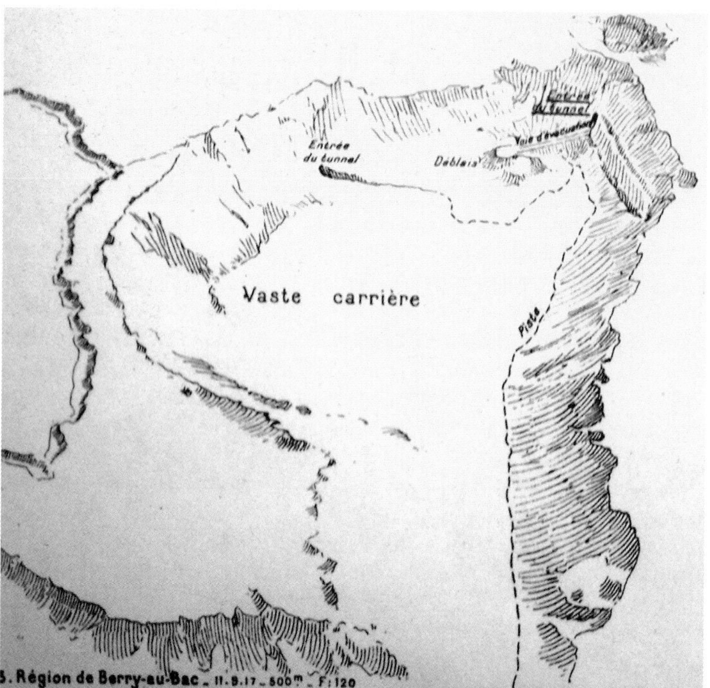

The companion line drawing shows the complex entrance was connected to a railroad network. (*Étude et Exploitation des Photographies Aériennes*)

history of coverage of the area to determine the width of the bands. Vertical coverage was not effective in determining the extent of the wire. A better glimpse of the barbed wire belts came from oblique coverage. The variety of barbed wire also contributed to the degree of difficulty of analysis. Concertina wire was difficult to detect. In winter, analysis was simplified as snow and frost highlighted the bands.

LIGHT RAILWAYS

Smaller, more accessible light railways (tramways) were a feature of the vast trench network that helped the interpreter better understand sector operations. Light railways were the logistical life blood of the trenches. The German forward railway systems were constantly being constructed or altered, especially during active operations. A single track built on an ungraded bed could be constructed at the rate of 8–16km in a 24-hour period. The load capability depended on the gauge of the track. Standard gauge track signified a heavy freight load that delivered larger quantities of troops and supplies. Smaller light railways extended the service to the front lines, supporting both infantry and artillery requirements. By the end of the war there were thousands of miles of light rail operating at the front.[1044]

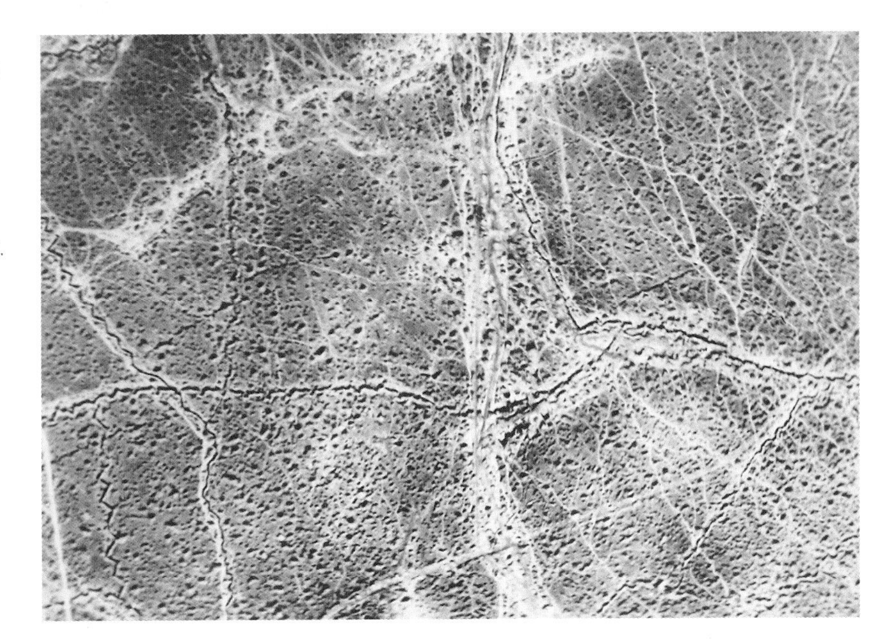

In-depth study of the disarray of No-Man's land revealed tracks and a light rail (tram) line in the front sector. (*Illustrations*, S.S. 550a)

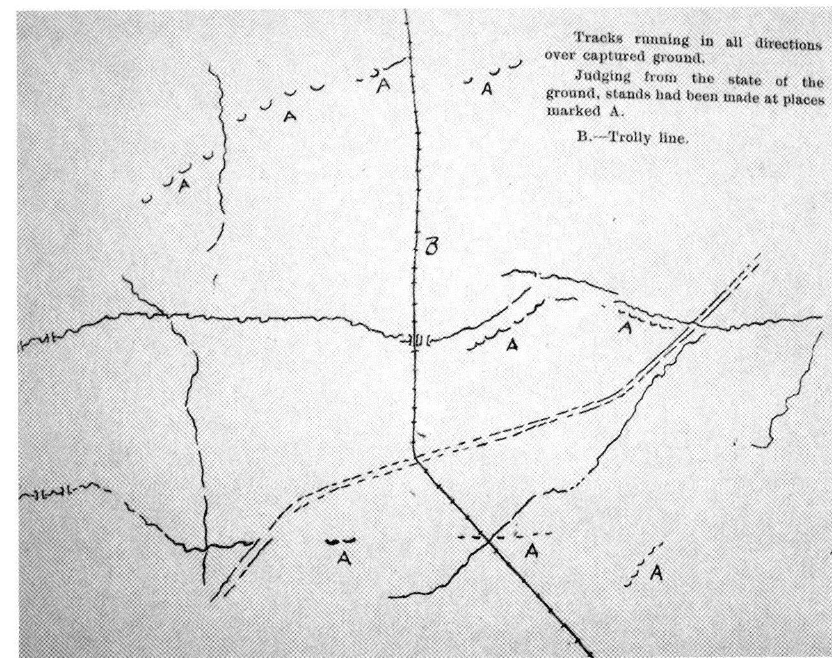

Companion line drawing highlights selected details of the front sector. (*Illustrations*, S.S.550a)

The light railway systems fell into three groups: the more common 1.00m and 0.80m gauge; 0.60m gauge; and 0.40m gauge. 1.00m and 0.80m lines existed there before the war. As a general rule, new 1.00 gauge was built only where lines of that gauge already existed, and where the use of a different standard would cause needless transshipment. During the battle of Verdun the Germans used the 1.00m tracks to move up their heavy artillery (420mm howitzers) to assist their offensive. The 0.60m and 0.40m gauge (also known as the trench tramway) came into existence at the beginning of trench warfare. Rail of 0.60m tracks was extended to the very front lines, serving the logistic needs of all major trench networks. Analysis distinguished them from paths in the area. Rail lines formed regular curves, generally following roads along the valleys or contours. These two narrow gauge tracks were only found in the forward areas. The starting point was generally a railway station. The 0.40m gauge, restricted to the more forward trench tramways where mechanical traction was rarely used, showed as a fine line on aerial photographs.[1045]

Light railways were distinguished on aerial photographs by a visible straight track leading into the trench network. Once amonst the trenches, sharper curves were observed. The German forward railway system was altered continuously to meet the logistical requirements of the moment, especially during active operations. Tracks were removed as soon as they were no longer needed, and were re-used to construct a new line. Photographic interpretation focussed on signs of deterioration and disuse. The mark of the track remained for some time after the rails had been removed. Aerial photographs could detect additional activity from footprints of men walking between the rails. A 1st Army AEF intelligence summary discusses light railway activity in the following report:

> Photographs of the region to the north of this show a new narrow gauge railroad running in a general southeasterly direction from about 302.5-282.5, to 303.3-281.7. These two points are in the vicinity of military camps and it is probable that the railroad serves not only as a communication between these camps, but also to supply the new trench system.[1046]

TRACKS AND TRAILS

The central focus of photographic analysis of enemy activity during the war involved the study of tracks and trails, the main thoroughfares of the front. During battle, they helped determine whether new trenches, shell holes and other defences were occupied. Footpaths provided the greatest insight into what the enemy accomplished on a daily basis. Photographic interpretation manuals continually stressed this feature.[1047] The most important purpose for analysing trails was to spot activity such as working parties that correlated directly to the state of readiness of a given sector.[1048] The interpreter could distinguish a well-trodden pathway by gauging the volume of activity on a respective trail as well as determining which routes were ignored. Tracks and trails also led the interpreter to key features within the sector such as dump locations, battery positions, headquarters, gaps through the wire, patrol paths and critical enemy observation posts. Tracks and trails also helped determine traffic patterns within a village, leading the interpreter to enemy operational centres and advanced listening posts.[1049]

Infantry movement within fire and communication trenches was continually analysed. Patterns of activity constituted targets worthy of disruption.[1050] Footpaths monitored from aerial photographs could not only gauge the level of activity in a sector, they also provided the analyst with an ability to determine sector boundaries; an invaluable insight when planning the battle. Infantry regiments and battalions used an assigned line for resupply. Oblique coverage provided the best photographic angle for determining the principal lines of traffic. The aerial photographic interpreter had to refine his knowledge of the front line infrastructure both by identifying the primary and secondary lines and eliminating those tracks that no longer supported operations.[1051]

SOIL

Aerial photography maintained constant surveillance of any new construction in a sector. Trench improvement, saps extending forward into No-Man's Land, or further excavation for shelters were examined on each photograph. In positional warfare, changes noted on the aerial photograph gauged the volume of soil brought to the surface. An abnormal level of earth meant digging deep for expansive shelters. Such activity was normally found near the primary communication trenches to shelter support elements or reserves. During the first year of combat, analysis of the amount of earth removed exposed a sector's weapons and ammunition depots.[1052] Spotting excavated earth prompted closer ground observation of a given area. Once the decision was made on the extent of the underground activity, the location became an artillery target.[1053]

It became a fact of life for all combatants that the aerial photograph detected even the slightest disturbance of the topsoil. The greatest aid came from the first images, providing a base by which change could be measured.[1054] The Germans soon modified their field procedures to cover the evidence. Working parties followed instructions to distribute the earth from excavations all over the landscape.[1055] Light railways proved invaluable in redistributing the soil. Front line mining operations were a notable feature of positional warfare with excavations penetrating as deep as the terrain allowed. The presence of large amounts of soil excavation became a signature. To compensate for the large volume, the soil was frequently put in sandbags to be taken to the rear and discarded.[1056] German successes at Verdun thanks to Stollen concerned the Allies for the remainder of the war. Stereoscopic coverage was continually desired. In the absence of viable coverage, the demand for prisoner interrogations increased.[1057]

Knowledge of geological formations soon had an impact on the thinking of operational planners. The war clearly demonstrated that protection against modern, high-power artillery could not be guaranteed by surface structures, no matter how strong the construction. Reliable defence against artillery was possible only by the excavation of deep works protected by earth or rock. Permanent fortifications built prior to the war had the luxury of time to explore the underground conditions through excavations and test borings. Once war commenced, there was no time for such consideration. Geology governed the time allotted to constructing the defence. Geological analysis assumed an important role toward the later stages of the conflict. The experiences on the Western Front prompted greater attention to terrain analysis and the geological make up of the battleground. After receiving orders to construct the appropriate defensive position, the engineer in charge reviewed the topographical conditions within the zone determining to what extent excavation was possible. Positional war focussed on these factors, determining a lethality opportunity cost for every man-made object below the surface. The commander now had to understand the geologic makeup at the front in order to assume the correct defensive posture.[1058]

SNOW

The climate of Northern France created a harsh environment for combatants in the winter season. For aerial photographic interpretation, the climate was a

Snow provided outstanding detail for aerial photographic interpretation. (*Étude et Exploitation des Photographies Aériennes*)

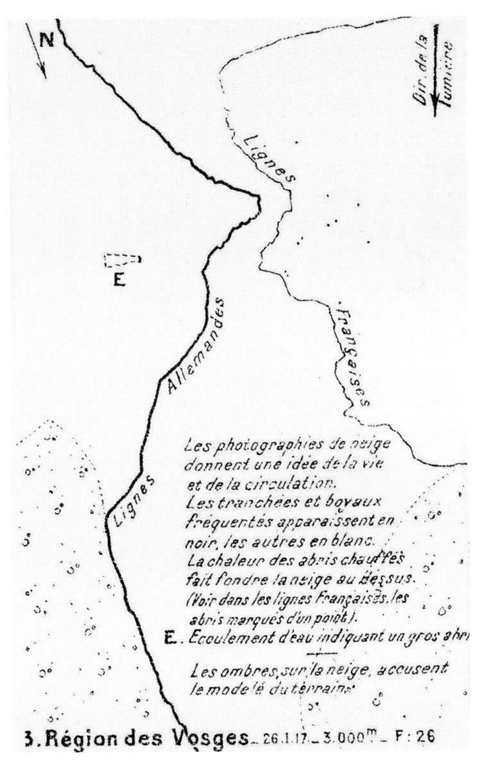

The companion line drawing clarified the extent of both German and French trench lines in the Vosges region. (*Étude et Exploitation des Photographies Aériennes*)

boon. Photographs taken immediately after snowfall provided breakthroughs in discovering the extent of activity. Black and white marks became the interpreter's standard for gauging the extent of recent activity. Wire, ditches, occupied shelters and camouflage became exposed with the clear contrast. Artillery was equally at risk, leaving a distinct black blast mark with every round fired.[1059]

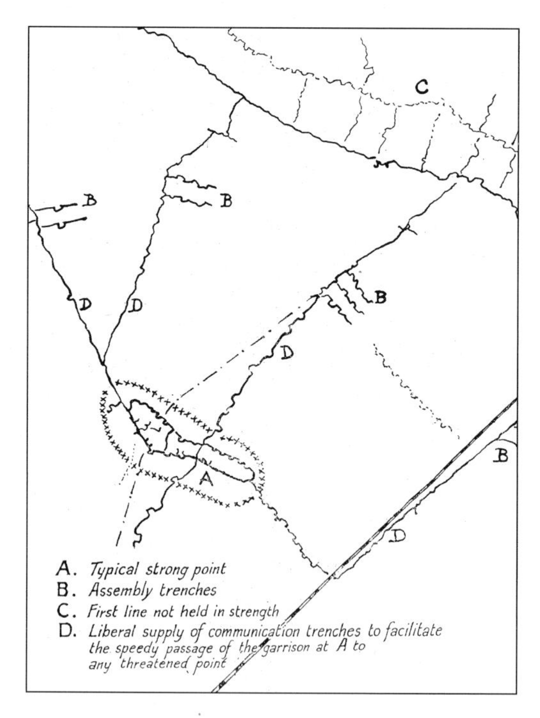

Companion line drawing highlights the *Bereitschaft* assembly area. (*Illustrations,* S.S. 550a)

BEREITSCHAFT (GERMAN RESERVES IN READINESS)

German reserves in readiness remained an ongoing concern for the Allied planner and for intelligence. Bereitschaft reinforcement helped sustain German positions. Trench network design called for communication trenches interspersed with shelters for rapid movement of rear-based reserves to any endangered sector. Bereitschaft movement was further supported through a network of rail centres for rapid transfer of personnel from other sectors as necessary. Supply depots were positioned to reinforce any counter-movement with weapons and materials. Beyond aerial coverage, prisoner interrogations proved invaluable for determining the extent of Bereitschaft in a given sector.[1060]

British aerial photograph showing front lines, assembly trenches and a strong point designated by the Germans as the *Bereitschaft*. (*Illustrations,* S.S. 550a)

MACHINE GUNS

Machine guns were the most sought-after target for aerial photographic analysis. Artillery could be detected through the variety of collection resources available besides aerial photographic intelligence. German weapons manuals attested to the machine gun's value, as reported in a 2nd Army assessment:

> *In General.* All kinds of machine guns isolated or combined at close or medium range, are an absolute success, if backsight and good observation of cone of fire are correct. Any attack will break down under the flanking fire of cleverly placed machine guns of any kind, served by resolute men.
>
> *In action around shellholes and in fights for strong-points.* Great morale effect of high machine-gun sweeping fire in connection with Artillery.

A superb oblique shot of the German lines at Gommecourt in 1917. The interpreter was able to assess possible locations for machine-gun nests along the forward trench. (*Illustrations,* S.S. 550a)

> Preparation and while flanking positions. Cones of fire keep crews of trenches and strongpoints down. Especially adapted for repelling counter-attacks. Therefore machine guns of all kinds are to be employed on the front lines. Machine guns of the defence in strongpoints are to be taken generally only by storm troops with hand grenades, smoke bombs, flame throwers and strungout hand grenades by flank or rear attack.[1061]

The machine gun's compact size and outstanding manoeuvrability resulted in a greater lethality being applied anywhere along the front. Machine guns could survive a blistering barrage while contained deep below in a shelter, resurface and quickly commence fire against the oncoming adversary. If one machine-gun team was destroyed, another surfaced and continued the mission.[1062]

Aerial reconnaissance and photographic interpretation emphasised annihilating the machine gun as the highest priority. 'The discovery of machine guns is of the utmost importance. It is essential that, in case of our attacking, every hostile gun should be destroyed by our bombardment. A single one remaining intact may be the cause of very heavy casualties.'[1063] Logical steps were taken to understand all facets of the machine gun's role and apply a methodology for detection and follow-up elimination through artillery targeting. In the first year of positional warfare, the German machine-gun emplacements were only prepared when the trenches were occupied. Given that a field of fire for a machine gun was generally 400 or 500 yards, the weapons were emplaced to cover the first line of trenches. Machine guns were placed farther to the rear if the slope of the terrain or shelter from houses with cellars was available to provide more cover.[1064]

The experiences of 1915 taught the French that the Germans maintained their machine guns in completely isolated shelters at a considerable distance from the trenches. This was done to avoid artillery. The shelter was made as inconspicuous as possible. Excavated earth was carried away. The machine-gun crew was provided with several days rations to sustain itself in place without detection by footpath. When possible, additional cover was provided by branches.[1065] A review of captured German documents dated September 1916 revealed to the photographic interpreters that at least 50 per cent of the machine guns were moved to rear echelons, observed in the second or the preferred third-line trench. The emplacements were selected to 'have a very wide field of fire or ... at the very least [to] flank the position'.[1066]

With each photo examined, special attention focussed on any abnormal configuration of the trench line. Any traverse that departed from normal size, shape or angle was scrutinised. Oblique coverage provided another angle to ascertain anomalies for further scrutiny.[1067] Covered emplacements became passé by the last year of combat; by then, mobility of the guns was a priority to sustain the defensive and increase survival. Only the quiet sections of the front retained a configuration that included fixed emplacements.[1068]

Locating machine guns was a difficult task, fraught with frustration for all concerned. Analysis of aerial coverage of the front was constantly cross-checked with the policies cited from captured German documents to predict machine-gun locations within the sector. Capitaine De Bissy, remarked on the challenge: 'One is ... compelled to give up almost entirely any attempt to discover the exact position of a machine gun from a photograph, but after examining the photograph, if one cannot say definitely there is a machine gun at such and such a point, one can say, if there is a machine gun in this sector, it is almost certainly at such and such a point.' [1069] As a rule, machine guns were found in the vicinity of communication trenches, allowing for rapid retrieval to the second line. They were sometimes placed between traverses to increase protection from enfilade fire. The interval between the traverses included only the shelter for the machine gun, its ammunition and the machine gunners.

By 1917 locating machine guns via aerial photographic interpretation was deemed almost impossible. 'The search for machine guns is more and more difficult.'[1070] At the commencement of combat, the machine guns were retrieved from shelters and placed at designated locations within the trench network. For machine-gun location, the stereoscopic photograph was important. Wire entanglements served as a guide to the intervals and the obstacles to be swept by machine guns.[1071] In 1917, the focus for discovering machine guns shifted to the rear trenches. Recognition features on the photograph included niches, observed indentations on the trench line, salients where the machine gun was a few yards to the rear, shelters that appeared to be strongly defended (calling upon stereoscopic coverage to reveal any spoil) and loopholes at a certain distance from the trench and within easy reach of a communication trench. Since the machine gun reinforced the flank, any associations with that role were also borne in mind during the examination of the photograph.[1072]

In the last year, as the fixed emplacements became rare, unique armour and turret configurations sometimes appeared at the head of a long winding sap running from a firing trench and surrounded by wire entanglements. Like the rest, these bunker-type guns were constructed with a view to providing flanking fire.[1073] Potential signatures for analysis included 'V' shaped marks in the forward edge of a parapet to permit the traverse of the gun. At the back of the 'V' a small dark spot similar to a shelter entrance provided a possible signature. What emplacements were left took the form of a square tray or concrete platform that varied in appearance with the sun angle, showing up as a white mark with a dark edge or as a comparatively dark square.[1074] Attention to all sources of intelligence was required to locate and destroy them. Prisoner interrogation concentrated on specific details that covered location, training and doctrine. Ground observation focussed on location and direction of fire. Machine guns were usually discovered by deduction from all available evidence.[1075] Once a position was determined, it was highlighted on the Plan Directeur map.[1076] The following 1st Army intelligence summary revealed a host of machine-gun emplacements in a given sector:

> Study of photographs taken Sept. 2.
>
> M.G. emplacements: 67.98–36.90; 67.90–36.86; 68.03–37.03.
>
> New wire at 69.05–37.20 with M.G. emplacement in corner of M.G. emplacement at 68.80–36.88. Tr. Du Cri evidently used considerably at night for reliefs, rations, etc.
>
> Path, much used, running N. from edge of woods parallel to Boyau des Releves, along E. edge of quarries with camouflage erected along E. side. Probable quarry entrance at 73.92–38.24 and 73.92–38.28.
>
> Covered. M.G. emplacement at 69.6–36.7. 2 M.G. emplacements at 68.80–36.86.[1077]

MINENWERFER (TRENCH MORTAR)

The minenwerfer provided continual fire at short range. The following German description of minenwerfer emphasised their flexible firepower:

In General. The field grenade fire, in close range high angle fire (up to 1300 m.) is mainly effective against living targets. Supports Artillery during the preparatory bombardment, often even after the beginning of the attack. Effective fire with rapidity up to 20 rounds a minute.

In action around shellholes and in fights for strong-points. In action against Infantry (Riflemen). As a consequence of great mobility especially adapted for the quick breaking up of enemy resistance in combat for positions and strong points. Accompanying Stormtroops and storm Infantry, firing from flat trajectory carriage, effective against living targets, also against prone riflemen at ranges up to 1,000 m. It replaces Field-Artillery the Ammunition supply requires special attention.[1078]

Light minenwerfers were encased within an armoured compartment or casemate, having a photographic signature of a whitish or grayish blot representing the hole of the tube. These holes were either round or square and their depth was determined by examining the shade created within the tube.[1079] A typical minenwerfer firing pit was akin to a sap along the communication trench. Medium-calibre minenwerfer were sometimes identified in shelters that had the roof removed. Larger minenwerfer were adjacent to supply lines for easier access to munitions stockpiles.[1080] Another approach to discovering minenwerfer locations was through a simple formula developed by the British. Thanks to years of experience, they knew that the weapon was only as effective as its range. Minenwerfer emplacements were identified within the trench network following a line drawn parallel to the British front line at a distance of about 500 yards to 700 yards. German observation posts were usually linked to the minenwerfer squad to determine the effect of fire.[1081] One challenge to interpreting minenwerfer positions was the similarity to the latrine position. Each was often mistaken for the other. The adjacent sap gave a clue. The sap leading to the minenwerfer was irregularly constructed in a zigzag pattern while that to the latrines was generally a straight path.[1082]

CHEMICAL MUNITIONS

The aerial reconnaissance search for chemical warfare continued throughout the war. Chemical weapons made the battlefield experience even more hideous. In the final months of the war, the Allies' Supreme War Council alluded to the possibility of finding enemy gas projector positions through aerial reconnaissance.[1083] It remained a priority intelligence target. Post-war discussions on aerial photography in the popular media alluded to the fact that gas attacks were prevented by aerial photographs revealing the presence of gas cylinders in the German front lines. The media also highlighted the fact that post-battle photography also aided in discovering and rescuing troops isolated after a gas attack.[1084]

Tactical coverage of the Western Front gave new meaning to the role of intelligence. Thanks to the stabilised front line the aerial photographic interpreter could hone skills over well trodden ground and provide the commander with as precise a depiction of the battlefield as possible. Such precision was best symbolised by the ability of the interpreter to detect individual footpaths along the trench network. Intense scrutiny forced the enemy to become active at night or work underground. The machine-gun teams, as the best example, would seek cover until the last moment.

The Western Front did not end with the third line of trenches. The tactical sector of the front line gained its strength from the strategical rear echelon. A new array of targets and weapons challenged intelligence to define the Western Front in terms of sustainability. The array of targets in the rear echelon proved equally significant as combatants constantly sought a breakthrough and a return to battlefield mobility.

The use of chemical weapons in the First World War still shocks today. The aerial photograph shows chemical munitions in use by the French. (*National Geographic*, January 1918)

CHAPTER 14

Strategical Targets in the Rear Echelon

Capitaine Paul-Lewis Weiller recognised that strategical targets were important for the commanders' understanding of the total battlefield. When his Weiller Group focussed on aerial reconnaissance of targets well beyond the front, he institutionalised an intelligence process that remains to the present day.[1085] Strategical analysis determined the enemy's ability to sustain a major operation; going beyond the immediate front sectors to cover an 'objective at a great distance' followed by the ability to 'ascertain the damage done'.[1086] (The term 'strategical' became shortened to 'strategic' post-war.) Its significance remains today. For the combatant of the Great War, the strategical rear echelon covered a broad range of very important targets to include rail networks, depots/ dumps, long-range artillery batteries, aerodromes and landing ground, ports, industries and population centres.[1087]

RAIL NETWORKS

Strategic sustainability was accomplished through an extensive array of rail networks. Intelligence priorities included determining the extent of activity from the amount of rolling stock in the yards or nearby supply depots (dumps) to ascertain if activity was normal or abnormal.[1088] Both tactical and strategical analysis required a significant assessment of enemy operations, through close study of railways and trench tramways. Besides the normal gauge lines, which were mostly built before the war, the Germans made an extensive use of the light railway system in the occupied portions of France and Belgium. They constructed new tracks near the front on a large scale. Personnel, stores and material were brought up by side tracks and branch lines, which ran to nearly all important points and villages in the front line system. The battery groups and single heavy batteries were almost entirely supplied by this method. Supply depots were located at the breaks of gauge as well as at various points along the tracks. Railway stations had unloading platforms for transferring supplies to the narrower tracks and switches. Each division employed a central supply station to support the reinforcement. The depots had an overhead appearance of large squares or rectangles separated by alleys and garage ways. Depots

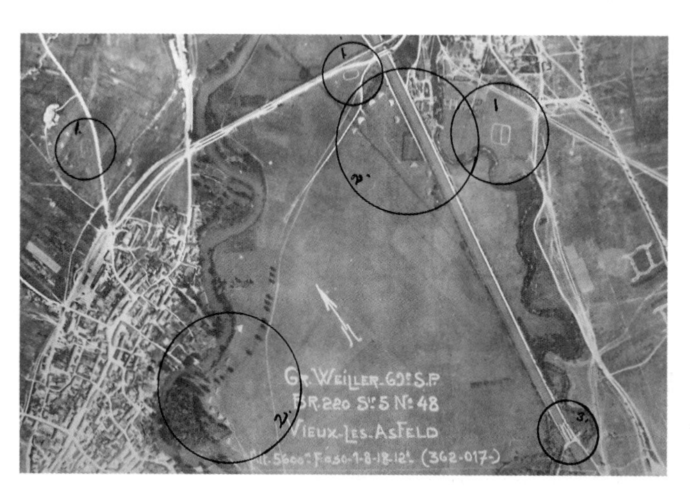

Weiller Group escadrille Br 220 aerial photograph of Vieux-lès-Asfeld, 25km north of Reims, taken on 1 August 1918. Photographic interpreters from SPAé 69 accomplished the analysis. Annotations show strategical interest targets of (1) training areas, (2) aerodromes, and (3) canal lock. (RG 120, NARA)

Strategical reconnaissance continually monitored rail activity for indications of increased support to the front sector. (Library of Congress, Brigadier General William Mitchell Papers)

supporting food supplies showed extensive pedestrian and vehicle activity.[1089] Train movements were a high priority target for aerial observation and intelligence reporting. This was reflected in the British policy that mandated lower echelon squadrons supporting Corps sectors to report on train movements as quickly as possible to higher echelon Army Branch Intelligence. The reports were consolidated with information from adjacent armies into the daily summary.[1090] Here is an example of strategical reconnaissance reporting of activity at a German-controlled rail station:

> Photos of September 14th taken at 11 hours show the following: The more northerly of the two large stations at PAGNY-sur-MOSELLE has been completely destroyed. No especial destruction is shown among the tracks, nor in the material dump on the platform, which is larger than usual. The big central building (machine shop) in the MONTIGNY station at METZ has been hit, and a big slice cut off from its southwestern corner. The tracks in front were also hit and a few cars have been wiped out.[1091]

DEPOTS/DUMPS

The insatiable demand for supplies created an extensive infrastructure of depots/dumps. The needs of infantry and artillery dominated the inventory of goods delivered forward. Supply dumps containing less precarious material were stored without attention to any particular arrangement. Camouflage was applied when available. Volatile goods demanded greater attention. Projectiles were stored in pits separated by thick splinter-proof earth shields designed to limit the effects of a partial explosion. Explosive powder charges were spaced at intervals under cover in smaller protected shelters.[1092]

Photographic interpreters searched narrow-gauge railway networks to co-locate any corresponding depot. Analysis showed points of convergence along the routes, tracks and railways of narrow and normal gauge as well as parallel road networks required for movement of the munitions.[1093] Large munitions depots were generally situated at rail-heads or convenient rail systems. Smaller depots in the vicinity of reserve trenches were quite common, generally taking the form of open dugouts for projectiles with separate accommodations for explosives.[1094] The volatile nature of munitions storage required precautions. The arrangement became a typical signature for large munitions depots. RNAS reporting of large coastal artillery reflected the standard: 'Numerous wooden huts separated by a considerable distance from one another, and surrounded by earth ramparts in order to localise the effects of a possible explosion are placed alongside the road or railway.'[1095] The following is an American intelligence summary concerning depots and dumps:

> The ammunition dumps of Bois du CHAPELET and the one north of CHAMBLEY show no change since September 2nd. These dumps were reported burning two days ago. The ammunition dump south of MARS-la-TOUR shows no change since September 2nd, and that of St JEAN les BUZY no change since May 19th. The materiel dumps at the last two towns show no change since September 2nd and July 19th respectively. The indications in this area do not suggest any withdrawal by the enemy beyond the Hindenburg Line.[1096]

HEAVY ARTILLERY

One of the highest priority strategical targets of the First World War was the long-range artillery, classified by the French as the Heavy Artillery of Great Power (*artillerie lourde à grande puissance* (ALGP)). The most notorious, the 210mm naval gun, with a range of 132km (82 miles), was a strategical threat and a weapon of terror against the French people.[1097] The gun provided a random threat to population centres. Paris, Dunkirk, Châlons and Nancy were the victims of the German long-range terror weapons. Such weapons acquired the soubriquet of the city that they targeted (ie the Paris Gun). The 210mm was derived from worn-out 380mm (known as the Long Max) railroad guns.[1098] The 380mm gun was equally impressive. It weighed 154 tons. Set on a cradle weighing 26 tons, the gun carriage was massive. The entire emplacement weighed 400 tons, rotating on 96 eight-inch steel roller balls.[1099] The numbers of this type of artillery were not substantial enough to contribute to any advance or success at the front. The 380mm gun had a distinctive signature for the photographic interpreter. A branch of normal gauge railroad supported the weapon system, providing the ammunition required. Should the gun be moved, the rail network was used. Constant surveillance of the railway network was maintained to identify the weapon before it was put into action. Its setup took a long time. The entire complex was positioned in woods and extensively camouflaged. This was, except for the visible gun barrel, typical of heavy batteries. The larger the gun, the more camouflage was employed. The blast of the gun was so great that it often killed some of the vegetation in the vicinity, which became an additional signature for finding the weapon system. The gun was mounted on a concrete platform about 20 to 25 metres in diameter and protected by a shield. Before the gun was fired, it was contained within a rectangular hangar to conceal its construction. The gun complex was surrounded by shelters and supplied through established logistic networks.[1100]

To complicate Allied intelligence efforts to locate the 380mm gun, the Germans co-located up to 30 batteries

A 380 mm gun photographed in the woods to the upper right of the frame. Note the craters from Allied counter artillery. (*Étude et Exploitation des Photographies Aériennes*)

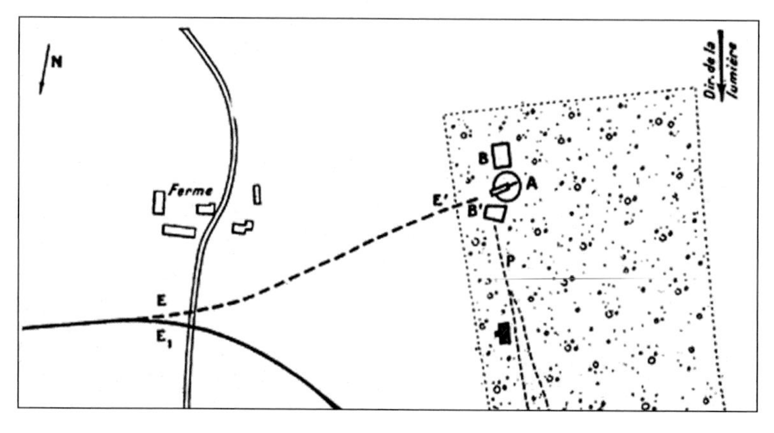

The companion line drawing shows the orientation of the gun. (*Étude et Exploitation des Photographies Aériennes*)

of various calibres in the region. As the gun fired, a simultaneous barrage by all the units occurred; masking the gun's actual firing location and complicating any observation from flash and sound-ranging units, captive balloons or aerial observation. The most fruitful sources of information came from aerial photography of the sector or from radiogoniometric detection of the gun batteries' supporting radio units.[1101]

The photographic interpreter developed lists of signatures to identify a specific German gun type. Besides the 380mm, the Germans had a variety of large artillery that was identified by proximity to rail networks and the gun carriage. The 21cm and 24cm resided on a platform at the bottom of a trench of 12 by 16 metres and about 1.25 metres in depth, bordered by two sections of parallel railways. The 355mm cannon used an analogous disposition with larger trench dimension of 15 by 17 by 3 metres. The 28cm and 305mm cannons were used almost exclusively on the sea coast and mounted on massive foundations of concrete. 305mm and some 420mm mortars were transported over roads by tractors. The long cannon of 21cm and 24cm and the long range 420mm mortars (16,000 metres range approximately) fired from trucks or platforms. 280mm, 305mm and 355mm cannon fired from a platform connected by a railway. The photographic interpreter was reminded when searching for these weapons that dummy emplacements and artillery were also located within the area to complicate the targeting.[1102]

AERODROMES AND LANDING GROUNDS

Unoccupied aerodromes were considered as important as occupied for surveillance. As one lecture to aircrews put it, 'The German aviation is marvellously mobile.'[1103] Key to the interpretation was identification of the aeroplanes and the type of hangars that provided

Aerial coverage of Marville landing ground. (NARA, RG 120, Box 5778)

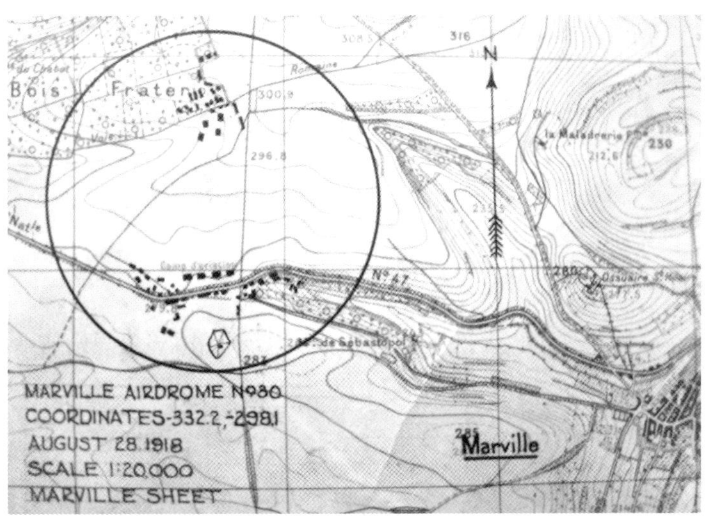

1/20,000 scale map shows the aerodrome hangars in relation to the aerial photograph. (NARA, RG 120, Box 5778)

protection and maintenance.[1104] As signified by the term 'Landing Ground,' the original airfields were temporary facilities. Many airfields included smaller canvas hangars near houses that served as billets.[1105]

Aerial photographic analysis established a routine for aerodrome operations. If observed activity noted new construction of aerodromes or hangars at existing fields, the enemy 'probably intends to make some move on the ground' to allow for aviation to increase to 'conceal his preparedness'.[1106] Construction schedules included developing the landing ground and erection of hangars and tents. The first accommodation consisted of small canvas hangars followed as quickly as possible by sheds. As new hangars were built, a corresponding decrease in the rear echelon aerodromes was observed.[1107] Aerial photography also verified the type, locations and numbers of aeroplanes at the field. One aerodrome feature visible on aerial photography was the Landing 'T,' a signalling device used to alert the pilot to the wind direction.[1108] Over time the Landing 'T' was replaced by smoke from a well-placed and controlled fire. The Allies learned from prisoner interrogations that German aircrews preferred a less conspicuous symbol at the airfield since aerial observation could detect the smoke originating from the field.[1109]

The shape of aerodrome hangars varied with the type of aeroplane. Small canvas hangars accommodating two scouts or one two-seater machine appeared semi-circular in shape and of a light colour. The smaller pursuit planes used rectangular shaped dark sheds. Twin-engined bombers had large canvas hangars for protection. The rectangular 'bessonneaux'-type hangar provided cover for several aeroplanes. Aircraft parks where extensive maintenance was undertaken usually had large wooden hangars or lattice girder structures.[1110]

The British published extensive information on aerodrome location and activity primarily based on aerial photographs. A 'Weekly list of Hostile Aerodromes on the British Front' described aerodromes of interest to the respective sector. It covered the numbers of observed sheds and hangars at each aerodrome. Additional sources on aerodromes came from prisoner interrogations, agents, and repatriated citizens.[1111] Whenever an aerodrome was discovered, the Branch Intelligence Section generated six copies of all aerial photographs covering the target. For previously discovered aerodromes, they generated four copies.[1112] The Americans also produced their own weekly assessment on aerodrome status.[1113]

French intelligence established a targeting plan that required all German aerodromes be identified and

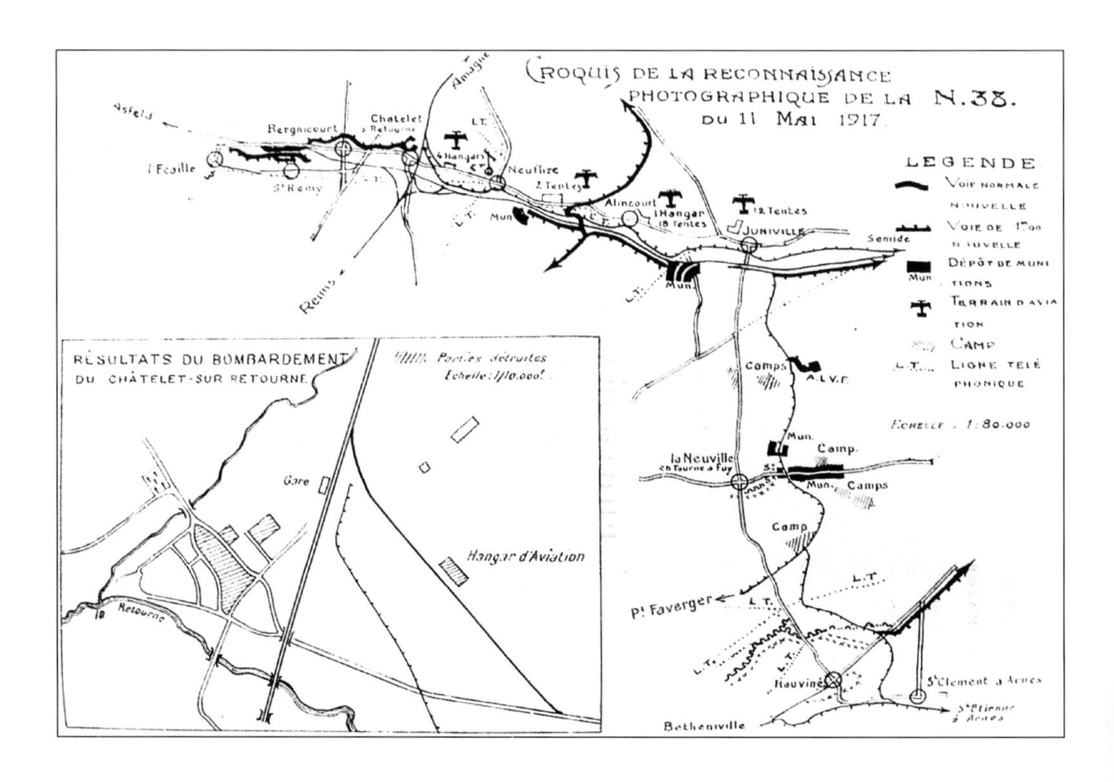

Croquis de la reconniassance derived from aerial photographs from Escadrille N 38 flying both Nieuport 17 and Spad VII in support of Groupe de combat 15 showing key targets in the Chatelet-sur-Retourne sector. (Eugène Pépin Papers)

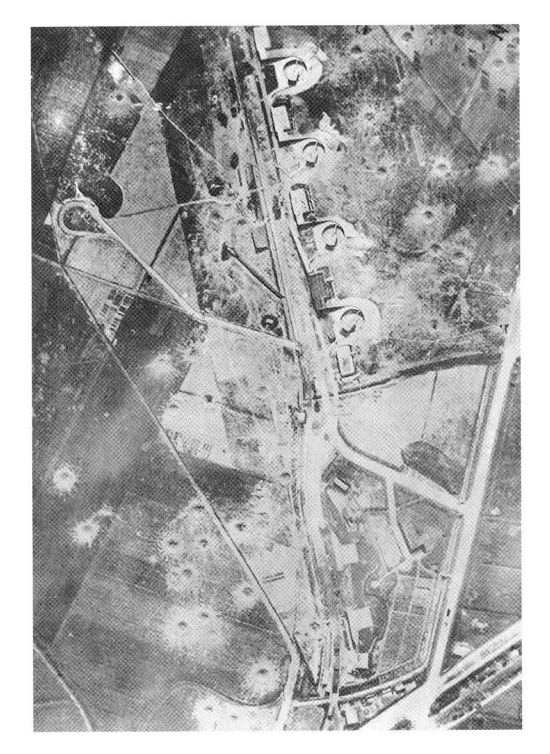

Left: Coastal artillery emplacements based on the Battery Hindenburg model were closely examined by French and British naval analysts for direction of fire. Battery Tirpitz was especially important, for the guns were aimed towards the vital port city of Dunkirk. (*Étude et Exploitation des Photographies Aériennes*)

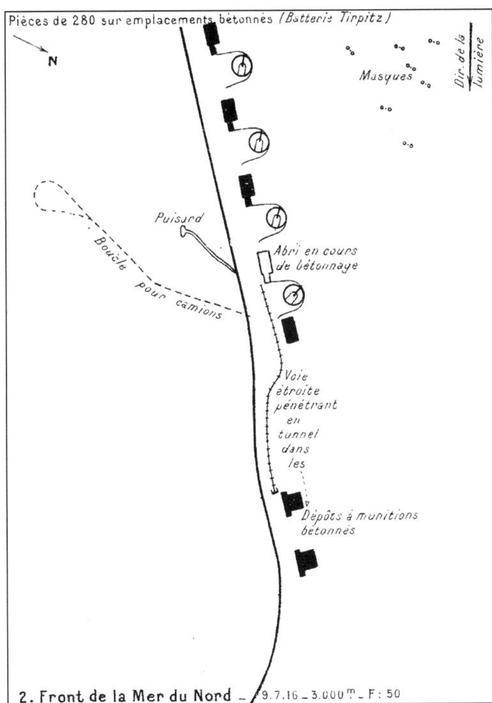

Below left: The companion line drawing illustrates the munitions lines to the artillery. (*Étude et Exploitation des Photographies Aériennes*)

photographed. Target value was based on the number of tents and hangars identified on the aerial photograph. The French established six regions for German airfield analysis; the Somme, Oise, Reims, Meuse, Moselle, and the region south of Strasbourg. Other 2e Bureau branches studied the entire German air force; focussing on order of battle based on aéroplane type, German aerial daytime tactics, night signalling operations and the use of ground beacons for aéroplane routes. A 2e Bureau liaison from GQG was assigned to each French bombardment group to provide the aircrews with any updates and to forward information gleaned from the bombardment sortie.[1114] An American intelligence summary illustrates key features:

> Photos of September 15 show two planes on the east field and four on the west field at BATILLY. Both of these airdromes have hitherto been reported as unoccupied. Many trucks appear on the field, probably indicating that air squadrons are moving in from vacated airdromes nearer the front.[1115]

Along the coastline, seaplane bases were a priority target for RNAS and French Navy aerial reconnaissance. During 1916, only one German seaplane base existed on the Belgian coast. German seaplanes soon proliferated, accomplishing pursuit, torpedo, aerial bombardment, and reconnaissance missions. 'Reconnaissances have enabled us to ascertain that the enemy in 1917 used his seaplanes for scouting over the sea, without needing to use his land machines.'[1116]

PORTS, INDUSTRIES, POPULATION CENTRES

Intelligence monitored the German threat against the vital sea lanes connecting France and Great Britain. Aerial coverage of the entire coastline from the Dutch border to Calais was acquired by British and French maritime aviation. Target descriptions covered submarine shelters; naval order of battle to include destroyers, torpedo boats, and submarines; ports; large artillery batteries in defence of the ports; and associated aeroplanes and seaplanes. The primary means of photographing these targets was by seaplanes using hand-held cameras.

The success of the Zeebrugge raid (albeit temporary) was verified by aerial photographic coverage. (Barrie Pitt, *Zeebrugge*)

RNAS analysis of submarine shelters along the coast showed six bases, two each at Bruges, Ostend, and Zeebrugge. Aerial photography confirmed the use of floating docks to repair the submarines. This high-value target was constantly scrutinised through all available intelligence. 'Different reconnaissances made in 1917 show that there were in the ports on the Belgian coast and at Bruges at least 15 submarines at anchor or on the stocks, without counting those that were submerged at sea or in the shelters.'[1117]

Aerial photographic analysis supported naval order of battle assessments and port activity. French reporting accomplished on 21 November 1916 showed extensive activity: 'enemy boats alongside the Quay, T.B.'s [torpedo boats] under steam, moving trains, rows of trucks, new railway lines, new buildings, numerous dugouts, smoke, etc.'[1118] When the port at Ostend was struck by aerial bombardment in June 1917, destroyers and torpedo boats diverted to Bruges.[1119] Aerial photographic analysis covering the vital Zeebrugge Mole was constant. Coverage pinpointed the location of minefields and confirmed the access channels. Additional trenches and barbed wire along the shore showed the extent of German efforts to protect against possible amphibious landings.[1120]

German long-range naval artillery was always an intelligence priority. French and British naval analysts shared each other's coverage to better determine the threat. One study by a French aerial photographic interpreter using RNAS coverage provided an in-depth assessment of Battery Hindenberg on the Belgian coast. Aerial photography provided descriptions of the guns' rotation and operating configuration, and measurements of the gun battery spacing, approximately 35 metres apart. Intelligence on the battery acquired from a deserter gave analysts an approximate size of the gun barrels (280 mm; about 11 inches). Gun rotation was estimated to be 165°, with the zone of fire 'confined to the sea'.[1121] Subsequent RNAS reporting in 1917 described ongoing construction of 'batteries of large calibre' to have deep emplacements and concrete shelters for ammunition and the gun crew. The distance between the guns was increased with the large calibre gun batteries divided into two groups of two guns. Around several batteries trench works provided additional defence against an infantry attack.[1122]

In April 1917 RNAS photographed Ghent and Antwerp four times at an altitude of 20,000 feet. They subsequently photographed Ostend, Bruges, and Zeebrugge using a 'big camera'. RNAS reports said the sequence of coverage 'helped understand certain works in connection with naval and submarine warfare'. Subsequent coverage helped the British analysts understand construction underway on coastal artillery, aerodromes, anti-aircraft batteries, and munitions depots. Strategical analysis benefited when RNAS merged with RFC into the RAF. Now experts from both services could apply their analytical skills to ongoing activities along the coast as well as establish precedent for subsequent strategical targets.[1123] When the Royal Navy conducted its historic raid on the German naval base at Zeebrugge, Belgium, on 22–23 April 1918, the process of in-depth strategical aerial photographic analysis was well established.[1124] Though it probably saved no lives in that suicidally courageous attack, at least the men knew what faced them.

ANTI-AIRCRAFT ARTILLERY

Anti-aircraft artillery began with an anti-balloon gun prior to the start of the war. Formations of anti-aircraft batteries were integral to the front line and the

German antiaircraft artillery was set on a circular platform to allow fire in every direction. (George P. Neumann, *Die Deutschen Luftstreitkraste im Weltkriege*)

strategic rear. The initial photographic signature in 1915 was a round pit with the pivot in the centre (identified by its circular platform) which allowed the gun to fire in every direction.[1125] From the beginning of 1917, the Germans employed small-calibre (20mm, 33mm, 35mm) anti-aircraft cannon to defend against low altitude raids on the front line trenches.[1126] The circular pattern was retained throughout the war and was easy to recognise. The normal configuration was two, three or four guns per battery. The breastworks for the guns were constructed in several formations throughout the defended area to allow for rapid transfer of the weapons to established positions.[1127] Anti-aircraft gun emplacements did not necessarily signify occupation. The guns left no blast marks on the ground.[1128]

RNAS analysis of anti-aircraft artillery along the coastline showed various signatures. Anti-aircraft was in place to protect the large-calibre artillery that threatened Allied naval operations, as well as to protect the munitions depots and aerodromes in the vicinity. 'The tendency has been not to place anti-aircraft batteries close to the objective it is desired to protect, but at a certain distance from them.'[1129]

TARGETS IN THE GERMAN HEARTLAND

The French prioritised German industrial complexes as the primary target regions. Four primary target areas, Mannheim-Ludwigshafen, Mainz, Köln, and Saar-Lorraine-

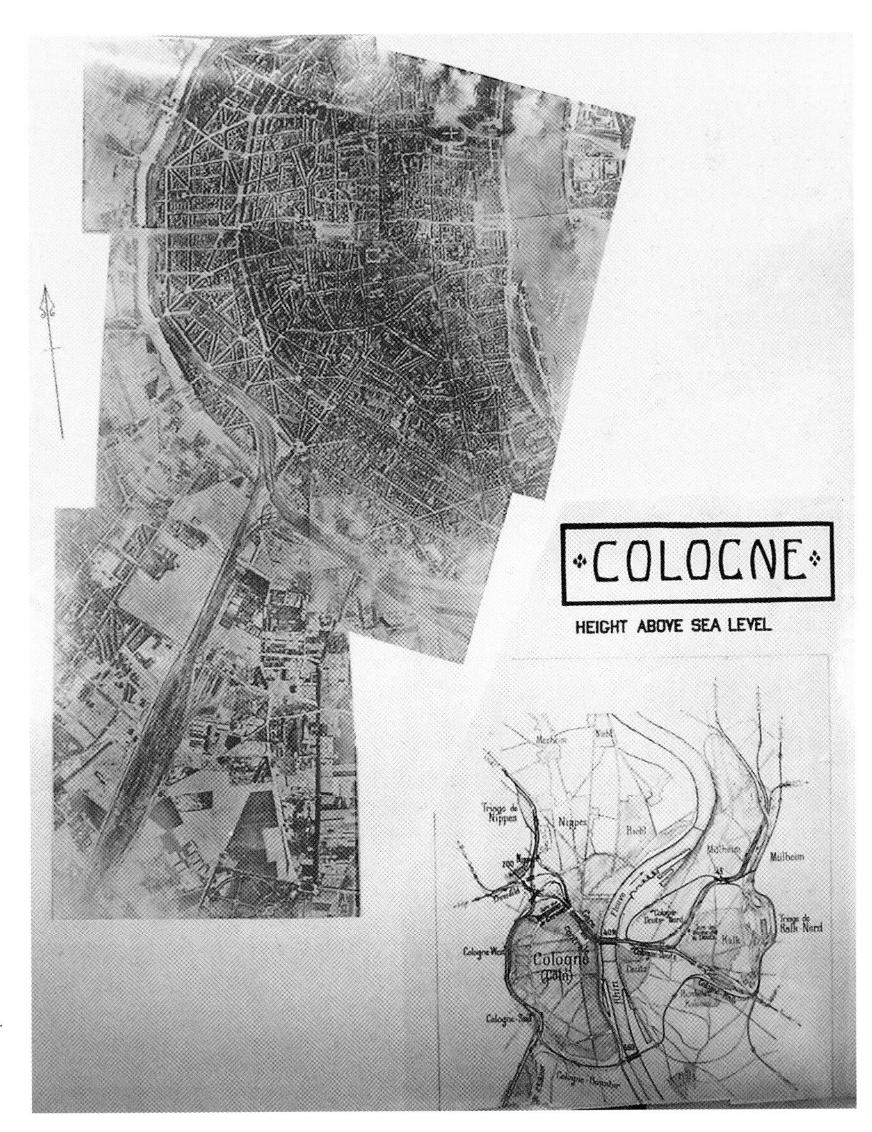

Detailed aerial coverage of the largest German city on the Rhine taken by RAF No. 55 Squadron. (Eugène Pépin Papers, MA&E)

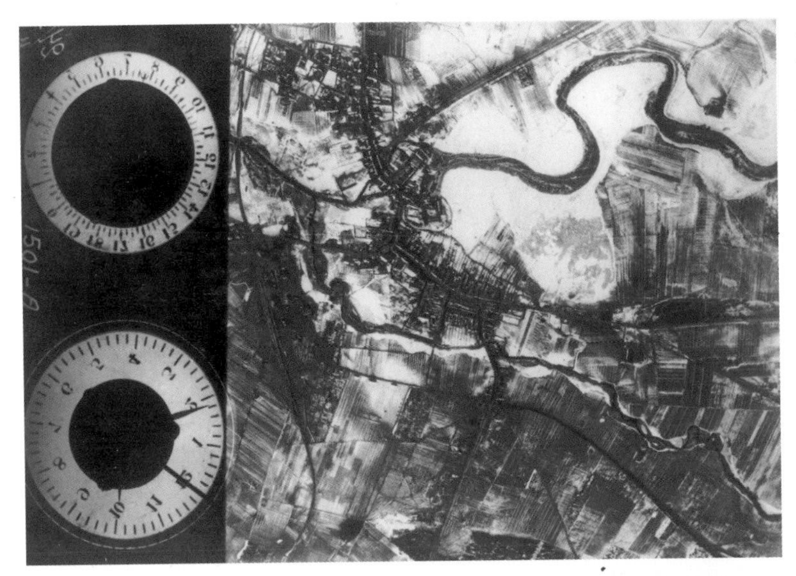

A superb example of strategical aerial photography taken by the 100cm Ulyanin aerial camera on board the Il'ya Muromets. (Marat Khairulin)

General Aleksei Brusilov personally inspects EVK leadership and the latest Il'ya Muromets Type Ye in 1916 prior to the most successful Russian offensive in the war. (Marat Khairulin)

Luxembourg were selected. Of the four, the French decided that the closest target area, Saar-Lorraine-Luxembourg, was their primary objective for strategical aerial bombardment. Their goal was to damage German industry. The impact on German citizen morale was not a priority for the French.[1130] They did attempt aerial bombardment against German civilian targets in retaliation for German attacks on French cities.[1131] The British considered the effect on morale as a secondary 'bonus'.[1132] By October 1917, the French had narrowed their assessment on strategical bombardment targets to isolation, or what they termed, 'internal blockade'. Isolation made it difficult for the Germans to move critical raw materials. The important Metz-Luxembourg rail links were a focus of much strategical bombardment.[1133]

Had the war continued beyond the Armistice, the Allies were prepared to conduct a more aggressive strategic bombing campaign against targets deep within the German heartland. British long-range aerial reconnaissance conducted by No. 55 Squadron covered many German cities, to include Frankfurt, Trier, and Köln.[1134] A unique strategic targeting occurred when the Austro-Hungarian Empire fell on 8 November 1918. The Allies planned to use bombers flying out of Prague to strike key German cities, including Berlin, Munich, Leipzig, Nürnberg, Frankfurt an der Oder, and Breslau. The British planned to fly long-range Handley-Page bomber missions against Berlin. The Armistice cancelled the plan.[1135]

The strategical targeting legacy defined aerial reconnaissance for the remainder of the twentieth century. Sustainability governed many decisions on the Western Front. Military planners were dependent on any body of knowledge that could describe the German's ability to conduct an offensive or defend against an Allied thrust. Most intelligence information was a tactical summary. The beginnings of strategical analysis were based on determining the strategic makeup of the rear echelon. The ability and intention to wreak havoc against vulnerable population centres carried a new message that modern warfare was total war.

CHAPTER 15

Artillery Organisations

Artillery, the 'King of Battle', ruled throughout positional war, determining the limits of conventional lethality. Artillery focussed on trench destruction and wire cutting prior to the infantry's advance. It also served as the primary means of disrupting enemy communications and countering the enemy's own artillery potential through counter-battery. Calculating the numbers of artillery pieces in a given sector was a priority throughout the war. Any discernible changes were a clue to enemy intentions.[1136] Artillery batteries were located along the entire front. An American estimate at the time showed the Germans had one field gun for every 19 yards of front, a light field howitzer for every 57 yards, a heavy field howitzer and a 6-inch gun for every 128 yards, a medium siege howitzer for every 256 yards, and a heavy siege gun for every 512 yards.[1137] German statements about artillery's role illustrated the offensive focus: 'Main object is to help the Infantry break quickly through the enemy strong points, to keep the attack going even in the face of strong enemy resistance, and to increase the element of surprise.'[1138]

Extensive analysis within the city limits reveals a detailed layout of German artillery concealed by buildings. (*Illustrations,* S.S. 550a)

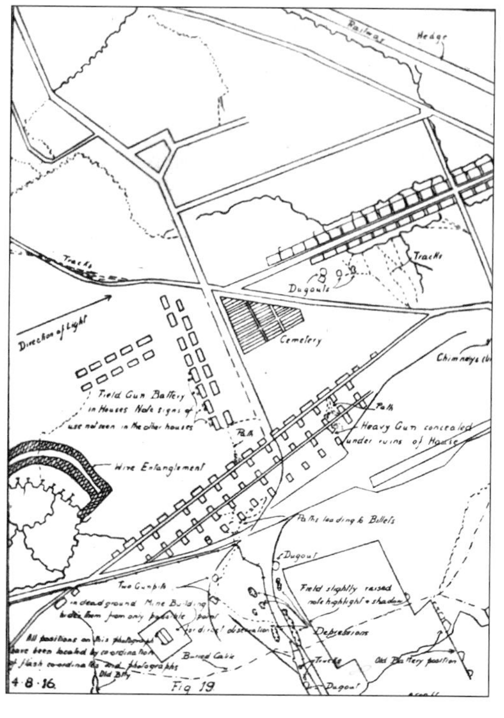

The companion line drawing clearly shows the details of artillery within the area. (*Illustrations,* S.S. 550a)

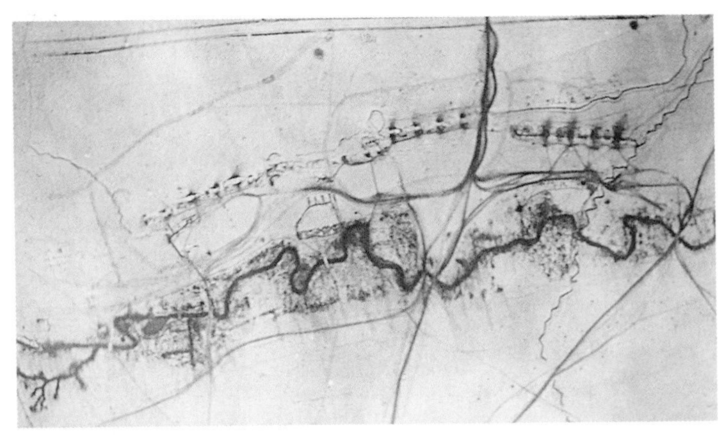

The search for artillery batteries was assisted by the black blast marks discernible on the snow. (*Étude et Exploitation des Photographies Aériennes*)

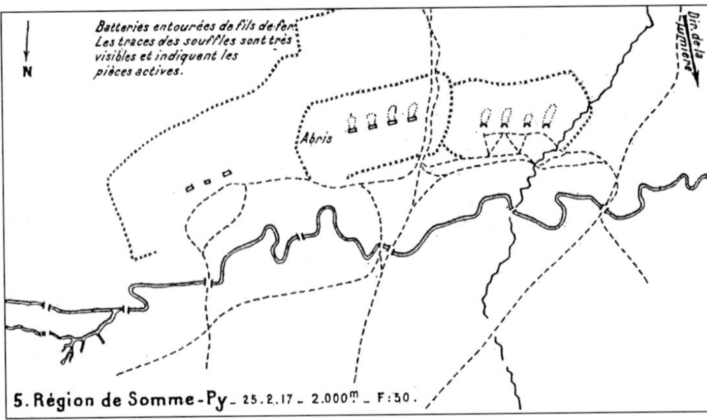

The line drawing shows the battery configuration. (*Étude et Exploitation des Photographies Aériennes*)

Artillery was also delivered chemical munitions. A captured German field directive confirmed German employment of gas shells against Allied batteries.

> The general appreciation of the value of the gas shell has now risen very much, as is proved by the high demands. Although the proper recognition of the useful nature of gas-shell bombardments is very gratifying, the general principle of economy has in this case to be considered, to a special degree, as the supplies are limited. It is therefore of special importance to have a sufficient amount of gas shell in the battery positions, in order, at the right moment, when the enemy is starting attacks on a larger scale, to neutralise the enemy's artillery and thus, at the same time, nip the enemy's infantry attack in the bud. Small harassing bombardments are to be confined to special cases.[1139]

LOCATING ENEMY ARTILLERY

Aerial observation placed a high priority on locating enemy artillery in a timely fashion so that acquired data could help achieve accurate counter-battery fire. When a major offensive was planned, aerial observation was airborne and on-scene to assist the first lethal rounds. Artillery operations were continuous, regardless of weather or darkness. A steady flow of information on enemy artillery was the daily standard. If the flow was disrupted for any reason, the squadron commander was held responsible for informing the artillery battery.[1140] Guidance to aerial reconnaissance was part of the planning process. When the British IX Corps during the Messines offensive considered the need for annihilation of key targets, the planning showed accurate fire was critical:

> The bombardment will be carried out not with the view of total destruction of the enemy's trench system but in order to destroy machine-gun emplacements, strong points, headquarters, and other selected points – by the destruction of which the enemy defensive system will be thoroughly disorganised. The further object of the bombardment, in addition to killing as many of the enemy as possible, is to thoroughly demoralise and starve out the troops holding the trenches. This will

> be accomplished by barraging continuously day and night the enemy communications and lines of approach and by shelling his billets and dumps.[1141]

Early in the war, emplacements (*épaulements*) provided sheltered protection for the guns and the support wagons. The photographic signature showed an alignment of four to six emplacements.[1142] By 1917, the emplacements were abandoned to increase artillery survivability through mobility.[1143] Targeting enemy lines of approach and known intersections was called upon when reconnaissance data was lacking. When the artillery resorted to mobile operations, especially in the last few months of combat, the batteries generally fired in the open without protection for guns or personnel. Under these circumstances natural features were utilised to the fullest extent to escape photographic detection.[1144] The interpreter looked for any signs of tracks, where a sudden stop or modification to the track suggested a nearby battery position.[1145] Once the artillery was discovered and coordinates determined, the perishable intelligence report had to be forwarded as quickly as possible.

Topography not only shaped the battlefield but the role of artillery. Northern France's topography included numerous quarries, villages, hedges and closely wooded terrain. Houses were particularly favoured for providing artillery batteries with cover and potential camouflage.[1146] A primary signature for detection was the blast mark. The force of the shock wave created visible erosion on the surface that was discernible on the print as a white scoring.[1147] To avoid blast marks being discovered, the battery sought cover from forests, houses, hedges, orchards and ditches.[1148] The hilly terrain of Northern France helped. As artillery was put into position, the Germans sought a reverse slope slightly behind the crest of a ridge to cover artillery operations from ground observation and ballon observateur.[1149] Such tactics did not work against áeroplanes.

TARGETING ARTILLERY

Intelligence support to targeting artillery was continual. Flash spotting and sound ranging operated day and night, whenever conditions permitted. Intelligence

The interpreter's world was complicated by the deception of dummy batteries. (*Illustrations,* S.S. 550a)

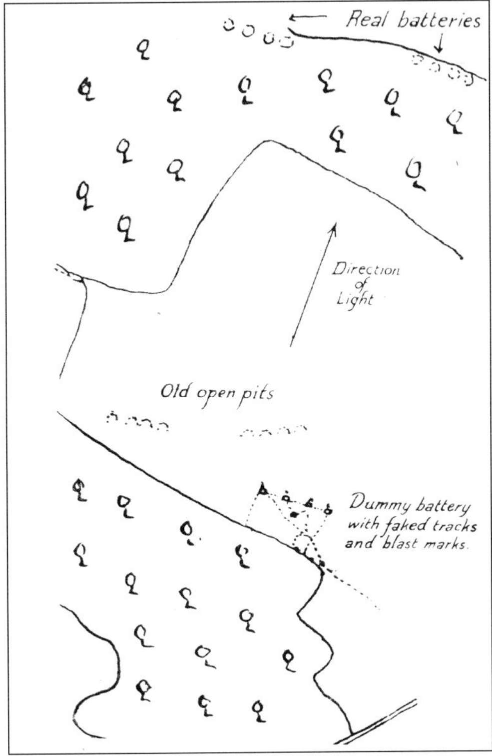

The line drawing depicts the dummy configuration. (*Illustrations,* S.S. 550a)

garnered from prisoner interrogation was only as good as the source.[1150] 'The best means of obtaining information is by photography, for the interpretation of aerial photographs affords information which would not be noted by an observer during a reconnaissance.'[1151] The French used triangulation to direct counter-battery fire. Using aerial photography, the enemy battery location was associated with prominent landmarks on the map. The convergence of the triangulated lines became the targeted area.[1152]

Aerial coverage provided a sense of control and purpose to artillery staffs. With it they possessed information that defined their target sets, to include observation posts, dimensions of the artillery battery area, and traffic. To their delight, the coverage also provided them with an image of ongoing activity that included ground traces, rail terminal points, and telephone lines.[1153] Liaison between the photographic interpreter and artillery unit was strongly encouraged for the mutual benefit that came of understanding each other's craft for one thing, it helped the battery understand its vulnerability from the adversary's aerial camera. The interpreter gained a better appreciation of the realities that confronted artillery in the field. Capitaine Pépin reminded all that '... in most cases photography alone will remain powerless.'[1154]

The British took extensive measures to manage the stream of information available on enemy artillery. In 1915, the British established a special topographic section known as the Compilation Section to monitor and track enemy artillery. Officers assigned to the function were sometimes flooded with over 100 aerial photographs a day covering known and possible enemy battery locations.[1155] A year later at the Somme, the British established the counter-battery staff officer (CBSO) to coordinate reporting from the battery and the squadrons to verify success or failure of targeting.[1156] The CBSOs also focussed on rapidly changing German artillery tactics.[1157] The CBSO managed the Counter-Battery Office operating in a given sector. A detailed battery database was established, containing known battery history, location, all ongoing and subsequent activity, Allied counter-battery activity, and the latest available aerial photograph. The counter-battery database proved very successful, eliminating waste and managing British artillery employment. Reporting on artillery batteries by the observers was given a scale of accuracy: 'A' accuracy signified that the observer could distinguish the separate pits of the battery under observation; 'B' Accuracy could distinguish the front and flank of the battery; and 'C' Accuracy only provided its approximate position. A British Counter-Battery Zone map was created paralleling the Plan Directeur, with updates reprinted and distributed to all concerned.[1158]

Artillery ruled supreme within the confines of positional war. It was the most lethal and effective means by which a commander could annihilate the enemy. Instrumental towards achieving this objective was a well-established liaison network throughout the front linked by an array of available communications. Through this elaborate and extensive network, aerial observation and the intelligence from photographic interpretation were effectively distributed.

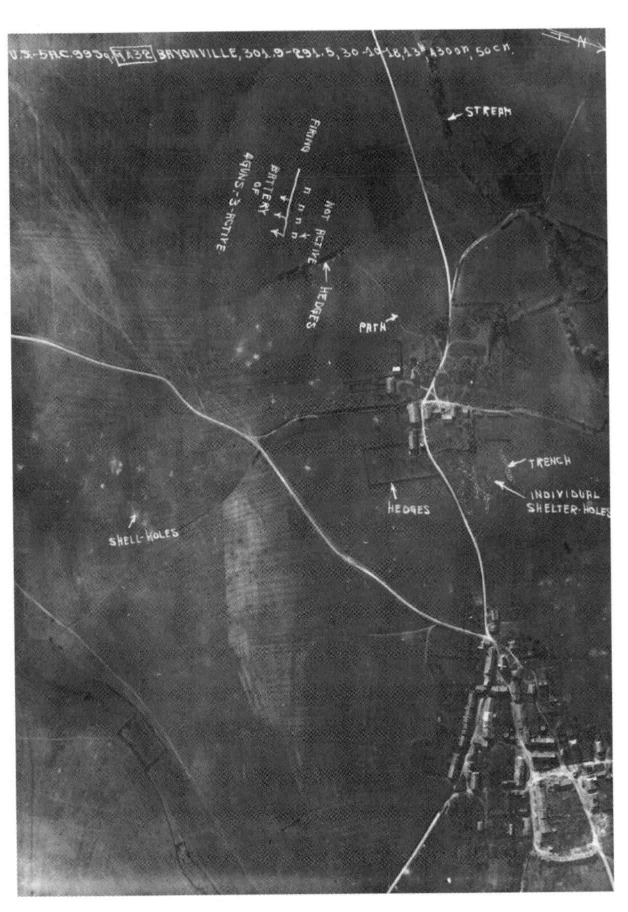

In the absence of a companion line drawing, annotations were effective at showing the location and strength of the artillery battery. (USAFA McDermott Library Special Collections (SMS57))

CHAPTER 16

Command Liaisons and Coordination of Artillery

Despite the recognition of the aerial photograph as the primary intelligence source, photography was recognised as part of the intelligence whole. *Étude et Exploitation des Photographies Aériennes* reminded the practitioners: 'It is not possible to reconstitute the whole organisation of the liaisons of a sector by the mere study of the photographs.'[1159]
It was through liaison that information flowed throughout the front. Liaison was the human-in-the-loop of information transference. French discussion conveyed the critical need for effective liaison:

> The purpose of liaison is to enable the Command to keep itself constantly informed of the situation of the Units under its orders and to supply it with the elements of its decisions, secure, between adjacent Units and among various arms, the transmission of orders, requests, reports, information and in a general manner, of all necessary communications in order to obtain the convergence of efforts, and in particular the intimate co-operation of the infantry and artillery.[1160]

Despite radio's infancy at the start of the war, it quickly became integral to operations. Both sides employed a host of alternative communication sources to ensure awareness of events underway and assist in decision making. From the start of the war a network of radiogoniometric intercept units was established to monitor enemy transmissions. During the war's first battles, signals intelligence revealed critical information on command location and enemy intentions. Triangulated radiogoniometric intercepts of radio emissions helped pinpoint critical nodes in the command and control network that comprised the PC. An added benefit from intercept was identification of units and construction of the enemy force's order of battle. This was essential in positional war's assessment of enemy capabilities.[1161]

Besides radio, both sides employed a variety of methods to communicate. Transmissions between key points such as headquarters and artillery batteries were primarily accomplished through telephone and telegraph.[1162] Portable searchlights communicated with aérostats using pre-arranged signals. Rockets (fireworks) were used as signals to order artillery barrage fire. Audio alarms – bugles, hand sirens (hand-operated klaxons) and oxygen sirens (compressed air klaxons) provided an additional means of communication.[1163]

POSTE DE COMMANDEMENT (PC)

PCs or Postes de Commandement (the British term was Command Post) were the commander's domain; the decision location for the respective zone of action. They were intended to maintain an air of permanence with liaisons to cover any event, any time. Ground observers attached to a particular PC signalled any assumption concerning enemy preparations, reporting changes through TPS. The PC staff received observation reports and combined them with the other sources, to include aerial photographic interpretation. PCs were normally in close proximity to artillery batteries, yet sufficiently distant to avoid destruction if the battery was shelled.[1164] There was no visual standard for PC shelter construction on both sides of the front. It varied from concrete to log cabin.[1165] As the information hub for the given sector, the PCs location was critical, requiring accessibility for the combatants within the sector, yet secure from enemy fire. It was the aerial photographic interpreter's challenge to understand the network as well as be an expert on the surrounding terrain to determine the PC location.[1166]

In the second half of the war, the main defensive line shifted from the front line firing trench to the rear trenches. The forward trench were transformed into 'listening units' (also referred to as listening posts by the British, the French term was *section de repérage par le son* (SRS)) linked to the respective PC of the sector.[1167] They were observed as small excavations either under or just behind the first belt of wire. Listening units became an integral design feature of the formidable Siegfried

Post of command buildings did not follow an established design pattern. This PC was overrun by the Americans at the Meuse-Argonne. (USAFA McDermott Library Special Collections (SMS57))

Stellung; reinforced by co-located machine guns, making their discovery a priority.[1168]

German observation posts served as the focal point for determining the situation facing the sector. Normally the post was constructed with concrete. It appeared to be rising just above the ground surface. Loopholes and periscopes provided access for sighting. German observation posts were constructed apart from the trenches, employing concrete for protection. Other posts were at the entrances of dugout/mine gallery shelters. These facilities exploited wooded areas and terrain features to further mask the construction. Observation posts were a challenge to detect, the entrances being 'camouflaged with care'. When the terrain permitted certain artillery observation posts were placed in front of or in the vicinity of the batteries they served. Observation posts on elevated ground served either the PC or artillery. The more distant observation posts in the rear echelon were noted for the construction, generally wooden or metallic, of towers 20 to 30 metres in height.[1169] The towers (also called pylons) were detected by the shadow of the support poles.[1170] Aerial reconnaissance was continually searching for tracks or saps leading to enemy observation posts. Stereoscopic coverage was ideal as it gave clarity to any footpaths in the vicinity.[1171] It was standard for both sides that observers had to be able to enter and leave their observation post unobserved day or night to avoid detection from the air.[1172]

COMMUNICATION NETWORKS

There were two primary targets for identifying German telephone communication networks: buried cables (cable trenches) and telephone aerial lines. The third target was visual signalling.[1173] 'Telephones form the nervous system of the trenches.'[1174] Telephone line networks were invaluable aids in finding headquarters, a telephone centre contained within a shelter, camouflaged batteries, observation posts or any significant location where enemy orders was disseminated.[1175]

At the beginning of the war, both sides realised that they needed more communication links to enable the commanders to conduct critical conversations with key elements engaged in battle. Command and control was first employed to acquire answers to urgent questions about the ongoing operation. Over time, communication links extended to all elements both at command centres and to key organisations throughout the sector. When the front stabilised during positional war, the requirement for more effective communications between the PC and the artillery battery became obvious and lasted for the remainder of the war. Artillery fire control became refined so that each group or battery was linked through a variety of means to ground and aerial observation stations.[1176]

Most telephone lines were buried across the front. Artillery barrages frequently broke the cable, requiring constant excavation to repair the damage. A temporary solution was to lay the telephone lines on the surface. Cable trenches containing underground cable lines for telephones converged toward locations where either visible or concealed shelters existed. They maintained a comparatively straight course and narrow construction and had the appearance of either a covered sub-aqueduct (a white path slightly in relief) or an uncovered channel (the earth being thrown out only on one side or alternating on both sides). The description for the pathway was described in one manual as having a somewhat 'woolly' appearance.[1177] Photographic analysis required frequent monitoring since the ground signature faded over time. It was an important part of the daily photographic interpretation routine to mark both buried cables and telephone aerial lines when freshly dug, before the traces disappeared.[1178] Cable trenches were generally visible on aerial photographs as long, straight, narrow light marks, similar to trails. Open cable ducts appeared as straight white lines with ragged borders with a dark line in the centre. The convergence of these lines towards a given point provided a significant signature of either an important headquarters or a telephone central. In the latter case, most of the cable trenches radiated towards the front.[1179]

Aerial (air) lines continued the communication feed outside of the trench area.[1180] The connecting telephone aerial lines appeared as a series of regular white dots from the displaced earth created from the poles set to hold the wire. The white dots were not the poles themselves, but the disturbed earth from digging postholes. Foot tracks from work crews walking from

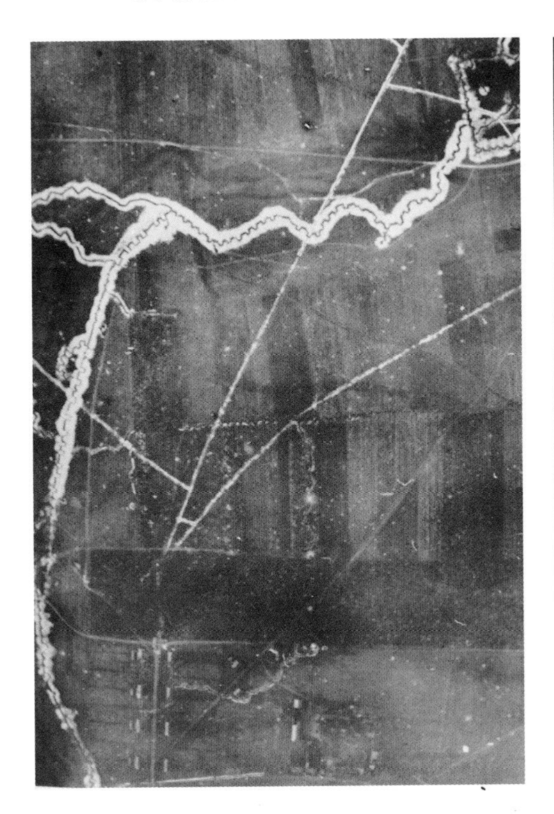

The detail derived from the photograph was a testament to the skill of the interpreter. The cable line helped determine the command and control network in the given sector. (*Illustrations*, S.S. 550a)

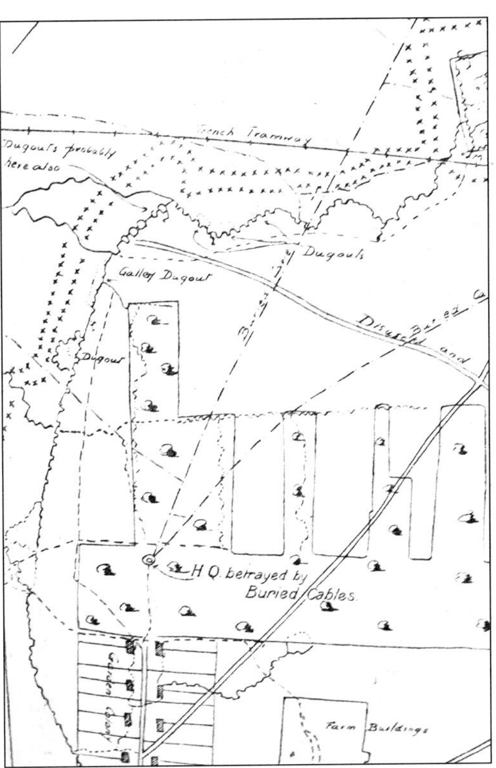

The line drawing includes the interesting assessment: 'H.Q. betrayed by buried cable'. (*Illustrations*, S.S. 550a)

pole to pole were visible. The interpretation sometimes benefited from shadows cast by the poles.[1181]

German policy discovered through captured documents provided further insight on the positions of telephone junctions. A 23 June 1917 OHL document ordered units to camouflage the surface lines and put the junction at a distance of about ½ km from the PC. Photographic interpreters found the search for PCs along the Western Front to be less challenging. The facility was strongly protected employing various communication links; telephone, radio, and footpaths for runners. The primary signal intelligence role, radiogoniometric analysis, was employed extensively and integrated with the photographic coverage to confirm the location.[1182] The search for survivable communications within the front sector led to employment of TPS ground telegraph lines. Conductivity of the terrain (illustrated on geological maps) was a factor in the employment of the ground path. Despite being susceptible to intercept, TPS was a great improvement for communicating within the front lines at a distance.[1183]

The final means of communication was through visual signalling stations employing two variations of search lights (signal lamps). Searchlight posts were installed and normally located at higher elevations in the rear of the sector. Searchlight locations were supported by a network of saps.[1184] Searchlights were capable of signalling up to 15 miles during the day and 45 miles at night. Smaller searchlights were effective from 3 to 12 miles during the day and 6 to 30 miles at night. The use of visual signalling was largely dependent on the nature of the ground. Sometimes the unit resorted to modifying the immediate terrain to improve the receptivity of the signal. Such modifications showed up quite plainly in a photograph, especially through wooded terrain.[1185] The notches in the soil created a 'luminous pencil of lights'.[1186]

COMMUNICATIONS AND ARTILLERY

The ensemble of the German batteries within a division sector was divided into regiment-size units designated as a close combat group (ie, the field artillery and sometimes a few heavy (short range) batteries under command of the Division) and group of Distant Combat (heavy and especially long range artillery serving the German Regimental Foot artillery staff). The Germans further subdivided their artillery into corresponding battalions of dismounted artillery and field artillery. It was difficult to align observed batteries to a particular unit because German artillery sometimes intermingled. Aerial photographic analysis could not determine the German artillery table of organisation. However, the composite of PCs, observation posts, artillery tracks, telephone links uniting the various batteries, and support infrastructure provided a fairly good estimate of the artillery configuration within the given sector.[1187]

The French saw communication as effective only 'if it is in intimate liaison with the Infantry on behalf of which it works.'[1188] The commander viewed liaison as integral to the artillery's program, emplacement and coverage of fire, the extent of munitions to be expended by the artillery unit, proportion of shells with each battery, battery movements, and priority of information required for the artillery operation. In turn, the artillery looked to liaison to alert them to the Infantry's planned manoeuvre, prepared signals, and the future locations for the advancing PCs. The liaison was required to thoroughly know the sector from a study of the Plan Directeur, relief models and aerial photographs. Liaisons communicated with the artillery through the telephone, backed up by visual signals and carrier pigeons. Two telephone circuits were maintained, one by the artillery and one by the infantry.[1189]

The lessons of the First World War showed that effective communication links were of supreme importance for gauging an adversary's ability to continue the battle. In the years that positional war dominated, command and control applied a host of new technologies such as radio and TPS to better manage application of force. All provided the commander with an ability to reach the entire battleground from the rear echelon. Aerial photography's role in this transformation contributed to, but did not dominate, knowledge of enemy command and control. It provided an additional source that reinforced the decision maker's understanding of the battle underway.

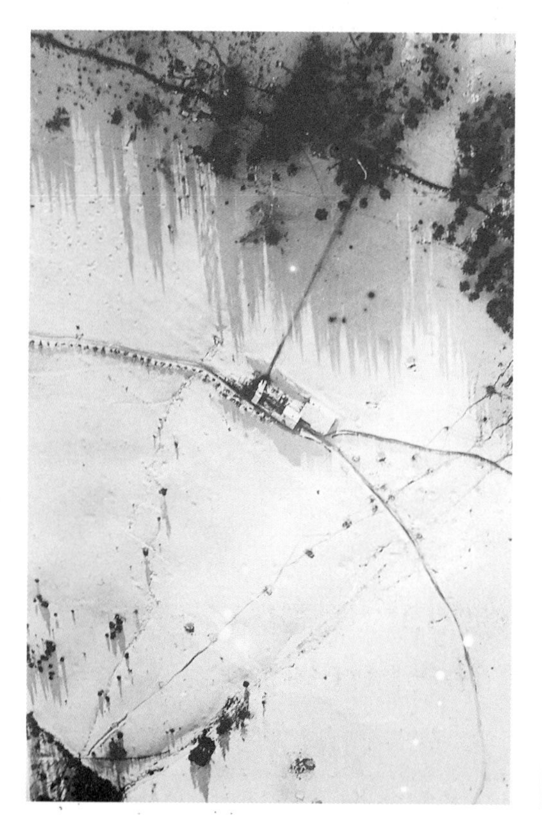

Aerial lines provided communications in a degraded trench network. (*Étude et Exploitation des Photographies Aériennes*)

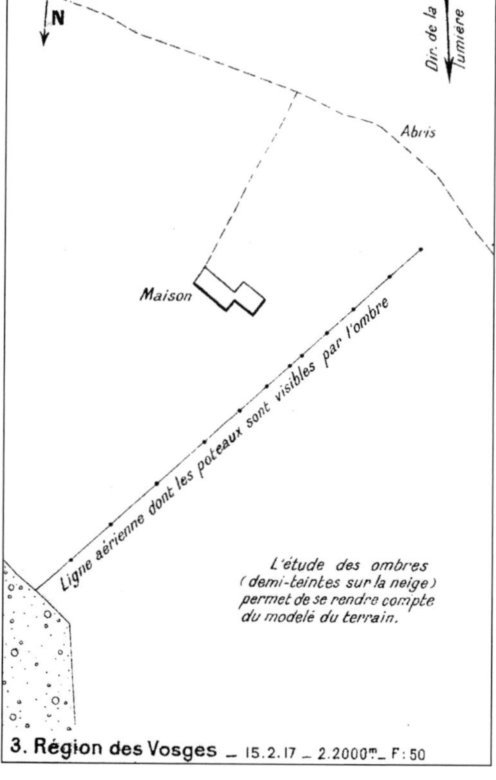

The companion line drawing illustrates the aerial line network. (*Étude et Exploitation des Photographies Aériennes*)

Challenges that Shaped Aerial Reconnaissance and Photographic Interpretation

CHAPTER 17

During the Battle

Recollections of participants of aerial reconnaissance making a vital difference in battle point to the importance of its overall contribution. Moore-Brabazon described a serious situation where last-minute reconnaissance delivered the impending attack from possible catastrophe.

> On a subsequent occasion, at another part of the Front, an attack had been planned to take place at dawn but the last photograph taken on the previous day revealed on examination a new position which proved to be a strong nest of machine guns. Had this not been discovered by photography, undoubtedly the attack would have been held up. As it was, orders were sent out to postpone the attack until the artillery had been brought to bear on the position. During the bombardment photographs were taken to show how completely it had been destroyed and the attack was then opened with complete success.[1190]

The ultimate test of aerial photography was its role in support of the offensive. The French experience at Fort Duouamont on 22 May 1916 demonstrated aerial photography's influence on commanders, even at the eleventh hour. Real-time access to information came with the aircrew's ability to arrive at the landing ground, download the camera and photographic plates, develop in the mobile laboratory, and quickly interpret the image. Tactical coverage provided a sense of assurance prior to the battle.

PREPARING AGAINST AN ENEMY OFFENSIVE

In positional war the upper hand was held by the defenders if they were ready to counter any attack. Surprise strengthened the chance for success in the campaigns. Reaction time was often the determinant of whether forces could withstand the destruction imposed. It was the same for both sides.

> At the moment of the enemy attack, the missions were identical to the French offensive. The uncertainty of the hour of attack complicated reconnaissance employment to effectively cover enemy actions exactly at the moment desired. This required the reconnaissance forces to maintain a greater vigilance and generate sorties with minimal delay. Reporting on the exact situation at the end of the day and in the morning remained a necessary requirement.[1191]

Aerial photograph from Escadrille BR 45 assigned to the Weiller Group shows a trench network northwest of Reims designed for training German infantry on raids and offensive operational techniques. (RG 120, NARA)

First and foremost, the interpreter had to ensure timely interpretation immediately after each aerial reconnaissance sortie. Intelligence was perishable. The interpreter prepared for the battle by studying the battleground's key features. Any inaccurate interpretation not only increased the risk to the aircrew acquiring the photographs but also to the infantry defending the front.[1192]

The first indications of an enemy attack were ascertained from coverage of the rear areas. The signatures were observed in this sequence: (a) The re-enforcement of anti-aircraft defences; (b) The construction of new routes of railways, normal and narrow track, and camps, etc; (c) The appearance of new battery emplacements; (d) The appearance of new works for infantry shelters, communications, minenwerfer emplacements; and (e) The contraction of sectors.[1193] Photographic coverage of rear echelons occasionally discovered training grounds for soldiers to practise raid techniques or offensive operational strategies.

An additional striking feature of the impending enemy campaign was the discovery of a number of Red Cross signs on the roofs of buildings. It became apparent that the enemy's medical arrangements were on a larger scale than the standard required for the normal allotment of divisions occupying the front line.[1194] German field hospitals marked with a red cross appeared as early as a month prior to commencement of the operation.[1195] '2 new red-cross signs are seen W. of the Bois Gerard (66.5–40.8), not far from the 17 centimetre naval gun lately reported active.'[1196]

German aerial activity increased markedly prior to an upcoming battle, acquiring more coverage of the battle lines, and refining last-minute planning.[1197] Aerial support for the offensive required aerodromes to be located as close to the front as possible. The last visible signature of the forward deployed aerodrome was hangers under construction. Canvas structures were first established, followed by larger and more permanent sheds. This process allowed the aerodrome to provide immediate cover as necessary, evolving to more permanent structures as the battle proceeded. It also freed up the supply of canvas to establish new forward aerodromes as the forces advanced.[1198]

The orchestration of balloons along the front also added to indications of enemy intentions. The saying:

When troops witnessed aerial observation balloons being inflated and launched, it alerted them to the possibility of an impending attack. Hence, the idiomatic expression, 'When the balloon goes up.' (USAFA McDermott Library Special Collections (SMS57))

A US captive balloon on station. (USAFA McDermott Library Special Collections (SMS57))

'When the balloon goes up' became an integral part of the soldier's vernacular anticipating the firestorm.[1199] Aerial observation from Allied aeroplanes and balloons monitored enemy balloons to determine if they were up or on the ground. Locations and status were provided to Allied aerial pursuit or artillery for targeting. Occasionally, the Germans moved balloons around the sector to confuse Allied assessments of enemy aerial observation locations. They were known to attach a partially raised balloon to a motor vehicle and drive from point to point.[1200]

Distinct indications such as a sudden increase in artillery activity prompted the worst feeling that an attack was imminent. Additional barbed wire at key locations signified a shift in the enemy's priority for defences, encouraging further intelligence analysis to support follow-on targeting by artillery. New configurations within the trench network such as construction of new saps, traverses or shelters meant the offensive had to contend with new obstacles. Improvements observed within the German communication trenches suggested easier movement of troops from the rear echelons to key forward sectors. New saps constructed forward of the first trench and hastily linked up provided additional protection for the advance guard in last-minute manoeuvres. Finally, commanders and staffs closely watched for any change in the logistical infrastructure. A rapid increase in the number of light railways sent a signal that the trench was being readied for protracted operations.[1201]

Analysis of ongoing battle planning had to be 'pushed in depth' to the rear echelons of the sector. The search for staging areas, areas for potential counterattacks, camps, communication flows, observation centres and artillery configurations was constant. Such information was critical in the final hours prior to the advance. Whenever preparation was detected the aerial observer conducted rapid calculations. Any column of troops in the rear echelons was analysed for unit size, location, time, and direction. A simple formula used by aerial observers to prepare the intelligence staffs for subsequent analysis estimated a column of squads with four men abreast as two men per metre, or 2,000 men per 1,000 metres. A column of two men abreast equated to 1,000 men per 1,000 metres.[1202]

Combatants on both sides of the Western Front alerted aviators to hospital locations with a plainly visible red cross. Newly identified sightings prompted intelligence to examine the possibility of a buildup of enemy operations. (USAFA McDermott Library Special Collections (SMS57))

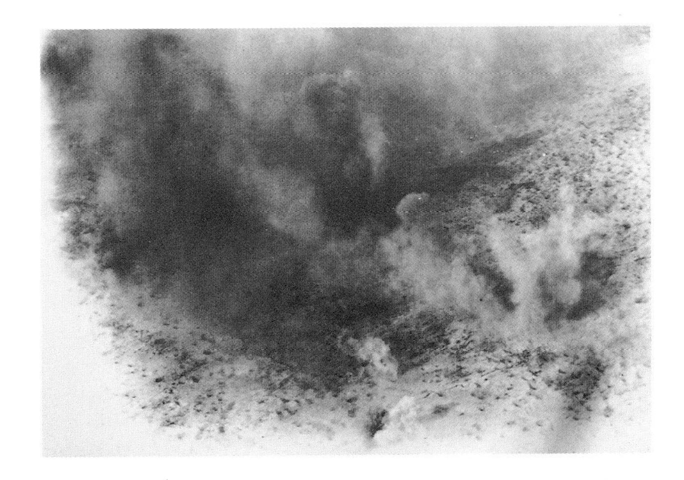

The drama (if not the close-up horror) of continuous bombardment of the front line was captured through the lens. (Pépin Papers, MA&E)

THE ALLIES ADVANCE

The dynamic environment of the attack made it necessary to take frequent photographs at the last minute – the French called it 'hasty organisations'. As intelligence received aerial photography in the closing minutes prior to H-Hour , the aerial photographic interpreter quickly studied the planned battleground for any last-minute changes. Since time was of the essence, the interpreter reviewed plate negatives instead of waiting for prints to be developed.[1203] The final preparation called for exact determination of whether or not the enemy trench was occupied. Early morning aerial reconnaissance was conducted with longer focal length cameras to acquire enough large-scale photographs to confirm enemy soldiers in place within the trench lines. Actual signs of occupation included tracks of supply, personnel relief, and activity at enemy PC locations. Enemy rear echelon areas were also monitored for last-minute activity.[1204]

The Germans disciplined themselves to employ as few trails as possible to limit any evidence of activity, since the trench lines were constantly reviewed. When Allied artillery commenced its barrages, aerial observers photographed the entrances to communication trenches or to temporary shelters to assist assessments of whether or not the shelter was still serviceable. Such analysis also aided in estimating casualties.[1205]

As the Allied attack proceeded, aerial reconnaissance focussed on reevaluating the enemy areas for new firing areas, footpaths, storage and staging locations.[1206] Detecting any signs of potential counterattack remained an immediate priority. The trails were constantly monitored for the footpaths that could lead the analyst to new headquarters and reserve areas.[1207] Oblique photographs of the front lines were highly favoured prior to the actual offensive for analysing ongoing activity and finding additional enemy shelters with reinforcements. When the H-Hour drew near, an exhaustive series of photographs, maps, and plans were made and issued confidentially to commanding officers.[1208]

As the battle commenced, intelligence and planning staffs concentrated on locating new artillery targets.[1209] They were confronted with a chaotic situation. Massive artillery barrages cut up the ground, obliterating trenches. Sometimes a new trench evolved from a line of fortified shell holes.[1210] Locating and destroying machine guns was a priority during an attack. Machine guns were found outside of trenches to conduct flanking fire against the advancing infantry. The photographic interpreter looked for a small white or black dot in a shell hole suggesting the location of weapons. Stereoscopic coverage was ideal. Not only did it provide more detail; it gave the photographic interpreter the best tool to conduct the necessary terrain analysis for the commander.[1211] Seeking active artillery batteries was an additional priority for the interpreter throughout the battle. Warning of counterattacks required an ongoing thorough examination of shelters. Throughout the battle any discernible tracks leading to the battleground were examined for any new enemy approaches.[1212]

As an Allied division moved forward into the battleground, it was assigned a route of march and a zone of action. Aeroplanes reconnoitred this zone and transmitted information by projectors, signal lights, dropped messages, or radio. Reception centres were established successively along the planned route of the advance to ensure continuous reception. Information centres were co-located at the landing ground, employing an array of communication means through telephone, radio, carrier pigeons, additional aeroplanes and automobiles.[1213]

Photographic interpretation during the time of the offensive included the first use of battle damage assessment (BDA; the French term was Register of

Destructions). This time-sensitive assessment was accomplished without delay. It involved special analysis of the plate or negative at the earliest possible moment (normally 15 or 20 minutes after the return of the aeroplane). Register of Destructions located the parts destroyed, breaches made in the barbed wire fences, areas which were hit but not demolished, and an assessment of untouched areas. This study compared successive conditions of the target area from both direct analysis and the study of new shadows. The French process was to provide a simple multi-coloured sketch or diagram (solid blue representing untouched target area, red hatching representing semi-demolished, and solid red a completely destroyed target area). The sketch was forwarded immediately to the GQG and lower echelon headquarters so they could make immediate use of the information. A limited number of prints were produced for the headquarters and artillery to verify sketches and other battle information. The photographs improved the effectiveness of fire by providing information about fire adjustment and destruction of strategic targets (batteries, railway stations, etc).[1214]

French artillery employed its SRA to focus on relevant targets being struck during the barrage. The database was a series of battery index cards containing relevant information to help the observers 'follow' the targets.[1215] SRA coordination with artillery extended to recommending the calibre of the round to either destroy or render ineffective any potential remaining threat, such as concrete shelters uncovered by earth shifting from the bombardment. The information provided to both artillery and infantry at the final moment prior to H-Hour was a critical facet of barrage analysis. A map of the destroyed targets compiled from aerial photographic analysis was perceived as the best means to convey to the infantry commander what obstacles

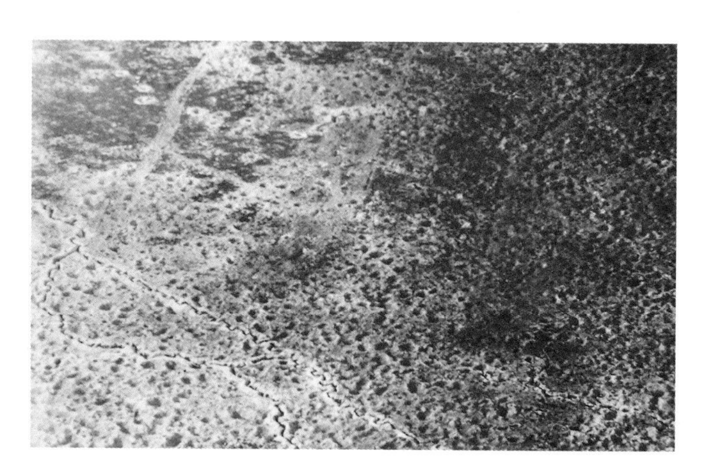

The worn battleground at Moronvilliers north of Soissons showed extensive destruction. Interpretation required detailed analysis of each shell hole. (*Étude et Exploitation des Photographies Aériennes*)

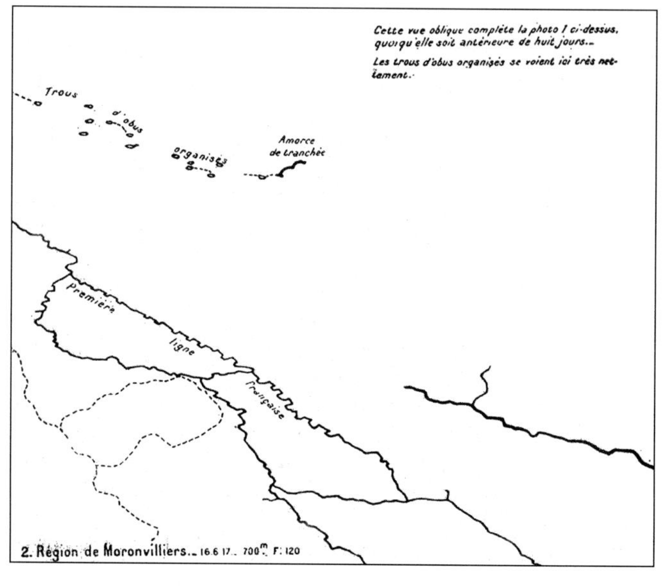

The line drawing described the remnants of combat capability in the area. (*Étude et Exploitation des Photographies Aériennes*)

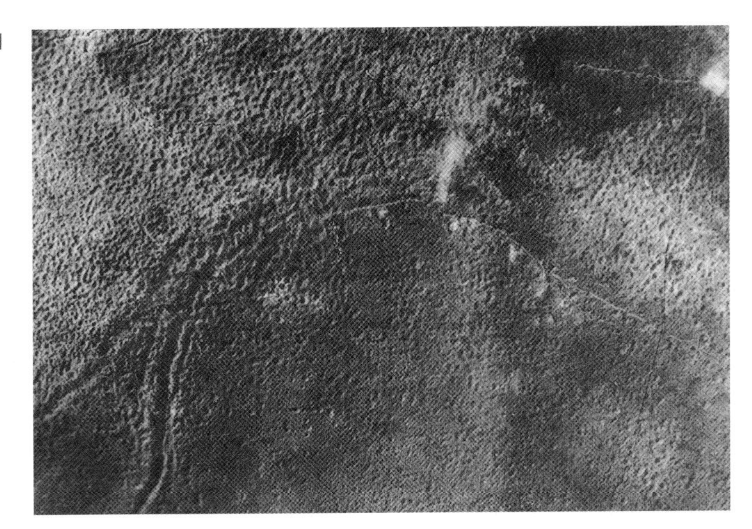

Craonne was a key battleground during the Nivelle offensive commencing in March 1917. (*Étude et Exploitation des Photographies Aériennes*)

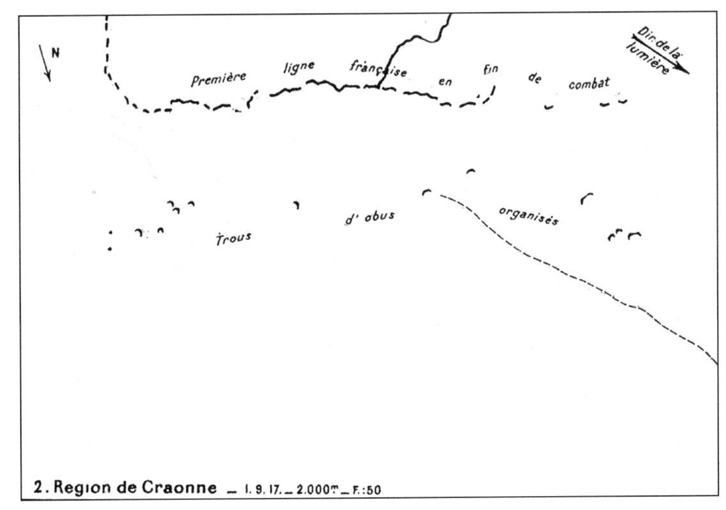

The companion line drawing highlights the French front line and trail activity at the end of the Craonne battle. (*Étude et Exploitation des Photographies Aériennes*)

could be in the way when the advance commenced.[1216] To assist the counter-battery fire, each French division employed a 75mm battery in reserve. The battery was readied to quickly respond with a barrage should intelligence indicate a German counterattack was underway.[1217] The French had an experienced aérostier assigned to the staff to assist in planning. Constant communication was an asset for balloon observers, particularly through telephonic contact.[1218]

Plan Directeur maps issued at the commencement of the battle required constant updates to confirm subsequent artillery targets. The French described its purpose: 'The Plan Directeur map of destruction which is made out daily during an offensive, will give at once the result of our fire of counter-preparation and of the enemy's fire of destruction.'[1219] The Plan Directeur provided the location of the Allied advance as well as an update of enemy defences.[1220] Along with the standard information portrayed on the Plan Directeur, the SRA contributed detailed information of the smallest calibre required to destroy the target.[1221]

Prior to the battle, the landing ground was organised to accommodate the entire communications network. When possible, the division commander generally had his artillery commander with him, moving successively from one information center to another as the offensive moved forward. The division artillery commander used the information centre to communicate with escadrilles working with the division to task infantry contact patrols. Aeroplanes designated for such duty (in theory one allotted per division) reported to the information centre of the division, either from the air via radio or after landing, to find which battalion they were to work with and the objectives. Once assigned, the battalion placed the identification panels to complete the communication with the aeroplanes.[1222]

Determining exact aeroplane objectives and assignment of duties was very important during the battle. Auxiliary landing fields were designated near the division information centre or in areas where effective communication was ensured. The division artillery commander kept in close contact with aerial observation, sending instructions through an auxiliary landing field. Artillery observers served as liaisons between the artillery staff and the aircrew, meeting and coordinating with the aircrew upon landing.[1223] Intelligence offices at CA were notified as soon as possible. Visual reconnaissance reports were telephoned and followed up with a written report. Photographs were developed as soon as the aeroplane landed. Two hours later the prints were dropped by courier aeroplanes to the CA battle headquarters (two prints, one of which was for the artillery); and the same for divisional battle headquarters andd armée headquarters. Simultaneously, the 2e Bureau (the British source refers to the equivalent BIO) and the CA GCTA (British Corps Topographical Section) commenced interpreting the photographs. A few hours later the annotated prints were forwarded on. The Topographical Section simultaneously revised the battle maps with the photographs. This provided commanders and infantry with a more specific understanding of the terrain and supplemented existing information on the German positions.[1224] More detailed analysis from aerial photography was generated by a daily report; a sketch map showed the results of the day's observations, and a sketch map for the artillery showed batteries observed in action. These reports and sketch maps were submitted for final review at the daily meeting of the 2e Bureau.[1225]

The French emphasised timely delivery of developed prints to support the ongoing battle planning. However, the prints were initially produced without the accompanying detailed interpretation. Armée and CA headquarters received the initial distribution. The remaining prints were sent with the report (when coming from a CA escadrille) or in the course of the evening (when coming from an armée escadrille). Additional prints required for ongoing analysis were sent as quickly as possible. Interpreted prints were usually the last to arrive for distribution.[1226]

THE END OF THE BATTLE

At the end of the battle, aerial photography assisted in determining the exact position of the new battle lines. Photographic confirmation of panel flags and smoking Ruggieri pots placed along the forward edge of the battle line provided an understanding of the new sector area. The panels were clearly seen by the photographs, even with a short-focus 26cm camera. Aerial photographic interpreters annotated any coverage showing the infantry advance as 'historical documents of great value'.[1227]

The classic signs of occupation were loopholes for firing, organised shell-holes, obstacles in the trenches (boyau), outlines of trenches, rifle pits, tracks, and frequently the men themselves. Photographic signatures of the advance included tracks from the Allied side ending at shell holes or fragments of trenches, obstacles remaining in the trenches, refitted enemy trenches for Allied use (niches in the fortifications [*parados*] or in shell holes), and the direction in which the earth was thrown. In times of Allied advances the Germans dug a hasty line to reinforce their defensive positions. Traverses were reversed to accommodate new strong points and movement of machine guns and minenwerfer. In-depth stereoscopic analysis determined potential machine-gun emplacements, and kept in mind new alignments for flanking fire. Minenwerfer and *granatenwerfer* (grenade launchers) were also discovered in new shell holes. Small paths were detected to the new locations. It was hoped that the verification of the Allied advance justified the offensive. Such activity was accomplished a few hours after an attack or shortly thereafter the following day. Large-scale photographs were produced and extensive analysis was undertaken with the magnifying glass. The photographs became the new adjudicators for terrain arrangement and the emplacement of the lines. As soon as the study was accomplished the photographs were distributed to produce map updates with new trench lines.[1228] Analysis continued between the aerial photographic interpreter and the intelligence staff. The 2e Bureau remained responsible for disseminating intelligence updates through intelligence summaries and directives based on new targets or techniques discovered.[1229]

Full-scale battles in the war were moments of great intensity; but they were not the norm. Quiet sectors covered most of the Western Front. When battle commenced, it was usually the culmination of a highly planned operation within a particular sector. Timely and critical battle decisions required accurate data. Despite the reliance on aerial photography in this war, it was clearly limited in its ability to define activity in motion. RFC speed tests conducted prior to the Battle of the Somme reflected an urgency to acquire aerial photography for the commander to make the correct decision to influence the battle. Timely and accurate coverage remained a worthy goal for every aerial observation unit. However, aerial coverage as a tactical asset in battle was limited. Strategical coverage of the rear echelon held a greater sway in determining ultimate success. Understanding the adversary's ability to sustain the ongoing battle became a focus for intelligence. As such, in the post-war environment aerial photography better served strategical analysis rather than the combatant.

The legendary Commandant Antonin Brocard cut his teeth in Deperdussin TT monoplanes flying reconnaissance for escadrille D 6 in the first months of the war. (By Henry Farré, courtesy USAFA Library, AF Art Program)

Capitaine René Hubert Roeckel defined his own artillery support role: 'Like a god, he commands the lightning and projects it from his batteries to whatever point he will.'[1842] (By Henry Farré, courtesy USAFA Library, AF Art Program)

A press image of the state-of-the-art Caudron G.3 reconnaissance aeroplane in 1914. The magazine praised its sterling service in the Battle of the Marne, portraying a brilliant azure butterfly. (Courtesy of Tony Langley)

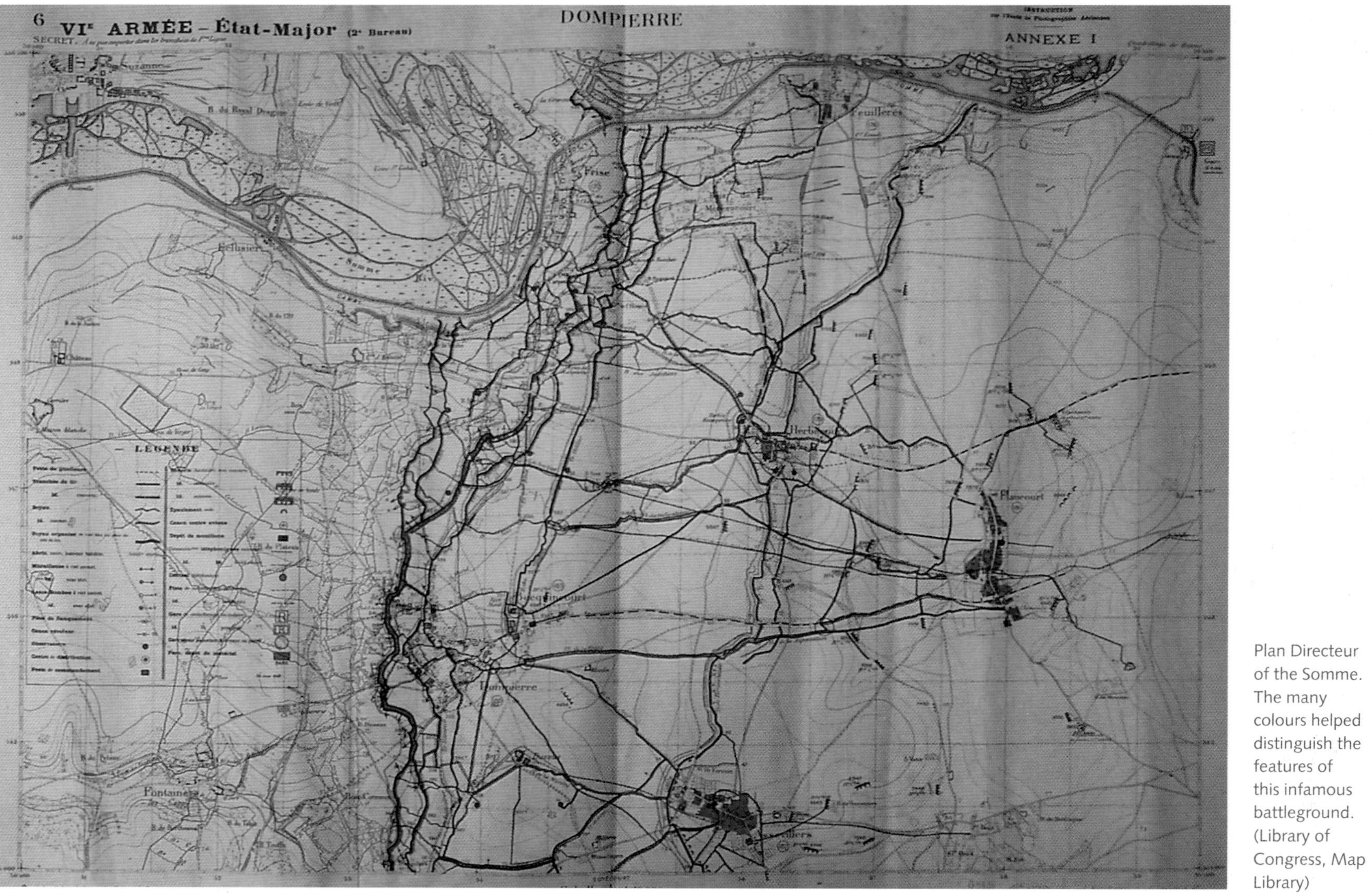

Plan Directeur of the Somme. The many colours helped distinguish the features of this infamous battleground. (Library of Congress, Map Library)

Excellent quality coverage of Dixmude by the legendary Belgian reconnaissance team of Jaumotte and Wouters with a rare annotation format on clear acetate. The city was first attacked on October 16, 1914. At the end of the month the Belgians opened the gates of the Yser river and the flooded area remained part of the front line until the end of the war. The ruined city was rebuilt in the 1920s. (SHD)

Two standard size prints produced by the French – 13 x 18 cm and 18 x 24 cm.

1/10,000 scale map of the key city of Ypres. (Library of Congress, Map Library)

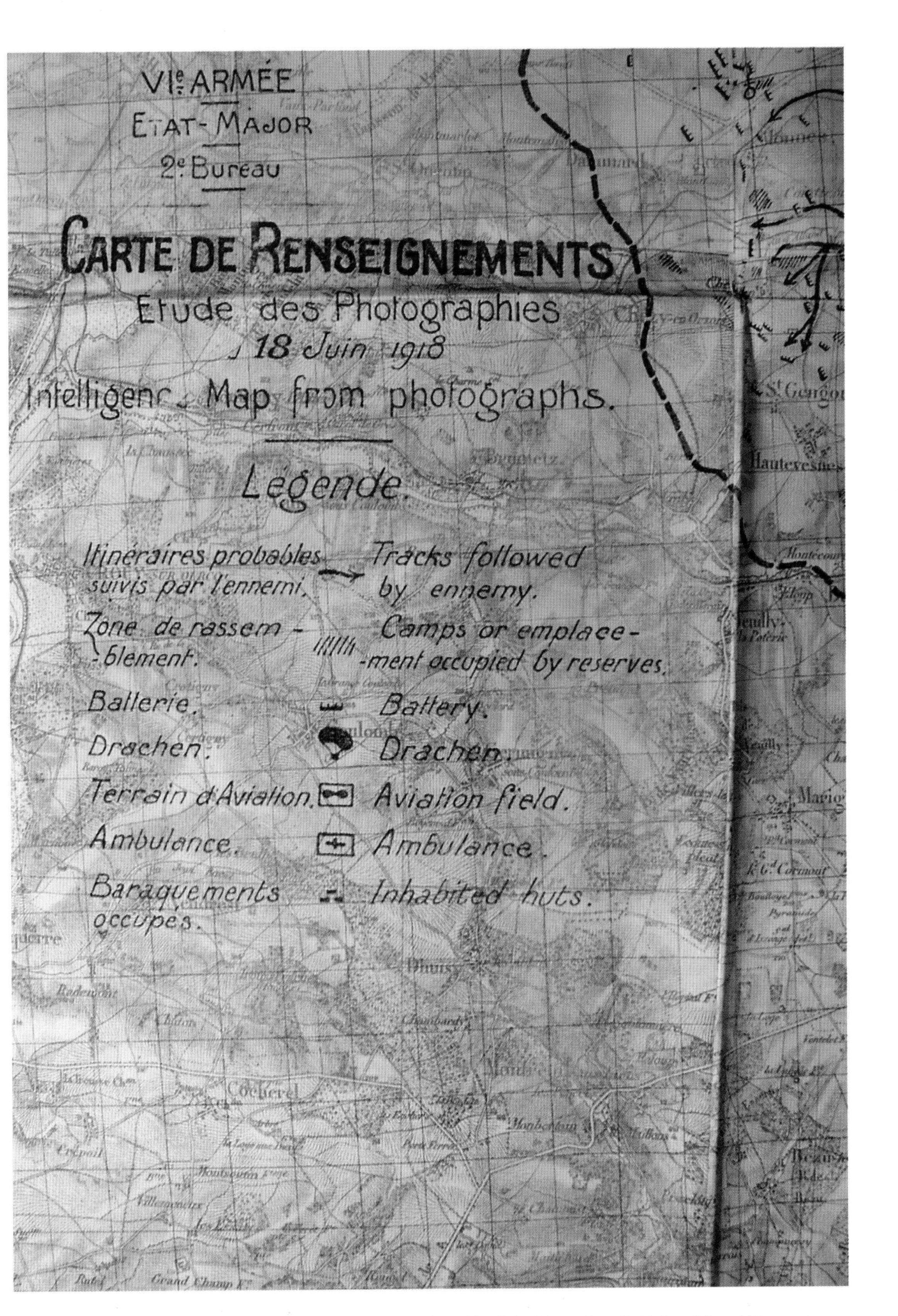

Legend details German targets portrayed on the map in French and English. (Massachusetts National Guard Museum)

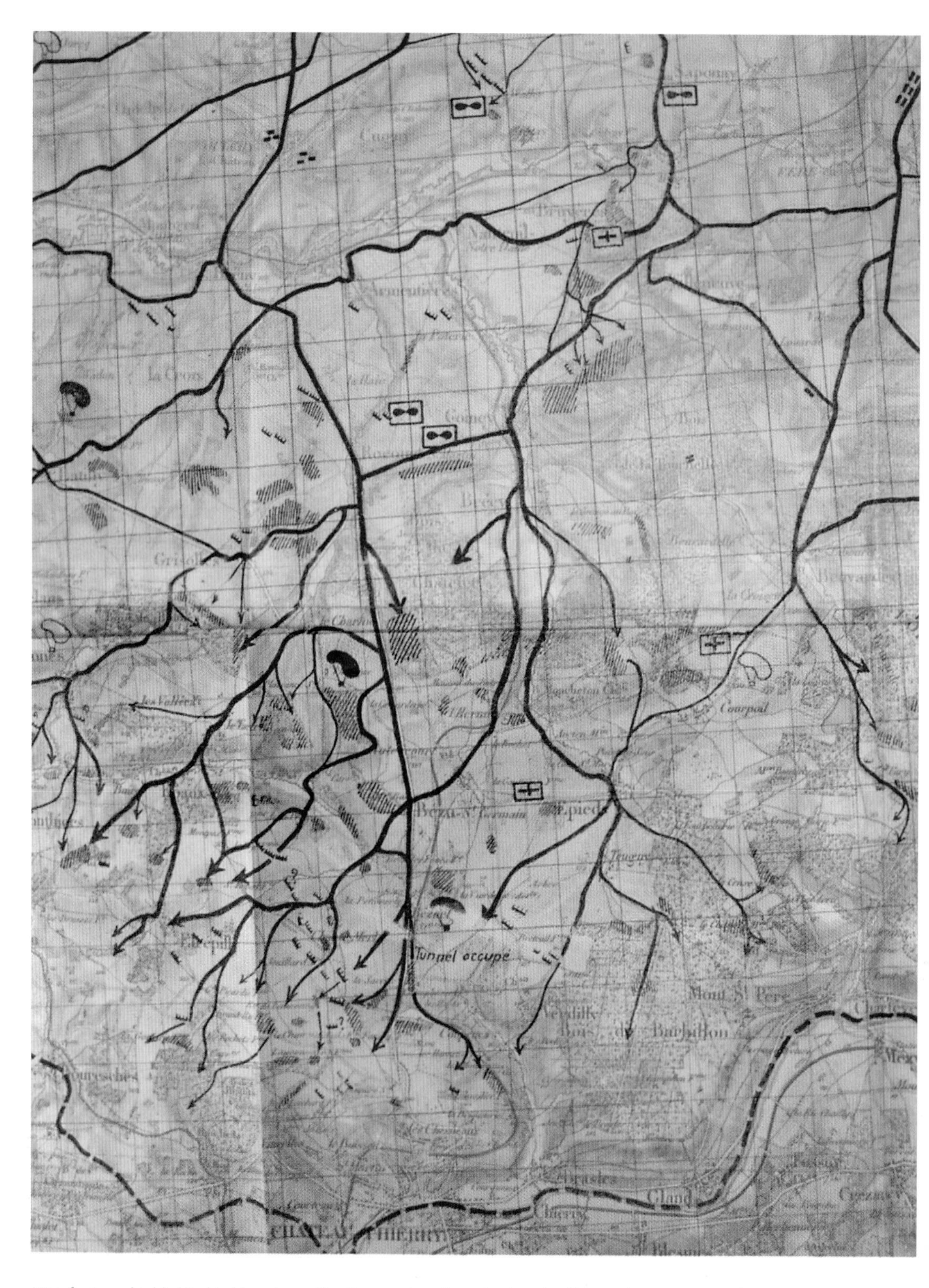

Map features highlighted in blue shows the German targets and networks. (Massachusetts National Guard Museum)

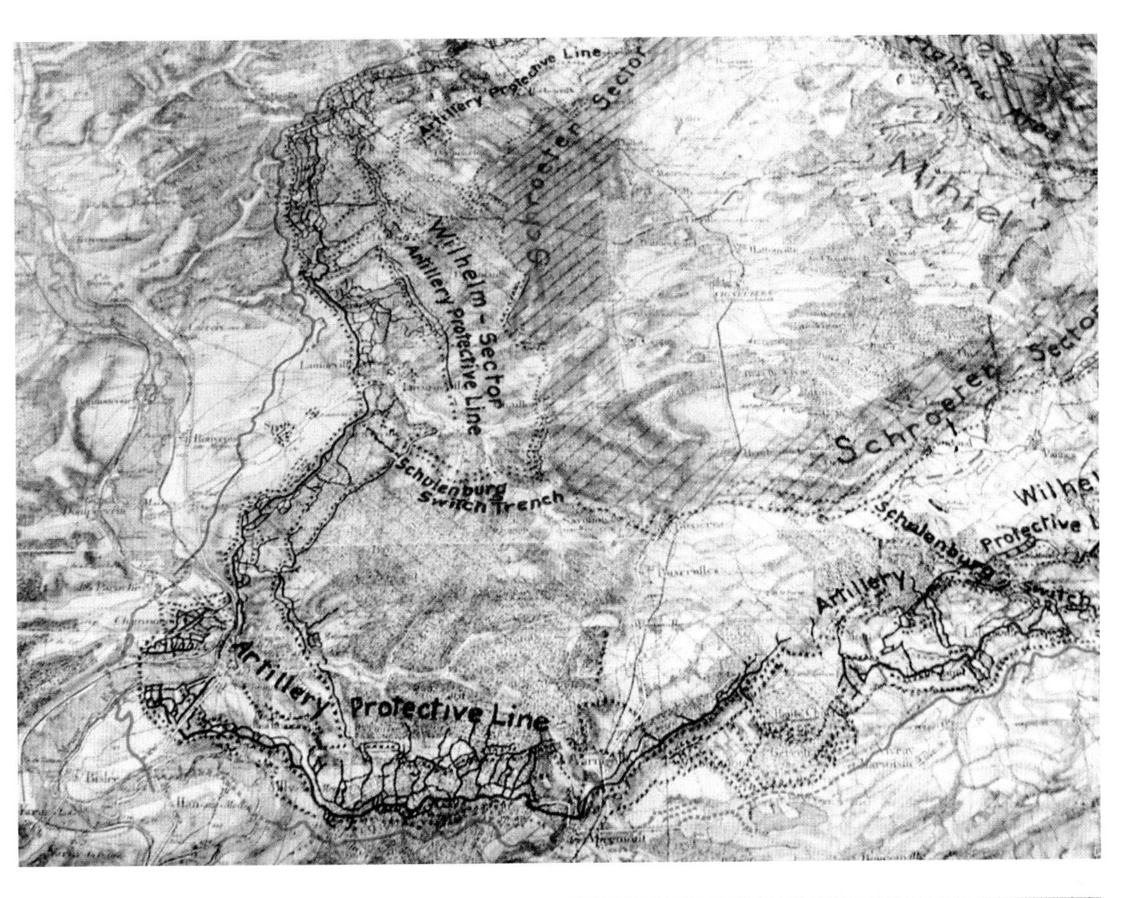

Detailed 1/80,000 scale map of the St Mihiel battleground used by the US 26th Division for planning. The German trenches are outlined in blue. (Massachusetts National Guard Museum)

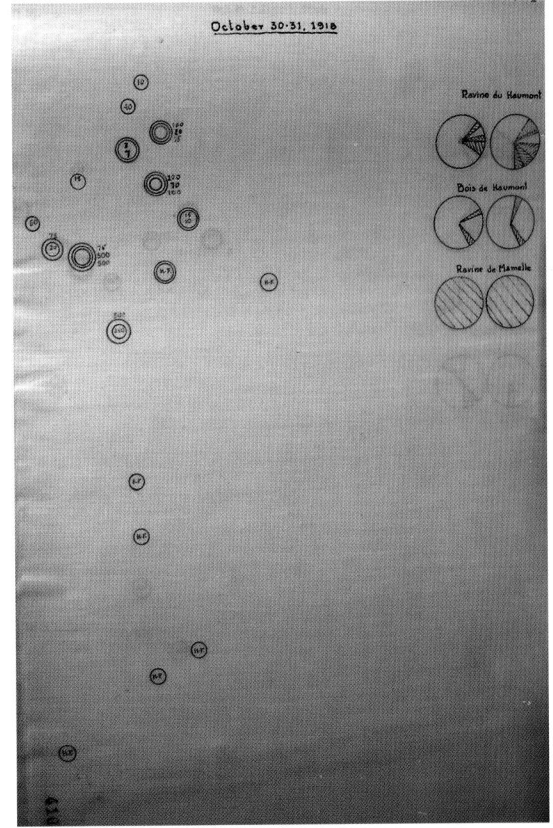

Details of artillery round impacts were plotted on maps. Here the US 26th Division overlay shows a variety of calibre weapons plotted. (Massachusetts National Guard Museum)

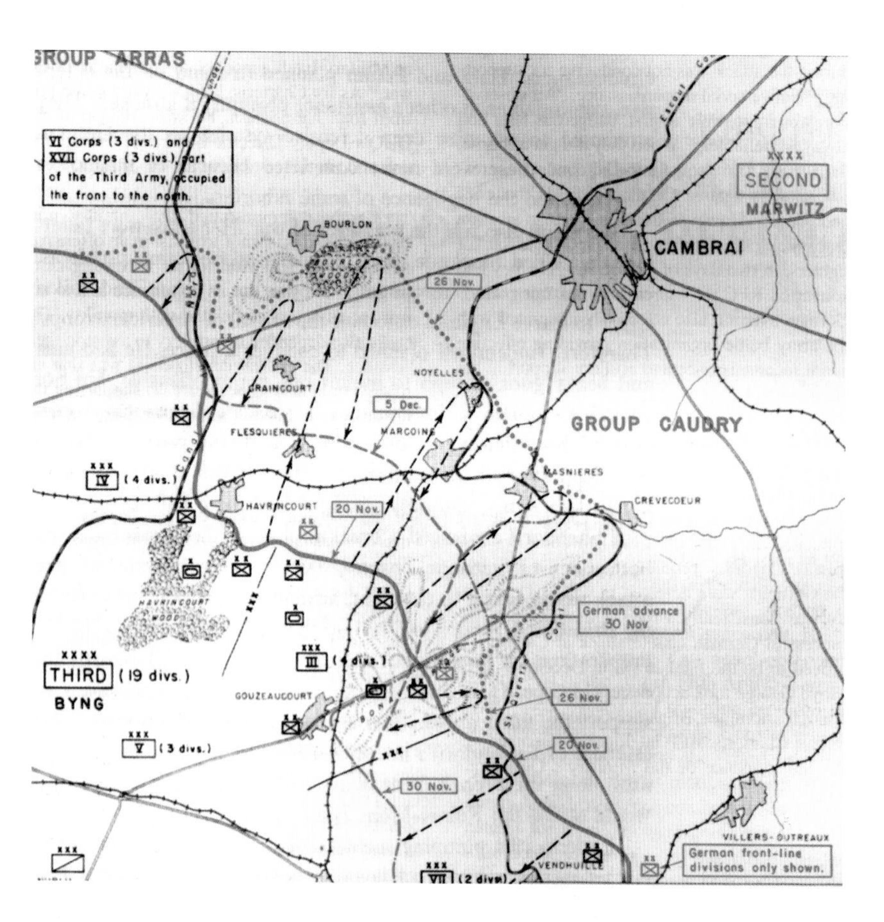

Cambrai. See page 93. (*The West Point Atlas of American Wars*)

'Farman [F.40] Regulating Fire' by Henry Farré. In the words of Capitaine Roeckel: 'The army corps plane looks toward the earth; he looks at the game like the artilleryman, and he controls it.' (Courtesy USAFA Dept of History, AF Art Program)

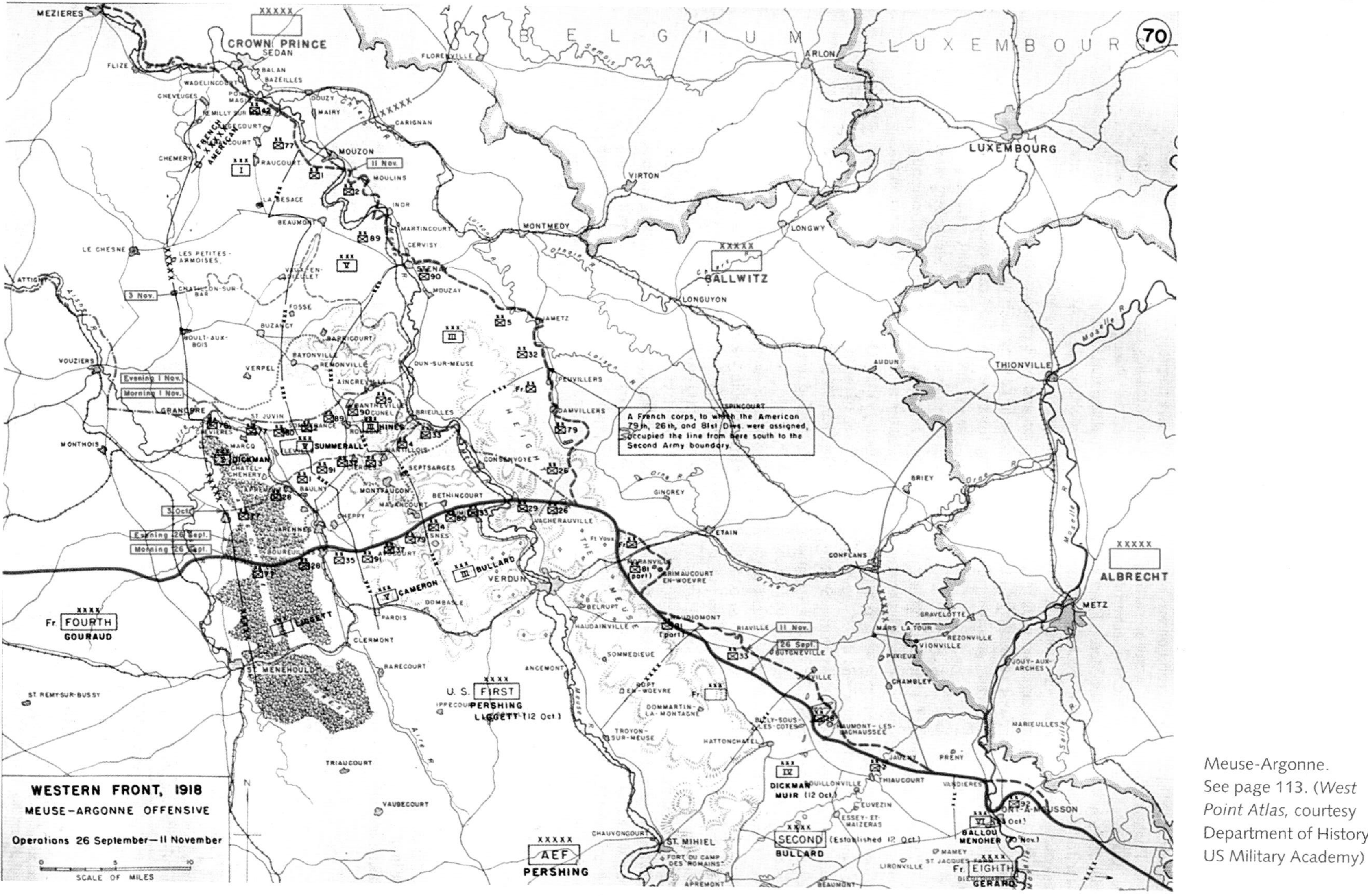

Meuse-Argonne.
See page 113. (*West Point Atlas*, courtesy Department of History, US Military Academy)

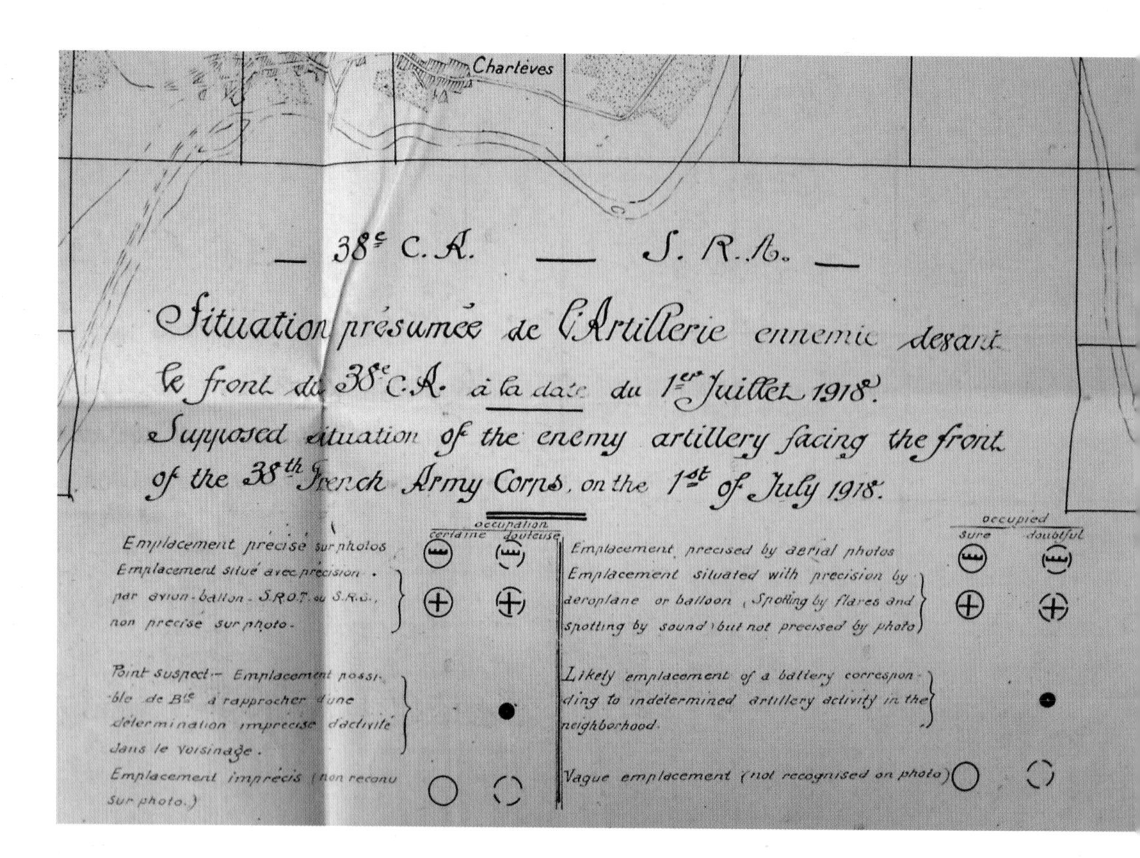

The map key used in 1918 by the French working with the US 26th Division illustrates the source and certainty of the information behind the target. (Massachusetts National Guard Museum)

'Caudrons [G.4s] Signalling'. Capitaine Roeckel: The point of aim – whether it is a trench, a battery, or a shelter – his will for destruction concentrates upon that.' (Courtesy USAFA Dept of History, AF Art Program)

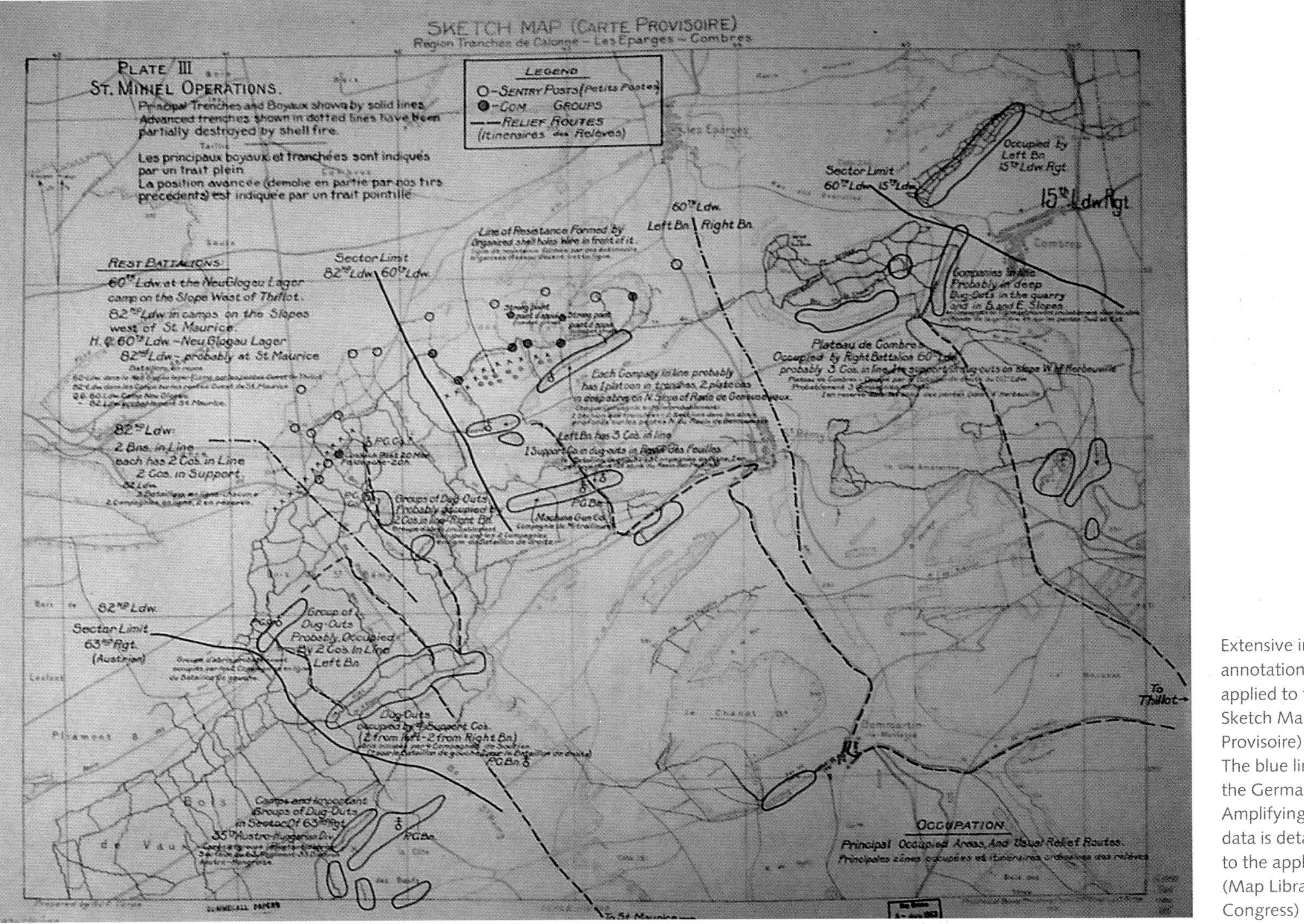

Extensive intelligence annotations have been applied to the AEF's Sketch Map (Carte Provisoire) of St. Mihiel. The blue lines depict the German trenches. Amplifying intelligence data is detailed in red to the applicable area. (Map Library, Library of Congress)

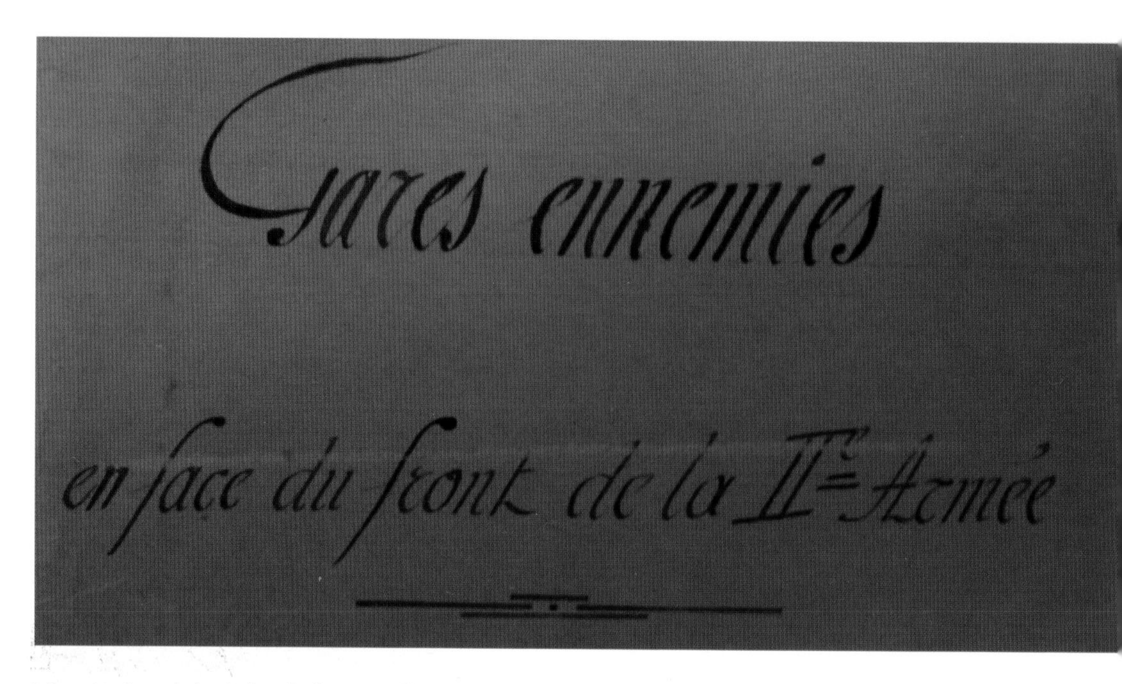

Gares ennemies

en face du front de la IIe Armée

Vibrant colours helped identify the target file. (Library of Congress, Brigadier General William Mitchell Papers)

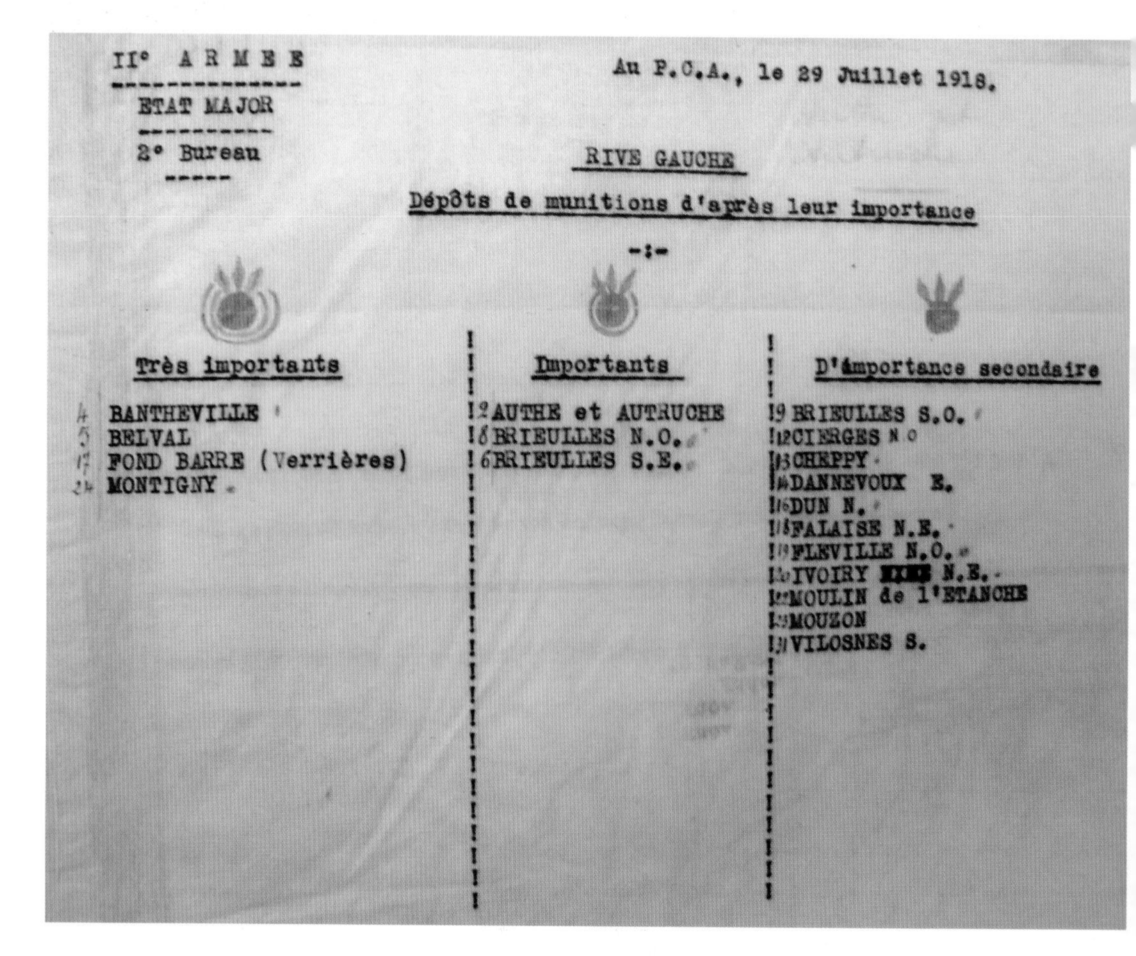

II° ARMEE

ETAT MAJOR

2° Bureau

Au P.C.A., le 29 Juillet 1918.

RIVE GAUCHE

Dépôts de munitions d'après leur importance

-:-

Très importants	Importants	D'importance secondaire
BANTHEVILLE	AUTHE et AUTRUCHE	BRIEULLES S.O.
BELVAL	BRIEULLES N.O.	CIERGES N.O
FOND BARRE (Verrières)	BRIEULLES S.E.	CHEPPY
MONTIGNY		DANNEVOUX E.
		DUN N.
		FALAISE N.E.
		FLEVILLE N.O.
		IVOIRY [illegible] N.E.
		MOULIN de l'ETANCHE
		MOUZON
		VILOSNES S.

Unique symbols show the order of priority for the target. (Library of Congress, Brigadier General William Mitchell Papers)

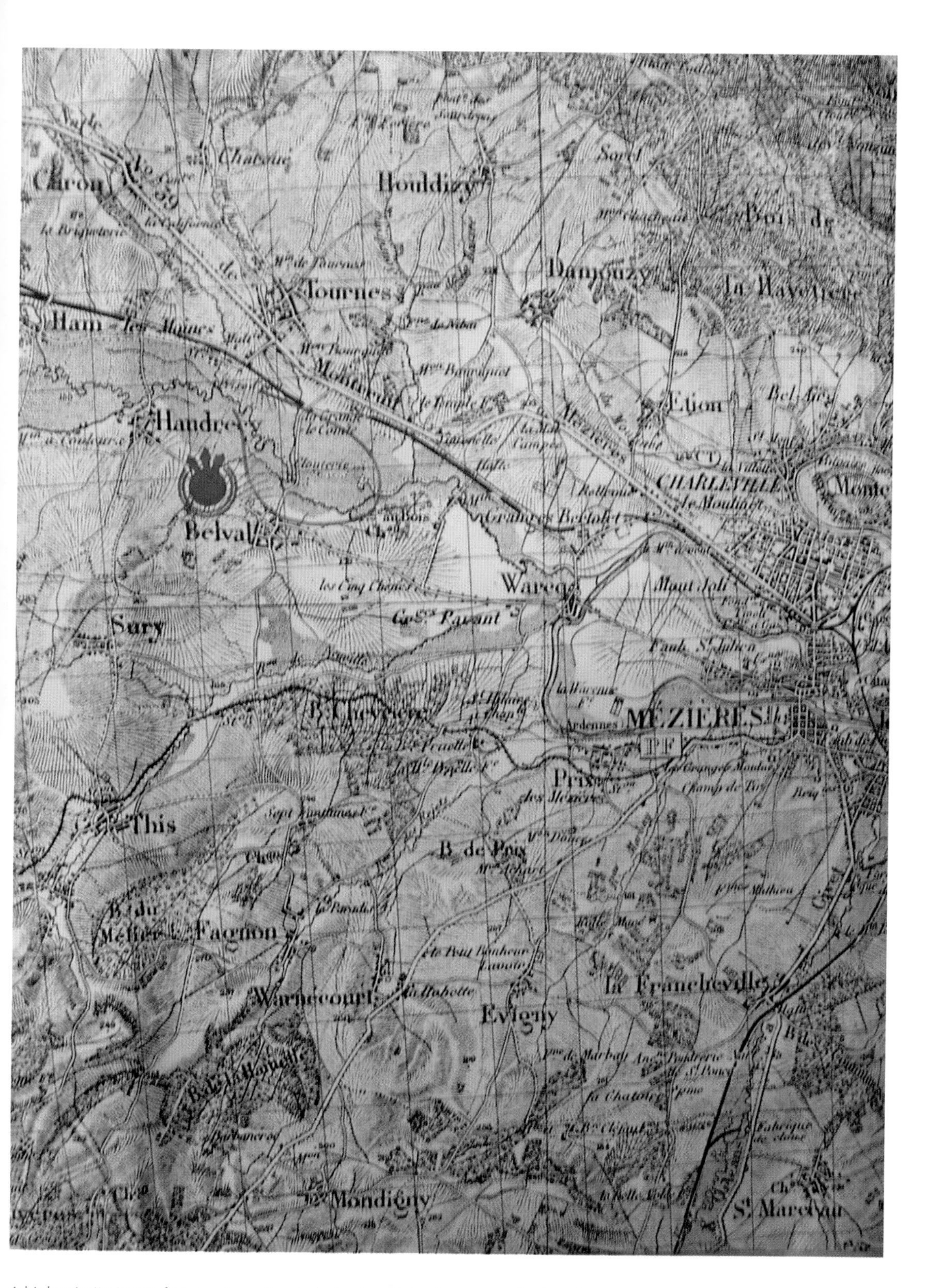

A high priority target for munitions depots in the target dossier was marked on the 1/80,000 map. (Library of Congress, Brigadier General William Mitchell Papers)

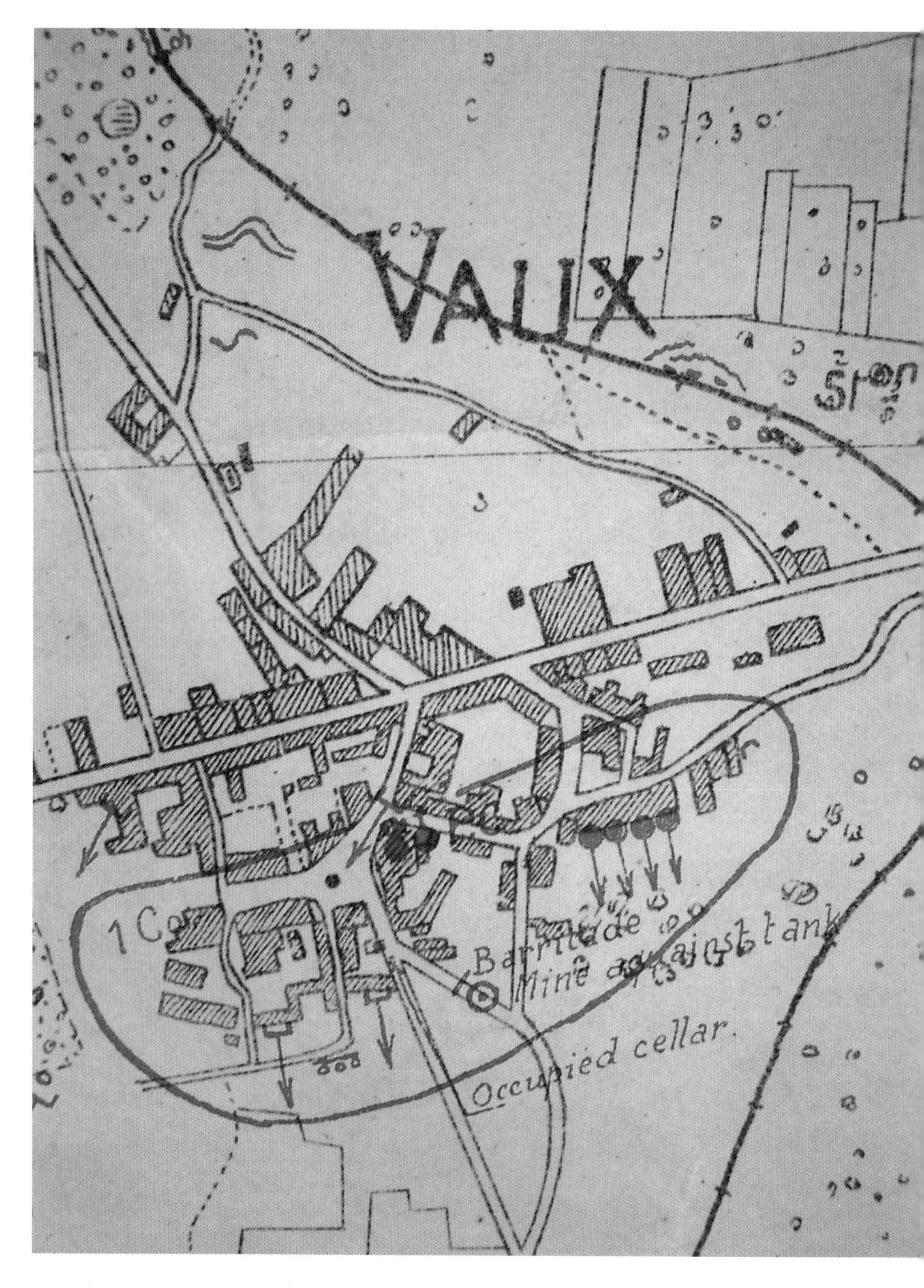

Detailed 1/5000 map of Vaux used by US 26th Division prior to advancing on Vaux. Aerial photography such as Steichen's provided clues as to the disposition of enemy forces in the area. See page 227. (Massachusetts National Guard Museum)

The self portrait by a member of the French camoufleur unit, André Mare, points to the tremendous cubist art influence on modern-day camouflage. (Graffin, *Carnets de Guerre*)

Aerial reconnaissance was dramatised and idealised through the media and the intrigue behind its analysis was amplified. (*Scientific American*, 26 January 1918)

CHAPTER 18

Study of 'Battlefield Destructions'

French interpreters considered the 'study of destructions' a continuation of battle analysis. Aéroplane and ballon observateurs served as the primary sources for determining destruction; yet both were limited in scope. An exact appreciation of the extent of destruction required ancillary information from sources such as prisoner interrogations. Destruction provided Allied commanders and planners with a degree of confidence that they could take the war to the enemy, be it artillery fired into the trench or aerial bombardment of the heartland. Annihilation represented progress on the battlefield.

DESTRUCTION AT THE FRONT

Massive artillery barrages transformed the entire region. The axiom of artillery was that 'sufficient fire could destroy any trench.'[1230] In many sectors along the front the destruction was so massive that a history of coverage was essential to determine the transformation. Coverage acquired after an artillery barrage was subjected to a standardised routine of analysis. Again, the first priority was placed on locating the most lethal weapons such as machine guns. Once an understanding of the immediate threat was accomplished, the trench network was examined to ascertain the flow of personnel and supplies reinforcing and sustaining the battle. This required a complete assessment of the shelters needed to maintain an existence in the forward area.[1231] Gauging the value of a particular trench network meant determining the effort underway to rebuild. Analysis included searching for signs of activity, new footpaths, and any materials that supported the defence.[1232] A standard template wasused so that all participants could quickly assimilate the acquired information. In the case of French photographic interpretation, the information was arrayed on the 1/5,000 map to provide as much detail as possible.[1233] Particular works being reconstructed after destruction, or preserved in spite of the bombardment, received special attention for analysis.[1234]

The target in the forward sector was assumed destroyed when its defensive value was reduced to that of connected shell holes. The trench was seriously damaged when it no longer could maintain the role that it was designed to accomplish.[1235] Functional trench works were examined for trafficability and utility. The extent of demolition was measured by the soldier's ability to exist within it. The trench was considered operational if shelter was still available.[1236] One French master of photographic interpretation, *Sous-lieutenant* Pouquet, provided his own methodology:

> A trench provided with large deep shelters is heavily damaged when half the shelters are unusable and when half the garrison is obliged to wait in the trench. An ordinary trench without shelter is seriously handicapped when it has been hit every 20 metres. A boyau is seriously affected when a continuous and sheltered traffic becomes impossible, for instance, when it is cut for 20 metres every 100 metres.[1237]

An artillery battery was considered demolished when guns, casements, dugouts, and the other targets in the area were destroyed.[1238] It was harder to gauge the extent of damage to auxiliary defence such as barbed wire belts or concertina wire. Analytical estimates on building and damage destruction considered structure composition, such as earth emplacement, masonry, cellars, and walls. The height of the rubble was derived from the length of the projected shadow.[1239]

The human toll from destruction was not easily discerned from aerial reconnaissance reporting. An examination of the hospitals in the area offered one clue. In the adjacent graveyard the interpreter could estimate the casualties from the number of crosses.[1240]

Artillery and aerial bombardment strikes against road networks near the front and in the rear echelons provided an indication of enemy priorities for resupply and movement of forces. New routes were created as Allied harassing fire limited the effectiveness of routes,

Left: There are no signs of life in the former town of Bixschoote in this aerial photograph. (*Étude et Exploitation des Photographies Aériennes*)

Below left: The French front line at the end of combat is depicted to the west of Bixschoote in the companion line drawing. (*Étude et Exploitation des Photographies Aériennes*)

N
Dir. de la lumière
Passerelle
Restes d'un village
combat
de
fin
en
P"
P
P'
A
T
P.C. probable
française
ligne
Première
A.A' – Déblais nettement rejetes du coté de l'ennemi.
P.P'P" – Petites pistes allant vers l'arrière.
T – Amorce de traverse
1. Région de Bixschoote – 12.8.17. – 1.600m – F.:50

light railways. Piles of material ('extensive dumping') were observed along the roads providing for rapid repair. The Germans built railheads to accommodate fast transfer of personnel and equipment to narrow gauge rail. When Allied artillery destroyed the facilities, interpreters detected alternative railheads being constructed. Once movement was reestablished, sightings of new and larger stockpiles became noticeable. When the supply dumps were built, signs of ammunition stockpiles became readily apparent. The identification of enemy priorities on the front and rear echelons guided interpreters to determine sector readiness for follow-on offensive operations.[1242]

Targets were classified either as temporarily out of action or completely destroyed. Analysis involved coordination with other experts to ascertain the degree of annihilation. The French had the artillery liaison assigned to the 2e Bureau confirm the calibre of the fired round to gauge the impact crater. The photographic interpreter correlated the calculations with stereo and oblique coverage to calculate the amount of soil unearthed from the barrage. Once the coordinated analysis was completed, the results were forwarded to the artillery unit to adjust fire as required. If the level of destruction was determined to be inadequate for the job, subsequent photographic analysis was conducted to ascertain if the artillery employed was of the wrong calibre or failing in some other way.[1243]

requiring ongoing map and aeroplane photograph study by the photographic interpreter to find those new routes. New tracks verified that the Germans were still operating in the sector and the intended Allied destruction was incomplete.[1241] The Germans gave a higher precedence to repair and construction of road networks than to

The final step in the French daily assessment on destruction was, that night, to modify the Plan Directeur for commanders' review. Photographic coverage in the morning and evenings provided the necessary material to gauge the extent of destruction.[1244] The photographic interpreter marked every observed strike and transferred the information to the Plan Directeur (1/5,000 scale) via an agreed-to colour scheme of marks.[1245] Once completed, commanders and staff prepared for the next advance.

AERIAL BOMBARDMENT

Strikes beyond the reach of conventional artillery were made by bomber aeroplanes. These extended well beyond the corps' sector limit of 10km (10 miles for the British). Not only did aerial bombardment provide a capability to destroy targets in the strategic rear echelon, it also served the Allies as the only force capable of taking the fight directly to the German heartland. Destruction by aerial bombardment was limited by payloads. Technology in aerial bomb construction was still in its infancy throughout the war. Accuracy was erratic, simply relying on the pilot dropping the bomb at the right place and time. Edward Steichen noted the sheer drama of aerial photography of bombardment: 'Vertical photographs made by the day bombing squadrons occasionally present a spectacular and dramatic interest in addition to their value as a record of the bomb raid.'[1246]

Oblique photograph of Vaux taken by Edward Steichen. Each building was analysed for any threat that could hide in the destruction. (McDermott Library, USAF Academy, Dwight Collection)

GQG's Aviation Militaire recorded its aerial bombardment campaign with extensive aerial photography and analysis before and after raids.[1247] The coverage not only confirmed the results of bombardment, it also established detailed maps of the bombardment routes. Rail stations and airfields remained a priority photographic target throughout the campaign.[1248] Upon return, the aerial photographs of rear areas were examined by the 2e Bureau for indications of activity and change in the target area. Any differences were marked on the 1/20,000 Plan Directeur with information embellished with notes.[1249] An RFC staff visit to a French aerial bombardment escadrille concluded that the French chose aerial targets that were either munitions factories or important railway junctions and stations.[1250]

In 1917, GQG's Aviation Militaire initiated 'a very remarkable photographic service', photographic aéroplanes were to accompany bombers on their missions.[1251] *Groupes de Bombardement* started acquiring intelligence officers and established intelligence facilities at the bombardment escadrilles. Targeting objectives and aerial photographs of the bombing routes were provided to the aircrews to assist in their long-range planning. Aerial photographic interpretation analysed enemy defences at aerodromes or anti-aircraft artillery detected along the planned bombing routes. The aircrews applied this to flight planning to determine appropriate defensive measures en route to and returning from the target area. The French commenced post-strike aerial photographic missions to determine bombing results, forwarding the assessments to commanders, and to try to improve

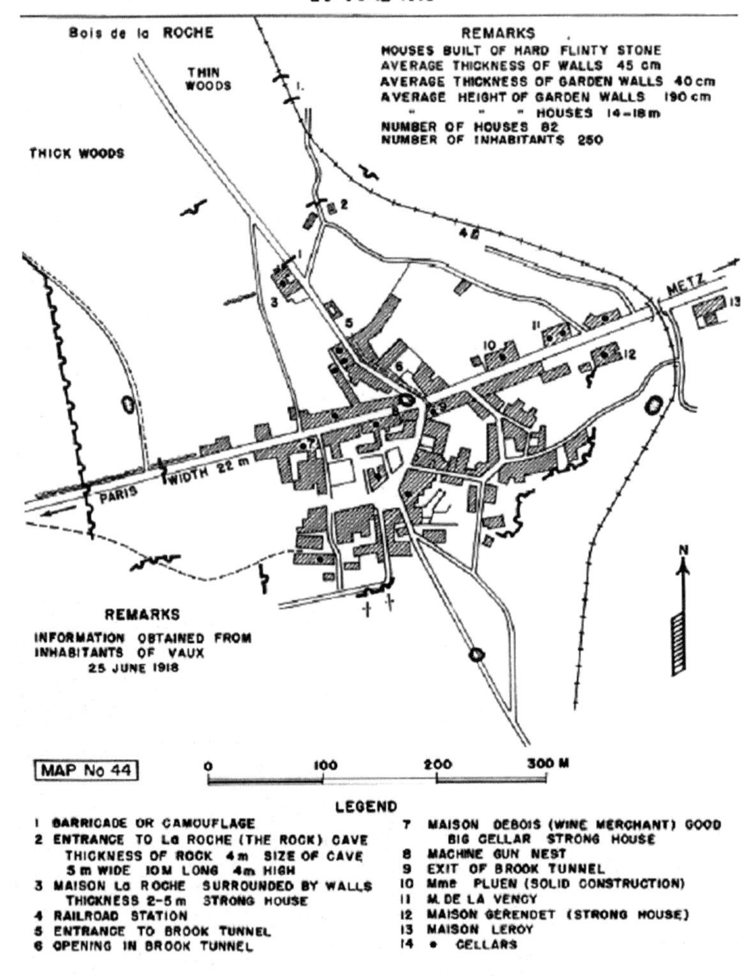

AEF G2 personnel compiled even more details on Vaux using both aerial photography and intelligence information from the inhabitants. See also colour section. (US Army, Office of Military History, United States Army in the World War)

subsequent sortie procedures. Aerial photography was rendering 'the greatest service, notably by permitting the unmasking of the adversary's systems of camouflage'.[1252]

The British also paid attention to aerial photography in their aerial bombardment campaign. Pictures of various kinds, maps, sketches, descriptions and all relevant information on targets were maintained in target folders. Strategic reconnaissance included the examination of all photographs for information relating to bomb targets and for the locations of aerodromes reported but not verified by photographs.[1253] Their squadrons' aeroplanes were equipped for both bombardment and aerial photography. A leading strategic unit, No. 55 Squadron, conducted extensive long-range operations against the German heartland, including coverage of Köln, Coblenz and Mainz.[1254] The squadron's War Diary described aerial photography's role with the bombardment sorties. The majority of aerial photographic sorties were unsuccessful. The diary simply stated, 'It was impossible to take any photographs.' However, no explanation for the deficiency was provided.[1255] Bombardment results did not escape the attention of Intelligence at GHQ. One very successful aerial bombardment raid and companion photographic mission in April 1917 was highlighted by Brigadier-General Charteris: 'Seventeen aerodromes, dumps, and railways bombed, eight tons of bombs dropped, 1700 photographs taken.'[1256]

American aerial bombardment training at the Flying Officers School at St Maixent emphasised aerial photography from the start. The photographic aeroplane was either in the lead of the formation or right behind in the No. 2 or 3 position. Lectures by the Headquarters, Air Service, Photograph Department to aircrews recommended that a lightweight automatic film camera be employed for taking photos of the bombardment.[1257] Exposure to aerial photography was not to create photographic experts, but 'rather to give the students practical knowledge of aerial photography that will enable them to carry out intelligently any photographic mission that may be assigned them.' Aircrews were reminded that aerial photography was necessary to 'bring back accurate and durable records of enemy activity behind his lines and to record the results of our own artillery and aircraft activities.' The discussion emphasised that the camera was 'the best observer there can be.'[1258]

A bombardment aeroplane's ability to conduct strategic strikes against the enemy heartland depended on the airframe and engine.[1259] French aerial reconnaissance went beyond the range of aerial bombardment sorties. In the course of the war, they accomplished the longest known Allied aerial reconnaissance, to Friedrichshafen and Essen (approximately 600km).[1260] Before the war ended, the French strategic bombardment was conducted by groupe d'armées (Army Groups (GA)) and armées with the understanding that aerial photographic reconnaissance was integral to the mission. In the northern sector, the Aviation Militaire serving IVème Armée controlled short-range missions. Armée-level escadrilles conducted photographic reconnaissance. The longer and more complicated missions such as night missions were planned by the Aviation Militaire serving the French *groupe d'armées du nord* (Army Group North (GAN)). It was specifically stated in the planning that long-range bombing operations ordered by GAN were to include aerial photographic reconnaissance. French Breguet bombers began targeting key railroad stations and enemy aerodromes. Guidance for their missions stipulated bombing in groups, maintaining a defensive posture in formation, avoiding any attacks on cities, and taking aerial photographs. Strategic targets in the February time frame included factories in southern Germany at Péchelbronn, Rothweil and Oberndorf.[1261]

British aerial bombardment underway by a DH 4 flight. The aerial photograph captures a high-risk moment for the DH 4 formation immediately below. (Courtesy of Colin Owers)

Enthusiasts for aerial bombardment in the final year of the conflict became the strongest advocates of aerial photography. American bombardment units were reminded that aerial photography validated their efforts. 'In fact, next to bombing, it will be their most important work.'[1262] The purpose of aerial photography cited in training for bombardment units was to bring back accurate and durable records of enemy activity from behind the lines and to record the results of artillery and aircraft activities. The senior leadership looked to aerial photographs to prove the value of their bombing squadrons:

> A Squadron may go out, drop its bombs with excellent results, and do a considerable amount of destruction, but if on returning it has no photographic record of the destruction done, the exact extent and amount of it, it can hardly be expected to receive credit for what it did. Obviously, it is in the interest of the pilots and observers concerned to perfect themselves in aerial photography.[1263]

Bombardment units recognised the limitations of the human observers, looking for enemy aircraft, flustered by anti-aircraft fire, and unable to see the details that an aerial camera provided. Photographic Officers assigned to the unit continually stressed the necessity of obtaining good photographic results so that the squadron received credit for a successful raid.

Aerial photography's role in a bombing raid was described by course lectures at the Air Bombardment school. The photographs were taken to verify the bomb bursts on the target area. If the photographic aeroplane led the formation or flew the No. 2 or 3 position, the aerial photographic pilot throttled down and waited 20–30 seconds before the observer exposed any plates. If the aerial photographic aeroplane flew back in the formation, the observer took photographs as soon as he was over the target and continued to do so as rapidly as possible until the target had passed from his

Aerial bombardment operations strongly encouraged aerial photographic verification of the strike location. The camera was considered the 'best observer there can be'. (US National Air and Space Museum, Garber Facility Archives)

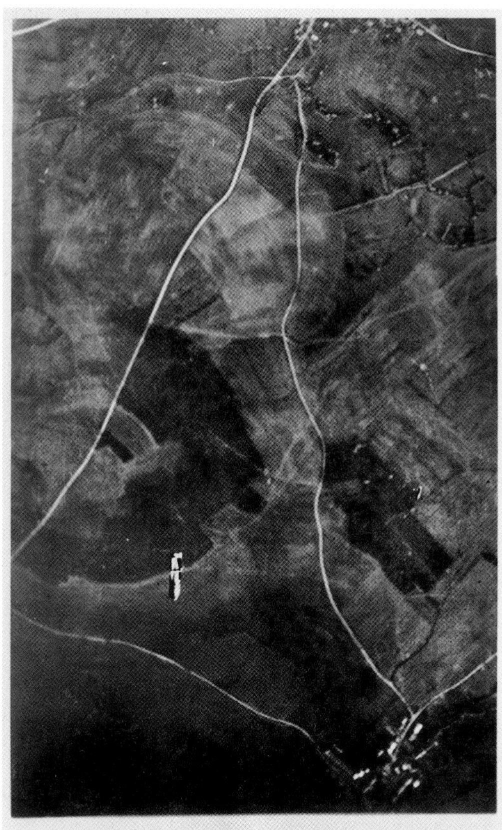

BOMB DROPPING

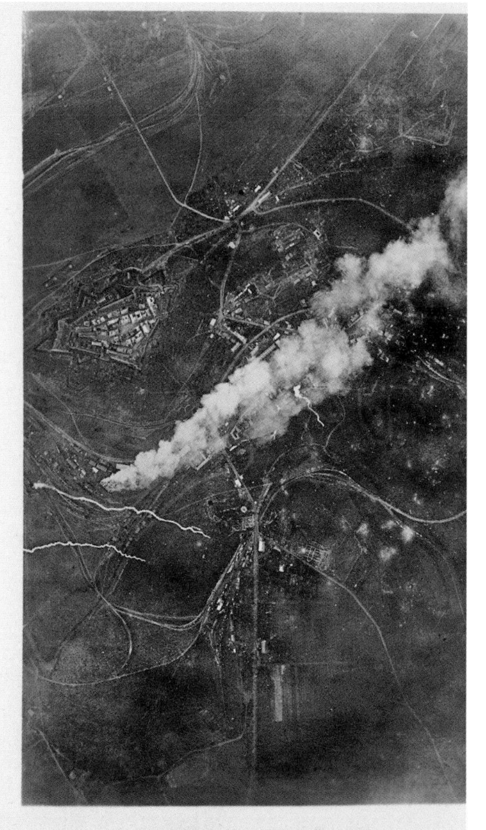

RESULT OF BOMB HIT

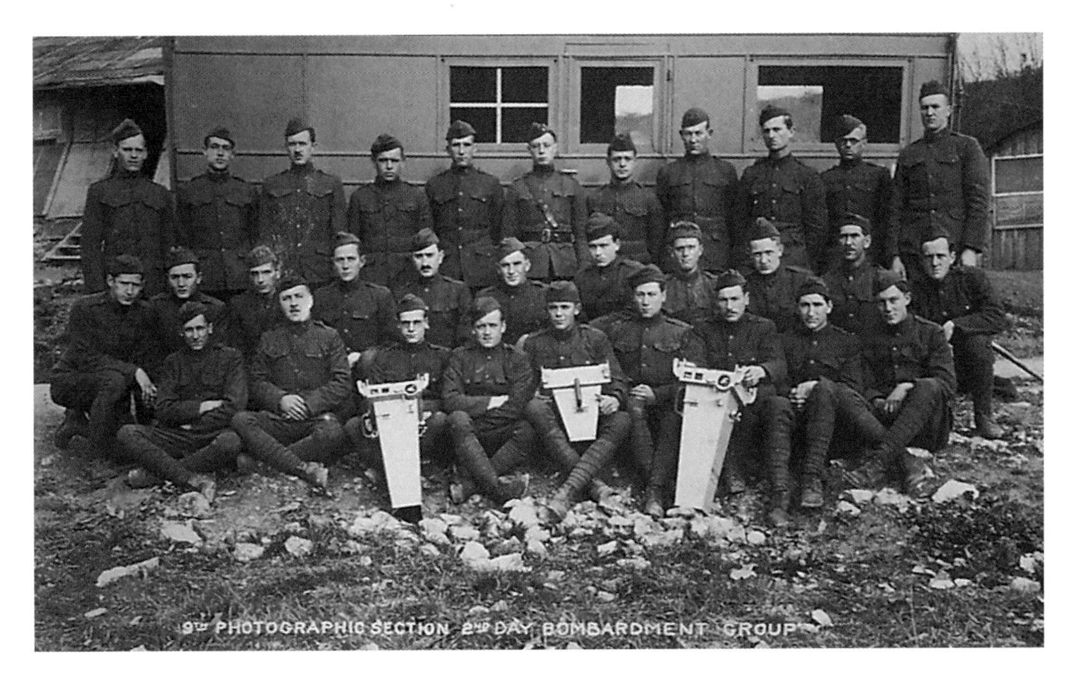

The 9th Photographic Section arrived just before the Armistice. Their assignment to the 2nd Aerial Bombardment Group illustrated the growing role of aerial photography within the operational units. Note two 50cm and one 26cm (Grand Champ) in their operational inventory. (Gorrell Report, NARA)

sight (and simultaneously kept an eye out for enemy aircraft). The aerial photographic aeroplane flew an erratic course to counter the threat from Anti-aircraft fire. This procedure was understood but contradicted the dictum that such missions were to be flown straight and level to acquire the photographs. There was also the possibility that the plates would fog up due to the Anti-aircraft bursts. However, that was considered remote. 'In fact, the less attention observers pay to such bursts, the better results they will get.'[1264]

Aerial photography did not stop with target destruction analysis. The return leg of the bombardment mission included aerial photography of additional areas requiring coverage. The 'camera machines' were withdrawn from the formation after the bomb run to cover the next target area. As was expected in cases of enemy fighter attack, the observer fired a prearranged flare, alerting the others in the formation to quickly regroup and help defend against the attack. After the threat disappeared, the formation proceeded back to base as quickly as possible. 'It must be realised that it is of more importance for the machines to return home with what information they have than to try to secure all their photographs and not reach home.'[1265] Any sighting of 'superior enemy planes' before reaching the target required the photographic machines returning home without engaging the enemy. 'Pilots and observers of bombing, photographic and reconnaissance machines should realise that it is tactically wrong for them to seek combat with enemy planes, and that it is more to their credit to return with reliable information regarding enemy activity having successfully dropped their bombs, than it is to return with no information but with a report of several enemy aircraft shot down in flames in the course of combat.'[1266]

The study of destruction remains one of the most important functions for intelligence, assuming the role of 'a veritable map of fire on which the shots may be counted, located, the effect appreciated and distinguished by comparison with previous documents, previous shots whose role has in general been nil, recent hits alone being effectual for the preparation of an attack.'[1267] Analysis of destruction was a discipline in itself. Its role in the strategic war was amongst the earliest justifications for aviation's potential in warfare, for it helped gauge the level and effectiveness of destruction by the artillery. Destruction analysis supporting aerial bombardment in the First World War measured the success of strategic strikes against the enemy heartland. The intelligence discipline study of destruction evolved into the Battle Damage Assessment that aided analysis throughout the remaining wars of the 20th and 21st centuries.

CHAPTER 19

Camouflage and Deception

A DUEL OF INTELLECT

Positional war established a new military duel, between a technician by training, determining any change through interpretations of minute differences on a film base, and an artist, using the battlefield terrain as a canvas. Their war was a struggle to shape perception. 'Remember that camouflage is intended in the absence of complete invisibility to render powerless enemy observation on earth, aerial or photographic, and must be conducted with extreme caution.'[1268]

To the participants, 'it was an interesting game.'[1269] One French photo expert equated it to a new cultural phenomenon despite the art of concealment 'being contrary to French nature'.[1270] Unreliable interpretation was to be avoided, for it created uncertainty that gave advantage to the enemy.[1271] 'The enemy, who is fully aware of the importance of the aerial photographs, will do everything in his power to make the study of film difficult. Skillful camouflage of organisations in large numbers and fake works are all means which permit the enemy to mislead our investigation.'[1272]

Throughout the war, the battle of minds adhered to the dictum, 'The eye only sees what it knows.'[1273] Both aerial photographic interpreter and combatant required knowledge of the technical possibilities of camouflage. 'Art' was used as if by a painter or sculptor who appreciates the part played by light, shadow, and tone.[1274] The French adopted this theme in their assessment of camouflage in 1918. 'To conceal oneself is now an art an art all the more difficult now that the means of observation are being constantly perfected.'[1275]

Camouflage practitioners were masters of the art of illusion. As early as February 1915, the French assembled a camouflage section from specialists at each command echelon to build dummy tree trunks or mannequin-like men and horses.[1276] The camouflage discipline soon included talent from the art world. One French artist, Lucien-Victor Guirand de Scévola, (1871–1950) was acquired from the artillery branch and given the challenge to expand the camouflage program.[1277] In 1914, as a telephone operator in the Pont-à-Mousson sector, he observed how vulnerable weapons and ordnance were at the front. To counter that vulnerability, de Scévola led development of a camouflage cover – a net splashed with earthen colours in a disjointed, 'cubist' manner. Shortly after the battle of the Marne, Guirand de Scévola communicated with both Général Joffre and President Raymond Poincaré about his discovery, leading to the first *section de camouflage* for the French armée. Armed with this expertise, the armée began to systematically develop camouflage for its soldiers and equipment. De Scévola's artistic experience included the newly developed Cubist technique, established by Picasso around 1907. He described his approach: 'In order to totally deform objects, I employed the means Cubists used to represent them – later this permitted me, without giving reasons, to hire in my [camouflage] section some painters, who, because of their special vision, had an aptitude for denaturing any kind of form whatsoever.'[1278] By 1918, the French camouflage section employed three thousand *camoufleurs*, including another well-known French artist, André Dunoyer de Segonzac.[1279] By the end of the war, de Segonzac served as chief of the French IIIème Armée's camouflage section and was awarded the Croix de Guerre.[1280] The unit was designated *La section des caméléons*. The camoufleurs had the distinction of carrying an insignia emblazoned with a chameleon.[1281]

In December 1915, the British military staff selected a camouflage enthusiast, Solomon J. Solomon, Royal Artillery, to work with the French and learn from their experiences. Solomon proceeded to establish the first British detachment to develop camouflage techniques.[1282] He sought out skilled artists, trained in designing imitations of nature within the 'world of the photograph'. Art became a military tool to counter aerial observation, creating a new class of 'war babies', camouflage and aerial photographic interpretation.[1283]

Military camouflage practitioners paralleled the art world. Camouflage influenced perception through

Decoys. Two 'tanks', British heavy and French Renault, were constructed in wood to convey the presence of armour in a given sector. (Gorrell Report, NARA)

shape, aspect, shadow and relief, regularity, and colour and its relative values. Camouflage required attention to shadow, highlighting the face of the object or a neighbouring object. Abnormal shadows gave away early attempts at camouflage and were avoided. They sought colour harmony with the surrounding objects.[1284] Colour served as another indicator of camouflage, including analysis of the spectral composition of incident light, the state of the surface of the coloured matter, observation distance and direction.[1285] These practitioners of the art of camouflage directed that it be 'less as a question of painting than as a question of sculpture, ie treat it by reliefs rather than by colours.'[1286] Behind the science of the camouflage art, aerial photography became the discriminator, reproducing the distinct spectral compositions beyond what the trained eye could determine.[1287] In his post-war reflection on camouflage art, Solomon summarised the thinking of the military practitioner:

> Even when art may be used to effect surprise it is doing no worse, but practically the same thing as the general who is pitting his brain against his opponents. That is known by the name of strategy or tactics. It is clear then, that the art which merely replaces the defensive or offensive concealment which nature in earlier wars afforded the needs to be developed to a high pitch of efficiency; for the side which neglects it will be unduly exposed to the onslaughts of an adversary who will be concealed and whose whereabouts cannot be detected.[1288]

TOOLS TO COUNTER CAMOUFLAGE

The aerial photograph became the discriminator, validating the ability of camouflage to provide adequate cover. Not only was the aerial photograph employed against the Germans' use of camouflage, it testted Allied camouflage throughout the war. Photographs were taken vertically and obliquely, in varied light, different sun angles and at various aeroplane altitudes to simulate the enemy observer. Prior to a military operation, French planners tasked aerial reconnaissance escadrilles to photograph sectors in order to evaluate the infantry or artillery unit's ability to mask any identifiable signature.[1289]

Aerial photographic equipment employed against camouflage included various panchromatic plates and colour filters. The aerial photographer had to differentiate colours at a level beyond the limits of human eyesight. The colour-filter lens resolved the hues of the spectrum into their parts and imparted a definite and characteristic shading on the photographic plate, disclosing a spot of 'impure' colour. Foliage green, for example, resembled red; the camouflage remained green or turned black on the print. This assisted the interpreter to diffentiate between real and artificial foliage. A plate susceptible to green was employed. Camouflage blurred on the image.[1290]

The most successful means of detecting camouflage was to exploit the history of coverage of a given sector.[1291] The interpreter could trace changes at given points in time thanks to a collection of aerial photographic coverage of the same location. The collection sometimes detected construction activity in progress. History of coverage allowed the interpreter to detect new works, extent of destruction, as well as explaining enemy intentions. However, once the camouflage was in place, it became difficult to interpret what was underneath.[1292]

Optics facilitated aerial photographic analysis of camouflage and deception. The most sought after coverage was stereoscopic, the dual optics providing a three-dimensional portrayal of the terrain. The stereoscope endowed the photographic interpreter with perspective and depth, a capability that could sometimes penetrate German efforts to camouflage vital installations. The stereoscope became the necessary instrument for discerning badly camouflaged works or evidence of 'too much' debris.[1293]

Despite access to tools, the interpreter inevitably did not penetrate all the enemy's camouflage. The Germans established the function of 'security officer' as a constant reminder to enforce rigid observance of the camouflage discipline. Some were attached with balloon units to monitor their own lines for discrepancies.[1294] Another challenge was operational ignorance about the basics of aerial observation and photography. Few laymen knew how to look at an aerial photograph. Airmen had the reputation of never looking at a photograph with a magnifying glass. The British camouflage expert Solomon later remarked:

> How they could expect to find incriminating clues, unaided, passes comprehension. There was ever the same incredulity right up to the end, in spite of the illuminating fact that bombing airmen were warned not to confuse an Earl's Court city with its real counterpart on the Rhine put up by camoufleurs to fog them.[1295]

THE INSTITUTION OF MILITARY CAMOUFLAGE

The best way to combat ignorance was to educate the combatants on camouflage construction and application techniques, by revealing the secrets. Camouflage was defined as 'The art of deceiving the Intelligence service of the Enemy'. The target was the enemy intelligence service, which was 'responsible for discovering and appreciating the facts as to our movements and intentions; therefore it is that service which it must be our first object to deceive'.[1296] Systematic camouflage included a positive and negative approach. Negative camouflage concealed the true facts from the enemy's intelligence service by preventing them from discerning the important organisations, to include concealment of an intended attack by monitoring for saps, trenches, loopholes for observation and firing, machine-gun emplacements, shelters, accessory defences, and 'fitted up shell holes'.[1297] The other technique was to create 'the impression of life'.[1298] By imitating reality they created positive camouflage, suggesting that an attack was to take place where no attack was intended; demonstrating the apparent existence of troops, trenches, batteries, and the like. The intention was to engage 'in the deepest mystery'.[1299]

All combatants took necessary precautions against aerial or ground observation. Given the stasis of the battlefield, the combatants had to adapt to life under constant scrutiny. Integral to success was the discipline required to avoid detection by constantly applying positive and negative camouflage techniques. Despite extensive camouflage efforts, the simple carelessness of a single soldier could lead to detection.[1300] American staffs quoted extensively from the experiences of Captain F. T. Colby, a Belgian artillery camouflage expert: 'Camouflage discipline should exist and be enforced as much as sanitary discipline is enforced'.[1301] Soldiers were directed to keep traffic to a minimum and to mask shiny objects on clothing and weapons. Creature comforts were minimised by suppressing smoke emissions and fire at night. Specifically, 'the trench should be as invisible as possible to observers on the ground or in the air'.[1302]

A camouflaged trench appeared on an aerial photograph as a slight blur, either lighter or darker than the surroundings, caused by insufficient or too much shadow in the texture of the material used.[1303] Within the trench, positive and negative camouflage could be applied to the construction of enemy observation posts and by depositing the soil so that it appeared as trench improvement. Debris from large excavations could go into sandbags dumped at a dummy emplacement or location of little consequence to draw fire.[1304] A good camouflage maxim was 'Make your dummies at the same time or before your real positions'.[1305] Instead of the normal 2-metre depth, a dummy trench could be established with as little as 40–50cm of depth. An aerial photographic study of shadows was the only way

Netting covering a light rail (tramway) in the front sector. (Gorrell Report, NARA)

to ascertain the true depth.[1306] Protruding saps were difficult to hide from aerial photography due to their trace on the landscape. Again, depth was determined through stereoscopic analysis of aerial photographs. Trench camouflage concentrated on covering fire-steps, observer loopholes and shelter entrances. Machine gun emplacements were 'practically impossible to discern' on the aerial photograph. Ironically, camouflage led to suspicion about an emplacement. 'An excess of precautions can be injurious.'[1307] History of coverage remained the best approach to counter any development or changes within trench networks.[1308]

Various techniques were created to challenge photographic interpretation. 'Umbrella camouflage' employed netting on posts with canvas strips attached. The umbrella covered large objects like ammunition dumps and spoil from mines.[1309] Fishnets were placed on wires to confuse analysis of trench defences after a heavy bombardment.[1310] A deceptive colour scheme for building and roofs was developed. Scrim, a material used for curtains or upholstery lining, disguised buildings by altering the outline, shadows and colouring. The British modified a house used as a corps headquarters to look like a ruin to aerial observation.[1311] Decoys were another component to deception. For example, false ammunition dumps were created with camouflage. The real dump might be in the connecting cellars of a ruined and demolished village, or in the excavated embankment of an abandoned railway.[1312] Constructing dummy field hospitals was suggested by planners in the last year of conflict, but was rejected as a violation of the Geneva Convention.[1313]

Camouflage was not confined to individual equipment. It expanded over the course of the war to cover large areas of the terrain. Roads were screened to thwart enemy observation. The first attempts used canvas sheets painted to represent the ground. Heavy to erect and difficult to maintain in position, on the suggestion of Solomon the canvas was replaced by lighter and more manageable material such as old fish nets or wire netting, garnished with tufts of painted or dyed fibres from the raffia African palm tree. In 1916, prior to the Somme offensive, British engineers wove thousands of rolls of wire netting with the fibre. In the course of time all British artillery batteries were equipped with the fibre covers. As the demand increased and the supply of raffia became inadequate, less flammable canvas strips were eventually substituted.[1314] Major George Patton, AEF, provided a brief description of road camouflage in a letter to his wife in May 1918.

> The Maj. and I left here at 7:30 and went by motor to with in about 1,500 yds of the front line. Most of the way the road was screened with camouflage. Sacking about 12 feet high on frames, it keeps the enemy from seeing what is on the road so he can not fire on it.[1315]

The art of camouflaging artillery was considered a vital element of artilleryship, 'a knowledge of which is of as high importance for the battery commander as the preparation and conduct of indirect fire, the knowledge of battery transport, or of any other of the basic principles of his education.'[1316] Artillery operations and camouflage were intertwined. Camouflage concealed the battery position, communications, and observation posts. It also helped direct enemy artillery towards a false battery.[1317] Flash screens were employed at night to hide gun flashes from the front and flank view. Eight-foot-high screens running out about eight yards to the front were placed to the right of the muzzle of each gun.[1318] Some artillery units disguised their gun positions by preparing several emplacements for each battery of four guns. The guns fired from one position for

Netting across the roads became a standard camouflage in the war. (*Scientific American*, 26 January 1918)

a time and then moved, generally at night, to another. This pattern was repeated amongst various positions. Small detachments at the vacated emplacements fired off flares, resembling the muzzle flashes of guns while at the same instant the real guns fired from another position. A combination of stereoscopic and ordinary photographs countered the deception by recording the blast marks from the artillery discharge.[1319] Elaborately camouflaged artillery was often detected with ease when the concealed object cast a noticeable shadow. History of coverage of a given location helped detect any attempt to add artificial cover.[1320] The French configured their artillery camouflage to the phase of the operation. During manoeuvre, camouflage was not a priority. However, when the battery was established for fire, an appropriate cover was applied.[1321] It was indicative of a commander's attention to camouflage that the forces were reminded to take appropriate measures. Such was the case during the Allied summer offensive in 1918, when Général Vincent, artillery commander of the 38ème CA, alerted both subordinate American and French artillery commanders of concerns over appropriate camouflage for operations. Vincent reminded the commanders to have their batteries photographed through aerial reconnaissance regularly to 'perfect the camouflage'. Reinforcing the policy was the directive that the first aerial photographs were to be 'rushed' to the commanders of division and heavy artillery.[1322] Artillery camouflage attempted to avoid the repetition of the battery signature. The geometric form of the employed battery required covering any discernible angles where the camouflage joined the soil. That required attention to what the shadow revealed. Tracks were conspicuous, requiring the unit to lay down turf in irregular fashion. The best camouflage for tracks was 'obtained by discipline'.[1323]

Artillery employed dummy configurations, notable for overcompensated features, providing the aerial photographic interpreter with conspicuous blast contours and artificial tracks. Additional tricks included smoke from false hutments to suggest life in the area. An operational technique to reinforce the site's deception was to quickly manoeuvre artillery into the dummy emplacement at night. Rounds fired gave the impression that the location contained an active unit.[1324] Unoccupied emplacements served as false batteries. One German document instructed that four emplacements should be constructed: a forward, spare forward, withdrawal and a spare withdrawal emplacement. The battery could manoeuvre between those locations, conveying an environment of non-activity.[1325]

Cooperation among the planning staffs was essential. Elaborate precautions for moving troops by road or rail during the night were useless 'if the construction of large and obvious dumps [and] new hutments, aerodromes and Field hospitals are springing up like mushrooms long before an attack is to take place'.[1326] In the rear echelon areas, dummy aerodromes and aeroplanes were a positive technique employed to confuse order of battle analysis.[1327] By 1917, it was obvious that all facets of aviation were vulnerable to observation and aerial bombardment. The French began camouflaging their áeroplanes that October. Colonel Edgar S. Gorrell, in a memo to the chief of the US Signal Corps, mentioned the transformation. 'Within the last three weeks there has taken place a great change in the method of doping French airplanes. All the French airplanes are now being doped a sort of silver gray colour underneath ... the sides and tops are camouflaged in a somewhat similar method as the camouflage of the Bessonneau hangars'.[1328]

Aerial photography during spring and summer had to break through the leaves and natural cover. However, interpreters considered camouflage with fresh branches and undergrowth a simple matter to discern.[1329] Cultural practices of the region also complicated analysis. In late 1914, an RFC aviator circulated a memo describing the lessons learned from observation in Northern Europe. Mistakes in identifying objects on the ground risked determining the true force disposition incorrectly. Rows of straw shelters used by farmers for drying tobacco or herbs were erroneously reported as enemy bivouacs. Some farmers' practice of piling beets or turnips created the misleading image of trenches where none existed.[1330]

CAMOUFLAGE IN CONFLICT

Both sides paid extensive attention to camouflage and deception, with mixed results. Ironically, the British had demonstrated a propensity early on to use the technique, but ignorance amongst the senior staff slowed the rigorous application of the art. In 1912, the British Army conducted manoeuvres with aeroplanes, providing aerial observation for the first time. One British division was so successful in hiding itself that the press labelled it 'The Lost Division'. Lieutenant-General Sir James Grierson, commanding the division, when asked by King George V how he accomplished the feat, replied, 'The men put their waterproof sheets over their heads and made noises like mushrooms'. In reality, the troops acquired camouflage material from tree branches, while straw and waterproof sheets covered guns and other essential items. They took the extra step to blend in of bivouacking close to hedges, walls, houses and woods.[1331]

By March 1915 French aerial observation reports emphasised that German artillery batteries had commenced taking evasive action and hiding from

Combatants on the Western Front complicated the photographic interpreter's analysis by creating fake artillery pieces. The artillery being examined was a German decoy. Similar but cruder dummies were deployed in the American Civil War, known as 'Quaker guns'. (Gorrell Report, NARA)

the ongoing Allied aerial reconnaissance.[1332] 'Keep everything low' was a camouflage instruction found on a German prisoner.[1333] Ernst Jünger, a future *Sturm Truppen* commander, recalled the precautions taken by the infantry: 'On account of the danger from aircraft we had to keep ourselves close in the overcrowded place.'[1334] By the final year, operational emphasis on the issue was clear.

> Concealment from aeroplane observation is more important than all other considerations in the choice of a location for an encampment. Carefully conceal beaten paths, trample down no paths through grain and clover. Columns should leave established roads under cover. Do not park vehicles in rows, but individually under trees or carefully concealed, never on the street. Keep at least ten metres inside the edge of the woods. It is the permanent duty of officers of General Headquarters, of balloon and aviation officers, to see to the carrying out of these orders.[1335]

The Germans became expert in simulating rail and road movements to confuse aerial observation and intelligence analysis. They marched back and forth on the same roads, employing a team of men on bicycles with smoke-producing devices to imitate clouds of dust on sheltered roads. Smoke-emitting automobiles were placed on railroads to simulate rail activity. The deception increased as aerial observation was forced to fly at higher altitudes.[1336]

Four years of positional war clearly confirmed the proposition that surprise was critical to successful conduct of the campaign. The Germans developed underground facilities and useed them effectively in major operations. Their most notable success was staging their 1916 advance attack in the Stollen underground facilities at Verdun. By 1917, German camouflage included construction of passages between shell holes. The excavated earth was thrown into neighbouring holes or spread on the ground. Tunnelled dugouts were designed to appear like shell holes so that aerial observation could not determine an entrance.[1337] In the final year, the most brilliant accomplishment was the strategic masking of the German forces staging in the Somme region prior to the *Michael* offensive.[1338] The Germans successfully coordinated their deception operations. The French proved inflexible in applying their cryptological and radio intercept strengths to deceive and counter the German buildup. One French général reflecting on the inadequacy of the French approach to camouflage and deception goals, summarised the situation: 'Camouflage ... is an anachronism.'[1339]

The British oversimplified German intentions prior to the *Michael* offensive based on over-dependence on order of battle information. This misperception helped the Germans' overall deception mission. Author David French described the British conviction that *Michael* was against the Channel ports, causing them to move an extra division and 20 batteries to reinforce the British Second Army guarding the coast.[1340] British offensive doctrine and strategy went counter to these lessons. By 1917, British policy had concluded that preparing for a large offensive was impossible to hide from the Germans. British military strategists determined that the best approach was to overwhelm the Germans with a superior force. This doctrine failed in the Cambrai offensive that took place that November. The subsequent disaster from *Michael* showed acute Allied vulnerabilities in misjudging the German strategic camouflage program and the latter's ability to mislead estimates of German force strength and intentions. Such a shortfall was almost fatal for the Allies.[1341] Partial blame can be pinned on a less rigorous application of preventive measures through aerial reconnaissance. The Supreme War Council in the spring of 1918 concluded, 'The failure to make the necessary photographic tests is the only explanation of much bad camouflage work which has been done in the past.' If Allied camouflage was to succeed, it required greater coordination between the camouflage and aerial photographic interpreter. As late as May 1918, the Inter-Allied Aviation Committee meeting at Versailles to plan the strategic campaign for the coming year cited camouflage as an area requiring more attention.[1342]

Art remained in conflict with tradition. For example, French operational and tactical systems were too inflexible to learn from camouflage developments that supported German attack methods. By 1918, the French command had not developed techniques equal to German camouflage and deception. 'It must be admitted that our military education has not been developed in this direction before the present events, the very idea of concealment being distasteful to our character.'[1343]

Allied commanders preparing for the final offensives prior to the Armistice did consider camouflage and deception. However, the experience did not always meet expectations. The innovative American aviator, Brigadier General Mitchell, personally took measures to complicate German analysis prior to the September 1918 St Mihiel offensive, ordering aerial units in the sector to construct fake hangars and aeroplanes at various locations close to the impending battleground. In the days before the offensive, fabricated aeroplanes were moved around the airfield suggesting minimal activity at the location. While this charade was enacted, real aeroplanes arrived at the forward aerodromes prior to dawn. The camouflaged hangars were replaced overnight by functioning hangars of similar dimensions. For the most part, the effort was successful. However, it only takes one mistake and the operation can be compromised. In Mitchell's case, one aero squadron made the critical mistake of arriving in broad daylight and commencing aerial reconnaissance of the strategic rear echelons near Metz. Their increased aerial activity alerted German intelligence to the impending operation.[1344]

Making camouflage and deception integral to success in positional war required extensive resources. The Americans quoted a captured German report to make clear the level of effort required:

> On account of the considerable labor involved and the great quantity of material necessary to produce the desired effect, complete camouflage, preventing aerial photography, can be supplied only to small works, cover for single guns, machine guns, etc.[1345]

When the Americans arrived, a US Camouflage Corps was established. Their average monthly consumption was an illustrative example of commitment to the art. In order to provide camouflage cover for the American sector, approximately 4,328,000 square yards of burlap, 200,000 gallons of paint, 7,700 fish nets, 50,000 pounds of wire and 2,160,000 square yards of poultry netting were required to conceal trenches and artillery battery emplacements. A single square mile of the front line was estimated to contain approximately 900 miles of barbed wire, 6,000,000 sandbags, 1,000,000 cubic feet of timber, and 36,000 square feet of corrugated iron.[1346] Had the conflict continued into 1919, the Americans could have been the de facto leaders of the Allies in applying camouflage to their operations, thanks to their industrial might.

Deceiving the enemy in modern warfare covered all facets of intelligence and information acquisition. Trench warfare created the right environment for testing the limits of perception, for the combatant was trapped in a stationary world.

The inability to manoeuvre beyond the gaze of aerial reconnaissance and the camera complicated the operational reality of combatants used to manoeuvring at will. The camouflage and deception challenge extended beyond the visual to other arenas such as communication.

The war demonstrated the enemy's ability to intercept and create false transmissions despite advances with radio technology. To operate in a world clouded by misperception required discipline and an unfailing belief that daily operations and procedures would ultimately ensure success.

CHAPTER 20

Arranging Aerial Reconnaissance

The Plattsburgh Manual, a popular army training manual employed by US forces at the time, illustrated the American leaders' confidence in the role of aerial reconnaissance:

> Airplanes will move far out, perhaps hundreds of miles, in front of our most advanced cavalry for the purpose of gathering information on large bodies of the enemy's forces. This is called Strategic Reconnaissance. Other airplanes do more local scouting. They go but comparatively short distances from the firing line for the purpose of determining the location of trenches, supports, reserves, artillery positions, etc. This is called tactical reconnaissance. They give their artillery commanders information as to where their projectiles are falling. During siege operations (as in Europe, where some trenches have remained in about the same place for long periods), photographers go up in airplanes each morning and photograph the enemy's trench lines. Blueprints are made of these lines. By comparing these with the lines of the previous day it is easy to determine the changes that have been made during the night. Other airplanes are detailed for the purpose of combat. They prevent opposing airplanes from gathering information.[1347]

The manual's summary of aviation highlighted the obvious: aviation's role was aerial observation. Combat was mentioned in the last line, and even that was about the information-gathering function. The companion *French Manual on Trench Warfare* made it clear that, 'In stationary trench warfare, observation is continuous as in combat.... The aeroplane and balloon are employed to obtain information (observations, photographs).'[1348] The French were the most sedulous in applying aerial observation to positional warfare. Their specialised system for the observation squadrons emphasised photography, infantry liaison and adjustment of artillery fire.[1349] They led in acquisition, integration, and analysis of aerial photography. In the early stages of the war they recognised that aviation 'furnished nevertheless very valuable intelligence, but this intelligence lacked the indisputable confirmation which photography affords.'[1350]

Voices from the aviation community of this era made it clear that the reconnaissance mission was the most important mission. One of the aviation pioneers for the US Air Service, Lieutenant Colonel Lewis H. Brereton, remarked: 'The basis of all war-time aviation is the observation air service.... The air service observation is the most important intelligence agent of an army .... Information regarding his [the enemy's] intentions ... is derived largely from aerial photography.'[1351] Such a feeling was echoed by the American aerial fighter legend, Captain Eddie Rickenbacker, who summarised his views on aviation at the time: 'I believe this function of "seeing for the army" is the most important one that belongs to the aviation arm in warfare. Bombing, patrolling, and bringing down enemy airplanes are but trivial compared to the vast importance of knowing the exact positions of the enemy's forces and "looking before you leap".'[1352] This attitude permeated throughout the ranks. One poignant comment was written by a leader within the US Air Service ranks, Major Philip Roosevelt, concerning the death of his aviator cousin Quentin (President Theodore Roosevelt's son): 'If Quentin has given his life we know that he gave it with a high heart in the performance of duty of prime importance, for the photographs which were obtained played no small part in enabling one French army to know with some precision the nature of the attack which sixteen hours later was launched against it.'[1353]

Aeroplanes were clearly recognised as a valuable asset as the war progressed, yet military commanders defined their purpose on traditional lines, applying their unique capabilities to existing military operations. The aeroplane's ability to roam the battlefield and beyond was a definite asset for both tactical and strategic operations. In early 1915, US attaché Colonel Squier wrote, 'For strategic and tactical reconnaissances the aeroplane is at present simply indispensable. In the present form of trench warfare the aeroplane is used to watch, sketch and plot the development of the enemy's trenches day by day, and in most cases it is the only method of keeping informed of the progress of their preparations.'[1354] Besides observing the trenches on a daily basis, aerial reconnaissance covered the immediate battle area to locate and examine gun emplacements, reserves, and railheads, chiefly to satisfy the corps or divisional commanders with timely information of the immediate threat. The longer-range strategic reconnaissance extended beyond the realm of the immediate tactical battleground to the adversary's ability to sustain combat operations. It provided insight into enemy intentions.

Aviation's evolving role in positional war caused leadership to slice the airspace into zones of responsibility. The analysis, acquisition, and transmission of information became institutionalised. By the end of November 1914, the RFC under the temporary command of Colonel Sykes described an aviation organisation that allotted the area of operation to the wings for 'reconnoitering and offensive duties for the units in that area'.[1355] Sykes' subsequent 'Notes on Air Reconnaissance' (Appendix C) defined four primary aviation missions: (a) Destructive; (b) Artillery Observation and Reconnaissance in the Immediate Front of Troops; (c) Tactical Reconnaissance for Armies and (d) Strategical Reconnaissance undertaken by Order of General Headquarters.[1356] Missions designated as 'tactical work' were assigned by the command echelon requiring information within their sector. Sykes' instructions elaborated, 'Normally, strategical reconnaissances will be ordered from R.F.C. Headquarters and only tactical reconnaissance be required by Corps Commanders.'[1357] Tactical reconnaissance had to be fast: 'The nearer the landing ground can be to Artillery H.Q. the better.'[1358] Taskings for RFC aerial reconnaissance by the end of 1914 showed a strategic emphasis with 'Special attention to be paid to rolling stock at all stations.'[1359]

'For infantry, order is above rapidity,' was a prevailing mindset in French military thinking.[1360] Whatever was to be accomplished by aviation ultimately had to benefit the fighting soldier. Doctrine conveyed through French training manuals clearly showed aérostats and aéroplanes as subordinate to the respective armée or CA. To meet the needs of the commander, aerial reconnaissance coverage encompassed three categories. First, aérostats provided the general observation and observation of fire against objectives visible from that vantage point, with ground observation , though limited, complementing this category. Second, views not accessible due to terrain were to be acquired through the aéroplane. Finally, aéroplanes could be longer-range resources, seeing far beyond the ballon captif and ground observation. The service Aéronautique commander serving armée or CA artillery objectives assigned duties to aérostat and aéroplanes as required.[1361]

BALLOON OBSERVATION/BALLON OBSERVATEUR PARAMETERS

Balloon observation (ballon observateur*)* was a prized resource for artillery, providing artillery spotting precision fire against targets beyond the limits of ground observation. Aérostat/ballon captive observation depended on the altitude of the balloon, the atmospheric conditions and the terrain. Rules governing French aérostat/ballon employment stated they were to ascend no closer than 5km to the front.[1362] Two observers could ascend to 1,200 metres while a single observer could reach 1,600 to 1,800 metres. When terrain obstructed the view, the aérostat/ballon commander prepared a sketch for each point of ascension and forwarded it to his headquarters command and the artillery commander.[1363] Balloon observation target assignment came from headquarters, artillery units and Service Aéronautique reporting.[1364] The French also directed division-assigned balloons through direct telephone contact to the division PC.[1365] The telephone connection to the artillery unit, providing instant communication for rapid targeting, was one advantage of ballons over aéroplanes.[1366] In 1917, French ballon captif companies were equipped with a radio receiver to enable them to receive TSF transmissions from aéroplanes flying in their area. Aerial observation was enhanced with aérostat and aéroplane working together against targets in the sector.[1367]

The Americans followed the lead of the French and British in aligning their balloon units to the army echelon requiring support, attached to a 'Balloon Group'. The balloons were then reallocated to serve the needs of the lower echelons. The Americans assigned their most reliable observers to support the corps sector to ensure the most accurate direction of fire for medium calibre guns. Balloons supporting the division helped regulate divisional artillery fire as well as provide surveillance and liaison for the infantry divisions. The number of assigned balloons usually corresponded directly to the number of divisions in the army corps. Balloon reporting covered enemy infantry and artillery activity, aeronautical activity, movements on road and rail, and sightings of explosions.[1368] Aerial photography from captive balloon platforms was considered but the lack of photo sections precluded it from becoming an integral component of the force. The US Air Service policy established that if photographic officers desired oblique coverage from balloons they were allowed a balloon ascent. Once acquired, the prints were processed by the photo section and not the balloon unit.[1369]

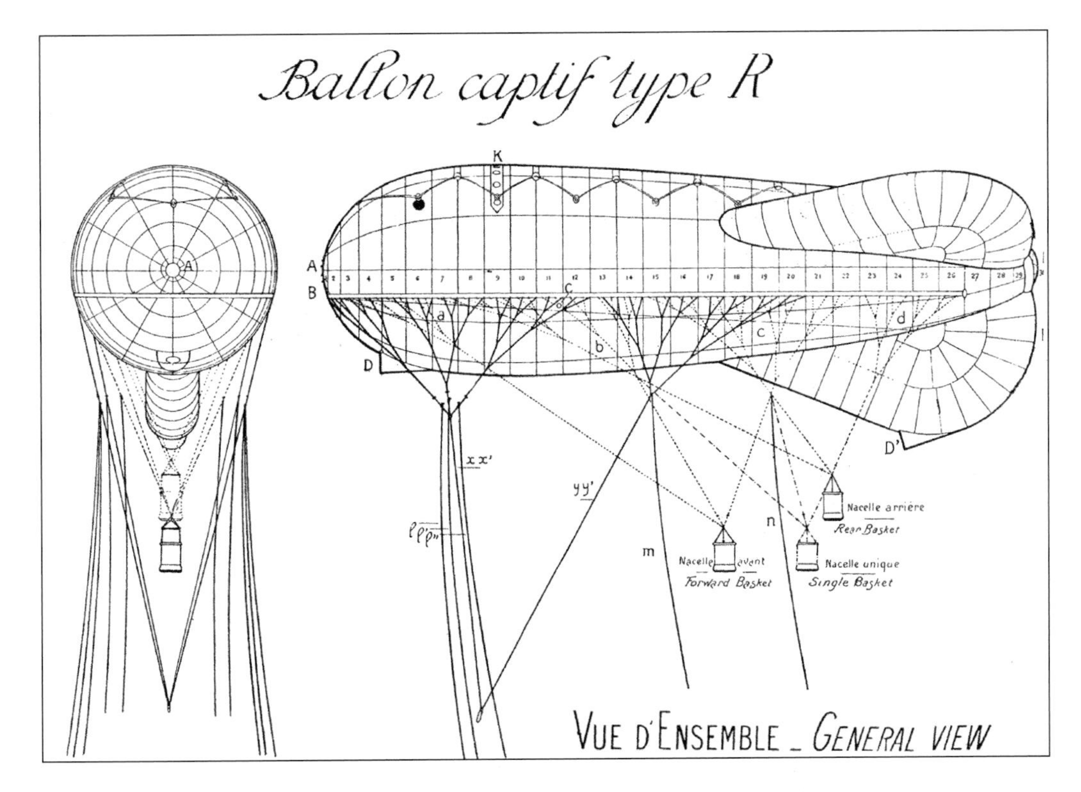

A detailed configuration of the ballon captif of the First World War. (*Instruction sur le Ballon Captif Allonge Type R*)

French aérostiers preparing for an ascent. Note the moving picture camera. (Gorrell Report, NARA)

AEROPLANE TACTICAL RECONNAISSANCE

Tactical reconnaissance extended the limits of captive balloon coverage as it covered both corps and division sectors. The division sector defined the smallest area of coverage (Appendix F). The Divisional Observation Squadron worked close to the front lines providing direct support to the respective division. The squadron only reconnoitred the areas opposite its division's sector to a limit of 5km.[1370] French CA service Aéronautique and the equivalent US Army Corps Observation were limited to less than 10km from the front line (the British operated their Corps squadrons out to 10 miles). In practice aéroplanes assigned to armée escadrilles were allowed to reconnoitre within the CA zone as required without prior approval.[1371] Army Corps Observation zone of action corresponded to the Army corps' artillery range.[1372]

The British 'Corps Wing' accomplished the more intimate work of tactical reconnaissance to include aerial photography, artillery spotting and patrolling near the line. This allowed the British Army to employ dedicated aerial resources to tactical missions. The formation of wings foreshadowed further decentralisation aimed at preserving the identity of the Corps, while at the same time distributing its operating units amongst the various commands of the Army.[1373] The Corps Wing contained up to four squadrons and all were equipped for aerial photography, operating from landing grounds close to the headquarters. The number of squadrons varied with the requirements of the Army's assigned sector. As a general rule the Army squadrons' landing field was farther back from the front line while Corps squadrons were in close proximity to Corps headquarters. These Corps squadrons conducted their sortie operations at an altitude of 8,000 to 12,000 feet, employing an 8- or 10-inch focal length lens, with a 20-inch lens used for 'special occasions'.[1374]

STRATEGICAL RECONNAISSANCE

Strategical reconnaissance accomplished visual and photographic reconnaissance over the front of the entire army to a depth of 50km behind the front lines. It was done by escadrilles and squadrons assigned to the respective army. Strategical reconnaissance gave 'an idea of the occupation of suspected regions and of the extent of zones of containment'.[1375] Strategical reconnaissance missions determined potential threats to the front lines from long-range artillery and aéroplanes. Order of battle analysis was conducted through it. Once that was determined, the potential for offensive operations or withdrawal was analysed. The French recognised that strategical intelligence collection required a fusion of all sources and the situation could not be 'perceived in one single glance'.[1376] Aviation complemented intelligence analysis at the headquarters level. Reconnaissance groups worked day and night.

A leading French advocate of strategical aerial reconnaissance was Paul-Lewis Weiller. He found that coverage acquired from a rigorous flying schedule of the rear echelon provided successful intelligence on German intentions in a given sector. The best formation, Weiller reasoned, was a group of large-scale reconnaissance flights committed to photograph the rear echelons to a depth of 100km as frequently as conditions permitted. The Breguet 14 A2 aéroplane fitted this strategy. It was rugged and could perform well against enemy pursuit aeroplanes. The combination of Weiller Group aerial coverage and 2e Bureau intelligence analysis proved to French commanders such as Général Foch that it was possible to estimate as far as one month in advance where the Germans were planning to attack or abandon a sector. Prior to this intelligence estimates relied on prisoner interrogations, which yielded vital

Thanks to the key map of the Weiller Group strategical reconaissance work, analysts had a clearer idea of activity in rear echelons. (de Brunhoff, ed., *L'Aeronautique*)

information only hours before the impending operation. Foch became the strongest advocate of strategical reconnaissance. He launched real as well as simulated counterattacks based on the Weiller Group's key maps. Most importantly for the conduct of the campaign, Foch directed maximum effort against those sectors that were identified through strategical reconnaissance to be abandoned. The enemy in the last months of the war became victim to the Foch strategy, one supported by in-depth strategical reconnaissance.[1377]

The heavy attrition of photographic reconnaissance prompted the French to reconfigure single-seat pursuit fighters such as the Nieuport 17, Spad VII and Spad XIII to photographic reconnaissance roles. Formation flying ending in dramatic dogfights continued to take a toll on reconnaissance assets. The independence of the single-seat fighter was an intriguing feature that aerial strategists figured would benefit aerial photography. However, photographic collection demanded straight and level flight, increasing the vulnerability of the fighter/reconnaissance aeroplane.[1378] The French persevered. Their pursuit aéroplanes were configured with a semi-automatic camera such as the Italian-made Piazza operated by the pilot through a Bowden cable. These sorties acquired high-altitude aerial photographs (4,000 to 6,000 metres) and oblique coverage.[1379] One escadrille, SPA 49, was renowned for flying as far as 200km behind enemy lines, taking 6,000 photographs during September 1918 alone.[1380]

The British Army Wing focussed on strategical reconnaissance, bombing and aerial photography. This echelon comprised better aircraft to accomplish these missions. Additional missions included pursuit and aerial superiority.[1381] The Corps Wing worked the tactical requirements of the corps assigned within the army. The British assigned two squadrons equipped for photographic work to each Army wing. The final number of squadrons varied according to local requirements. The essential functions of Army and Corps remained distinct. However, an Army squadron might be called upon to do work normally falling to the Corps squadrons. One Army squadron was equipped for photography. As the war progressed, the British intended to have every squadron equipped for aerial photography.[1382] Since Army squadrons were intended to fly at higher altitudes (12,000 to 18,000 feet), the aeroplanes were configured for cameras with 20-inch focal length lenses.[1383]

ORGANISING THE AERIAL FORCE

Organising aerial observation to accomplish expanding missions helped establish and refine the French escadrille and British squadron. Better control of aéroplane resources required assigning similar aéroplane types together for more efficient maintenance. The French escadrille was commanded by an aéroplane pilot capitaine. Escadrille personnel normally comprised ten pilots, eight observers and three machine-gunners in addition to officers and men entrusted with the maintenance of flying material, rolling stock, radio telegraphy and armament.[1384] The aéroplane could serve a diversity of roles, be it aerial observation, pursuit or bombardment. Modification came with new requirements. French long-range reconnaissance gravitated to multi-engine aéroplanes to acquire strategic information well beyond the front lines.[1385] The

French emphasis on faster and higher flying aerial reconnaissance led to employment of the existing pursuit inventory for reconnaissance. The 50cm aerial camera behind the cockpit allowed the single-seat pursuit fighter to do the job. (SHD, 87.2859)

French also made a significant change in supporting heavy artillery targeting. Elite reconnaissance aircrews were assigned to Artillerie Lourde (AL) escadrilles to detect and photograph significant targets. AL escadrilles were numerically designated from 201 through 228. The aerial reconnaissance specialists were experts, forwarding essential information to longer-range heavy artillery units capable of delivering more lethal firepower.[1386]

As the demand increased for more aerial photographic coverage, the burden had an impact on the entire Allied aerial configuration and stretched available resources. For example, AL escadrilles supporting CA artillery were exempt from assignment to other duties.[1387] More aeroplanes and aircrews were needed to meet the demand. By the end of the war, the French had organised their aerial reconnaissance force according to capability. A long-range reconnaissance group consisted of one escadrille of Spad single-seaters (10 Spad VII or 10 Spad XIII) for fast reconnaissance, one Breguet escadrille (10 Breguet 14s) and one Voisin escadrille (10 Voisin bombers) for night reconnaissance.[1388] The RFC increased the squadron size from 12 to 18. The British GHQ envisioned a force comprised of two squadrons of two-seater fighter reconnaissance machines, two long-distance reconnaissance machines with eight hours fuel capacity, one squadron of single-seat fighters, and six bomber squadrons.[1389] US Air Service aero squadrons assigned to aerial observation roles in 1918 also had 18 aeroplanes assigned. The daily schedule had two flights of six aeroplanes accomplishing aerial reconnaissance. During active operations all flights were operational.[1390]

As the war entered its third year, the British recognised that the aeronautical forces needed to be more effectively integrated with the army. The RFC was subsequently reorganised to include brigade-level formations that were attached to the respective army at the front. Similarly, the RFC Brigade placed the headquarters within the vicinity of the army it served. These brigade-level units, commanded by a brigadier general, directed both corps- and wing-level resources specifically to support an assigned army sector. An RFC Brigade comprised a total of 108 aeroplanes, with its headquarters located near the headquarters of the army it was supporting, approximately 14 miles behind the front line. Within this force, the Brigade could allocate aeroplanes either to the army or the subordinate corps.[1391]

FORMATION FLYING

By 1916 the traditional array of aeroplane sorties confronted an increase in aerial combatants, causing the missions to become dependent on formation flying. Discussion of the aerial campaign at that time reflected the growing numbers. 'All reconnaissance, ordinary patrol, and bomb raids are now done by numbers of machines varying from three for line patrol, to bomb raids of as many as 20 machines, ie, all work is done in formation.'[1392] Aerial observation became extremely vulnerable to the evolving German pursuit aeroplanes. Major-General Trenchard, RFC commander, commented, 'Information can no longer be obtained by dispatching single machines on reconnaissance duties. The information has now to be fought for, and it is necessary for reconnaissance to consist of at least five machines flying in formation.' The evolving aerial landscape fully recognised that if aerial observation was to survive, it required the increased firepower offered by additional observation planes configured for aerial combat, or pursuit aeroplanes.[1393] The attacks of the German fighter scout were likened to 'guerrilla tactics, hanging on the flanks and rear of the formation, ready to cut off stragglers, or attacking from several directions simultaneously...'[1394] There was (greater) safety in numbers, no matter what the mission. As one British aviation lecturer in 1916 commented, 'On reconnaissance a straggling machine will probably be attacked by superior numbers, though the formation itself will be left alone. If a pilot therefore finds that he cannot keep up and is certain to be left behind, he should turn back at the line at his own discretion. This applies only to reconnaissance.'[1395] When the Americans commenced aerial observation, they adopted the standard of two protection planes flying with a single aerial photographic plane.[1396]

Techniques in formation flying were refined each year. By 1916, British instruction paid attention to the role of flight leader, working the rendezvous, flying at a reasonable speed that all could maintain, and keeping the rest of the aeroplanes in the same direction. The flight leader was distinguished by streamers attached to the wings. The formation circled the landing ground until the flight leader was satisfied that all aeroplanes were ready to proceed. While in formation for reconnaissance, it was advocated that hostile aircraft should not be attacked unless 'a good opportunity offers'.[1397] Aerial photographic formations were a triangle design, with the three scout aeroplanes in the lead carrying aerial cameras. The rear aeroplanes provided cover while the aeroplanes carrying cameras could concentrate on the photographic mission at hand, 'knowing that their tails are being protected.' The scouts were there to provide defensive protection, since the primary objective was to obtain the required photographs.[1398]

Air superiority became a more dominant objective as the war progressed. By the conclusion of the conflict improvements in combat potential had surpassed aerial

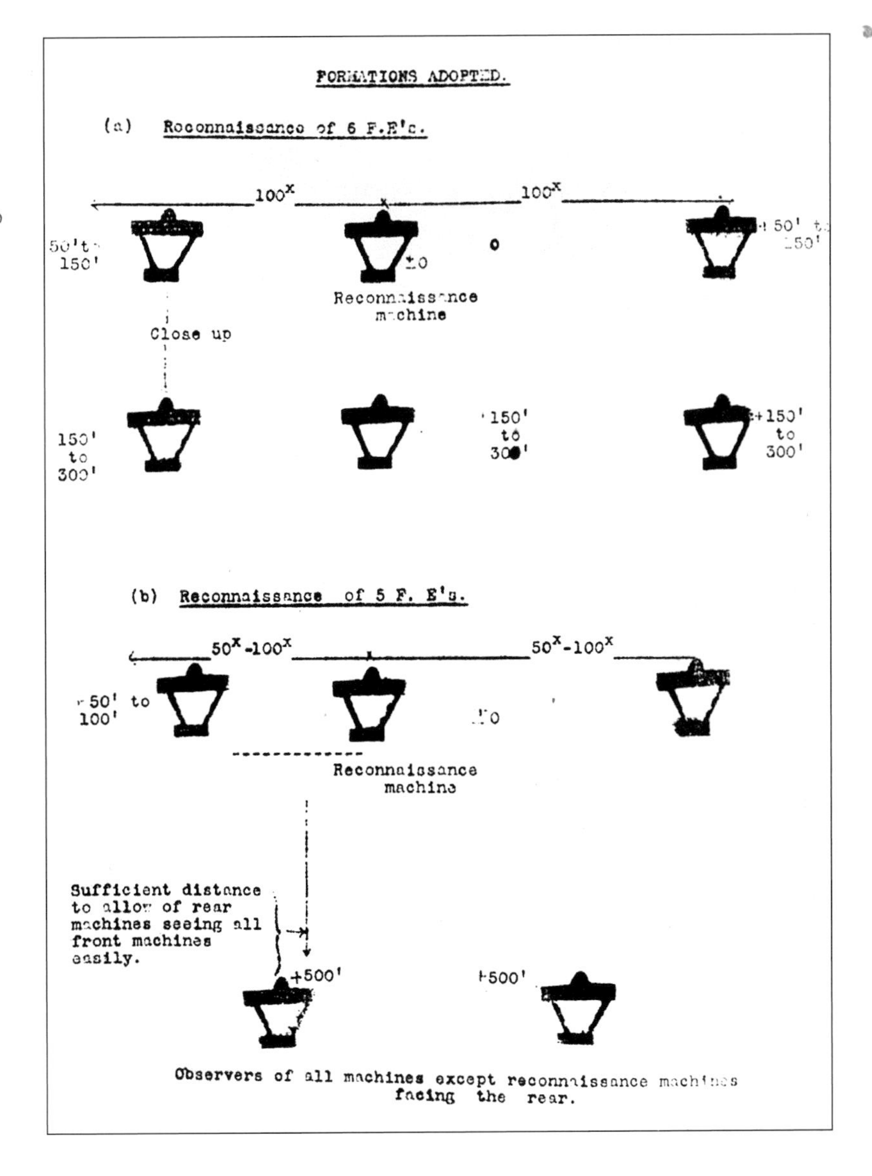

The evolution of formation flying emphasised protection of aerial reconnaissance. A British training manual portrays two FE 2b formations. (TNA: AIR 1/1625/204/89/7)

observation developments. 'Air superiority ensures you two things: a screen for your own movements and a knowledge of the enemy's. Such knowledge eventually leads to but one conclusion, the ultimate superiority in arms of the aerially superior forces.'[1399]

Aerial photographic formations retained the basic configuration to the end of the conflict. American guidance on photographic missions recommended the triangular formation, with the plane with the camera being in front and the other two on either side, slightly behind and above, giving protection to the camera man with their machine guns. Pursuit support to photographic missions was retained throughout the war, and sortie planning was inclined to include pursuit resources in the formation.[1400] Ironically, a formation of several aeroplanes providing protection for the aerial photographic aeroplane caused so much attention that it often led to an attack by enemy formations. Several aerial observation teams finally preferred to go alone on these missions, trusting to the pilot to spot any 'Huns' and feeling that if they were attacked it was easier to take care of themselves than to keep in a formation and fight a pitched battle.[1401] Inevitably, the issue of what constituted the best use of reconnaissance resources in this precarious environment led to the genesis of aerial reconnaissance performed by 'a fast machine capable of flying at a great altitude'.[1402] Such a preference carried on long after the war. It would become the standard for the Second World War and subsequent Cold War missions. It is the definition of the SR-71.

DEFENDING THE RECONNAISSANCE

Aerial warfare revolved around escort duty for reconnaissance. For some it was a monotonous experience. The 1916 diary of Canadian Lieut. John Brophy portrayed the routine where the bait was the 'camera machine': 'I was on escort duty to FEs who were going over to take pictures. I stuck just above the camera machine and followed him around. As we came to the lines we saw four huns, who made off when they saw us coming for them. Flew about while the FEs took pictures.' Brophy commented a few days later, 'Went out again at 11 to escort FEs who were taking photographs. The FE arrived late and in the meantime four of us were tearing up and down the whole Somme front looking for him. Dropped four bombs on Le Transloy for old time's sake and finally found FE after dodging archies for over an hour. Sat above the FE while he toured about.'[1403]

The demand on aviation resources to protect aerial reconnaissance soon prompted designers to suggest changes in reconnaissance aeroplanes. The mission required the aerial observer pay attention to manipulating the camera rasther than watching out for threats. It was comparatively easy for an enemy patrol of pursuit planes to come within striking distance before the observer was aware of their approach. The greatest advantage an enemy pursuit plane could have in the attack on the observation plane was surprise. Even with other planes in formation spotting an incoming attack, the inability to effectively communicate with the aerial photography aeroplane made the mission difficult. Sometimes the only indication of danger was machine-gun fire. By the end of the war, aerial reconnaissance turned to improving the camera mechanics so they operated automatically. The first semi-automatic cameras such as the French De Ram and the British 'LB' Type were attempts to free the aircrew from totally concentrating on the camera and taking the photograph.[1404] The pilot assumed responsibility for taking the photographs while the observer both looked out for enemy fighters and conducted visual observations to supplement the information derived from the photographs.[1405]

Aeroplane evolution paralleled armament, employing weapons of greater firepower and versatility. The prevailing view throughout the conflict was that reconnaissance aircraft should emphasise collection over protection. However, after the first year of combat, protection was a part of aircraft design. By 1916, mounted, synchronised machine guns allowed aircraft to engage in a frontal attack. The observer also became an aerial gunner. Despite such efforts to improve armament, aerial reconnaissance aeroplanes were vulnerable to the single-seat pursuit fighter to the very end of the war.[1406]

The other threat that prompted aviation to ascend to higher altitudes was from ground anti-aircraft weapons. At the outset of the war, before anti-aircraft guns increased their accuracy and range, observation planes flew at lower altitudes and took satisfactory aerial photographs with rudimentary equipment. However, anti-aircraft became more effective, forcing the planes higher. It culminated in many improvements in the aeroplane, motor, and aircrew flying gear.[1407] The

The German 88 mm anti-aircraft was a formidable threat for two world wars. 'Archie' was a continual threat throughout the war, prompting aerial reconnaissance to operate at extremely high altitudes. (Neumann, *Die Deutschen Luftstreitkraste*)

Friends in the area protecting aerial reconnaissance contributed to the evolution of the air superiority role in aerial combat. A Nieuport 17 pursuit fighter on patrol with aerial reconnaissance aeroplane. (Pépin Papers, MA&E)

The evolution of aerial reconnaissance in the final year of the war created a formidable fighting machine. The Breguet 14 A2 was well armed fore and aft. Enemy aircraft would not be quite so ready for their 'close-up'. (Gorrell Report, NARA)

tendency of aerial reconnaissance to fly a standard straight line course at a constant altitude gave anti-aircraft, referred to by British pilots as 'Archie,' as perfect a target as they could desire.[1408] The name 'Archie' is said to have originated from the character of the bad boy Archibald in the comedy of 'Mamma's Boy,' produced in London in 1915. The catch line of the comedy was an ineffectual shot followed by the reply, 'Now, Archibald!'[1409] One aviator recommended a gentle slideslip to a lower altitude as an effective way of getting away from Archie's dangerous fire.[1410]

MODES OF COMMUNICATION

It took extensive cooperation between aerial resources and the traditional combat arms to determine the missions that defined aerial observation for the war. The first months of positional warfare honed the requirement. Communication made observation's role more effective and important in the first year of combat rather than camera systems. The one-way radio Morse key transmission from aircraft to artillery battery became a fixture for the entire war. It provided observation depth beyond balloon range that meant greater flexibility for the artillery in selecting targets. Two-way communication was achieved by patterns of cloth strips laid out by ground units to the aircraft. It primarily linked to artillery batteries to provide definitive updates on the impacts.[1411] Communication improved in both range and effectiveness over the years. By 1917, jamming from German radio intercept units and friendly interference were countered by a 'clapper break' that allowed the Morse key pitch to vary with each aircraft. The British Sterling radio transmitter was the commonest unit, providing a range of 8–10 miles, sufficient for supporting Corps-level operations. The transmitting aircraft extended a 120-foot aerial from the fuselage to conduct the artillery support mission. By the end of the conflict, radio transmissions extended over the strategic sector, up to 70–80 miles to the ground receiver.[1412] In 1915, a new pattern ground-strip for communication in the form of an arrow was developed. The ground signal was displayed by battalion headquarters when the advance was held up by enemy resistance. The arrow gave the direction, while bars were added, each symbolising 200 yards that the obstacle was distant from the arrow. The British historian, Sir James Edmonds, considered this the first established organisation of air cooperation.[1413]

The aeroplane transmitted via radio/radio telephone, projector, messages (or photographs) dropped in bags or by signal light. The ground receiver responded by displaying white cloth panels (that highlighted certain objectives gained), smoke signals, radio/radio telephone, or lamp. The Allies used a simple code, spelling out letters in Morse that were sent by a signalling panel – a shutter of lathes like a Venetian blind, exhibiting white on one side and dark green on the other. The observer reported location, designated objectives, and results of fire. They were supplied with special lithographed maps for marking the positions of the Allied and enemy troops. Radio was the main communication link for artillery spotting, used for urgent information on position of friendly troops, requests for fire, extension of fire, and the like. The Allied infantry communicated with the aviators by setting off flares within the trenches so as not to attract further attention from the enemy. The advanced troops received signals from the aerial observers by message bag, lamp, Klaxon horn, and/or white Very light.[1414] Very lights generally sent four signals: 'Where are you?'; 'Understood'; 'Counterattack'; and 'Areas I am flying over are unoccupied by enemy.' The signal lights were fired above the German lines so that the lights fell in front of the troops.[1415] As a general rule, visual signals were repeated and panel signals left in position until the aeroplane replied, 'Understood,' preferably by light signal.[1416] The use of the Very pistol had its drawbacks for the aircrew. A member of the Royal Artillery, Lieut.-Col. Neil Fraser-Tyler, flew with the RFC over enemy lines to gain a better appreciation of the artillery spotting mission. He described a moment when they attempted to signal to the batteries. 'On firing coloured lights from a Very pistol the battery engaged pre-arranged targets, but it was a squally day of thunderstorms and hail, making observation very difficult. The Hun anti-aircraft guns always get very fussy when a plane starts firing lights; apparently they think it is a signal for something desperate to happen.'[1417]

MISSIONS

Significant roles were artillery spotting, infantry contact, aerial photographic and long-distance reconnaissance, reconnaissance of particular zones or localities, and exercises. All special missions required extensive preparation with the ground liaison. The group or squadron operations officer preplanned the mission and named the aircrew in advance.[1418]

Artillery spotting remained the priority for aeroplane aerial observation for the duration. Aerial reconnaissance developed techniques that helped gauge potential destruction of the battery emplacement area. Active and inactive gun positions were determined through direct observation of guns firing. The seasons were employed to reinforce the estimate. During the winter observers learned that dark smudges in fresh snow marked recent firing. Even hidden artillery could be revealed by this and associated trails. The ideal time frame for artillery reconnaissance was at dusk when the flash could be detected.[1419] Avion de réglage also constituted another mission in support of artillery. Aerial photographs were taken in advance of trial fire by a designated artillery unit to determine accuracy.[1420] One recollection of the British process for artillery spotting describes the work of No. 3 Squadron, Royal Australian Air Force:

> After take off, the RE 8 would wait for the Battery to lay out a giant white linoleum-formed letter 'L' on the ground to signify the Battery was ready to commence firing. The RE 8'd then cruise at about 85 miles per hour toward the target area and, using the Standard Sterling Transmitter in the RE 8 with its 120-foot-long trailing aerial, they'd key, in Morse Code, the letter 'G' to let the battery know they could let 'go' with their salvo. On the ground, near each Battery location, a Wireless Operator would receive the RE 8's signal on his crystal detector receiver using a 'cat's whisker' tuner, headphones and an aerial pole about 30 feet high. Incidentally, the 'cat's whisker' had to be continually re-tuned to the point on the crystal which gave the best signal reception because, as the shoot progressed, the receiver would be constantly jolted by the concussion of the Battery's guns and the enemy's counter-bombardment. Nevertheless, the RE 8's signals, relayed to the Battery, would start each salvo and the RE 8's job was to note where each salvo fell in relation to the clock-face on the photograph and to then signal the Battery with the clock-code-letter designating the distance the salvo fell short of the target. Of course, the Battery would keep correcting its aim, always trying to hit the target, while the RE8 would continue to fly in large circles between the Battery and the target to report changing circumstances.[1421]

Infantry contact patrols provided an overview of the situation at a given sector at a given time, especially during an attack. This was difficult because of the hostile environment. Observations included indications of fresh digging or newly taken positions.[1422] The infantry contact patrol closely followed the movements of the attacking force and transmitted the results by the quickest means available. The attacking infantry informed the infantry contact patrol with banners or rockets. The infantry contact patrol relayed the information to the headquarters by radio or message drop. Upon landing, the infantry contact patrol made an immediate telephone report to the CA battle headquarters and the divisional headquarters.[1423] The French allocated one infantry contact aéroplane (*avion d'accompagnement de l'infanterie*) per division to accomplish the mission. While flying over the given sector, the infantry contact aéroplane was responsible for observing the signals of the PC and transmitting to the général commanding the division. Additional roles included transmitting orders of the division's commanding general to the infantry as well as providing information quickly to the Artillery Command. These division subordinate aéroplanes were to fly above the advanced elements at an altitude not to exceed 1,200 metres or below 500 metres.[1424] The infantry contact mission provided some of the best information on locations of friendly troops as well as enemy movements.[1425]

Infantry contact was one of the most demanding of missions, requiring the aircrew to fly the aeroplane at dangerously low altitudes. Lt. Katcher, an AEF Air Service observer before the start of the Meuse-Argonne offensive, reflected on the infantry contact mission:

> An observer, to be successful, must bring up to the minute the information he has obtained before leaving the airdrome. He will get this by flying over our batteries' zone, before visiting the advanced

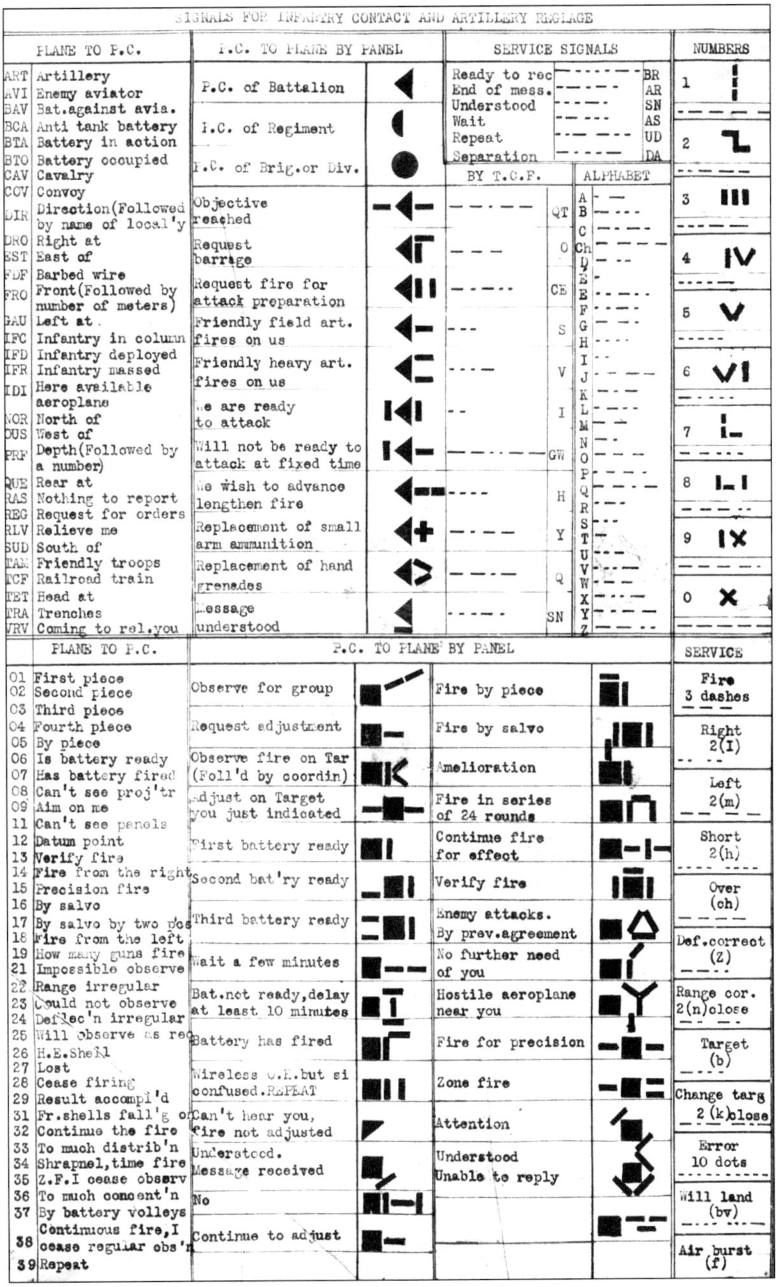

Ground panels allowed infantry to communicate with aerial observation and infantry contact aeroplanes. The 99th Aero Squadron accomplished infantry contact missions through the use of 'smart cards'. (USAFA McDermott Library Special Collections (SMS57)

SIGNALS FOR INFANTRY CONTACT AND ARTILLERY REGLAGE

PLANE TO P.C.

Code	Meaning
ART	Artillery
AVI	Enemy aviator
BAV	Bat.against avia.
BCA	Anti tank battery
BTA	Battery in action
BTO	Battery occupied
CAV	Cavalry
COV	Convoy
DIR	Direction(Followed by name of local'y
DRO	Right at
EST	East of
FDF	Barbed wire
FRO	Front(Followed by number of meters)
GAU	Left at .
IFC	Infantry in column
IFD	Infantry deployed
IFR	Infantry massed
IDI	Here available aeroplane
NOR	North of
OUS	West of
PRF	Depth(Followed by a number)
QUE	Rear at
RAS	Nothing to report
REG	Request for orders
RLV	Relieve me
SUD	South of
TAM	Friendly troops
TCF	Railroad train
TET	Head at
TRA	Trenches
VRV	Coming to rel.you

P.C. TO PLANE BY PANEL

P.C. to plane by panel	By T.C.F.
P.C. of Battalion	
P.C. of Regiment	
P.C. of Brig.or Div.	
Objective reached	QT
Request barrage	O
Request fire for attack preparation	CE
Friendly field art. fires on us	S
Friendly heavy art. fires on us	V
We are ready to attack	I
Will not be ready to attack at fixed time	GW
We wish to advance lengthen fire	H
Replacement of small arm ammunition	Y
Replacement of hand grenades	Q
Message understood	SN

SERVICE SIGNALS

Ready to rec	BR
End of mess.	AR
Understood	SN
Wait	AS
Repeat	UD
Separation	DA

ALPHABET

A B C Ch D E É F G H I J K L M N O P Q R S T U V W X Y Z

NUMBERS

1 2 3 4 5 6 7 8 9 0

PLANE TO P.C.

01	First piece
02	Second piece
03	Third piece
04	Fourth piece
05	By piece
06	Is battery ready
07	Has battery fired
08	Can't see proj'tr
09	Aim on me
11	Can't see panels
12	Datum point
13	Verify fire
14	Fire from the right
15	Precision fire
16	By salvo
17	By salvo by two pcs
18	Fire from the left
19	How many guns fire
21	Impossible observe
22	Range irregular
23	Could not observe
24	Deflec'n irregular
25	Will observe as req
26	H.E.Shell
27	Lost
28	Cease firing
29	Result accompl'd
31	Fr.shells fall'g on
32	Continue the fire
33	To much distrib'n
34	Shrapnel,time fire
35	Z.F.I cease observ
36	To much concent'n
37	By battery volleys
38	Continuous fire,I cease regular obs'n
39	Repeat

P.C. TO PLANE BY PANEL

Observe for group	Fire by piece
Request adjustment	Fire by salvo
Observe fire on Tar (Foll'd by coordin)	Amelioration
Adjust on Target you just indicated	Fire in series of 24 rounds
First battery ready	Continue fire for effect
Second bat'ry ready	Verify fire
Third battery ready	Enemy attacks. By prev.agreement
Wait a few minutes	No further need of you
Bat.not ready,delay at least 10 minutes	Hostile aeroplane near you
Battery has fired	Fire for precision
Wireless O.K.but si confused.REPEAT	Zone fire
Can't hear you, fire not adjusted	Attention
Understood. Message received	Understood Unable to reply
No	
Continue to adjust	

SERVICE

Service
Fire 3 dashes
Right 2(I)
Left 2(m)
Short 2(h)
Over (ch)
Def.correct (Z)
Range cor. 2(n)close
Target (b)
Change targ 2 (k)close
Error 10 dots
Will land (bv)
Air burst (f)

infantry and observing attentively the aerial activity, the artillery battle, the grenade fights, etc. When he has grasped the situation and sees a relative calm, he will then come over the infantry and ask for the line. And, even then, he cannot always see bayonet fighting during the course of which the infantry cannot display their panels, therefore, causing his mission to be unsuccessful. To keep the command well informed as to the situation of our advanced line is the first and most important duty for the observer of an infantry contact plane; the difficulties that he will have to overcome to be successful do not permit those in command to expect him to give regularly other information or have him undertake different missions. However, after the location of the advanced line has been reported, or, in open warfare, if the troops are advancing rapidly, the observer may render other services...[1426]

Post-war reflections on the impact of infantry contact emphasised that success required commitment by both the aviator and the infantry combatant. During the Meuse-Argonne offensive, US Air Service support to the First American Army's 5th Corps was unable to find the front line through lack of panel markings being displayed by the infantry. Panels were never detected while the attack proceeded. During lulls in the battle, markings were misleading because panels were spread out well behind the actual front. Infantry contact was most successful when the aircrews flew down to 300 metres and actually spotted the brown uniforms of the advancing American infantry. The German's blue-green uniforms were cited by the aircrews as having a superior blending quality that proved more difficult to spot, even at such low altitudes. The best solution for determining the front line was to locate friendly troops at a given point and fly over the length of the line while the observer marked the position. The confirmed reference point helped orient the observation to other friendly pockets of infantry in the advance.[1427]

The US Air Service had a different term for infantry contact. Known as the 'cavalry reconnaissance patrol', American observation aeroplanes flew at altitudes to accomplish observation of the terrain immediately in front of the advancing Allied infantry, locating the position of machine-gun nests, strong points, and all other hostile defences likely to retard the progress of the attack. Information was transmitted to the immediate front-line troops by dropped written messages. The infantry was kept informed at short intervals of all developments in the area immediately ahead and were given information of great value in aiding their advance. This avoided the time constraints of aerial photographic interpretation, but lost the verification process that went with it.[1428] In addition to reconnaissance, the US Air Service observation planes were sometimes called upon during the low altitude sorties to drop newspapers and cigarettes to the advancing troops at the front lines.[1429]

THE DAILY SCHEDULE

A sector designated as an area of interest was usually covered daily by the observer aircraft. An aerial photographic sortie occurred about every three to five days, depending on the aerial photographic collection requirements of the intelligence staff.[1430] To gain the maximum benefit from each sortie, reconnaissance sought to surprise an exposed enemy as well as allow for the ideal sun angle for acquiring shadow on the aerial photograph. One post-war summation showed daily operating hours for aerial photography occurred between 0800 and 1600 hours in midsummer, 1000 and 1400 hours during the spring and autumn, and 1130 and 1230 hours during the limited hours of winter.[1431] When the season permitted, routine sorties fell between 0900 and 1200 hours, 1400 and 1700 hours, and a final flight shortly before sundown.[1432] An opinion on the time for reconnaissance was provided by one French aviator: 'The whole sector to which the aerial photo section has been assigned should be photographed daily; sometimes it is even worthwhile to do this twice daily, morning and evening; the pictures should be developed very quickly.'[1433]

Weather constantly challenged successful aerial observation. Weather permitting, aerial reconnaissance sorties operated before sunrise. Normally, about 45 minutes to an hour before daybreak the aéroplane took off to the desired altitude and flew back over the enemy's line in order to be over the rear echelon areas at the break of day.[1434] The ultimate goal was to acquire coverage at any time under any conditions. The French realised that photographic reconnaissance was necessary regardless of what the climate offered. 'We must be thoroughly convinced of this principle; photographs must be taken at all hours and in all kinds of weather'[1435] When navigating a shoot, aircrew commonly had to repeat the orientation when coming in contact with a thick bank of clouds.[1436]. The sortie would be flown at very low altitudes (100 metres or less) when the ceiling was between 200 to 300 metres. This took the enemy by surprise, gaining information or inflicting damage before quickly ascending to the cover of the cloud bank. Another advantage gained in flying during inclement weather was enemy chase 'did not have to be reckoned with on these missions'. [1437] One of the first British officers in the RFC tried to put a brave face on things. 'The war has taught us to laugh at weather.'[1438]

Aerial observation's established routine became les effective over time. Thanks to the enemy manoeuvring primarily under cover of darkness, night reconnaissance was needed to support commanders with updates on enemy activity. Observers on the early morning sortie often saw extensive movement across the sector.[1439] Prior to dusk or dawn, aerial observation capitalised on the start and finish of activity.[1440] Evening reconnaissance was conducted for the sole purpose of spotting active hostile batteries. The flash from the guns was clearly observed at dusk. The sorties took off approximately one hour before dark and flew at an altitude of 2,000 metres inside friendly lines. The missions gave excellent results. Not only could the location be ascertained, but the calibre of the artillery piece could be determined by the frequency of fire. The last vestige of daylight allowed the aeroplane to return to the landing ground. The field would light flares for the returning aircrew.[1441]

Night reconnaissance meant judging the amount of light available. The four types of night flying conditions were defined as bright moonlight, half moon, quarter moon and dark starlight. During bright moonlight sorties, the pilots were told to fly at 2,000 feet or under. Lectures on the mission described troops or transport observed on the road in these conditions as 'dim patches'. Parachute flares were employed to light up a quarter-of-a-mile-square area. Dropped from an average altitude of 1,800 feet, the flare burned for about four minutes.[1442] Night aerial photography tests using flares were undertaken by the British right up to the Armistice.[1443]

Night reconnaissance required strict operator discipline, with the French selecting their best aircrews for the mission. The focus was on rail traffic, activity in the industrial centres and aerodromes. Traffic density had to be addressed, with the possibility of over 100 aéroplanes operating simultaneously in a particular region. Each unit was required to establish clearly defined air routes and altitudes to avoid potential collisions.[1444]

Night operations ensured ongoing reconnaissance of rail and road activity to determine the extent of reinforcement. French aviation soon gravitated toward night operations for aerial bombardment. In 1915, bombardment escadrilles collected information while on the sortie. Upon return any relevant observations were collated and forwarded to the 2e Bureau staff.[1445] Such activity gained importance as the Germans employed the cover of darkness to mask their major operations. By 1918 this was literally the order of the day everywhere. In July a German general mandated that 'Large troop movements and marches incidental to a change of billets will, as a rule, only take place at night. When troops on the march are taken by surprise at night by parachute flare dropped by a hostile aeroplane they will stop immediately and not move.'[1446]

Allied night reconnaissance paid off in the final months. French escadrilles specifically set up for night reconnaissance were recognised by Général Pétain for their continued efforts. American aviation soon created its own night reconnaissance flight. Prior to termination of hostilities, the 9th Aero Squadron became the sole American squadron with night reconnaissance responsibilities.[1447]

PHOTOGRAPHY ASCENDANT OVER THE AERIAL OBSERVER

In the final summer of the conflict, the US Air Service through its Office of the Director of Military Aeronautics provided lectures to newly arriving aviators on the benefits of aerial photography over observers. The discussion was to the point. Human frailties did not measure up to the power of technology. One report observed:

> Let us consider point by point the advantages possessed by the lens and the plate over the eye and the memory. It covers a given subject in a thorough manner far more perfectly than the eye, which is disturbed by many causes within and without, seeing as it does only one thing at a time, while the lens sees all at once with perfect impartiality. It eliminates any element of imagination, for the lens is neither dramatic nor poetic and sees things as they are. The lens can cover a territory wider in scope, and in a shorter time, than the eye and the brain and memory could cover even under the most ideal conditions; and its whole maze of detail being grasped and recorded simultaneously...
>
> The camera has been adapted to act as an interpreter extraordinary ... The camera, therefore, has the advantage of a calm and keen perception and a perfect memory, to say nothing at all about the speed with which it gathers and records all in view with positive finality that leaves no question of a doubt. Possessing these extraordinary powers of perception, together with the power of retaining and recording matter gathered, it, therefore, constitutes an asset which is invaluable to the director of any military operations, whether they be contemplated on a large or on a small scale. The pilot and observer, therefore, should realise this and feel that, in the camera, they have the best aid to the proper discharge of their duties, without which their reconnaissance, no matter how diligently and courageously performed, would be minimised or even nullified.[1448]

Observation had peaked in 1916. By 1918, aviation was an integral part of the battle, preceding the infantry, breaking counterattacks, and pursuing a retreating enemy. The role played by the aerial observer was still relevant and useful, but it did not compare to photographic technology. Observation was not advancing as an integral part of combat in the last year of the war. The quality of the observation force could not equal the advances in photography.[1449] Three years of war refined the photographic process so that its intelligence contribution to daily operations was a standard fixture of daily life. The front lines were well monitored throughout. 'Observation aeronautics did not yield the results which it gave in 1916. For this we may attribute as the principal reason the decline in the value of observers.'[1450]

The British 'Notes on Observation from Aeroplanes' in February 1918 reviewed the prevailing attitude among

A ground panel signal is laid out for the infantry contact aeroplane. (Maurer, Ed., *The U.S. Air Service in World War I*)

the forces to aerial observation. Ignorance about the mission was still prevalent among the 'officers of other arms'. The Notes explained that 'Fighting upon the ground is conducted upon one plane. Aerial fighting is complicated by a third dimension.'[1451] The challenges involved in artillery spotting were underlined. 'It must be remembered that the artillery machine is heavily laden, but has poor speed and climb relatively with fighting machines, and that its attention is concentrated upon work on the ground ... Aeroplanes cannot do the work of cavalry.'

MOBILE WARFARE

In 1918, with the return of mobile operations, lower altitude coverage was demanded. Short-focus lenses and wider viewing angles, required for quick reconnaissance to identify newer terrain features facing the advancing forces, now dominated the information battlefield.[1452] During this time, the French increased their information networks to ensure that vital reports reached the mobile combatants. American post-war review described how the information was processed as forces advanced:

> During movements, each large unit, generally the division, is assigned a route of march and a zone of action. Airplanes reconnoiter this zone and transmit information by projectors, signal lights, dropped messages, or radio to information centers. These are established successively along the route so as to secure continuous reception. Information centers are in communication with the landing field of the airplanes by visual signals, telephone, radio, carrier pigeons, airplanes, or automobiles. An officer of the Air Service of the army corps or division reconnoiters if possible a suitable landing field near the information center. The necessary sentinels are assigned to it and communications organised. The division commander generally has his artillery commander with him, and moves successively from one information center to another as the movement continues.[1453]

The architecture had come full circle. The landing zone techniques that applied in 1914 became policy in the final advance of 1918.

It is a testament to aerial reconnaissance that it met the initial scepticism of a well-entrenched military establishment with conclusive results, results that placed intelligence on equal terms with combat arms. Four core missions had been successfully accomplished by aerial reconnaissance. First, artillery spotting ensured effective targeting for the most destructive weapon in the inventory. Second, visual reconnaissance ensured the most effective employment of weapons at the commander's disposal. Third, the aerial photograph proved to be the most effective technology for confirming both strategic and tactical assessments. Finally, greater liaison between the air arm and ground forces created the infantry contact mission that improved awareness of enemy vital areas and targeting. The four core missions were eventually supplanted by two additional tasks, tasks that became the inevitable focus of military aviation. Aerial combat would take precedence over aerial reconnaissance. Preventing the adversary's access to the 'higher ground' as well as protecting one's own reconnaissance ability became the new raison d'être. Aerial superiority became the final goal, as armed forces sought total dominance of the sky.[1454]

CHAPTER 21

The Spectrum of Technology

Aerial photography as an intelligence source changed the very nature of the Great War's battleground. Access to aerial reconnaissance information governed many decisions on the Western Front. Camera technology – the apparatus, film and photographic plate medium and techniques associated with acquiring and developing an image – was in place before the war and changed little throughout. The magnitude of aerial photography's role did not immediately alter the science of photography. Charles Gamble, the RNAS leader on aerial photography, wrote in his post-war reflections that application in a combat environment was photography's key to success:

> It would be difficult to find any form of energy either visible or invisible capable of influencing the sensitive material whose effects had not been registered by the camera. The problem then that presented itself to the pioneers in war aerial photography was simply a problem in photography and the proof of this lies in the fact that having stated it to be a practicable thing they proceeded without any more ado to put the belief into practice.[1455]

THE CRITICAL COMPONENT

The heart of camera technology was the lens, the most critical component governing the aerial camera's ability to image the target. The lens determined the aerial reconnaissance altitude and the required ground track. Intelligence and information gleaned from aerial photography was at the mercy of lens quality, and lenses in quantity were needed to meet increasing demands for aerial reconnaissance coverage. Many issues complicated the lens problem: a lack of standardisation among the Allies, limitations caused by airframes, and probably most important, a serious shortage of raw material to make high-quality optics. The lack of a standardised lens for Allied aerial photography was serious, since the many variations in commercially produced lenses forced the use of inconsistent, makeshift cameras. When aerial photography gained a foothold in British military operations, the Ministry resorted to advertising for and paying premium prices for many critical photographic components, especially lenses.[1456] Limitations imposed by aeroplane designers determined camera size and specifications. British aeroplane designers built 'stop-gap' focus lenses configured to the aeroplane designer's specification and space available.[1457] Moore-Brabazon reflected on trying to work through the industrial hold on camera design in relation to aeroplane specifications: 'The lesson that sticks out from these difficulties is that a great machine having one small part missing throws the whole scheme out of gear. Every part has to be planned and allowed for beforehand if disaster is to be avoided.'[1458] The final restraints on camera technology were the lack of raw materials and the industrial processing of the lens itself. In 1914, only two manufacturers in the world made the heavy crown barium glass necessary for building high-speed, high-power lenses. Germany's Carl Zeiss firm (located in Jena) and France's Parra-Mantois factory were the world's leading manufacturers of high quality optics. Parra-Mantois thus became the sole supplier of heavy crown barium glass for the Allies.[1459]

Other French firms produced a lesser quality glass, including Hermagis, Lacour-Berthiot, and Krauss. What little they produced was sufficient only to meet the demands of Aviation Militaire.[1460] The British had no manufacturer. Their only link to Parra-Mantois was that they were the source of high-grade coal required for achieving the temperatures necessary to fire the glass.[1461] Once the aerial camera was produced in early 1915, the British War Ministry staff attempted to gain access to a reliable source of lenses. British attempts to acquire lenses from the French Lacour-Berthiot firm were unsuccessful owing to lack of available stock. The French Service Géographique strictly controlled their indigenous lens output and made it clear to the British that they were unable to supply lenses in a reasonable time because of the lack of glass.[1462] The British staff resorted to extraordinary measures, visiting every

Enlist Your Lens in the Air Service

IF you have a powerful photographic lens, put it to work for our men "over there;" let it disclose from the skies of France hidden machine-gun nests waiting to spread death among advancing American troops; let it save hundreds of American lives from being snuffed out in the trenches by shells from concealed batteries. An official report calls the situation "critical," brooking no delay.

What is especially desired at the present time are lenses of from 7 inches to 24 inches focal length and with speeds of from F 3.5 to F 7.7. Practically all lenses of this type will be purchased as soon as they can be found. The following are some of the foreign makes wanted: Carl Zeiss Tessars, Bausch & Lomb Tessars, Voigtlander Heliar, Euryplan, Cooke, Goerz, Bush, Ross, Ross-Zeiss, Krauss, Krauss-Zeiss, Steinheil-Isostigmar, Rodenstock. In addition, matched pairs of stereoscopic lenses, with speeds of F 4.5, focal lengths of 4½, 5, 5½, 6, 6½, and 7 inches, are needed.

If you are in doubt as to the value of your lens, ask the nearest photographer.

Remember that you can probably replace your Anastigmat lenses with others just as serviceable for you but not adaptable for the army. If you have a lens such as your army needs, send at once its description and the price you think fair to

SIGNAL EQUIPMENT No. 33
WASHINGTON, D. C.

A remarkable admission of a critical resource deficit in time of warfare. Like the British in 1915, the Americans faced a chronic lens shortage in 1917, culminating in a call for lenses from the public through popular publications. (*National Geographic*, January 1918)

second-hand lens dealer in Paris. It was of little use, for the quality of what they found was very poor.[1463] As the war progressed, the optical supply worsened for the British. British optical glass manufacturing remained practically non-existent until 1918.

The British military complained to the optical branch of the British Ministry of Munitions that a shortage of high-quality optical glass was handicapping aerial reconnaissance. The Ministry subsequently announced a compulsory return of lenses from the public to serve RFC purposes and commenced to work closely with British industry to solve the optical glass shortage. They became acutely aware that wartime demands far exceeded contributions from the British public. Advertisements were inserted in trade and other papers for lenses of 8- to 12-inch focal length. People owning such lenses flocked into the Air Ministry to offer their lenses for use in the field. Despite such support, an industrial surge to produce high-quality glass remained the only solution to achieve and maintain the level of critical optics necessary. They reached this goal only in the final months of the war.[1464]

All the Allied participants felt the strain. In post-war analysis of the crisis, former US Photographic Section staff officer Captain Herbert Ives commented on the impact on the Allied aerial reconnaissance effort:

> It is important that the historical aspect of this matter be well understood by the student of aerial photographic methods, for the use of these odd-lot lenses reacted on the whole design of aerial cameras and on the methods of aerial photography, particularly in England and the United States. Almost without exception the available lenses were of short focus, considered from the aerial photographic standpoint; that is, they lay between eight and twelve inches. This set a limit to the size of the airplane camera, quite irrespective of the demands made by the nature of the photographic problem. Lenses of these focal lengths produced images which, for the usual heights of flying, were generally considered too small, and which were, therefore, almost always subsequently enlarged. Such was the English practice, which was followed in the aerial training of aerial photographers in America, where exactly similar conditions held at the start with respect to available lenses. French glass and lens manufacturers did succeed in supplying lenses of longer focus (50 centimetres), in numbers sufficient for their own service, although never with any certainty for their Allies. The French, therefore, almost from the start, built their cameras with lenses of long focus, and made contact prints from their negatives.[1465]

The British solved demand for a longer lens (20-inch; 50.8cm) using a variety of sources. Moore-Brabazon reported: 'While we were waiting for the first British production, 50cm. lenses were captured from the enemy almost every day for three weeks. This saved

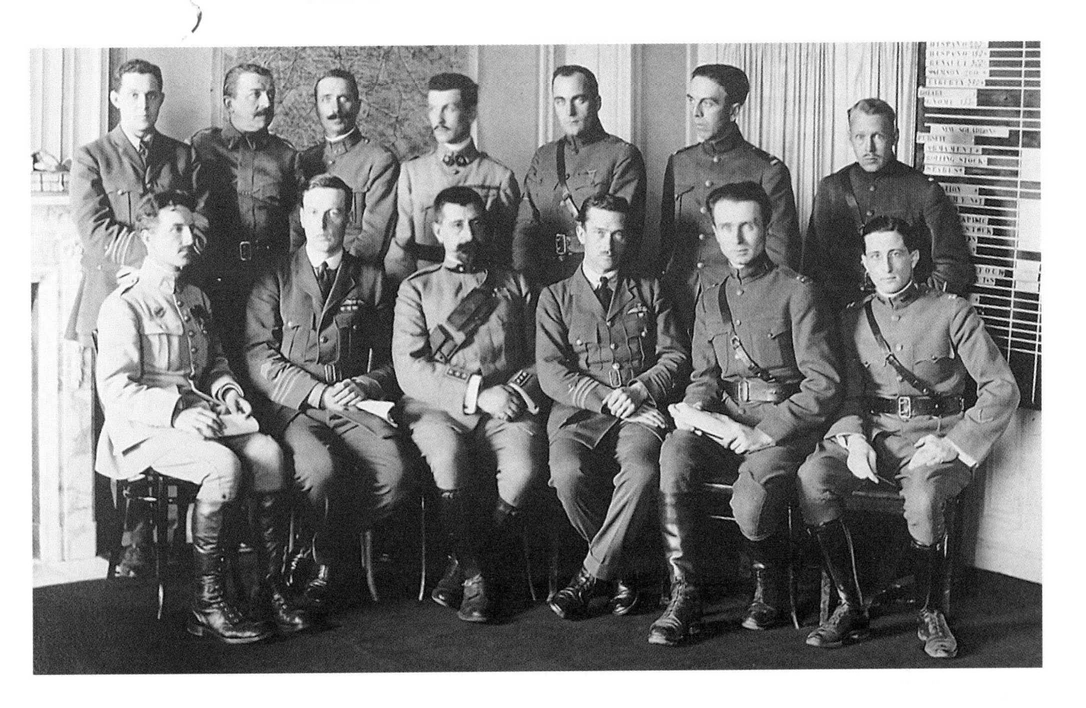

The Allies' leading experts on aerial photography met in conference to discuss a variety of critical issues, to include plate design specifications and lens distribution. The photograph includes Major Toye (back row, far left), and from left to right on the front row Lieutenant Labussière, Lt-Col. Moore-Brabazon, Colonel Tardivo (Italian Army) Major Laws, and Captain Steichen. (Gorrell Report, NARA)

the situation by giving us the very lenses we required most.'[1466] Every lens was employed. The British even developed the 'B' Type camera – at their repair depots in France – using recycled lenses of downed German reconnaissance aircraft.[1467] 'We never knew from day to day where we were going to get our next lenses, and the situation often became tense – indeed it was at times heartbreaking.'[1468] Another post-war review of the situation by one of Moore-Brabazon's counterparts, Major Edward G. Toye, recalled, 'Looking back on the glass situation, it is remarkable how quickly the prejudices and difficulties with regard to the production of high class optical glass were overcome, although at the time the progress seemed to be maddeningly slow.'[1469] It was soon evident that French lenses, as well as French glass, had to be a shared resource. In the summer of 1917, the issue was addressed by the highest French authorities and the Inter-Allied board. Under the direction and guidance of the French *Service Géographique, Section Optique*, lens production was soon increased and the output placed on an allocation basis satisfactory to Aviation Militaire.[1470]

The demand for higher quality optics increased as the need for intelligence on the battleground grew. Intelligence analysts insisted on detail, demanding at a minimum 20-inch lenses.[1471] Longer focal length resulted in more detailed photographs as well as increased aircrew protection. In the first three years of the conflict, improvements in air defence forced observation planes to ascend from an initial operational altitude of three to four thousand feet to 15,000, then 18,000 feet and higher.[1472] The evolving aerial observation mission required greater numbers of longer lenses such as the French 120cm.

American photographic experts learned of the lens supply crisis when they arrived in late 1917. 1st Lt. Edward Steichen notified the Signal Corps' Photographic Division in December that the French Service Géographique, Section Optique, was the 'central and absolute authority having control of all optical glass questions for the French Government'. Steichen concluded, 'Their word is final on all such matters and, as you can see, vague.' He advocated that the ultimate solution to the glass situation 'lies in the production of glass in America.'[1473]

This was not a trivial issue for the US Air Service. Colonel Raynal C. Bolling, the lead for advising both AEF and the War Department on critical challenges facing the US war effort, personally kept Brigadier General Benjamin Delahauf Foulois, Chief of Air Service, updated on the lens problem. In a 22 January

1918 memorandum, Bolling discussed the Service Géographique's position and the need for some independent action. The Parra-Mantois domination of heavy crown barium glass, plus the inability of the British and Americans to manufacture the required amount, were of serious concern. Bolling mentioned that Parra-Mantois' production was wholly inadequate to meet the combined needs of France, England and the US. Consideration was given to Parra-Mantois's profitability from increased production of the glass ('undoubtedly a precarious and unprofitable business'). The firm was concerned about the cost of setting up new production, and of creating a huge surplus when the war ended and demand subsided. Bolling even considered working out a deal to guarantee profitability for Parra-Mantois after the war. The Service Géographique insisted on the universal adoption of French lenses. At the time, speculation was that the British production authorities were undecided. Bolling concluded that the problem of adequate negotiation rested with the Service Géographique staff. 'The French Service Géographique are entirely unwilling to entertain this suggestion, but speaking frankly, their attitude not only in this but in some other matters has not indicated to us sufficient openness of mind and spirit of give and take to make us ready to recommend simple acceptance of their decision.' Bolling concluded that he had led the US effort up to that time on the issue, but felt it was time to have a more senior person working within the Committee on French-American relations to come to an agreeable conclusion.[1474] In the interim, the British established a capability and commenced their own glass production suitable for aerial cameras.

The American's advance aerial photograph team was quickly caught up in the controversy. Major James Barnes, 25 January 1918:

> The entire Allied Air Photography is dependent on the glass output of the firm of Parra-Mantois and this is controlled by the *Service Géographique*. Captain Laronde insisted that everything possible had been done that increased the output of the Parra-Mantois firm. He did not give any proof to substantiate this and our information leads us to believe the contrary. He decided that the only method of increasing production was by adopting one type of lens ... Captain Laronde's argument in favor of adopting French lenses as the desired type of lens is that these lenses are a safer manufacturing proposition, as smaller glass melts are needed. This is more of a laboratory than a military point of view.[1475]

Barnes added, 'American opinion favors lenses now made in England, as they are generally better and 50 per cent faster, consequently giving more day and longer hours for taking air photographs.'[1476] He noted that the American supply of lenses sufficed only for three more months. The shortage coincided with the operational employment of America's first observation squadron to support operations in the front. A solution was required as soon as possible. Major Campbell responded to Barnes on 30 January 1918, revealing that British demands were being simplified, suggesting they were going to rely on their own resources to meet the requirement for high-grade lenses.[1477] Barnes expressed his growing frustration when he wrote Moore-Brabazon on 17 February 1918 that 'The French are trying to get out of making any delivery of not only glass but I believe of the lenses they have contracted to deliver. They are exactly three months behind now and have not delivered any at all since the 29th of November.'[1478]

Toye wrote to Barnes that the British had solved the glass problem to their satisfaction. They decided to ask for lenses only, not glass. He also mentioned that Campbell had 'gone under' from overwork. The standing up of the RAF at that time was also consuming a lot of energy.[1479] The staffs worked feverishly to resolve the issue, which was elevated to the ministerial level. Brigadier General Foulois received word from the AEF staff on 26 March 1918 updating him on the precarious state of camera optics: 'The lens and glass situation with the A.E.F. is at present in a most crucial state. In spite of all attempts to arrive at some definite understanding with the French Government, nothing yet has been accomplished.'[1480] Confusion was being created by myriad offices sending different cables on the situation. A diplomatic crisis had evolved to the point that 'it [was] in desperate need of solution.'

AEF staff recommended that General Pershing send Steichen to represent the Air Service and work with the Director of the Technical Section in dealing with the French on the lens issue.[1481] At the subsequent Conference on the Glass Situation held on 25 April 1918, the American staff-leader, Major Gros, noted that the British required 120cm lenses. A production rate of 62 lenses per month was required. The French estimated monthly production of 200 of the 26cm, 145 of the 50cm and 17 of the 120cm lenses for 1918. At that time, the British representative mentioned they only had 12 of the 50cm lenses in the entire British Army. Captain Steichen placed the US need at a minimum of 50 of the 50cm lenses. The French responded that the only way to distribute lenses was according to the number of infantry divisions at the front.[1482] The 26th Division was the only American unit serving in an active sector. By August 1918, the Americans had 12 divisions on the front line. Steichen strongly appealed to the French to immediately provide the British with the number of lenses required to allow them to catch up.

Comparable Allied Aerial Camera Measurements – Lens Length

British Standard	French Metric
4 inches	10.2 cm
6 inches	15.2 cm
6¼ inches	15.9 cm
7 inches	17.8 cm
8 inches	20.3 cm
9.8 inches	25 cm
10 inches	25.4 cm
10.2 inches	26 cm
10½ inches	26.7 cm
12 inches	30.5 cm
14 inches	35.4 cm
15 inches	38 cm
19.7 inches	50 cm
20 inches	50.8 cm
20.4 inches	52 cm
27 inches	68.6 cm
47 inches	119.4 cm
47.2 inches	120 cm
72 inches	183 cm

Allied lenses were of differing focal lengths, to meet varying publication capabilities and needs. (Author analysis)

The French representative insisted upon not drawing on the reserve stock of lenses, 'which he states are kept for a possible rainy day.'[1483] That remark drew the following comment from Steichen: 'Lenses would be much better stocked at the front, where they can be used, than kept in reserve at the Service Géographique.'[1484] Discussion ensued about the word 'stock'. The conference ended with the agreement that the allocation of 50cm lenses would be: French (55), British (55) American (11).[1485] The French agreed to provide the vital, heavy crown barium glass, and the lenses for both countries were to be made in England, in accordance with a British formula, which accelerated the production.[1486] The final allocation for this critical resource was made in April 1918.[1487] Lieutenant-Colonel Moore-Brabazon described subsequent discussions in June 1918 at an Inter-Allied Conference on Photography. He observed that the French Parra-Mantois firm remained the only Allied industrial centre that could produce raw glass of the quality required for photographic lenses.[1488]

The solution to the lens shortage crisis came in the final year of the war. British glass manufacturers commenced a major effort to produce optical glass equal to the German standard. Moore-Brabazon:

> This was a challenge to the British manufacturer who at once took it up. Lenses were soon being sent in for test and amongst the first was the Cook Avair by Taylor & Hobson. I believe their second effort to be equal to the best German Zeiss that we could procure. The tests were made from a height of 9,000 feet in Cameras fitted with two lenses side by side photographing exactly the same object, the Zeiss on the one side and the Avair on the other. The tests were very encouraging. Then came the question of production. No serious delay was experienced. Following the Avair came the Ross Express and the Ross Airo, both wonderful lenses and soon we had more lenses than Cameras to put them in ... They produced lenses with wonderful covering power and very fast. Large numbers of these were at once put on order and we were soon well ahead of our requirements.[1489]

PLATE VERSUS FILM

One tool of First World War aerial photographic reconnaissance, the photographic plate, has fallen into obscurity. The plate on which the image was captured consisted of glass with a silver bromide emulsion on one side. Plates were packed with the emulsion side facing each other and were separated by cardboard.[1490] Film technology existed, but problems with cameras and the film itself prevented it from becoming the primary image base. The film option was tried out early in the conflict by Lieutenant Grout and Lieutenant Pépin, whose 300 photos acquired from a 6cm x 200 metre film strip covered the Verdun sector using the Douhet-Zollinger camera. This camera proved too difficult to handle and its short lens limited the operational altitude for acquiring readable coverage.[1491]

The British also experimented with film in a combat environment. As the war entered the second year, they developed a film camera, the Aero Camera or 'F' Type Camera. The Russians used this camera on the Eastern Front to support their cartographic production. The 'F' Type Camera was designed to take a continuous series of pictures on a roll of film 5 inches wide and 25–50 feet in length. A 50-foot roll provided coverage for 120 exposures. The 'F' Type Camera was uniquely configured for flight, being positioned on a bomb rack. The automatic camera was driven by air pressure from a small four-bladed propeller. Every 300 propeller revolutions triggered the camera to take a photograph, providing a continuous stream of coverage.[1492] Despite experimentation with this medium, the Allies stayed with plates until late in the war. Film also had its difficulties owing to electrical discharges portrayed on the emulsion in cold and dry environments such as those found in the upper atmosphere.[1493]

The true champion of film for aerial photography throughout the war was George Eastman, chairman of the Eastman-Kodak Company of Rochester, New York. From the start, Eastman monitored development of the aerial camera, occasionally meeting with the British on the subject. In 1915, he observed, 'The military and naval authorities are taking considerable interest in photography for aerial observation ... Experiments have been going on lately at the military flying grounds in the use of automatic photography on airplanes.'[1494] The war provided Eastman with an outstanding opportunity to develop and promote an aerial film camera. He aggressively promoted his automatic camera, reminding everyone that plates were not as effective as his film:

> We are especially devoting ourselves to aerial photographic work, and have already devised special emulsions and apparatus for the purpose, some of which are very promising. Among other things we have an automatic film camera which will make fifty exposures, six inches square, in succession, which are equal to anything that can be

A glass photographic plate showing detail in the light. (SHD)

> taken singly with glass plates.... The Allies are using plates almost exclusively, because they have been unable to get the photographic results required on films. This is partly because they have been trying to use the ordinary camera films, which are quite unsuitable for the work. They also have been unable to handle the films mechanically.[1495]

Kodak's subsidiary, Folmer & Schwing, came through with the state-of-the-art aerial film camera during the war. It was radically different from anything that had been seen in Europe up to that time. The aerial film automatic camera, known as the 'K' type, carried film capable of 100 exposures (18 x 24cm) per loading. The 40-pound 'K' was light for an aerial camera. The film rolls were easy to handle and also light (4 pounds). The most novel and ingenious feature of the 'K' camera, the 'vacuum back,' held film flat. The filmstrip passed over a flat perforated sheet, behind which a partial vacuum was created by suction from a Venturi tube extending outside the body of the aeroplane. It was driven either by a wind turbine of adjustable aperture or, in warplanes, by electric current from the heating and lighting circuit. The 'K' camera allowed the observer to start the machinery going and regulate its speed according to the rate of travel of the aeroplane. The final result was a series of pictures that formed a continuous mosaic photograph of the aeroplane's ground track.[1496] Eastman persisted in making the design available for government inspection. He told Secretary of War Newton D. Baker that: 'Mr. Folmer will at once proceed with the construction of experimental models of cameras to be submitted to you, and we will place blueprints of the drawings of these models at your disposal to enable you to obtain tenders for their construction from other firms as well as ourselves.'[1497] The 'K' camera design was a winner, and the US War Department promptly adopted it, but it never saw combat.[1498] The problem for Kodak and the film camera was that the Allies favoured plates. When the AEF arrived, they quickly accepted what was available and did not debate the issue. Despite this setback, Kodak continued to refine their aerial film camera. Their persistence paid off in the long run. Film became the mainstay for American aerial photography shortly after the war.[1499] Eastman remained the 'K' camera champion throughout the war. In a letter to the War Department, he made his case:

> We have tried the camera out twice in the air at the field of the Curtiss Company near Buffalo and it worked perfectly from the start, and the second trial was made only for the purpose of correcting the focus of the lens. It operates very simply, using the air pressure generated by the plane itself. The pictures are sharp and in every way satisfactory. The speed at which the exposures can be made is variable and the starting and stopping of the camera and the speed regulation is done with one lever so that the instrument is a one-man affair, to be operated by the aviator, which is a very important thing considering the lives it will save. The film has a special emulsion which overcomes all the objections you have noted. The reloading of the camera can be very easily done in the air, but in that case it would probably have to be done by an observer.[1500]

Aerial plate cameras had mixed reviews throughout the war. Vibration was always a concern. Instructors cautioned that the observer not touch the plane while taking pictures. The magazine was difficult to operate and required sensitive handling. That the plates were glass also contributed to their limitations.[1501] One combat veteran observer still advocated plates, commenting on film's limitations:

> It is a curious thing that films can't be used in the air with any consistent success. The centre of a film photograph is almost invariably out of focus. The aerial photo covers so much ground that a film is never flat enough and apparently can't be stretched flat enough, to preserve the focus. Britain and France discovered this after unwearying investigation and a constant succession of failures, and America was wise enough, for once, to profit by somebody else's experience. Since aerial photographs are worthless unless they are clear and full of detail, the question of bulk and weight was also set aside and plate cameras were universally adopted. This meant a lot of extra work to ensure success for the photographic observer had to shift every plate through the magazine at least once before the aircraft left the ground to ensure the plates fit.[1502]

The deliberation on the role of film continued late into the war. An argument raised by Major Vincent Laws at the Inter-Allied Photographic Conference of 19 August 1918 pointed out that taking photographs on films could not be satisfactory no matter how good a camera was evolved. The question of dealing with films from a development point of view was, in Laws' mind, the impasse. As of August 1918, he was not aware of any method that could be looked upon as satisfactory in the field.[1503]

Despite the dispute amongst the experts, the Conference revealed the emergence of film's role in providing broad coverage for strategical missions. Aviation Militaire adopted the principle regarding a new role for the film camera. Armée escadrilles were to discard the use of plates in favour of film. CA escadrilles were to continue to work with plates for shorter-range missions. The French also ordered that 'old style' plate camera production was to cease.[1504]

FILTERS

Despite their shortcomings, and in part because they were already in use and a familiar item, the British and French stayed with plates for the duration. Americans followed suit as the dependent newcomer. Film was on the cusp of acceptance, coming into its own immediately after the war.

In northern Europe lighting posed a challenge most of the year. The atmospheric conditions made photographing a target a formidable undertaking. As the threat from pursuit aeroplanes and air defence artillery forced photographic missions to operate at higher altitudes, the issue of haze became more urgent. Above a certain height – varying depending on the haze – a normal lens could not image the target. After extensive research the solution was found in the use of special colour filters containing a yellow cast, which obscured the bluish light characteristic of haze. Filters of new materials specially adapted to aeroplane use were made available as a result of study.[1505] The visual effect of haze did not always translate to the photographic plate. When the view was almost clear and no filter was used, photographic results were found to be flat. The rule of thumb was 'the greater the density of haze the deeper the filter should be.'[1506] Filters cut through the haze of northern Europe, but not through cloudbanks. Weather still determined the success of the mission.[1507] By war's end, panchromatic plates were in use. One British observer recalled: 'In our day, the sensitive material was at best orthochromatic, such as the Wellington Anti-Screen plate of honoured memory, of a speed not more than 300 H&D. We generally used a X2 yellow lens-filter, fitted as a disc of gelatine between the lens-components, when the light was good.'[1508]

Filters also conferred a critical tactical and strategic advantage to intelligence. The panchromatic plates and filtration of the light helped identify enemy attempts at camouflage. The aerial photographic interpreter had to differentiate between colours that ordinarily looked alike but were actually of different colour composition. A colour-filter lens was developed that separated the hues of the spectrum into their parts and imparted to each a definite and characteristic shading in the photographic plate, disclosing a spot of 'impure' colour. Those areas that were camouflaged came out blurred on the image.[1509] Filters complicated the ongoing strategy of positional war. Units such as artillery depended on camouflage to boost battery survival; the photographic image refined through the filter increased the uncertainty of preparing the battlefield.

TECHNOLOGY TRANSFER

French and British photographic experts expected the arriving American forces to come to the battleground appropriately equipped to conduct aerial photographic reconnaissance. The Americans were recognised as leaders in film technology. However, such expertise did not serve the well-established French and British aerial photographic infrastructure. Instead, for the remainder of the war, both had the additional burden of supplying Americans with critical photographic resources.[1510] As a result, each Ally tried to establish the priorities for production and development. One senior AEF staff officer told colleagues that the French were trying to influence the direction of the talks on the sensitive issue of simplifying lens construction. The chief of the Service Géographique wrote a letter to the GQG to prevail upon the Americans and the other Allies to reduce the size of the plates, thus simplifying the fabrication of the lenses. The American officer wrote, 'To my mind this reads like the man who tried to persuade himself to cut off his nose so that he would not have to wash it.'[1511] The suggestion was disregarded and lens standards remained as they were.

The Allies influenced the evolution of US photographic organisation. Photographic instruction in the United States followed the British model. The Americans followed French practices when they flew aeroplanes of British design. Adding to the complicated scenario, the Americans configured their British-built aeroplanes with French cameras.[1512] A more rigid structure had to be imposed. The Americans gained agreement from the British and French to standardise aerial photographic plate size and lens size. Since French and the British plate sizes varied, it consumed valuable time to achieve a standard. If the French used British photographs or maps, a photographic enlargement was required. Likewise a reduction was necessary if the British used French images. The lack of standardisation increased the time spent on handling the different size plates, so delaying analysis. By 1918, all agreed to a format of 18 x 24cm. Additional standards included a 50cm lens and British dimensions for camera cone and attachments.[1513] Steichen remarked in a memo to the Signal Corps Photographic Division that the standards had more impact on the British because their cameras had to be enlarged to handle the new photographic plate size.[1514]

All things considered, the French had made the greatest effort during the war to provide photographic resources. US Photographic Section records show that five per cent of materials came from England, eight per cent came from the United States, and the remaining 87 per cent from French stocks. Not a single American-made negative or photograph was at the Front prior to the signing of the Armistice.[1515] The first train of

American plates and paper was en route between Paris and Columbey-les-Belles on 11 November.[1516]

The technology exchange throughout the war reflected a strong camaraderie amongst the experts. However, warm relationships were not as prevalent amongst the military in general. Moore-Brabazon, for example, amiably resolved a technical question with peers, but suffered the political consequences: '[I]n photography nobody could help me – nobody else knew anything about it.' He went to the French technical experts at the *Section Technique* in Versailles. The discussions were totally open and promoted a sharing of techniques and capabilities. However, upon return, Moore-Brabazon was 'soundly told off' by Brigadier-General Trenchard for making the trip to Paris. About two weeks later, Trenchard was present at an inspection conducted by the French on one of the British squadrons. To his delight, the French had asked Trenchard for permission to 'copy one of our gadgets in photography' for use in their own machines. Trenchard was magnanimous and agreed. However, he was totally ignorant as to what the 'gadget' could do or where it came from. Moore-Brabazon, when approached, revealed that not only was the device of French design, he had received Section Technique permission for incorporating it into the British camera system: 'I cannot tell you how delighted the General was to think that within a fortnight of my visit to Paris the French were asking our permission to use something they had invented themselves!'[1517]

Inter-Allied conferences became the best medium for Allied photo experts to reach compromise on technical issues. The breadth of exchange on photographic research was almost breathtaking. At a conference one month before the Armistice, Steichen described British attempts to establish a breakthrough on nighttime photographic reconnaissance. 'The British representative submitted a report of their experiment with night photography, and showed sample prints made. These photographs are the first concrete results in night photography that have come to the attention of the Allied Photo Services. It was decided that all information of this nature should be promptly distributed among the Allies.'[1518] In early November 1918, Moore-Brabazon described efforts up to that time for developing the Photographic Flare to be used to illuminate railway stations. The flare was designed for photographs to be taken by night from a height of 1,000 to 2,000 feet. The flare was dropped from the 20-lb. bomb carrier and timed to explode at a height of 400 to 500 feet above the ground. The effective part of the flash lasted about 1/100th of a second. The photographic plate was kept exposed during the whole of the flash. The successful tests occurred during the last month of the war at the RAF's Armament Experimental Station, in Orfordness.[1519]

Another interesting insight into ongoing photographic research conducted by the Italians was discussed at the October 1918 conference. Italian research on the photographic spectrum had discovered a unique application. Colonel Tardivo, Photographic Section, Italian Army, described the theory of infrared photography for submarine detection and signalling, but no written data was furnished.[1520] Aerial photography was quickly moving into new territory.

AERIAL PHOTOGRAPHIC GROUND SUPPORT TECHNOLOGIES

Post-war popular publications discussing the lessons of the First World War entertained their readers with elaborate descriptions of behind-the-scenes processes that contributed to the Allied victory. One source described with great imagination what aerial photographs revealed about enemy intentions. The key technology behind the entire process was the magnifying optics employed by the aerial photographic interpreter, to include the basic magnifying glass for single frame shots and the more elaborate and highly prized stereoscope. Without the ability to see and interpret detail through stereoscopic analysis, aerial photographic interpretation would have been severely handicapped. The stereoscopic legacy remains an important component of imagery analysis, even if it does not live up to its reputation in some circles as 'the deadliest weapon of the war'.[1521]

Intelligence analysis at the landing ground was possible because of the photographic laboratory technology. Once the aerial photographic plates were in the photo laboratory, the tank method called for plates to remain in the developer until they just began to fog. Too long a period in the developer created flatness on the surface of the negative, a lack of contrast. The prints were dried in alcohol. For time-sensitive coverage of targets of high value, the French were able to produce prints 30 minutes after the plane had landed. The normal development time was two hours.[1522]

Mobility was integral to SPAé, Photo Section and Photographic Section operations. Each Ally maintained a camion, lorry, or truck assigned to the section. Photographic processing mobility increased intelligence value by shortening acquisition, processing and interpretation time.[1523] The basic model required two men to drive and maintain the lorry, run the charging set to batteries, and help in the dark room as needed. Four personnel mixed chemicals, developed, washed prints, fixed, enlarged, dried, trimmed and handled photos and negatives. NCOs were responsible for all the work done in the lorry. They assisted the indexing,

The French were the first to develop a mobile photographic facility. The laboratory was self-contained and transported on the back of a French Camion (truck). (SHD, from *Vues d'en Haut 1914–1918*)

'We never had running water.' Developing aerial photographs in the mobile laboratories took advantage of whatever the terrain provided. (Courtesy of Michael Mockford and the Medmenham Collection)

Interior view shows space allotted to interpretation and filing. (Gorrell Report, NARA)

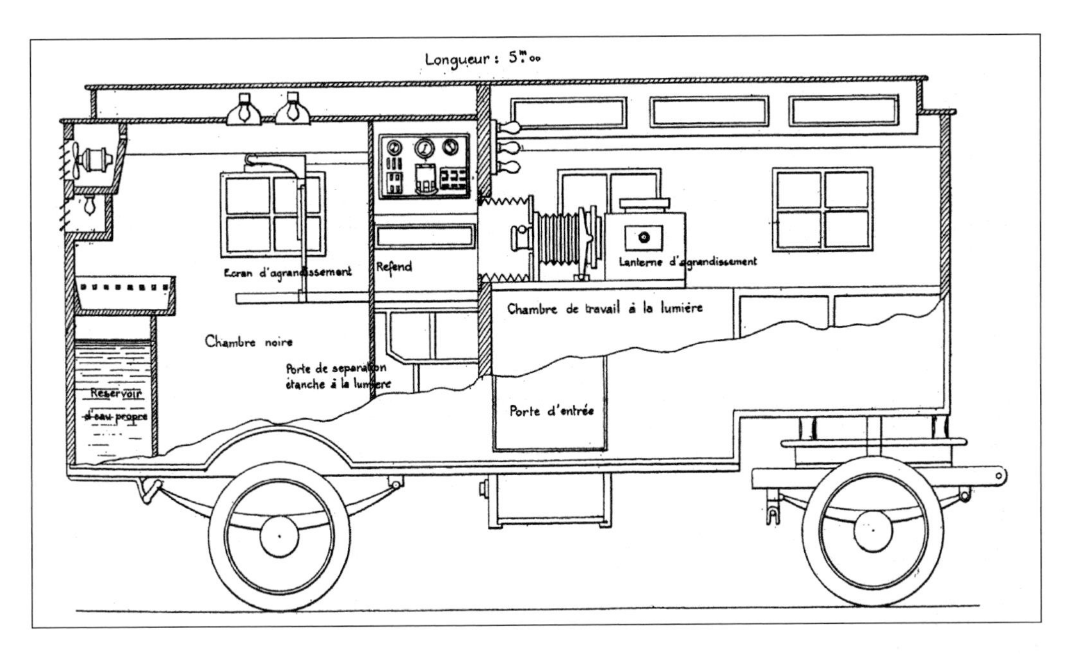

The French also employed a photographic trailer. The line drawing shows the details within the confined spaces. (Brunoff, ed., *L'Aeronautique*)

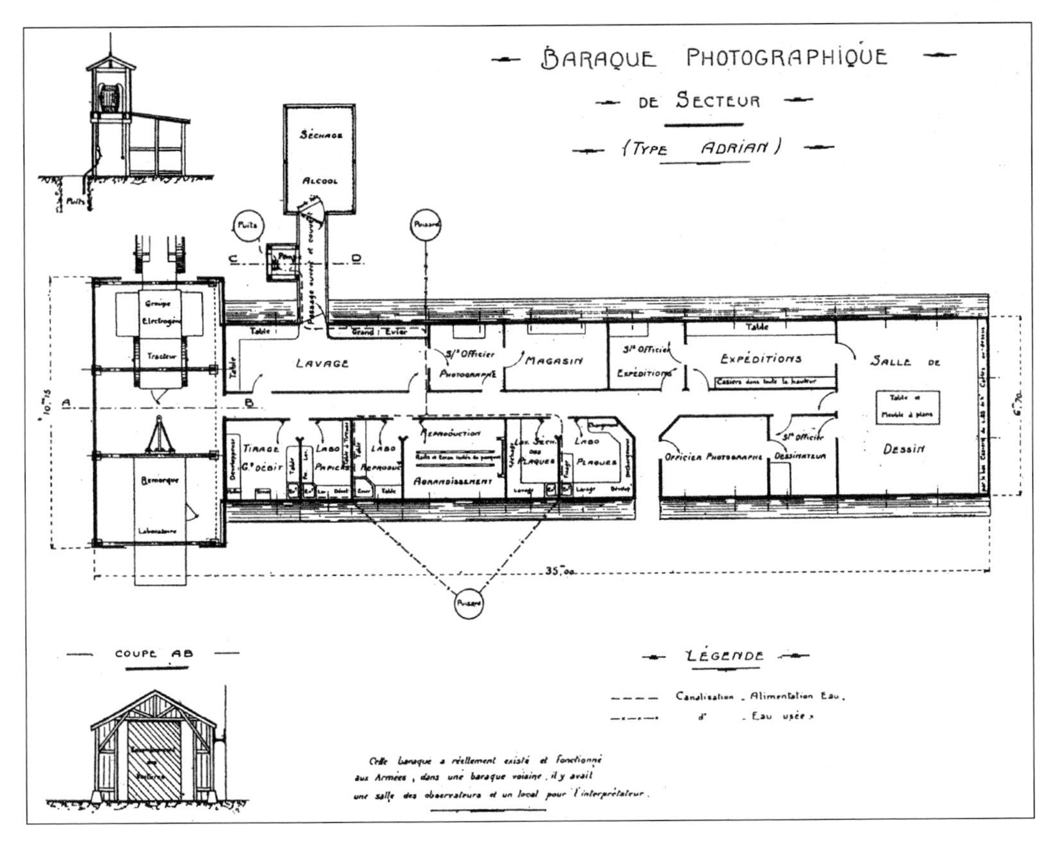

Accommodations for photographic processing and interpretation were increased by the fixed Photographic Baraque (Cabin). (Brunoff, ed., *L'Aeronautique*)

numbering, filing of plates, prints, and other tasks. The NCOs also kept a tally on all stores and ensured stock was maintained. The officer ensured the collection of photographs was kept safe and available on request. The photos were catalogued by map squares. The officer also instructed observers on the camera and its use.[1524] However, it was a flexible operation. 'Even in the most unfavourable of circumstances, the lorry can be drawn up near a house of some description where a table and chair can be placed at his disposal upon which the work of identifying the photos, and getting them ready for issue can be carried out.'[1525] The only thing needed was access to a water source. One photo lab technician remembered: 'Water was the most essential, but the scarcest of our needs, and had to be trucked in GI cans from the rivers, which sometimes were several kilometres away. We never had running water.'[1526]

The basic mobile photographic laboratory was compact. British specifications resulted in a cramped environment that all photo processing and interpretation personnel had to endure. 'The Dark Room Lorry was to be 12 feet long by 6 feet 6 inches in width. It was divided into five compartments: small store, officer's desk enlarging and printing room, dark room for papers and dark room for plates. The dark room was initially envisioned for six men or more to work in if necessary with a place for the OIC to write and watch his stores.'[1527] French laboratories possessed additional equipment. The Photographic Camion consisted of a tractor, a laboratory trailer which could accommodate eight personnel and a generator to provide power for the laboratory. They also constructed a wooden portable laboratory known as a Photographic Cabin (*baraque*) to accommodate larger production requirements.[1528] An entourage of Photographic Camions could be assembled to support an operation on demand. As one French officer recalled, 'In the sectors of attack, several sections are brought together and assigned to a large photographic barracks where the automatic machines are installed.'[1529]

When the AEF set up operations in theatre, they worked with what was available and also commenced development of an American version of the mobile laboratory. Their mobile photographic truck contained all the equipment necessary to rapidly produce prints in the field. The truck body was equipped with a dynamo for the electrical current required for lights and drying fans. Each unit contained an acetylene generator for emergency use if the electrical apparatus should break. The mobile dark room carried on the trailer of each unit was equipped with tanks, enlarging camera, printing boxes, and other necessary apparatus. Some 75 complete units were built in the US and shipped.[1530] Technological questions held centre stage as procedures were worked out for efficient aerial reconnaissance. Deliberation went on amongst the Allies to find the best techniques to process and analyse photography. Perhaps even more important was the allocation of critical resources such as the high-quality glass and determining the extent to which each Ally should commit to sharing trade secrets. The arrival of the Americans only increased the urgency to work out how to share limited resources.

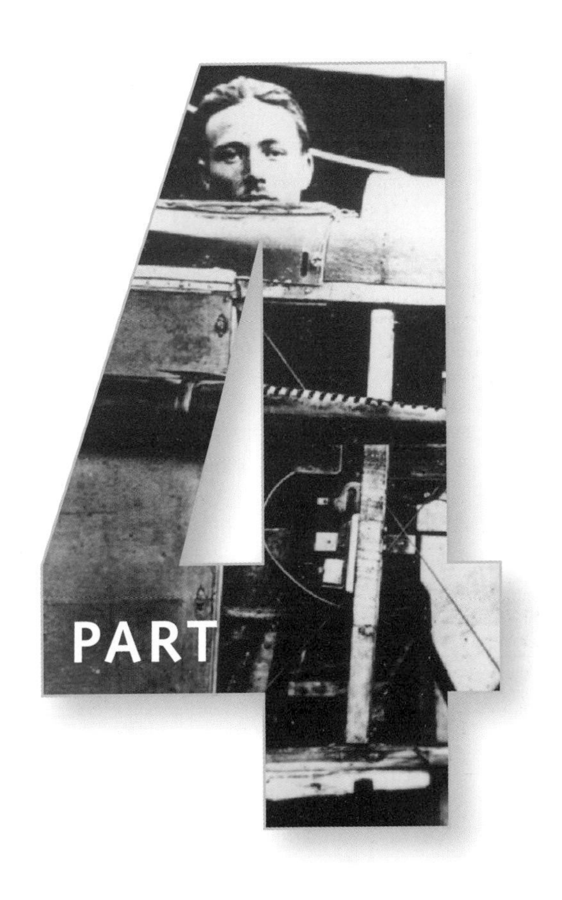

The Enduring Technological and Human Legacy

CHAPTER 22

Operational Aerial Cameras

Camera design and photography had been evolving for 70 years prior to the war. By war's start, photography was limited only by the imagination of the viewer. It offered a tremendous outlet for both the professional and the novice. Buyers enjoyed a selection of cameras from the Kodak Brownie hand-held to elaborate movie equipment. However, camera systems that had been developed, such as aerial mapping cameras, were far too heavy for serious consideration by military aeroplane designers. As aerial observation gained value with military forces, efforts were made to develop cameras that could effectively photograph the battleground, with refinements to improve both reliability and ease of handling in a combat environment. Technology transformation occurred, not in the science of optics, but in the ability to stabilise the airframe and manipulate the photo plates so that a series of photographs could be acquired while maintaining constant vigilance against a rapidly growing threat.

Taking a photo with an aerial camera comprised three steps: set the shutter, release the shutter and change the plate. Aerial plate-camera design covered four principal parts: box, lens, shutter, and plate holder. The box was light but strong enough to support the mechanisms required to take the photographs. The inside of the box was black to prevent reflection on the plate. The lens was the critical component and was made of high-quality glass.[1531] Aerial cameras required the shutter to acquire a calculated amount of light on the photographic image while moving at a certain speed. Some of the aerial cameras used at this time were manually operated, creating a distorted image of the ground. Another challenge to aerial cameras was vibration from the aeroplane motor, which forced the use of a faster shutter speed of 1/125 of a second or less.[1532] The plate holder or magazine contained metal holders known as the septum that held both unexposed and exposed plates. From these basic components, camera design evolved to handle the rigours of flight.

THE AERIAL CAMERA INVENTORY

Shortly after war was declared, the public rushed to offer the combatants suggestions on the camera's use in the conflict. In Britain, military staffs received novel ideas concerning the art of the photographically possible. One memorable offer came from the noted American explorer and photographer of the African continent, James Barnes, who later launched America's photo effort for the US Signal Corps in 1917. In August 1914, Barnes had just arrived in Britain as forces were mobilising. Having just extensively employed an array of still and movie cameras to record his latest sojourn, he was quick to point out to a British military senior at Westminster Town Hall the value of photographic reconnaissance, emphasising that military information by which an army acted, either on the defensive or the offensive, best came from the air. The only way one could achieve accuracy in this endeavour was using a recording 'flying eye' – the camera. Barnes and his explorer colleague together offered plans for a portable photographic laboratory – a lorry, fitted up with dark room, drying frames, everything necessary for the development of negatives taken automatically on continuous film. The prints were to be made in the field in the shortest time. The novel Barnes proposal included a 'guarantee to show to any general officer or the Intelligence Department a moving picture on the screen, of the country traversed by any air scout within an hour after descent from flight of inspection.' The resulting response was brief but gracious. The immediate priorities of mobilising forces and deploying to the combat areas were too overwhelming to take advantage of the offer.[1533] Barnes had his chance to apply his photographic prowess later in the war when he established the US Army's Photographic Division.

Aerial photography in the first year of fighting was accomplished with hand-held cameras. British and French efforts prior to the outbreak of the war were experimental. The early cameras were not well-suited for aerial observation. One hand-held camera that received considerable attention from the early British aviators was the Pan-Ross, a Goerz-Arshutz camera type used by press photographers.[1534] When the RFC set out for France in August 1914, five 'Penros' cameras were listed on the equipment inventory list, fitted with 6-inch Panocentric lenses.[1535] An unexpected problem cropped up. The bellows on the camera blew in from the air stream and the focus became permanently disturbed.[1536] Despite these shortcomings, hand-held cameras performed as well as early machine-mounted attempts.

The vast majority of cameras in the First World War were non-automatic. The ideal was the automatic camera, allowing the air crew to concentrate on defending themselves while flying the sortie. When the automatic camera operations finally became reality in the last year of the conflict, the responsibility for taking photographs shifted to the pilot, allowing the observer to take additional photographs with a more portable hand-held camera as well as look out for trouble. The combination increased the coverage and provided supplementary information that aided the photographic interpretation. Some single-seater fighter aircraft were configured with rudimentary controls to operate a semi-automatic camera; however, the mechanism was found to be unstable and unreliable.[1537]

The aerial camera of this era was categorised by the method of operation and the sensitive material (plates versus film) employed. Plates were the standard, despite the fact that film was equally effective. There were three types of plate aerial cameras: non-automatic, semi-automatic, and automatic. First World War aerial cameras were mostly fixed-focus box styles configured for plates. The average plate size varied with nationality. British cameras used a smaller 4 x 5 inch; French frames were larger at 18cm x 24cm. The French 18cm x 24cm plates produced contact prints whereas the British had to perform an additional step in the development process using an enlarging camera to produce 6½ x 8½-inch prints from the 4 x 5-inch negative. In all British aerial fixed-focus cameras the lenses were interchangeable so that either a short- or long-focus lens could be fitted to suit the requirements of the mission. Shorter lenses were used for broad coverage or for supporting map production. Long lens cameras produced the large-scale photographs that helped analysts interpret targets.[1538] Nomenclature for aerial cameras varied with each country. In 1918, both the French and British gave their cameras an alphanumeric designation. French cameras were listed with an 'F'. The British defined their cameras by medium and function. Plate cameras received a 'P' prefix; film cameras were labelled as 'F' and gun cameras as 'G'.[1539]

FRENCH RESEARCH ON AERIAL CAMERAS

The French government had long recognised the military value of photography. At Versailles, their *Section Technique* laboratory examined a broad array of technologies for military applications, including aerial photography. Section Technique successfully developed camera systems that served as the core of French aerial photographic reconnaissance for most of the war and into the immediate post-war period. The designs were consistent, with the aerial camera body in a pyramid shape and built of aluminium. The French established their nomenclature process for aerial cameras based on focal length – 26cm, 50cm, or 120cm. Later in the war, a standard nomenclature was developed such as F-1 for the 26cm camera. It was a tribute to the success of Section Technique that a post-war military assessment rated development of the aerial camera alongside other key military-related technologies such as the 'wireless telegraphic sending apparatus, machine guns firing by synchronisation through the propeller, machine guns in the rear turret, and bomb-throwing gear with shells in the wings'.[1540] Also instrumental in the aerial camera design were the Gaumont and Duran firms, which provided the key components for camera assembly.[1541]

The Douhet-Zollinger camera, which provided the initial breakthrough for French aerial photography of the front opposite the French IVème Armée sector in March 1915, was not actually French. The Douhet-Zollinger was built about 1913 in Turin, Italy, under the direction of Commandant Douhet and engineer Zollinger. The French acquired a prototype for testing when war was declared. The Douhet-Zollinger was one of the few operational automatic film cameras in the war. The camera body was aluminium, with a Tessar-Zeiss optical lens set to a focal length of 10cm. The shutter speed was set at 1/100 a second. This German lens of course was not available to the Allies after the start of the war. The film was 6cm by 20 metres, providing continuous coverage for 300 photographs. The film was perforated along the edge, driven through the camera by means of sprockets. This automatic camera was replaced by the more sturdy plate cameras such as the 26cm. The automatic function was not re-evaluated until the final year of the conflict.[1542]

THE BRITISH PROTOTYPE 'A' TYPE CAMERA

The British initiative behind developing aerial photography was fed by their own aviators and by introduction to the ongoing efforts of the French. It became apparent to RFC pioneers, Lieut. Pretyman and Lieut. Darley, that hand-held cameras yielded limited results. Moore-Brabazon remembered the tricky start:

Exterior view of the 'A' Type camera. (Nesbit, *Eyes of the RAF*)

'We were then asked to produce the same results [as the French]. Handicapped as we were for apparatus, etc., work was commenced with the five existing cameras which were very soon reduced to two owing to losses due to enemy action and cameras becoming unserviceable.'[1543] C.D.M. Campbell, a photographic expert prior to the war, knew which technical barriers had to be crossed to develop a functional aerial camera. On 16 January 1915, Campbell outlined the requirements:

> Simplicity and strength are of primary importance. The apparatus comprised the lens, camera, shutter, and a method to change plates. The lens at present used on the Aeroplex camera is the Telecentric by Ross of 12 inch equivalent focus, and has a back focal length of 6¼ inches. Optically, this lens gives fair definition of full aperture. Assuming 6,000 feet to be the lowest an aeroplane can fly over the enemy lines with comparative safety. This served as the height taken as a standard on which size of image was based. The lens gave a scale of 1 mile to 10.6 inches. Pictures were made to have a scale of one mile to 21.2 inches, or 1/3,000 roughly. The scale for trench work is too big and the small area made identification difficult. The team showed the images of varying size taken from one standpoint, concluding that a focal length of 8 inches would serve as the happy medium.[1544]

Campbell also calculated that a photograph taken from approximately 6,600 feet equated to a ready-made 1/10,000 scale map, ideal for restitution.[1545]

Regarding the next stage of development Moore-Brabazon recollected, '(We) were sent home to design and make a camera; this was done about February 1915. Six box-type cameras were evolved, this being the real R.F.C. camera, and from that moment photography improved a good deal.'[1546] Moore-Brabazon and Campbell returned to England to commence camera development. 'We started with a bellows camera – that is, a camera which pulls out – and you can imagine what a bellows camera would be like in an aeroplane.'[1547] It was soon understood that a camera had to be designed specially for the work, 'so the personnel then in the field consulted together and produced ideas which were conveyed home to Messrs. Thornton-Pickard who in a very short time produced what was known as the 'A' Type camera.'[1548] In February 1915, Mr Richard Hesketh, secretary of the Thornton-Pickard Photographic Manufacturing Company Ltd., met with Moore-Brabazon and Campbell and talked about British aerial reconnaissance basics, being careful not to violate the prevailing security laws. The discussion led to what was required to configure an effective aerial camera for the RFC. The 'A' Type design met the requirements. All significant issues were addressed, including protecting the fabric focal-plane shutter from the pressure or air suction and the position of the two handling straps.[1549]

Hesketh immediately followed up, sending a telegram to the Thornton-Pickard staff on the criteria behind the camera design. When Moore-Brabazon met Thornton-Pickard, Chairman of the firm a few days later, he was handed a completed camera.[1550] The 'A' Type camera was shaped like a conical box and constructed to withstand major vibration from the aeroplane. It was a folding-type that included a bellows. The lens was fixed in distance from the plate, an idea promoted by Campbell whereby the aerial camera focus was set at infinity. This served as the standard for the remainder of the war.[1551] The first 'A' Type camera was not easy to handle. The first exposure required eleven distinct operations followed by ten distinct operations for each subsequent exposure.[1552] Under the leadership of Campbell, the 'A' Type camera was modified to take 18 successive photographs by means of a changing apparatus.[1553] The effort to streamline the photographic process continued throughout the war. The 'A' Type prototype was rushed to France and flown over the German lines on 2 March 1915. The first 4 x 5 inch plate was exposed above the village of Fauquissart by an observer of No. 3 Squadron. Prints were made and personally briefed to the BEF General Staff. The RFC now possessed the technology and the organisation required for operational aerial photographic reconnaissance.[1554]

The War Office and the RFC were happy with the design.'The 1st Wing reported that the Camera had quite fulfilled expectations, and that the Squadrons ... are delighted with it.' The only early complaint was that slides carrying plates were apt to break rather easily.[1555]

The 'A' Type prototype became a source for modifying aerial cameras to better handle the challenges of flight. The 'A' Type was operated by the observer, who gripped the camera through straps or brass handles and leaned over the side of the aeroplane to take photographs. Unfortunately for the observer, the process of taking a photograph was very complicated, particularly in the harsh environment of open cockpit flying.[1556] One observer recalled the experience with the 'A' Type camera. 'On my first jaunt, taking pictures of a certain trench-line in the vicinity of Thiepval, I nearly lost my balance as I leant over the side with my two gauntleted hands firmly clamped in the weighty affair's straps. My safety-belt was lying idle on the floor of my nacelle, and I was reaching out to avoid a landing wire that was in my field of fire.'[1557]

Later that summer, the British design team developed the 'C' Type camera, a semi-automatic plate-changing mechanism to make the air crew's handling of the camera easier.[1558] The first steps were taken towards organising the supply depot for aerial camera supplies and component parts. A Photographic Supply was created within Britain's Military Aeronautics Branch of the War Office. This office was responsible for the dispatch of cameras and photographic stores to France and to the other theatres. It was not a smooth running operation. Eventually, the Directorship of Aircraft Equipment (DAE) was formed to purchase cameras and deliver the required goods to the photographic officer making the request. Post-war review mentioned the process had problems and that bookkeeping was 'dreadful'.[1559]

STABILISING THE AERIAL CAMERA

After the first fixed mounting of the camera in the summer of 1915, for the next six months coverage was marred by camera vibrations. The primary British photographic aeroplane of the time, the BE 2c, suffered from extensive vibration to the outside mounting that resulted in blurred coverage. Tests were done to find a solution to the vibration caused by the propeller draught. The ideal place for the camera was determined to be internal and behind the pilot. Research by Campbell led to a successful design for strapping the 'A' Type camera firmly to the side of the fuselage with leather straps and a wooden frame. This not only reduced vibration, it also increased the verticality of the photograph for restitution.[1560] In his post-war notes, A.V.M. Brooke-Popham identified RFC NCO pilot Scholefield as the first to fix the camera rigidly to the aeroplane fuselage and take photos.[1561]

Camera stabilisation was one of the successes during the war. Early attempts at photography through a hole in the cockpit floor tried a variety of measures to steady the camera for each exposure. Great care had to be taken in the manipulation of the camera so that it did not interfere with the control wires. Aerial photography had to contend with the vibration of the aeroplane motor and air flow which caused serious problems

The second operational British aerial camera, the 'C' Type, was similar in appearance to the 'A' Type, but possessed a semi-automatic device for changing plates. (Courtesy of Michael Mockford and the Medmenham Collection)

The 'C' Type was the first British aerial camera to be strapped to the outside of the fuselage. The aeroplane is an FE 2b. (Courtesy of Michael Mockford and the Medmenham Collection)

when the tiny metal screws of the camera system became loose, causing the camera lens to drop out. The French advocated that the camera be held between the knees whenever possible. They recognised that the body was the best shock and vibration absorber. Another unique shock-absorbing mechanism was four tennis balls within the suspension. However, they did not support the camera effectively when the aircraft landed. In the latter part of the war aerial cameras were mounted on a suspension framework cushioned with foam rubber.[1562]

Once aloft, the size of the camera determined if the observer could take photographs outside or from within the cockpit. Camera operations varied with the camera and the airframe. The American journalist Ralph Pulitzer described the operating space of a Caudron G.4 during one sortie in August 1915:

> In the floor of the little cockpit, right in front of my feet, was a little glass window through which I could watch the ground passing directly (though some thousand feet) underneath. Just behind this window, in the floor under my feet, was a little metal trap-door. By straddling my feet I could open this, for the purpose of either taking vertical photographs or of dropping bombs.[1563]

The third operational British aerial camera, the 'E' Type, had a metal frame for the camera. ((Nesbit, *Eyes of the RAF*)

BRITISH AERIAL CAMERAS

The British inventory of cameras charts the energetic application of technological improvements over the course of the war. Moore-Brabazon, Campbell and Laws continued to be the design team.[1564] The prototype hand-held 'A' Type camera was replaced in the latter half of 1915 by Thornton-Pickard's 'C' Type camera. The 'A' Type had a tendency to distort the image because of the angle at which the aerial observer held it. The 'C' Type was the first aerial camera designed to be strapped to the outside of the aeroplane fuselage to produce better quality vertical photographs.[1565] The 'C' Type employing the body and plates of the 'A' Type Camera could acquire more aerial photographs and use a longer lens. It was fitted with a semi-automatic device for changing plates from one full magazine to an empty magazine. Eighteen unexposed plates were placed in the position above the focal plane aperture. A handle at the top of the camera changed the camera plates. During the exposure period the slide projected into a compartment on the opposite side of the camera from the receiving magazine. The cycle continued until the 18 plates were exposed.[1566] The 'C' Type camera remained the standard RFC aerial camera until the spring of 1917. Its greatest shortcoming was that the wooden construction would expand and contract at different altitudes and temperatures, sometimes causing the lens to be thrown out of focus.[1567] Some of the remaining 'A' Type cameras were modified with the 'C' Type plate changer. This variant was known as the 'D' Type.[1568]

The US Army adopted a 'C' Type variant for training. The American-made 'C' Type was produced by the Folmer and Schwing Division of Eastman Kodak Company and configured with an 8½-inch lens that imaged the standard 4 x 5 inch plates. Like the original British design, each 'C' Type camera was equipped with two gravity-loading magazines with a total capacity of 18 plates.[1569] The 'C' Type and follow-on 'E' Type cameras served as the primary camera types for training the British and Americans.[1570]

The 'E' Type camera was developed in the autumn of 1916. It was the first British all-metal vertical aerial photography camera.[1571] It differed from the 'C' Type with a metal frame that included a flap over the lens, eliminating the 'C' Type sliding screen. An added capping shutter covered the focal plane shutter during resetting. Remote control was accomplished by cords (called the Bowden Brake Cable) attached to the plate-changing machine. Its most significant improvement, an adjustable lens focus, allowed broad or detailed focus with an 8 to 10½-inch lens.[1572] Both 'C' Type and 'E' Type cameras had one major design flaw. They were never built for plates larger than 4 x 5 inches or for lenses longer than 12-inch focus.[1573]

During the Somme campaign of 1916, the demand for aerial photography prompted the British to come up with a unique camera, comprised of German aerial camera components. This 'B' Type camera with a simple aluminium cone design was manufactured at the British Repair Depots.[1574] The 'B' Type camera was primarily used for oblique photographs. What made it unique was that the critical lens component was acquired

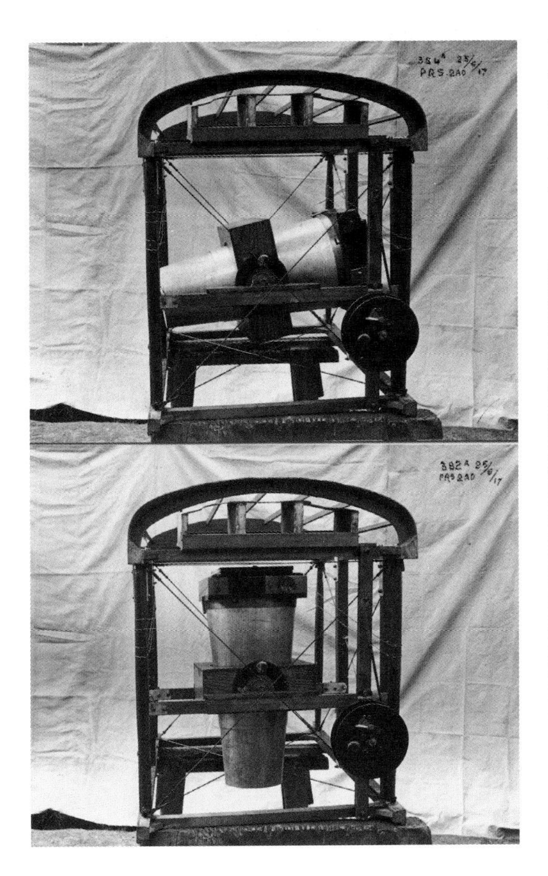

The 'B Type was designed in the British Repair Depots. It was primarily used for oblique shots and was unique in using captured German aerial camera lenses. (Courtesy of Dave Humphrey, Photographic Museum at RAF Cosford)

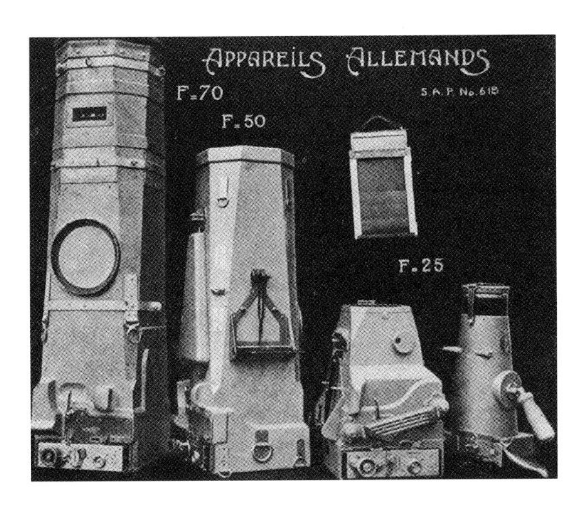

The German aerial camera inventory was impressive, with beautifully designed casings containing the world's best optics. (Herbert E. Ives, *Airplane Photography*)

The 47-inch frame camera was one of the longest lens cameras used by the British in the war. (RAF Museum, RAF Hendon)

from captured German aeroplanes. British intelligence knew that most German aeroplanes were configured to handle aerial photographic reconnaissance and any lenses acquired from captured or downed aeroplanes were a precious find.[1575] The German Zeiss or Goerz lenses were deemed 'superior to anything made in England.'[1576] The 'B' Type lens length was 20 inches, much longer than any lens produced at that time by any British manufacturer.[1577] They had a variety of lenses (10 in., 14 in., 20 in., 27 in.) and plate size (9cm x 12cm; 12cm x 18cm) and the German cameras were well designed for the reconnaissance mission.[1578] The 'B' Type used a larger 6½ x 8½ inch plate. Exposures were taken remotely by means of a Bowden cable. The value of German optics is proven by the urgency that they were sought after by both sides throughout the war.

The best British aerial camera of the war, the 'L' Type. See page 4 for use of the Bowden cable. (Gorrell Report, NARA)

The FE 2b was modified to contain a long lens aerial camera. (Courtesy of Michael Mockford and the Medmenham Collection)

The demand for more detailed photographs by intelligence personnel led to longer 20-inch lenses. The British produced a long lens variant in 1916 known as the 47-inch lens camera. It was designed to specifications similar to the French 120cm and configured to the front of pusher aeroplanes like the FE 2b.[1579] Moore-Brabazon recalled that the British acquired obliques for planning tank attacks using a camera with a 72-inch focal length. The extremely long lens was 'a remarkable feat.'[1580]

The 1916 Somme experience demonstrated to the British that large quantities of photographs were needed to assist both infantry and artillery. Later that year while at Training Brigade, the innovative photographic genius, Victor Laws, developed the 'L' Type aerial camera.[1581] Produced in greater numbers than any other line of British aerial cameras, the semi-automatic 'L' Type camera improved on the 'E' Type camera and became the most successful British aerial camera of the war. The semi-automatic arrangement of the plate changer made it possible for the pilot to operate the camera and freed the aircrew to defend their machine against an attacker. The photo plate magazine was made of metal and accommodated 24 plates. Like the previous 'C' Type and 'E' Type, the British continued to produce the standard 8½-inch lens for the 'L' Type camera configured for the 4 x 5 inch plate.

Lieut.-Col. Moore-Brabazon reflected on the transition to the 'L' Type: 'Improvement on cameras was again effected by the use of the wind vane, rendering the changing of the plates purely an automatic affair,

This 'L' type contained within a DH 4 fuselage frame was built in the US. (*Instructions for Installation of Type L Cameras on Airplanes*)

the Pilot having but to release the shutter.'[1582] The automatic mechanism changed the plate and set the shutter after each exposure. An air speed of over 50 miles per hour caused the cycle of camera operations to be accomplished within 10 seconds.[1583] When the 'L' Type was set to automatic, a locking device was put in place to keep the camera trigger from activating until the camera was reset.[1584] However, biases existed amongst the aircrew. The semi-automatic feature was usually disregarded in favour of hand operation. The aircrew did not trust the propeller (wind vane) designed to operate the camera, regarding it as another part that could fail. The 'L' Type camera drawbacks were that the upper magazine could not be closed or removed until all the plates had passed through, and the gravity feed for changing plates tended to jam. British observers dealt with the jams by turning the propeller backward or sometimes just shaking or 'thumping' the camera.[1585] Jamming the magazine was the fault of the operator, not the camera. A photo lab man remembered, 'The magazines ... would jam once in a while when the observer, in his excitement, would operate the pump change too rapidly and not allow the plate enough time for the springs to elevate it enough to move into the back position.'[1586] The 'L' Type was the first aerial camera produced in the US under contract. Edward Steichen recalled, 'The only aerial cameras received from the US which could be utilised in France were a small number of type L of British design, of American manufacture. This camera showed some minor improvement from the L type of British manufacture. The few we had were used with good results at various training centers, but were not used at the front.'[1587] The American equivalent took 24 plates.[1588] The US Air Service's Bombardment unit tried out various cameras (26cm; 50cm; 'L' Type) and came to the conclusion that the British camera, both on account of its focal length and of the ease with which it was installed in the planes, provided the best configuration for bombardment coverage. The rail surrounding the observer cockpit was too high to permit holding the 'L' Type over the edge. The bombardment unit used six of the 'L' Type cameras as an experiment.[1589]

The British 'LB' Type was a modification of the 'L' Type, designed by both Laws and Moore-Brabazon. The 'LB' Type was similar to the 'L' Type but the mechanism was simplified. The chief difference was an easily removed self-capping focal plane shutter, which could be replaced without disturbing the other parts of the mechanism.[1590] The greatest advantages over the 'L' Type were the possibility of using lenses of any focal length from 4 to 20 inches and having a detachable shutter.[1591] The 'LB' Type changed plates and shutter setting in a single operation by the action of an air screw in the slip stream or by hand. The exposure was made by a focal plane shutter with a separate capping device. The magazine containing the unexposed plates was placed

The 'LB' Type semi-automatic camera was designed by Laws and Moore-Brabazon. (Nesbit, *Eyes of the RAF*)

immediately over the exposure aperture on the body, the plates falling into position as required. The exposed plate moved along horizontally until it was opposite the receiving magazine.[1592] After the war the 'LB' Type was designated the P7 camera.[1593]

The final British camera to see service in the war was the 'B.M.' Type camera. It was becoming apparent in 1917 that, with the increasing altitude required for safe coverage from enemy pursuit aircraft and improving air defences, British designs incorporating a 10-inch lens and enlarged print from a 4 x 5 inch frame were insufficient to accomplish the mission. The 'B.M.' Type camera was developed to overcome this difficulty by employing a whole plate and a 20-inch lens. In the final year of the war, because the Inter-Allied Commission recommended a common photographic format that provided British aerial photographic interpreters with a more convenient medium, the size of the plate was changed after the manufacture of the first two models to the standardised Allied 18cm x 24cm format.[1594] The 'B.M.' Type was the largest and final camera type introduced by the British. It was metal, including the magazines, and each magazine held twelve plates of 18cm x 24cm. The 'B.M.' Type was enormous, weighing in at 82 pounds (competing with the 80- to 100-pound American version of the French De Ram). It served a

The 'BM' Type was the largest aerial camera employed by the British in the war. This camera is contained within the fuselage of a Bristol Fighter. (Nesbit, *Eyes of the RAF*)

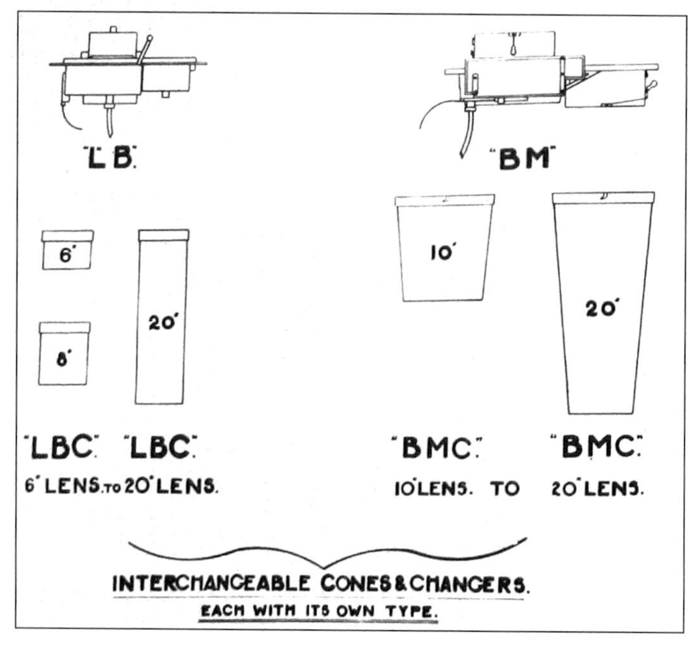

Diagram shows the interchangeable cones for both 'LB' Type and 'BM' Type cameras. (Laws Papers, RAF Museum, RAF Hendon)

variety of reconnaissance requirements with the ability to interchange four lens cones (7 in.,10 in., 14 in., 20 in.). The 'B.M.' Type mechanism operated under the same mechanical principles that powered the 'LB' Type. A wooden propeller was carried in a propeller bracket fixed on the side of the aeroplane. Power went to the camera from the propeller through a flexible shaft in a casing attached by bayonet couplings. The shutter release was triggered by a Bowden cable.[1595]

Early handheld cameras like the Pan-Ross were replaced by more sophisticated equipment. The 'A' Type hand-held camera served British reconnaissance for most of the war.[1596] Later in the war, the British would develop the 'P' Type metal body hand-held camera to replace the 'A' Type. The 'P' Type had a fixed cone holding a 8½- to 10½ -inch lens and a detachable plate-holder for 5 x 4-inch Mackenzie Wishart envelopes.[1597] Another RNAS aerial camera was produced by the Houghton-Butcher company and called the P18.[1598] Hand-held cameras were essential for RNAS seaplane operations locating submarines or photographing German targets on the coast. The seaplane could not accommodate an outside attachment for vertical shots. Subsequent RNAS photography was primarily accomplished with hand-held cameras shooting at an oblique angle. RNAS hand-held cameras had a variety of modifications and lens lengths (6 to 15 inches).[1599] The Americans brought with them an array of 'private cameras' to acquire short-focus aerial photographs. This lessened their dependency on the Allies, particularly in covering the normal field of view that lenses such as the 26cm offered. Since aerial cameras were a critical resource in the conduct of every campaign, the idea of an expiry date for an old aerial camera system did not exist. The

The RNAS had a variety of aerial cameras to conduct anti-submarine operations. One such camera was the Houghton Butcher frame camera. (Nesbit, *Eyes of the RAF*)

The 'P' Type camera was the most popular of the British hand held cameras. Here a RNAS officer demonstrates holding the camera. (Ives, *Airplane Photography*)

Aiming the 26cm for a shoot demonstrated by the aerial observer. All Allied aerial photography was governed by 'Aim towards Germany!' (Gorrell Report, NARA)

British prototype 'A' Type cameras were employed for the remainder of the war. Thornton-Pickard continued to manufacture the 'A' Type for commercial purposes for another decade. British aerial cameras supported Allied forces on the battlegrounds beyond the Western Front, including Gallipoli, Italy, Macedonia, Palestine and Mesopotamia. Each deployment of aerial photographic cameras came with an entourage of photographic interpreters and laboratories. The only difference was the interpreter's knowledge of the terrain and culture of the region.[1600]

FRENCH AERIAL CAMERAS

The 26cm (F-1) hand-held camera was the oldest camera variant used by the French, ideal for acquiring broad area photographic coverage with the short focal length. The 26cm was produced by the De Maria Lapierre firm with components such as the shutter provided by others such as Kopcic. The magazine held twelve 13 x 18cm photographic plates.[1601] The 26cm proved ideal for aerial mapping and provided the French with their first aerial surveys of the battleground. Subsequent roles for the 26cm included artillery observation and fire control. By the end of the war, the 26cm teste Allied camouflage. The broad area of coverage also served infantry contact (infantry aeroplane) missions locating the advance line of troops during combat operations.[1602] De Maria Lapierre's 26cm required the operator to point the lens upwards after the exposure. The plate was changed by pulling the inner body of the magazine out and then back into the camera framework. The operator then set the shutter. The final step was to point the camera at the target and expose the next plate by a gentle pull on the exposing lever.[1603] When the US Air Service's Bombardment unit tested the 26cm camera, they expressed frustration over the fact that it took up too much room in the observer's cockpit when trying to invert it to change the plates.[1604] When the French modified their cameras to the larger 18 x 24cm camera plates, the process was simplified, requiring each exposed plate to be pulled out after each shot.

Once broad area coverage was acquired by the 26cm, a subsequent aerial photographic reconnaissance sortie employed the 50cm (the F-2; also designated 52cm) to acquire a more detailed study of the terrain. The 50cm provided increased coverage ranging from four to six thousand metres.[1605] The F-2 went operational in 1916.[1606] Operating the 50cm required pulling the drawer out of the magazine after each exposure. The French design had the magazine on the top of the camera, a challenge to observers when it was attached to the gun turret rail.[1607] Like the 26cm, the 50cm employed the standard 13 x 18cm plate. Gaumont produced the twelve-plate magazine. Later the French reconfigured the 50cm to a larger photo plate of 18 x 24cm. The working area of the plate was reduced to 17cm x 23cm to compensate for the 0.5cm edge.[1608]

In 1917, the French upgraded their 26cm (F-1) with another 26cm hand-held camera known as the *Grand Champ* (the F-3, also known as 25cm). It covered a broad area for mapping reconnaissance, employing a larger photographic plate of 18 x 24cm. This larger plate size remained the standard for French reconnaissance for several years after the Armistice.[1609]

Finally, the 120cm (F-4), also designated the 1.20 metre, photographed particular points or a limited sector in order to secure a specific piece of information.

Above: Three photographs showing the various configurations of the 50cm within a fuselage. The unique tennis ball suspension at the top of the camera was complemented by the spring suspension at the bottom. (Gorrell Report, NARA)

Above: The 26cm 'Grand Champ' provided larger contact frame aerial photographs than its 26cm predecessor. A Breguet 14 A2 aerial observer demonstrates the aiming technique. (SHD)

Above: The 50cm fit snugly in back of the Spad VII cockpit. (SHD, 87.2856)

The 120cm camera was long, sometimes protruding as much as two feet below the standard Maurice Farman or Caudron fuselage. If the 120cm was not streamlined, the aeroplanes could 'lose' up to 20 horsepower. The use of this camera was limited to good weather owing to the small lens opening.[1610] The media labelled the 120cm the 'vest pocket' camera.[1611] The 120cm was built by the Gaumont firm. The photograph magazine carried twelve larger-format 18 x 24cm plates.[1612]

The majority of cameras used by the Americans in 1918 were of French design. By the end of 1917, the French were producing 25 aerial cameras a month for the American forces. Practically all aerial photographs taken by the Americans were 18 x 24cm format.[1613]

Simplicity often overrode technical advance in the thinking of many. One French observer commented:

Photograph shows 120cm camera size in comparison to a member of the AEF photo section. (Ives, *Airplane Photography*)

Sergent Charles Hallo of escadrille MF 63 demonstrates aiming the 120cm from the forward cockpit of a Maurice Farman MF 11bis. (SHD, 877/092)

A unique attempt to configure the 120cm was attempted by Escadrille SPA 42 by mounting the camera on the top wing of a Spad XI. (MA 1856 in Service historique de l'armée de l'air France. (*Les escadrilles de l'aéronautique militaire française:symbolique et histoire 1912–1920*)

> Aerial cameras may be automatics, semi-automatics, etc. Automatic cameras are proper for one-seaters, for the Pilot cannot let go of his controls to take the photographs, but they are not needed on aeroplanes with two seats. The best thing on two-seaters is a good plate camera such as the French and English use. Automatic cameras have complicated, fragile mechanisms, which get out of order easily, and in France, where the temperature is often very low in the air, we run the risk of having the oil in the mechanism freeze and the machine refuse to work at all; besides the operator cannot tell during the flight whether or not his camera is working, and this causes a great deal of trouble. For this reason we should positively not use an automatic camera on an aeroplane with two seats. Such a course would have bad points and no good points at all.[1614]

THE 'DE RAM' SEMI-AUTOMATIC CAMERA

In support of automatic cameras, Steichen spoke for many when he declared, 'During a mission over the lines the operation of this camera took up all the attention of the observer. From the beginning of our experience, it was clear that an automatic camera was not only desirable, but would eventually become a necessity.' Allied efforts to build an automatic camera resulted in several unsatisfactory prototypes. In the last month of the war the Americans fielded and operated 15 semi-automatic cameras over the lines.[1615]

The development and deployment of an automatic camera was a dominant theme as the Americans established their presence. They found something close to what they were looking for in the French-built De Ram camera, named for the inventor, Lieutenant G. de Ram of the French Aviation Militaire. The semi-automatic camera allowed the pilot and observer to concentrate on defending themselves while the camera took sequential photographs. A semi-automatic prototype was successfully designed by the end of the war. The De Ram came to the attention of AEF senior staff members in July 1917. It was a large, heavy camera mounted in the fuselage and operated by either the pilot or the observer. The camera weighed approximately 90 pounds when loaded with the maximum stock of 50 plates, an enormous number.[1616]. The large format plate (18 x 24cm) assisted the photographic interpreter. Operating the De Ram appeared to be a simple process. An exposure was created after the operator pressed the release, resulting in the plate changing and the shutter being set automatically for the next exposure.[1617] Preparing the camera before and after flying the mission required extra maintenance, since the entire camera assembly, not just the lens cone, was removed from the aircraft.[1618] De Ram first met with failure when his prototype was rejected by the French government. His disappointment was short-lived. The Americans were impressed.

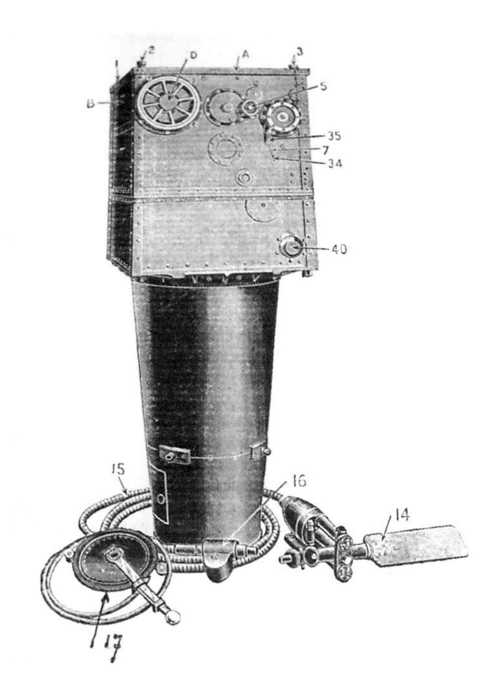

The largest semi-automatic Allied camera of the war was the De Ram. (Carlier, *La Photographie Aérienne Pendant la Guerre*)

In November 1917, Colonel Edgar S. Gorrell of the AEF staff contacted the senior US photographic experts in theatre, Major Barnes and Capt Steichen, to review the De Ram. Gorrell was sold on the prototype he saw and wanted the experts to confirm his impressions. Barnes and Steichen subsequently visited the De Ram workshop and provided positive reports on its configuration and operation. In addition to the 50 18 x 24cm plate capacity, they described the plate movement being controlled by a propeller or helix connected with the machine by a tube. They reported that the De Ram demonstrated an advantage over the rest of the camera inventory thanks to its movement being exactly controlled by a pilot or observer. Barnes stated that the De Ram's principal defect was weight and post-mission processing. The camera had to be removed from the fuselage and taken to the dark room to be opened.[1619] Besides the rotary magazine of 50 plates (changing box), the De Ram had interchangeable cones for the French focus standards of 26cm, 50cm and 120cm.[1620]

Leading photographic experts had the highest hopes for the De Ram despite the fact that the only Allied single-engine reconnaissance aéroplane capable of carrying the heavy load of camera, air crew and plates was the Salmson 2 A2.[1621] Steichen commented, 'if the De Ram Camera works out satisfactorily as a building proposition, it will very likely be adopted by the Allied armies as the *Standard Camera*.'[1622] Steichen's praise was exuberant:

> These cameras were ingenious instruments invented by Lieutenant DeRam and were built under his supervision in a small shop near Paris. The camera works entirely automatically; it can be set for any given interval of exposure and simply requires to be started when the airplane reaches the region to be photographed. It has a capacity for 50 exposures. In spite of the fact that this machine was new in principle and that these first cameras could scarcely have been considered final or complete, results were produced that were not only valuable in themselves, but pointed the way to new and hitherto impossible accomplishments in aerial photography.[1623]

The American acquisition process involved several layers of government. A model of the De Ram was sent to the US for evaluation, where it gained acceptance. Manufacturing rights to produce the De Ram were acquired. American requirements stipulated that it be

Far left: Assembling a semi-automatic De Ram aerial camera in the factory. (Gorrell Report, NARA)

Left: Lieutenant Georges de Ram, the inventor of the semi-automatique camera that bears his name. The famous aerial observer shot down four German aeroplanes while on reconnaissance sorties. (Bailey and Cony, *French Air Service War Chronology* 1914–1918)

semi-automatic so that the observer or pilot only had to release the shutter at will and a fresh plate would be in place.[1624] Delivery was scheduled for November 1918. At the time of the Armistice, 200 De Ram cameras were nearing completion.[1625] The discussions about the value of the De Ram lasted throughout the last year of the war. In one June 1918 memo from an Inter-Allied conference, the British delegation leader, Moore-Brabazon, reflected a concern with the American infatuation with the De Ram. 'The Americans seemed to be gambling on a supply of De Ram Cameras as their big type Camera, but they want to be helped out later by a supply of BMs [British semi-automatic aerial camera] if we could see our way to supply our own needs and have an excess.'[1626] As the war neared its end, Moore-Brabazon reversed his opinion, expressing the hope that the Americans could produce enough cameras to fulfil British requirements as well.[1627]

In the end, only a few French-produced De Ram cameras saw combat.[1628] The De Ram camera received its baptism of fire in the Argonne offensive. During one reconnaissance mission, seven German fighters attacked two Salmson 2 A2s flown by Americans. One of the aéroplanes was equipped with a De Ram automatic and the other with a hand-operated camera. The Americans shot down two of the fighters and drove off the remainder without interference to their own photographic work. The De Ram camera had successfully and automatically accomplished the photography required.[1629] De Ram gave the aircrews a taste of the aerial photographic future. Now the observer could increase his survivability by focussing on the enemy threat and operating the Lewis gun.

There are conflicting accounts of the De Ram's performance. A postwar account of aerial support to the First American Army stated that 'A French automatic (De Ram) camera gave better results than those hand operated, furnishing over 70 per cent of the later photographs taken on the Fifth Corps front.' [1630] Not all memories of the De Ram were that positive. The 1st Army Observation Group was not impressed, declaring the camera unreliable. The cameras were mounted and tested in September 1918. The De Rams worked fairly well at first. However, when assigned to operational missions, the cameras continually jammed or the plates were either unexposed or unevenly exposed.[1631] This feeling was echoed at squadron level. One photographic laboratory member with the 88th Aero Squadron remembered: 'It had a tendency to jam easily, so we did not use it much.'[1632]

THE ITALIAN PIAZZA

An Italian aerial camera known as the Piazza was employed by armée escadrilles for strategical reconnaissance 1917–1918. The focal length was 24cm.

An exceptional photograph showing an aero squadron's entire camera inventory: the De Ram, four 50cm, 26cm 'Grand Champ,' and the 26cm. (USAFA McDermott Library Special Collections (SMS57))

The camera employed a Bowden cable to allow the pilot remote access to the shutter. The Piazza was capable of imaging up to 36 13 x 18cm plates per sortie. The Piazza's smaller size and automatic features made it an ideal aerial camera for Spad aéroplane reconnaissance flying at altitudes up to 6000 metres.[1633]

AUTOMATIC FILM CAMERAS

In the final weeks of the war, the French introduced several film cameras for strategical reconnaissance. At the Inter-Allied Photographic Conference of 18 August 1918 discussion among the representatives focussed on a Duchatellier camera currently in production and configured with an 18cm film spool.[1634] The lens focal length was 24cm.[1635] The French demonstrated a new film with a thick base made by Pathé. Despite British concerns raised that the Pathé film did not lie flat enough to hold the film to the register, the French successfully employed the camera. Two other film cameras, the Lenouvelle 18 x 24cm camera and the Type R.A. 18 x 24 roll film camera with vacuum device were introduced as well.[1636]

Allied cameras in the First World War demonstrated an ingenuity in the design of rugged and reliable systems necessary to accomplish a difficult mission. Despite the challenges, aerial camera design was simplified to effectively operate within the constraints of the plate configuration. The French designs of 26cm, 50cm and 120cm showed the greatest consistency, partly because they were employed longer than other cameras. The British penchant for improving the technology resulted in the well-thought-of 'L' Type camera that produced most of the photographs in the final year of the conflict. Semi-automatic and automatic aerial cameras were examined because aerial reconnaissance needed to balance mission accomplishment with the ability to defend. Interestingly, film did not make its mark following the successful employment of the Douhet-Zollinger in early 1915. Had the war continued, film would have assumed a more important role thanks to the persistence of George Eastman and his influence on the Americans. That was to be the case in the post-war era, as aerial photography converted to automated film systems.

CHAPTER 23

Aerial Reconnaissance Platforms and their Configuration

As aerial observation missions expanded to serve increasing demand for information and aerial photographs, aircraft design improved. Each successive year demonstrated the need to improve on the ability to engage and defeat the enemy. More effective fighting machines with greater firepower and airframes designed to take the stresses of rapid manoeuvre were built. The early years showed that non-manoeuvring aerial observation platforms were easy prey. However, aerial manoeuvre hampered photography. An interim solution became aerial supremacy through formations. However, such formations drew the attention of enemy pursuit aircraft and anti-aircraft batteries. The solution lay in advancing aircraft design to meet the demands of combat and the reconnaissance mission. The next evolution was to ascend to higher altitudes to escape both anti-aircraft fire and the enemy pursuit fighters.

The limitations of vision required aerial observation to operate slowly and in low-flying aerial platforms. Operating above 1,000 metres stretched the ability of the aircrew to see activity. It also challenged the early aircraft designs; climbing to that level took over 15 minutes. The pusher plane with its unlimited range of observation had to give way to a more rugged and survivable tractor plane. Aeroplanes and aerial photography evolved simultaneously. The requirements of the war forced rapid development of aircraft specifically designed to facilitate photography, using cameras specifically designed to work on aircraft.

Critical to aeroplane design was the attention paid to the existing camera configuration required for successful aerial photography. The lack of long lenses in the British and American optics inventory resulted in camera frames of limited dimensions. British and American designers therefore developed aircraft with only limited space for an aerial camera. Americans employing French aéroplanes and adopting French aerial reconnaissance practices enjoyed the benefits of longer focal length cameras such as the 50cm and 120cm embedded internally. For most of the war the British had to rely on smaller cameras attached to the outside of airframes.[1637]

Aerial reconnaissance planes serving in the First World War covered the spectrum of capability from some of the best aircraft of the era to some of the worst. The ten most notable aerial observation aeroplanes to serve the Allies on the Western Front are described in the following sections.

BLÉRIOT XI

The Blériot XI aéroplane is best remembered for the first flight across the English Channel in 1909. It served as many European nations' introduction to military aviation, including France and Great Britain. Blériot XI's fulfilled an important purpose; they helped define the basics of flight and the role of aerial reconnaissance, to include aerial observation, artillery spotting, and as a platform for taking photographs with hand-held cameras. Blériot XI aerial observation roles were conducted at low altitudes and slow airspeed. The primary threat was the carbine, either from infantry on the ground or rudimentary attempts by German adversaries in the equally slow-flying Tauben. Blériot XI observation was hampered by the wide wingspan directly under the forward cockpit where the observer sat. After 1914, the Blériot XI served the Allies as a basic trainer.

Crew	1 or 2 dependent on variant
Average maximum airspeed	100kmph; 62.1mph (single-seat) 120kmph; 74.5mph (two-seat)
Climb to 1,000m	12mins
Ceiling	1,524m; 5,000ft
Endurance	3hrs (single-seat) 3½hrs (two-seat)
Aerial camera	hand-held[1638]

The Blériot XI 'Parasol' was modified to improve the aircrew's view without the wing obstruction. (SHD, B87.3513)

BLÉRIOT EXPERIMENTAL (BE) 2A–2F

The most recognisable British aerial observation aircraft for the first three years of the war was the ubiquitous BE 2. It arrived with the first BEF expedition to France in August 1914 and remained in later reconnaissance variants until 1917. Stability and safety were the driving requirements behind the BE 2's initial pre-war design, and stability would be retained for the rest of its tenure. It suffered from a broad range of critical vulnerabilities including a low-powered engine and an inherent inability to outmanoeuvre any threat. As an aerial reconnaissance platform, the BE 2 performed as well as could be expected. The observer sat in the front seat surrounded by a 'cage' of struts, wires and the engine that severely limited what little defensive firepower the aircraft had.[1639] The BE 2 was the pioneer platform for development of the British 'A' Type camera and carried the majority of subsequent types for the remainder of its tenure. Despite an incredible service record of providing aerial observation for British forces, comments on the BE 2 aircraft by the pilots that flew it were not flattering. It was an understatement when a 1914 British report on the plane stated that the BE 2bs were 'not liked by all pilots'.[1640] In retrospect, as one aerial observer who flew BE 2s surmised: 'The BE 2c aeroplanes were designed for drudgery, and carried weapons for self-defence – like old ladies never without their parasols.'[1641]

The Blériot Type XI-2 Artillerie – one of an eventual twelve variants of this famous aeroplane –which entered service with the French and Italians in 1911 to provide reconnaissance for the artillery. Note the first use of identifying roundels on the wings. (Spellmount)

A common sight for most of the war. The BE 2c acquired most of the RFC's aerial photography until replaced by more survivable DH4, Bristol Fighter and RE 8. (Courtesy of Michael Mockford and the Medmenham Collection)

Crew	2
Average maximum airspeed	115kmph; 72mph (BE 2c) 141.6kmph; 88mph (BE 2d) 144.8kmph; 90mph (BE 2e)
Climb to 1,064m; 3,500ft (BE 2c)	6mins, 30secs
Climb to 1,524m; 5,000ft (BE 2d)	24mins
Climb to 1,829m; 6,000ft (BE 2e)	20mins, 30secs
Climb to 1,981m; 6,500ft (BE 2c)	20mins
Climb to 3,048m; 10,000ft	45mins, 15secs (BE 2c) 82mins, 50secs (BE 2d) 53mins (BE 2e)
Ceiling	3810m; 12,500 ft (BE 2c) 3657.6m; 12,000 ft (BE 2d) 2743.2m; 9,000 ft (BE 2e)
Endurance	3¼hrs(BE 2c) 5½hrs (BE 2d) 4hrs (BE 2e)
Aerial Camera	hand-held 'A' Type 'B' Type 'C' Type 'E' Type 'L' Type[1642]

MAURICE FARMAN (MF) 11

Another pre-war design aéroplane with extensive service for the first half of the war was the MF 11. The MF 11 was a pusher biplane with a raised nacelle serving as a fuselage. The wide-open front design was superb for aerial reconnaissance. However, the observer sat in the back and manned the machine gun in early models. Later, the seating was reversed to provide better observation.[1643] When the MF 11 was assigned to photograph with the 120cm camera, a steel tube frame was installed to hold the camera in place at the front of the nacelle.[1644] The MF 11 could climb very quickly and land over a short distance, making it a good tactical aerial observation platform in the first years of the war. However, airframe construction contained several design flaws such that even the slightest shock could damage the landing gear and the nacelle.[1645] The MF 11 was known by the British pilots as the 'Shorthorn'.[1646] The follow-on Farman design to the MF 11was the Farman F.40, the mainstay of French aerial reconnaissance in 1916.

The MF 11 was a superb design for Allied aerial reconnaissance in early phases of the war. The proud maintenance crew poses with the 26 cm aerial camera. (Courtesy of Walter Pieters; *The Belgian Air Service in the First World War*)

Loading the 120cm aerial camera into the cockpit of an MF 11 was a challenge. The MF 11 flew for escadrille MF 24 in 1917. (Zinn/Rockwell collection, courtesy of Alan Toelle)

Crew	2
Average maximum airspeed	118kmph; 73mph (MF 11) 116kmph; 72mph (Shorthorn)
Climb to 2,000m; 6,561ft	22mins
Climb to 914.4m; 3,000ft	15mins
Climb to 1,981m; 6,500ft	21mins
Ceiling	3,810m; 12,500ft
Endurance	3hrs (MF 11) 3¾hrs (Shorthorn)
Aerial camera	(FR) (MF 11) hand-held, 26cm, 50cm, 120cm (UK) (Shorthorn) hand-held, 'A' Type, 'B' Type, 'C' Type (BE) 26cm, 50cm, 120cm[1647]

CAUDRON G.4

If there were a design that could represent Allied aerial reconnaissance for the first half of the war, it would be the Caudron G.4. It was one of the best second-generation Allied reconnaissance aéroplanes, entering into operational service in August 1915. The Caudron G.4's predecessor, the Caudron G.3, had helped define roles for aerial reconnaissance. The follow-on Caudron G.4 was designed with a second engine that increased the range and loiter time (five hours), and it gave the aerial observer an unobstructed forward view. One pilot recalled, 'The G.4 was more satisfactory, even though it was slower. It was more reliable. You could manoeuvre it at will and loop it and it would never go into a spin. It was about as safe as an aircraft can be, as far as flying performance is concerned.'[1648] In 1916, Caudron G.4 aéroplanes gave significant support to the French armées at Verdun, accomplishing aerial reconnaissance and avion de réglage missions. Additional Caudron G.4 escadrilles were assigned to support French AL.[1649] The British also produced Caudron G.4s under licence, primarily supporting RNAS coastal patrol.[1650]

Above Right: The Caudron G.4 equipped with a long lens 120cm provided some of the most detailed aerial photographs in the war. (SHD, from Christienne and Lissarague, *A History of French Military Aviation*)

Right: The Caudron line of aerial reconnaissance aeroplanes was the mainstay of French aerial photography in the first half of the war. 'Lulu' was assigned to escadrille C74, a Franco/Belge unit operating in support of the French armée along the North Sea coast. (Courtesy of Walter Pieters; *The Belgian Air Service in the First World War*)

Crew	2
Average maximum airspeed	130kmph; 80.7mph (sea level) 125kmph (2,000m; 6561ft)
Climb to 1,981m; 6,500ft	17mins
Climb to 2,987m; 9,800ft	33mins
Ceiling	4,300m; 14,107.6ft
Endurance	5hrs
Aerial camera	(FR) hand-held, 26cm, 50cm, 120cm, 26cm (Grand Champ) (UK) RNAS: hand-held (BE) 26cm, 50cm, 120cm[1651]

FARMAN EXPERIMENTAL (FE) 2B

The British Royal Aircraft Factory FE 2b was initially used as a reconnaissance aircraft, evolving to other missions, to include night bombardment. The observer had an excellent view, sitting far forward in the nacelle directly in front of the pilot. The FE 2b was one of the first aeroplanes to carry an aerial camera inside the nacelle.[1652] The FE 2b did engage in aerial combat, being equal to a Fokker Eindecker in aerial manoeuvre and credited for ending the 'Fokker Scourge' over the Somme battlefield.[1653] The FE 2b reconnaissance aircraft earned the reputation that it could fly extended sorties over the front 'to bring back its information as quickly as possible without incurring casualties.'[1654]

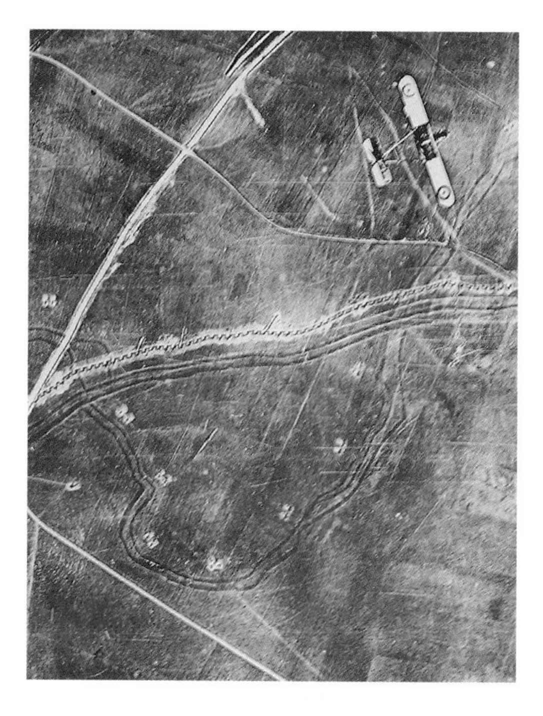

Aerial operations over the front as seen from an FE 2b. (*Illustrations*, S.S. 550a)

Crew	2
Average maximum airspeed	147.2kmph; 91.5mph (sea level) 130kmph; 81mph (1,981.9 m; 6,500 ft)
Climb to 914m; 3,000ft	7mins 24secs
Climb to 3,048m; 10,000ft	39mins 44secs
Ceiling	3,352.8m; 11,000ft
Endurance	3hrs
Aerial camera	'A' Type; 'C' Type; 'E' Type; 'L' Type; '53'[1655]

Above and right: One of the last pusher aerial reconnaissance aeroplanes. The FE 2b also served as a fighter and aerial bombardment aeroplane. (*Above*, Gorrell Report, NARA; *Right*, courtesy Les Rogers via Aaron Weaver)

SOPWITH 1½ STRUTTER

The Sopwith 1½ Strutter served British (RFC and RNAS), French and US squadrons from 1916 to 1918. It acquired its name from the arrangement of the upper wing centre section W-strut supports. It was first delivered to the RFC in early 1916.[1656] It had the distinction of being the first true tractor pull configuration two-seat fighter with an efficient synchronised forward-firing armament (Vickers gun) to see service in the war.[1657] The Sopwith 1½ Strutter's long range made it the inevitable choice for distant reconnaissance flights, participating in the campaigns of the Somme, Arras, Messines and Ypres.[1658] The 1½ Strutter's modest £842 per airframe price tag appealed to the Allies – this aircraft was significantly better and less expensive than the BE 2 series.[1659] The RNAS was the first to receive the Sopwith 1½ Strutter and initially assigned it to offensive patrols and fighter escort for French bombardment aéroplanes. The RNAS eventually employed the aeroplane to accomplish aerial reconnaissance of the coastal region. The French were very impressed with the Sopwith 1½ Strutter for its reconnaissance and bombardment capabilities. They subsequently acquired the licence from the British and built 4200 Sopwith 1½ Strutter airframes, calling their version the Sopwith 1 A2. When the Americans operated at the Front between May and July 1918, several units (88th, 90th, 99th Aero Squadrons) flew the Sopwith 1½ Strutter in reconnaissance sorties over quiet sectors.

Crew	2
Average maximum airspeed	167kmph; 104mph (sea level) 160kmph; 100mph (1,981m; 6,500ft)
Climb to 1,980m; 6,496ft	12mins, 40secs
Climb to 3,048m; 10,000ft	17mins 50secs
Ceiling	3,352.8m; 15,500ft
Endurance	4hrs
Aerial camera	(UK) 'A', 'C' Type, 'E' Type, 'P' Type (FR) 26cm, 50cm, 120cm (US) 26cm, 50cm, 120cm, 'L' Type, (BE) 26cm, 50cm, 120cm[1660]

Top: The British designed Sopwith 1½ Strutter accomplished a wide variety of functions to include aerial reconnaissance, aerial bombardment and RNAS submarine spotting missions. (Nesbit, *Eyes of the RAF*)

Centre: The French 50cm clearly fitted within the Sopwith 1 A2. (Gorrell Report, NARA)

Left: A post-war photo of Sopwith 1 A2 aircrew handing off the 26cm to the Photographic Officer for delivery to the photo laboratory via the motorcycle dispatch. (Gorrell Report, NARA)

RECONNAISSANCE EXPERIMENTAL (RE) 8

This RE 8 survived a strike by an artillery shell during a sortie over the front. (Courtesy of Colin Owers)

The RE 8, the most widely used British 2-seater on the Western Front, did most of the British artillery spotting, contact patrol and aerial photography for the last eighteen months of the war. One former BE 2e pilot commented on the RE 8, 'This machine has an adjustable tail plane for varying the speed and climbing powers. It can be moved in the air by means of a wheel and is a most fatal gadget if used at the wrong time ... I cannot say that I like the machine. It is too heavy and cumbersome, more like flying a steam roller than an aeroplane. Still, they do 105mph, which is quite an advantage over the old B.E. 2e.'[1661] Known as the 'Harry Tate' from the name of a swivel-moustached music-hall comedian (sounding like RE 8), this aeroplane was a stalwart of the RFC/RAF.[1662] The cameras were configured internally forward of the observer's cockpit. Designed to replace the BE 2, 4,099 RE 8s were produced during the war. Its worst moment was at Arras in April 1917, when on one day six RE 8s of No. 59 Squadron were shot down by von Richthofen's Jasta 11 of the elite Jadstaffel. Despite such setbacks, the RE 8 performed admirably in artillery spotting and aerial photography, supporting British offensives at Messines, Ypres, and Cambrai. A single RE 8 (A4397) accomplished an impressive 147 sorties over the Western Front, logging 440

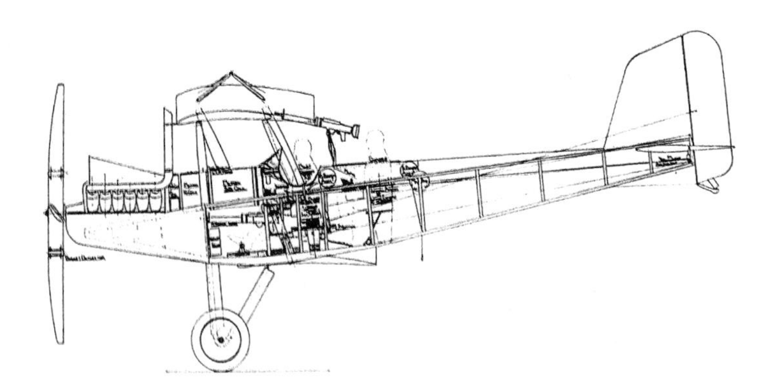

Early line drawing of an RE 8. Note the compact British aerial camera under the pilot's cockpit. (J.M. Bruce, *RE 8*)

The RE 8 withstood punishment from enemy pursuit aircraft and anti-aircraft weapons. (Courtesy of Colin Owers)

hours and 35 minutes of flying time.[1663] By the end of the war, fifteen RE 8 squadrons operated on the Western Front.

Crew	2
Average maximum airspeed	157kmph; 98mph (1,981m; 6,500ft) 148.8kmph; 92.5mph (3,048m; 10,000ft)
Climb to 1,981m; 6,500ft	21mins
Climb to 3,048m; 10,000ft	39mins 50secs
Ceiling	3,352.8m; 11,000ft
Endurance	4hrs 15mins
Aerial camera	'A' Type, 'E' Type, 'L' Type, 'LB' Type, 'P' Type[1664]

BREGUET 14 A2

The Breguet 14 A2 was one of the finest aéroplanes to see service in the war and post-war era. It was a remarkably rugged airframe, built of duralumin oxy-welded joints (claimed to be the first aircraft built in this way) that could absorb damage and allow the aircraft to carry enough firepower to inflict its own. The Breguet 14 was best known in its 14 B2 bombardment version. The companion Breguet 14 A2 assumed many reconnaissance roles, both tactical and strategic. Eventually, it replaced most of the French AL escadrilles supporting heavy artillery targeting. The Breguet 14 A2's ability to combine a rugged and defendable airframe with an impressive operating ceiling (7,600 metres) made it one of the most formidable aerial reconnaissance platforms of the war.

Crew	2
Average maximum airspeed	203kmph; 126mph (2,000m; 6561ft)
Ceiling	7,600m; 24,934ft
Climb to 2,000m; 6,552ft	6mins, 9secs
Climb to 3,000m; 9,842ft	10mins, 19secs
Climb to 5,000m; 16,404ft	22mins, 28secs
Ceiling	7,600m; 24,934ft
Endurance	3hrs
Aerial camera	(FR) 26cm, 50cm, 120cm, 26cm (Grand Champ) (US) 26cm, 50cm, 120cm, 'L' Type[1665]

AIRCO DE HAVILLAND (DH) 4

The Airco DH 4 was one of the most versatile aeroplanes designed and built during the war. It was a powerful aircraft for the time, both in power plant and armament. Aerial photographic reconnaissance was accomplished through internally mounted cameras. The DH 4's airframe interior was large enough to handle the standard line of both British and French plate cameras.[1666] This was the first time that a British-designed aeroplane could accommodate the state-of-the-art cameras internal to the fuselage, a critical improvement of British aerial reconnaissance design. The pilot was underneath the top wing and separated from the observer by an 85-gallon gas tank. Communication between the two crew members required the use of sign language.[1667] This multi-purpose aeroplane provided both British and American

Left: Breguet 14 A2 *Gabrielle* assigned to escadrille BR 202 flying missions in support of *artillerie lourde à grande puissance* (ALGP) (MA7180).

A DH 4 aerial observer receives the 26cm camera from the Photographic Officer. (Gorrell Report, NARA)

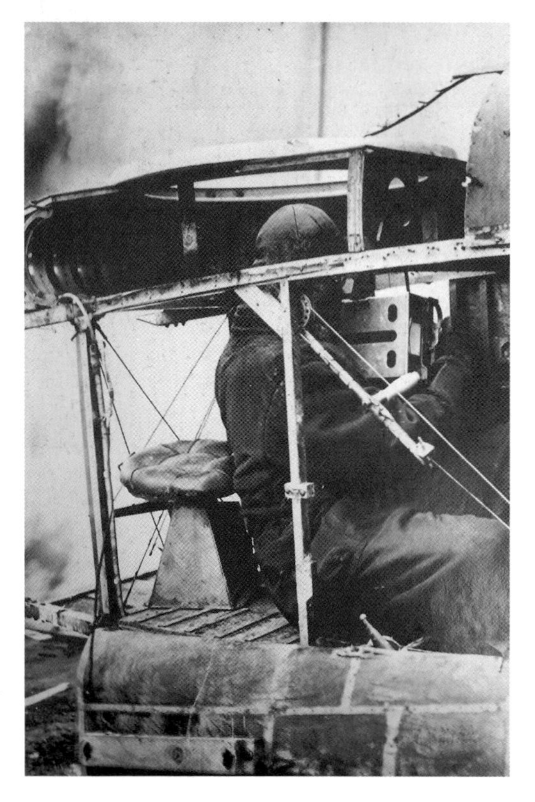

The versatile DH 4 carried a variety of British and French aerial cameras. The major shortfall in the design – the distance between air crew members – made communication difficult. This DH 4 was intended to accomplish long-range aerial reconnaissance of Northern Germany. (Courtesy of Colin Owers)

The cramped conditions of the observer's compartment complicated taking oblique photographs from within the fuselage. (Gorrell Report, NARA)

forces with a sturdy aerial platform capable of flying long-range sorties deep into German territory at high altitude.[1668] DH 4 capabilities prompted the RNAS to plan a long-range (14-hour sortie) reconnaissance against Northern Germany, but the sortie was never undertaken.[1669] American units flew the DH 4 powered by the American-built Liberty 12 engine (a subject of great pride for American aviation forces).

Crew	2
Average maximum airspeed	(Eagle VIII power plant) 202 kmph; 126mph (4,572m; 15,000 ft) (Liberty 12 power plant) 193 kmph; 120mph (4,572m; 15,000 ft)
Climb to 1,828m; 6,000ft (E VIII)	4mins 50secs
Climb to 3,048m; 10,000ft (E VIII)	9mins
Climb to 4,267m; 14,000ft (E VIII)	14mins 45secs
Climb to 5,486m; 18,000ft (E VIII)	23mins 30secs
Climb to 1,981m; 6,500ft (L 12)	8mins 30secs
Climb to 3,048m; 10,000ft (L 12)	14mins 48secs
Climb to 4,572m; 15,000ft (L 12)	33mins 42secs
Ceiling	6,705m; 22,000ft (Eagle VIII power plant) 6,096m; 20,000ft (Liberty 12 power plant)
Endurance	3¾hrs
Aerial camera	(GB) 'L' Type, 'LB' Type, 'BM' Type, 'P' Type (US) 'L' Type, 'LB' Type, 26cm Grand Champ, 50cm, 120cm[1670]

SALMSON 2A2

The Salmson 2A2 served extensively on the Western Front with the Aviation Militaire and the US Air Service. The camera pointed through a trap door in the floor of the rear cockpit. The Salmson 2A2 was the only single-engine aerial reconnaissance aéroplane to carry the semi-automatic De Ram camera in combat. The De Ram's heavy weight and bulk were too much for other single-engine platforms. The observer acquired additional oblique coverage by mounting a smaller camera (eg, 26cm Grand Champ) on the gun turret rail in place of twin Lewis guns.[1671] The Salmson 2A2 was a dynamic aéroplane with good performance and handling characteristics. When it became operational in late 1917, the Salmson 2A2 provided the Aviation Militaire with a reconnaissance aéroplane that could compete with German advances in aeroplane design. Its

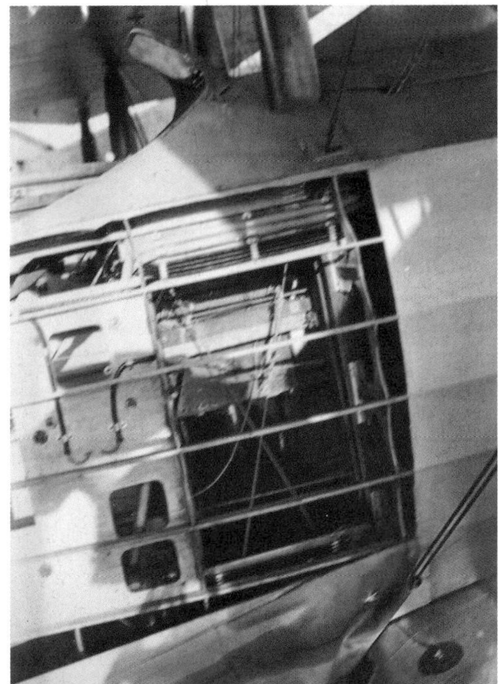

Above: The De Ram, largest aerial camera in the Allied inventory, was also carried by the Salmson 2 A2 in front of the observer cockpit. (Gorrell Report, NARA)

Left: An aerial view of the Salmson 2 A2 with camouflaged surface. (Courtesy of Colin Owers)

The Salmson 2 A2 was a superb aerial reconnaissance aeroplane, flown by both French Eescadrille and American Aero Squadrons. This aeroplane was assigned to the US Air Service's 1st Aero Squadron. (USAFA McDermott Library Special Collections (SMS57))

only design flaw was similar to the British counterpart, the DH 4. The distance between the pilot and observer caused communication problems, especially during combat. One US aviator, Lt. W.P. Erwin of the 1st Aero Squadron, destroyed eight enemy aircraft using only the front gun of the Salmson 2A2.[1672]

Crew	2
Average maximum airspeed	188kmph; 116.8mph (sea level) 181kmph; 112mph (3,000m; 9,842ft)
Climb to 2,000m; 6,561ft	7mins, 17secs
Climb to 4,000m; 13,123ft	17mins, 20secs
Climb to 5,000m;16,404ft	27mins, 30secs
Ceiling	6,250m; 20,505ft
Endurance	3hrs
Aerial camera	(FR) 26cm (Grand Champ), 50cm, 120cm (US) 26cm (Grand Champ), 50cm, 120cm, De Ram[1673]

The number of airframes committed to aerial observation throughout the war far exceeded any other purpose. As they evolved, not only did they fly at higher altitudes to compensate for the increasing threat of anti-aircraft guns and pursuit fighters, they were more effective, faster photographic platforms. With the evolution of the more sophisticated automatic cameras, aeronautical design had to compensate for the increased weight and provide manoeuvrability against potential threats. Such was the case with the Salmson 2 A2, certainly one of the finest aéroplanes to see combat in the war.

CHAPTER 24

Training the First Generation of Aerial Photographic Interpreters

INITIATION

> 'Cadet Goddard, you're a photographer.' 'Well, not really sir. I am an artist.' 'But you take pictures. You own a camera.' 'I suppose so, sir.' 'We're establishing an officers training school for aerial photographers. We need the right kind of people in it. It is a vitally new important field in fighting a war. The training course is three months, and you'll be a commissioned officer before most of your fellow cadets finish their pilot training.' 'Well that's fine, sir, but I joined the air service to be an aviator.' 'And I promise you, you will ... some day. But right now we need aerial photographers more than we need pilots.'[1674]

This rocky introduction to aerial photographic interpretation created a lasting memory for Brigadier General George W. Goddard, a leading American figure in the first half-century of photographic reconnaissance. The rush to acquire expertise to meet the demands of the war effort meant shaping a cadre of expertise in the photographic interpretation ranks from those possessing a recognisable background in photography. The call went out early on all sides for people who knew what a camera was, what was required to process the film, and what the picture represented. Very quickly, what was a hobby became a cornerstone of intelligence gathering, and the foundation was laid for imagery analysis. From hand-held cameras and 'happy snaps' came a new profession within intelligence, and many of the techniques and methods in use a century later were developed in the First World War by this first generation of aerial photographers and analysts.

A French post-war aeronautics assessment succinctly stated, 'We may say it was as the fighting went on that this science was evolved and developed.'[1675]. The new aerial photographic mission challenged the armed forces, particularly in the development of human resources. Personnel selection, training, and organisation of flying forces developed into an educational system on a scale much larger and more diverse than anyone anticipated. No nation had, up to the outbreak of the war, ever trained an aviator for aerial warfare. The knowledge gap covered every component of the aerial platform including radio communication, photography, aircraft maintenance, and other specialised skills. Each nation went through similar trial and error. All participants from the recruits to commanders had to come to terms with the aerial platform in their own way. What worked over time was extensive interaction between the traditional arms of artillery and infantry and the new force of aviators and photographic interpreters. The military institution needed a training regimen that ensured aerial photographic interpretation was equal to the challenges and that established a level of credibility with the combatants. The first years of the conflict put a burden on everyone developing the proper training regime. By the time the Americans entered the conflict, the value of aerial photography was clear and the training plans were well established.

THE EMPIRICAL BASELINE – LESSONS LEARNED FROM THE FRENCH AND BRITISH

Photographic interpretation came into its own after a year and a half of war. The leading photographic interpreters of France and Great Britain were called in to author the manuals that served as the basis for photographic interpretation in the twentieth century, validating the idea that the aerial photograph was a vital commodity. In late 1915, Capitaine Grout, the SPAé (Photo Section) pioneer, was reassigned to establish the complete course curriculum for training and certifying qualified personnel for SPAé.[1676] He established in March 1916 the Photographic school at Chalais-Meudon and subsequently moved it to Plessis-Belleville. The school focussed on the process of restitution and supporting Plan Directeur development. Besides interpretation, the students learned the art of developing mosaics from

Basics in aerial reconnaissance were provided to aviators such as this American aircrew on a Farman F.40. A hand-held camera is mounted forward. (Gorrell Report, NARA)

vertical and panoramic photographs and of analysing stereo photographs.[1677]

The British preceded the French by half a year with their School of Photography, Mapping and Reconnaissance (also called the RFC School of Photography) at Farnborough in September 1915. The British photographic pioneer, Flight-Sergeant Laws, took the initiative to stand up the operation.[1678] While working with Moore-Brabazon and Campbell, Laws had avoided a commission for financial reasons. As a Flight-Sergeant, Laws made more money than an incoming lieutenant, and he could avoid the bills that came with officer's mess membership. In November 1915, Laws accepted a commission in the RFC and became the school commandant. In September 1916, Laws was reassigned to the Training Brigade where he not only designed the 'L' Type aerial camera, but also acquired his aviator's wings.[1679] In December 1917, he took Moore-Brabazon's place as Senior Photographic Officer at GHQ and remained in that capacity for the remainder of the war. Moore-Brabazon left to manage photographic requirements at the War Office in London. Laws continued to serve in uniform for another 25 years until the close of the Second World War, retiring as a RAF Group Captain. He then employed his vast photographic experience to establish a company that photographed the finish line at greyhound and horse racing tracks.[1680]

Four months after opening the school, Farnborough produced five officers and 42 NCOs trained in aerial photography. Their training was equal, if not superior, to the overseas wings, who had access to recent aerial photography. A parallel effort with Wing photographic officers to train pilots and observers in aerial photography was not as successful. A contemporary report revealed 'it still leaves much to be desired.'[1681] By the autumn of 1916, training was not keeping up with the demand for photographic expertise. Aerial photographic training within the RFC had been entrusted to Moore-Brabazon. At the same time, photographic supply problems also were building, resulting in increased friction between the BEF and the War Office. The problems were solved by assignment of additional personnel to the photo staff. The RFC also increased awareness of aerial photography by establishing an Officer Training Brigade to train personnel at the Wings.[1682]

The RFC School of Photography was initially housed in two large disused plane packing-cases. Attendance at the school soon numbered in the hundreds and it developed into an up-to-date scientific institution, equipped with every modern device and latest appliance. In the model dark-rooms every candidate for acceptance as an RFC photographer had first to pass a test to determine his suitability for the work. He received a month's practical intensive training, with attention paid to the process of development and enlargement of negatives by artificial light. Much importance was attached to timely production of the enlargements, 'for the fate of a battle often depended upon the promptness with which large-scale copies of a vital subject could be supplied the Intelligence Staff.'[1683]

At the RFC School of Photography, the student received training in the principles of colour photography, lantern-slide making, the production of stereoscopic mosaic photographs, the rapid drying and finishing of prints, map reading and plotting, and other relevant subjects. Upon completion of the course work, the airman photographer was reassigned to a Service Squadron overseas and served a Photo Section working in support of an RFC Reconnaissance Flight.[1684]

The British complemented their aerial photography training with an Intelligence school at Harrow, providing a comprehensive understanding of the German military order of battle. Another school was located at the Horse Guards in London.[1685] For the novice in the intelligence arena, the school provided a good foundation on the organisation of the German military. By 1918, Americans started attending and found the course on intelligence good, but too focussed on learning the organisation of the British army. One recalled that subject was 'minutely taught for two weeks'.[1686] One student wrote to Lt Col Conger, AEF Assistant G-2, 'The course is good and planned entirely to fit the students for "fighting intelligence". Excellent lectures on all subjects and the aerial photography, which we have just started, promises much.'[1687]

Key to learning were the manuals, describing both the concept of positional warfare and fundamentals of aerial photography. The *Reference Manual of British Trench Warfare 1917–1918*, commented on battlefield preparation, including reconnaissance:

> Success in an attack on a line of trenches depends on the training of the troops and on the thoroughness of the preparations made. Units in occupation of a system of trenches must always consider their line and the enemy's defences from the point of view of attack. Constant observation and patrolling will be required to add to the information available. The ground between the opposing front lines must be accurately reconnoitred so that no unsuspected obstacles, such as sunken wire or ditches, may hold up an assault. Every effort must be made to locate machine-gun emplacements or strong points in the enemy's line, so as to be able to assist the artillery in the bombardment of the hostile defences. The whole of the enemy's system of defences over which the assault is to go must be made familiar to all ranks. A great deal of information is available from the excellent trench maps compiled from aeroplane photographs and a study of the photographs themselves.[1688]

Lieutenant McKenny Hughes studying aerial photographs prior to a mission. (IWM)

The manual also mentioned one of the best teaching tools: 'The use of a magic lantern to throw aeroplane photographs on a screen has been found of great value.'[1689]

Training on or familiarisation with aerial photography was also conducted at the front, even for senior members of the British Army. Lt. Thomas McKenny Hughes, a British army intelligence officer with aerial photographic experience, was assigned as an instructor in 'Interpretation of aeroplane photographs'. One class included the visit of a senior officer:

> After a bit of business getting the light to work, I got under way with my celebrated lantern exhibition and was getting along quite nicely when the door behind the screen opened and the Corps Commander and the BGGS [Brigadier General, General Staff] shuffled in, tripped over the electric light wire, put the light out and fused the arc light. The Corps Commander then fell over a chair and I felt it was time to pull up a blind, which I did. He then told me to carry on as if he wasn't there, which was difficult as we should have a light working if he hadn't been there. However, it was eventually got going again and I continued.[1690]

Hughes was an experienced aerial observer. He accomplished several aerial photographic missions in his time at the front. Finally, in February 1918, Hughes was shot down and killed while taking oblique photographs in the vicinity of Bailleul.[1691]

INTRODUCING AERIAL PHOTOGRAPHIC RECONNAISSANCE – THE AMERICAN WAY

The American experience provided a concise overview of how aerial photography entered into the military mainstream of the early twentieth century. They employed the best aerial photographic methods of the British and French, and proceeded to implement a program to employ photographic talent.[1692] Instruction covered three separate categories of students: observers to operate the cameras in the air, intelligence officers on the ground to interpret them, and technicians to aid in developing, printing, and enlarging photographs and keeping the equipment in serviceable condition.[1693]

The war engendered a sense of urgency throughout the US military. The greatest demand for specialisation was in the Air Service. At the start of the war there was no real experience in handling airplanes that were combat-ready. American instructors did not exist. The schools in the United States were technically unprepared to meet the demands. America looked to the Allies to assist them. The initial vanguard to develop the Air Service training program included then Major William Mitchell and Captain Joseph Carberry. Carberry commenced an orientation with French Flying Schools to gain an appreciation of what had to be established. The orientation quickly revealed that the United States had to establish training programs in France to acquire the true operational experience and become a more credible fighting force. Carberry was designated in July 1917 as Officer in Charge of Instruction under the general supervision of Colonel Mitchell, Aviation Officer of the AEF. In September, the Instruction Department was established with Carberry as Chief and Major Davenport Johnson as Assistant Chief. The initial staff included Private 1st Class Douglas Campbell, soon to be commissioned and destined to be the first American aviator 'ace'.[1694]

In December 1917, Brigadier General Foulois assumed direction of the US Air Service. Three schools were established to work aviation requirements. Colonel H. C. Smither was appointed Assistant Chief of Air Service. His assistant, Lt Col Marlborough Churchill, subsequently returned to the US to become the G-2 for the War Department; the first flag officer for intelligence in the US Army. In a reorganisation on 30 April 1918, the Training Section established an Intelligence Division to complement the existing training underway for pursuit, bombardment and observation. 1st Lt. L.A. Smith was appointed the first chief, and his duties included

> A collection of all information from activities in the Zone of Advance and of all technical information such as would be of interest in the training of Aviation Personnel. The preparation and edition of the Weekly Training Bulletin, a bulletin of information of value to the Training Schools and training students, and a proper distribution of such bulletins. A general supervision of Intelligence Organisations at the Aviation Training Schools, to ensure proper dissemination of all Air Service information published and distributed to the schools.[1695]

Along with the photographic discipline, the intelligence discipline had to be mastered. Nolan recommended Americans look to the example of the Allies and model the US intelligence function accordingly. Visits with the Allies provided a sounding board for the American novices. Training an efficient cadre took time. At the Canadian Corps Headquarters, Canadian intelligence officers recommended that Americans acquire older men for Intelligence work. The Canadians stated from experience that it took a man accustomed to responsibility and with a broad general knowledge of military affairs. The American intelligence function was primarily manned by young and inexperienced officers. Inexperience in warfare had been a serious handicap for the infantry, only slowly being hardened to the realities of the ongoing struggle. The Americans in general were already at a disadvantage coming into the struggle in the fourth year of the conflict. The combatants most likely paid little attention to the newly appointed, youthful intelligence staff.[1696]

AERIAL OBSERVATION

It was obvious to the AEF command at Chaumont that the French regarded air observation as essential, especially on a battlefield dominated by artillery. The AEF got their aerial observer manpower from the ranks of the ground forces of the AEF, with the focus on artillery officers well grounded in the requirements of artillery targeting. The observer, or the 'second man' in the plane, did not have to be a qualified pilot. In the summer of 1917 the AEF sought volunteers, but subsequently resorted to detailing artillery and infantry officers to aerial observation. Leaning on the French experience resulted in the AEF adopting French concepts. Pershing had French manuals for liaison between air and artillery translated into English and distributed to AEF Air Service trainers and observers. He also encouraged the War Department to accept and teach French tactical and operational concepts.[1697]

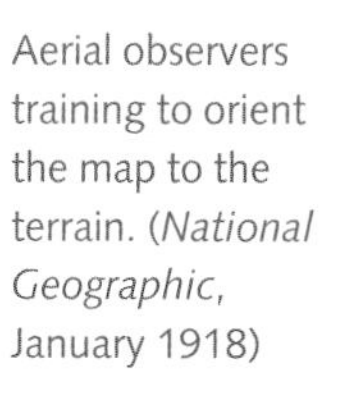

Aerial observers training to orient the map to the terrain. (*National Geographic*, January 1918)

US aerial observers receiving in-depth training using French aerial cameras such as the 26cm. (Gorrell Report, NARA)

Efforts to train an effective observation force met with challenges from the start. Aerial observers trained in the US arrived in France 'with most indefinite ideas regarding the principles involved in the performance of actual aerial missions over the lines. In some instances they had been taught obsolete methods.'[1698] Cooperation between the Allies was essential. French instructors gave photographic training to the pilots and observers at the AEF Training Centers. The shortage of instructors was constantly felt.[1699] Both the French and the British were magnanimous in providing whatever material was required.[1700] However, training objectives were out of sync. It was assumed that both aerial squadrons and the photographic laboratories would undergo parallel training and proceed to the Western Front at the same time. However, this was not the case, for training in photographic science took longer to master.[1701]

For many Americans aerial observers, the Air Service Observation School at Post Field, Fort Sill, Oklahoma was their introduction. The course of training generally under the heading of 'Theory' covered 19 hours of Artillery Contact, 6.5 hours of Visual Reconnaissance, 23.5 hours of Photography, 25 hours of Infantry and Cavalry Contact. Under the heading of 'Practice', training covered approximately six hours of Artillery Adjustment, five hours of Visual Reconnaissance, and one hour of Aerial Photography.[1702] Subsequent training in France was more realistic. One French instructor, Lieutenant Ledy, devoted much time to aerial photography, to include basics in interpretation. Lt. Gilchrist recalled: 'We learned that newly dug dirt photographs different from other dirt, that tracks reveal all sorts of things, such as gun positions although the guns themselves are hidden, that shadows tell the time of day.'[1703] The fully trained observer had to send and receive eight words a minute by telegraph code, to make 12 good aerial photographs on 18 assigned locations, to locate and direct artillery fire against enemy batteries, and to conduct a prearranged shoot without error.[1704] The most enduring lessons of course came with combat experience.

Despite recognition of the importance of observation among the aviation community, the quality of the observer force was always an issue with the field army units, the intelligence staffs, and amongst the aircrew themselves. This was especially true for the AEF Air Service in the final days. The rush to meet manning requirements resulted in lower training standards. One veteran observer, Lt. John Snyder, served in the 91st Aero Squadron for most of the final year of combat and was amazingly frank in his opinions, going beyond the standard lecture and describing in detail the concern that his colleagues were not 'on top of their game'. He thought the squadrons were being filled up with pilots and observers who had just completed 'a very meager training consisting of: Observations as seen from a class-room.'[1705]

As the war entered the final weeks, shifts in training were implemented. By November 1918, more attention to night reconnaissance was called on as the campaign became more mobile. Aerial observation was forced to develop methods to track night troop movements. Navigation was emphasised more in the curriculum. However, after the Armistice, the AEF discontinued training and transferred the functions to rear echelon staffs. The value of aerial photography on a mobile battlefield was becoming questionable.[1706]

A leader in US aerial reconnaissance was Lieutenant John Snyder. Veteran of the 91st Aero Squadron, he become an authority on aerial observation. He defined the aircrew role in aerial reconnaissance. (Gorrell Report, NARA)

ESTABLISHING AN AMERICAN PHOTOGRAPHIC TRAINING PROGRAM

The effort required to establish a credible aerial photographic interpretation capability was Herculean. Upon the entry of the US into the conflict, the Signal Corps still held sway over the role of aircraft. All units were burdened with organisation standup, supply, material, personnel, training (both aerial and interpretation) and production. Aerial observation was a 'small side issue'.[1707] Training of pilots and observers in photography at Air Service installations was usually entrusted to the French. However, establishing Photographic Sections meant homegrown resources.[1708] Fortunately for the Americans, they possessed a wealth of photographic talent throughout the nation, particularly from industry.

The first requirement was to assemble an initial cadre for aerial observation, aerial photographic interpretation and the photographic laboratory. Once the staffs of the Signal Corps recognised this, the call went out across the nation for those with photographic experience. All Army posts and recruiting centres were requested to have their commanders provide the Signal Corps' Photographic Division with a list of such enlisted men claiming to have photographic training of any sort. The recruitment phase was even supported through ads and articles in local papers. Manning shortfalls remained a constant challenge for the headquarters staff. Even as troops arrived in theatre, there was a constant review of any available talent to meet the demand. Enlisted personnel already in the AEF were screened. Talent was soon identified from amongst the volunteers and assigned to build the program.[1709] Orchestrating the

The first US school was established at Langley Field. Schools at Cornell and Rochester would shortly take over the operation. Training included operating a mobile photographic laboratory complete with camouflage coverage. (USAMHI)

training from headquarters was equally challenging. Establishing a Photographic Division to monitor ongoing developments required inputs from sources never previously considered. Ingenuity became the order of the day. The first priority was establishing laboratories at the schools where it was assumed the photographic training would take place. Though the preexisting training centres at Issoudun, Tours, and Amanty continued, many of the new instruction centres were not functioning until the end of February 1918, just in time for handling the arriving masses serving the AEF.[1710] The British and French had regular schools to instruct their officers. The British school was at St Omer, location of the British Headquarters. Americans attended these courses until their own service set up a course designed to support AEF missions.[1711] Members of US Photographic Section No. 5 remembered how the opportunity to join the photo ranks came about:

> 'The Air Service needs men with photographic experience' appearing in our daily papers caught the eye of not only thousands of professional photographers but of many others of all walks of life, professions and occupations, who had at one time or another done some kind of photographic work, at a time when they were seriously considering what branch of the service they were most fitted for to serve their country.[1712]

Accounts of the wonderful work accomplished in aerial photography, the prospect of early service in France, and the excitement of flying over the trenches caught the immediate attention of the recruits. This was further enhanced by the assurance of early promotions and commissions, which prompted hundreds to offer their services at once. The prevailing initial expectations of duty were a few weeks of training in the States and then across the water to photograph the manoeuvres of the Kaiser's army. Those selected went to one of three new schools for Aerial Photography: Cornell University or Rochester, both in New York, or Langley Field, Virginia, where they were simultaneously instructed in the principles of aerial photography and the art of becoming a soldier. The first American school of aerial photography was opened at Langley Field on 19 October 1917. The school had one camera, of foreign make, and was pleased if 50 per cent of the pictures taken at 6,000 feet turned out well.[1713] Developers and printers were trained at Rochester, and 680 had graduated by 30 June 1918. The officer's school at Cornell University taught map compilation and interpretation. In-theatre training was also offered at photographic 'huts' at the flying fields.[1714] After completing the work at those schools, some of the men were sent to various Aviation Fields for practice and official work while others continued to do 'Squads-Right' until they received overseas orders.[1715]

When the US declared war on Germany, George Eastman of Eastman-Kodak informed Secretary of War Baker, 'We offer use of any of our experts in an advisory capacity ... We shall be glad to provide school accommodations and instructors for training men for the photographic work of the Aviation Section here in Rochester ... We will also select some of our younger specialists and recommend them to you for service with the photographic division of the Aviation Section here and abroad.'[1716] The offer was appreciated but nothing was done for several months. The War Department

The Cornell University photographic school maintained an up-to-date aerial photographic map of the entire Western Front. (USAMHI)

initially declined Eastman's offer to establish the school. Eastman wrote Dr Richard C. Maclaurin, president of the Massachusetts Institute of Technology:

> We offered to establish a photographic school for aviators here and furnish the equipment and materials. The reason the Government gave for turning down our offer was that it could not send men away from camps for instruction. Of course we have facilities here that could not be obtained elsewhere, and we believe that we could teach the men much more effectively than could be done anywhere else, as we would have a staff of instructors such as they could not possibly get together in any other place ... We are especially devoting ourselves to aerial photographic work, and have already devised special emulsions and apparatus for the purpose, some of which are very promising.[1717]

On 10 January 1918, nearly seven months after Eastman first offered to assist the Signal Corps in the establishment of their Rochester School of Aerial Photography, the War Department asked Eastman to restate his proposal, which he did in detail.

> We offer, without rental, for a period not exceeding six months from the 1st of February, 1918, the use of the third story of our restaurant building No. 28, known as the Assembly Hall, for barracks ... The Government will have to feed its force of men but we offer the use of our kitchen, which has a capacity for feeding two thousand people at a time ... We are prepared to draw on our staff for any assistance you may need. We can loan you a certain number of specialists to assist in the instruction and to give lectures, etc., both from our scientific staff and from those sections of our employees who have to deal with the teaching of photography, a considerable number of our men being engaged on the instruction of photographers throughout the country. We would be glad to loan without charge the services of such of these men as you may require until you can replace them by men trained in the school.[1718]

The War Department made a concerted effort to channel more intelligence personnel through aerial photographic training. The demand for commercial or amateur photographic experience was modified to completion of a three- to six-month training program. Special instruction was necessary to explain the basics behind army photographic work, even when the work covered simple darkroom procedures or retouching.[1719] Major Barnes, one of the driving forces behind the US military's initial commitment to aerial photography in addition to Steichen, eventually commanded the school at Rochester in the summer of 1918. He agreed that the calibre of the officers and enlisted at the Rochester school was very impressive. The enlisted men were in the main, 'of a very high order of intelligence'.[1720] Barnes commanded about 3,000 men during his tenure. They possessed an esprit de corps that Barnes had never seen equalled. This was reflected in the post-war reports. The American Photographic Sections were judged excellent and their training received from Rochester was rated highly satisfactory.[1721] A few students who demonstrated a greater ability and intelligence for the discipline were sent to Cornell University for further training and commissioning as Photographic Intelligence Officers.[1722]

By June 1918, an overall training centre in theatre was approved to provide organisational orientation for incoming units and personnel. A baseline was critical to help all participants understand the mix of nationalities, echelons, new technologies and intelligence priorities that affected aerial photography. The school opened in August, and with each passing month more aerial photographic-experienced personnel arrived. The public were excited about the photography training organisation. The *Scientific Monthly* reported in April 1918:

> Plans have been completed for the great enlargement of facilities for training and equipping the aerial photographic force for photographing the German trenches from the skies and keeping up to the last-minute the large composite picture of the whole German front ... Aerial photography has greatly developed during the war. During the single month of September, British official reports stated that 15,837 aerial photographs were taken by the British alone. No new trench can be dug, no new communication system opened up, no new batteries placed but the ever-present and infallible camera above records it for the examination of the staff below. So piercing has been the work that camouflage has been developed as a protection, thus forcing aerial photography to even greater ingenuity. Every sector of the front is divided into plots about half a mile square, each one numbered and entrusted to a squad of photographers, who become fully familiar with it. As fast as the photographs are made, they are developed, printed, reduced, or enlarged to a standard scale, and then fitted into their place on the large composite photograph of the sector.[1723]

Allied cooperation was critical to developing US aerial photographic interpretation training both in Europe

Capitaine Carlier was one of the leading aerial photographic instructors at the Cornell University photographic school. (USAMHI)

and America. When the US entered the war, Great Britain and France sent many personnel to America to aid in organising the forces, particularly the air service and auxiliary units such as photographic interpretation. At first, it was an exercise in futility. Many handpicked Allied liaisons initially wandered around Washington with practically nothing to do. They approached the designated action officers such as Major Barnes and imparted wisdom on subjects that were still vague, but recognisable for their potential value. Self-study assisted the standup. Available literature and textbooks became the seed corn for the institution. Soon, heretofore untouched Intelligence Department files of British reports were scrutinised. The French Mission provided aerial photographs that served as the foundation for subsequent training.

Barnes requested that the War Department seek immediate British support. The British came through with a superb duo of experts. Three weeks after the Barnes request, the British aerial photographic pioneer Major C.D.M. Campbell, aided by Sergeant-Major Haslitt, arrived in Washington. Campbell's coming was a godsend, for his reputation was already legendary amongst people in the aerial reconnaissance field. The tour in the US was meant to prolong his already illustrious service, for Campbell was slowly dying of consumption (tuberculosis). His assignment was perceived as a means to lighten his load following three intense years at the British Air Ministry. He came armed with extensive supplies relevant to setting up the photographic training program: samples of all the British cameras in use, lantern slides for several sets of lectures, and the complete photographic equipment of an air squadron, from plates and paper to all the chemical materials required. Campbell also brought the Campbell Photographic Gun which, instead of shooting bullets, recorded on a film the precision of aerial gunners.[1724] His efforts were extensive and a great help to the newly established American program. He toured the training centres, gave lectures and showed movies and air pictures portraying the German lines and back areas. An old Standard training plane that included an attached English camera was used to demonstrate the photographic configuration. Training was accompanied by tea served to the attendees.[1725]

Aided by Campbell's reputation, Barnes and Steichen demonstrated to senior leadership the amazing potential of aerial photography. Campbell was a persuasive influence, using his entourage of equipment and the Standard training plane to deliver a photograph to a general on the ground barely fifteen minutes after it was made in the air. Sadly, the barnstorming tour did little for Campbell's failing heath. He was ordered home to England that spring and died on 3 September 1918. Major Campbell is still remembered as the godfather of American aerial photography training. His depth of knowledge was such that even brilliant members of the photographic community such as Steichen were willing apprentices.[1726]

THE SCHOOLHOUSE

Schools at Cornell University and the Eastman-Kodak plant were established within a few weeks. Cornell opened on 9 January 1918.[1727] Goddard remembered that the first class was of 40 students. At Schoellkopf Hill, classrooms accommodated sinks, tables, and partitions for the photographic laboratory, dark room and classroom facilities. This effort coincided with the arrival of the British and French instructors.[1728] As a result of its rapid establishment, the school soon consumed the chemicals and lenses required. The school had to resort to advertising in newspapers and mailing to all photographic studios to buy available lenses.[1729] Goddard recalled that the art of aerial photographic interpretation required a careful eye and quick mind. However, the students considered it a 'pretty tame business'. What was more exciting was the prospect of flying and taking the photos themselves.[1730]

The Cornell school maintained an up-to-date map of the entire battle area from the English Channel to the Swiss border. The coverage was located on a long, high wall in the classroom. The details covered not only the first, second, and third German trench systems and No-Man's Land, but the entire network of English, American and French trench systems. All of the front was covered and current photographs taken by French, British and American photographic units were forwarded to the schools every two weeks. Each day the students interpreted the various pictures with the assistance of the French and British instructors familiar with the particular areas along the battle lines. The students then revised the map with updates.[1731]

Both Campbell and Haslitt served at Cornell. Their depth of knowledge was critical to getting the American novices adjusted to the discipline of aerial photographic interpretation. Goddard remembered Haslitt as 'a tough little banty rooster, all spit and polish, who had served over the lines in France as an aerial photographer. He knew his business but he was highly critical of what he considered our deplorable lack of discipline and slovenly unmilitary dress.'[1732] He had some words of wisdom for the trainees: 'Once the bloomin' sun goes down and it gets dark, armies are blind as bats. Neither one knows what the other's up to till the blinkin' light's turned on again.'[1733]

At Cornell, French instructors presented an equally passionate view of the photographic art. Lieutenant Carlier was physically and mentally the opposite of Haslitt. Carlier gained an early reputation for brilliance assisting Capitaine Grout in establishing the French SPAé. After the war, Carlier produced *La Photographie Aérienne Pendant la Guerre*, the single most detailed assessment of First World War aerial photography and interpretation. His passion for the aerial photographic art rivalled that of Campbell and Haslitt. The only thing that was important was getting the photograph, even if it meant forsaking military discipline to achieve the objective. Haslitt and Carlier engaged in roaring arguments over their respective approaches, the result being a continuous contest for doctrinal dominance on the relative merits of British and French photographic methods and achievements over enemy territory.

Besides classroom instruction, the training included operations with aircraft. At Rochester's nearby Baker Field, aerial photographs were taken by one of the three photographic aircraft assigned to the school. The film was quickly downloaded, rushed by motorcycle to the mobile laboratory and delivered to the final judge (the Mayor of Rochester) within 30 minutes of landing. Barnes expressed wonder at the ability of the students to perform this task so well. 'To see laboratory workers dry a negative by pouring alcohol on it, ignite it without injuring its clearness in the slightest way, and then to see them do the same thing with the photographic prints was an exhibition of skill that was astonishing.'[1734] Rochester's curriculum included Copying Camera, Contact Explanation, Photographic Unit configuration, Maps, Mosaic Maps, Plates and development of Plates.[1735] Life was disciplined according to the standards of the day. However, it had an air of community. The school published its own newspaper, *The Airscout's Snapshot*, providing a glimpse into daily life. Major Barnes became the commander at Rochester, succeeding Captain Charles F. Betz as commandant of the United States Army School of Aerial Photography at Kodak Park on 31 July 1918. Barnes, the original member of the US Photo Service, served in this capacity for the rest of the war.[1736]

Major James Barnes would leave the AEF and assume command in July 1918 of the US Aerial Photography School at Rochester, New York. (USAMHI)

In retrospect, the Allied aerial photographic training regimen was an unqualified success. The US alone generated 25 Photographic Sections and their full complement of aerial photographic interpreters, photographic laboratory officers and staffs to provide an impressive depth to the ongoing requirement for intelligence and aerial photography. Had the war continued into the next year, American energy directed toward establishing a broad range of trained aerial photographic talent would have surpassed the well-established resources of France and Britain. Yet, as quickly as it became an integral part of the expanding Allied force structure, the aerial photographic talent would subsequently disappear as the forces drew down and returned to civilian life.

CHAPTER 25

The Human Experience of Aerial Reconnaissance

THE 'OFFICE'

'Reconnaissance work is not spectacular.'[1737] It involved a sortie that could last for up to three hours at a time, with interruptions at any time from atmospheric conditions or attack.[1738] Aerial observation missions were not glamorous. One British aviator compared aerial reconnaissance to a routine office job, taking the daily train to and from work and waiting while the board sat in debate over the profit and loss account.[1739] 'Photography again. I am getting thoroughly fed up with this job. It is the most difficult and dangerous job a pilot can get.'[1740] Commentary on the observer role was equally dolorous:

> I soon found that the main function of an observer, during the training period of an R.F.C. pusher aircraft squadron in this country, was to act as ballast. His other duties included the very simple one of starting the rotary monster, bearing the illogical name Gnome, imprisoned in its cage behind the pilot, on every occasion of a forced landing beyond the reach of the aerodrome staff, plus that of accepting full responsibility for any error in navigation on the part of the helmsman.[1741]

To be fair to the veterans of the discipline, not all reminiscence on the aerial observer role was negative. One British aerial observer reflected on the reward of accomplishing an aerial observation mission: 'There was something hugely satisfying, however, in both these pursuits ... To go over the lines and look vertically down on the enemy's most treasured and private property, and to know that you had it in your power to bring about its destruction or capture, that was in truth a job worth doing.'[1742]

Opinions varied on what made the ideal observer. One declared that the assignment demanded 'a steady, philosophical, and physically tough type. In vertical map-making he needed all these qualities, for despite shaking from Archie bursts he had to fly an undeviated course, exposing plates at finely calculated intervals so that they would overlap precisely to form a mosaic.'[1743] One post-war British assessment described the person behind the camera as 'a man in a state of stress and in the strained condition which ensues from any operation he may be called upon to perform in the course of a flight ... so that it may become to him almost an automatic action.'[1744] A successful observer had to develop the knack for both observing and remembering

The aerial observer, the unsung hero of the Great War. (SHD)

the aerial view, for he was liable for whatever was seen or not seen. Background experience in the military arts was necessary: 'The observer had to be a trained tactical officer, because in reconnaissance of this nature, an untrained person cannot interpret the military significance of what he sees.'[1745] Demands went beyond the physiological challenges of operating in an often cold, wet, icy, windy, open cockpit while someone is trying to kill you. The observer had to possess a mental discipline that demanded a precise concentration on what was observed: 'He must withstand the temptation to make conjectures, or to think that he has seen something when he is not absolutely certain of the fact, since an error in observing, or an inaccuracy in reporting may lead to false conclusions and cause infinite harm.'[1746] Succinctly put: 'The vital duty of the observer was to know exactly what his objectives were and how to get it on the plate.'[1747]

The initial criterion for becoming an aerial observer was rudimentary at best. One of the British RFC pioneers was William 'Sholto' Douglas, an observer in No. 2 Squadron, who described his initial introduction to aerial observation: 'The various observers in the squadron were canvassed with a view to finding out whether any of them knew anything about photography. I had had a camera as a boy, and had taken, developed and printed some very amateurish photographs. On the strength of this, I was appointed the squadron's official air photographer.'[1748] Douglas learned first-hand that the observer function required innovation to ensure success. Flying observer in the BE 2a, Douglas took the bold step of cutting an oblong rectangular hole in the bottom of his cockpit and pushed the camera through the hole to take a snapshot. 'It did not occur to us to fix the camera in the floor of the cockpit; but in any case this would have been impossible, as the camera was of the folding type with bellows.'[1749]

The observer had to know how to operate the camera and determine at what intervals aerial photographs were to be taken to ensure effective overlap. Once over the target, the observer 'was on his hands and knees in the "office" looking down through the little glass plate in the floor, releasing the shutter, changing the plates and counting out the seconds between exposures.'[1750] Douglas remembered: 'This procedure was not too easy in the cramped space available, especially as the weather was cold and bulky flying kit a necessity. Each plate had to be changed by hand, and I spoilt many plates by clumsy handling with frozen fingers.'[1751] Taking a successful photograph required tremendous patience and dexterity on the part of the observer. Pointing a camera through the floor was difficult because 'one could not see the ground at all,' except through the camera's aperture.[1752] Later designs allowed the observer to sight through a pinhole, releasing the shutter just as the target centred. What instruments that existed within the aeroplane were needed to fly the plane. One aero squadron ground crew recalled the cockpit display: 'A tachometer, oil pressure gauge, altimeter, air speed indicator, and a compass were about all they had. There was no gasoline gauge, and gas consumption was measured by time and the R.P.M. of the motor, which added up to approximately two and one half hours to a tankful.'[1753]

Later aeroplane designs included features to assist in the shoot. Two radial lines were painted on the outside of the fuselage in an area convenient for the observer's visual reference. The angle of the lines corresponded to the angle of the lens of the particular camera on board. This provided a reference for adjusting the camera at any altitude. Sighting took place along the first line toward a designated reference point. Once the aeroplane passed the reference point the observer either commenced counting or activated a stopwatch

Radial lines painted on the side of 1st Aero Squadron Salmson 2 A2 assisted the aerial observer in the shoot. The angle of the lines corresponded to the angle of the lens of the aerial camera. The time it took to fly between the marks served as an estimate for the coverage acquired by the photo plate. (Courtesy of Colin Owers)

Oblique photographs were improved when the aerial camera such as the 26cm was attached to the gun rail. This Salmson 2 A2 belonged to the 91st Aero Squadron. (Courtesy of Colin Owers)

to gauge when the second line passed. The flight time between the marks served as an estimate of the area covered by one photo plate. Dividing the time by two provided the proper time between exposures to allow for a 50 per cent overlap.[1754]

On the reconnaissance mission the observer not only maintained the camera but annotated a mapping board and notebook to retain information and the location of the print. These notes were invaluable both to the observer and to the aerial photographic interpreter.[1755] Observers had to learn the art of acquiring oblique photographs either through reconfiguring the French 26cm camera on the aerial gunner's machine-gun turret rail or through use of a hand-held camera such as the British 'P' Type camera. One American observer recalled the challenges in acquiring oblique photographs:

> We experienced a great deal of difficulty in our attempts to take oblique photographs. Our first attempt to suspend the camera was on the inside of the fuselage but this proved to be very unsatisfactory, in as much as the oil from the motor could not be properly guarded against causing the lens to become dirty and fogging the plates. With this method of suspension great difficulty was experienced in securing a sufficiently small angle with the horizontal. The photos did not give the proper perspective after several attempts at getting a suspension to fit the machine-gun turret [rail] so the photos were taken over the side where the observer could see what he was shooting. This later arrangement gave excellent results after a yoke had been made which absorbed all vibrations from the plane. However, besides being very hazardous to fly at very low altitudes over the enemy lines with this suspension instead of the machine guns it was also very difficult to manipulate due to the blast of the propeller coming full on to the camera, making it extremely difficult to swing the camera toward the side. In changing magazines difficulty was also encountered with the shutter of the cameras.[1756]

One aeroplane was readied for any short-range reconnaissance missions. Once the observer was notified of his assignment, he received instructions from the operations officer on the details of terrain, type of information desired, tactical situation, and communications data. Following that initial brief, the observer then conferred with his pilot.[1757] Prior to mission takeoff, the observer consulted with the intelligence staff on the target area. Those observers detailed to aerial photographic reconnaissance missions carefully studied the maps of the territory to be photographed, looked at previously taken photographs, and completed a thorough analysis of the ground to be covered, to include any important points associated with the mission. While the observer completed his final preparations, the squadron photographic officer ensured the mounting of the camera and accessories in the aircraft. The photographic officer also made sure that the observers using cameras knew how many photographs they were to take and how many plates they required in the magazines. Observers were told to expose all the plates: 'this is a rule, the importance of which cannot be over-emphasised'.[1758]

The aerial observer in the first year of conflict developed techniques for tracking the path of the aeroplane over the designated target area. The observer would direct the pilot to fly toward a prominent landmark. Getting to the exact point sometimes required the observer to lean over the side of the fuselage, signalling to the pilot from time to time with his hand so that the course would be exactly as required. The early technique sought flying from the landmark toward one of the cardinal points of the compass. Once over the landmark the observer commenced shooting the camera and marking the spot on tracing paper covering the map contained within the cockpit. The steps would aid the subsequent interpretation and restitution.

Upon return, the observer taking the aerial photographs assisted the intelligence staff in identifying the plates for interpretation. The mission was not complete until the interpretation report was completed. The veteran American observer, Lt. John Snyder, reaffirmed that aerial observation reporting, with or without benefit of the camera, required

> ... the observer make out his report immediately upon his return from a mission regardless of whether it is of a positive or negative nature, for it is just as important to get negative reports as it is the positive ones, but when an observer does make negative reports he must be sure that the report is accurate ... The observer must at all times report everything that he sees, regardless of how unimportant it may seem to him. It may verify some other observer's report or some report that the staff received from agents.[1759]

Lieutenant Snyder is handed the approved 50cm camera from the squadron's Photographic Officer for installation prior to takeoff. (Gorrell Report, NARA)

In the case of bombing missions, observers were strongly encouraged to expose all plates taken over the target to acquire as much information as possible from the sortie.[1760] Each observer had his routine prior to takeoff. Lt. Snyder shared his techniques with new students at the front:

> After getting into the plane and before starting on the mission, I make a very careful examination of the camera and see that it functions properly. I always remove the magazine from the camera, wind up the curtain then releasing it to see that it works properly, and then put the magazine back and work the drawer at least once to see that it does not bind. As for the setting of the lens and the curtain aperture, that must be left to the photographers who are specialists in that line ... When taking off I always held the camera with my hands, keeping it free from the wires, and keep the extra magazines between my feet so as to prevent them from falling over or bouncing, which would result in the breaking of plates.[1761]

The amount of time on a mission was limited and required the observer's total concentration. Snyder remembered it took about one hour to gain the altitude necessary to work Army observation missions. Once at cruising altitude, the aircrew adapted to their 'office' environment:

> After attaining the desired altitude and before crossing the lines I again try the shutter so that in case it would have been affected by either the cold or the vibration, we would not have to risk our lives unnecessarily, but would turn back and get another camera.[1762]

There were differences of opinion among the Allies as to the appropriate role for the pilot when it came to aerial observation. One British observer remarked: 'In French Corps Squadrons, the pilot is generally a mere chauffeur.'[1763] The pilot was actually an integral part of the aerial observation mission. The pilot had to know the elements, to include direction of the wind, so that he could make proper corrections for drift, altitude and speed of the aeroplane. He also had to keep to a carefully prescribed straight and level course, a process that required tremendous discipline, precision and steady nerves.[1764] To achieve mission success required the pilot be a top performer, possessing great skill and a refined sense of orientation.[1765]

The French assigned some of their best personnel to aerial reconnaissance. Each French escadrille included a photographic section composed of specially trained men.[1766] Many recollections spoke of the discipline required to be a successful reconnaissance aviator.

Upon return, Lieutenant Snyder hands the exposed plate magazine to the motorcycle dispatch. (Gorrell Report, NARA)

> Flying a photo machine is the most difficult and 'brainy' flying of all. The conditions are such that it demands the most of a Pilot, while it is of such importance that the enemy spares no expense or trouble to bring him down. His machine is always the objective of all enemy chase machines who can reach him, and a target for the anti-aircraft batteries, for he has to fly a straight and even course, always. He cannot vary his line of flight to avoid the inferno of anti-aircraft shells that are belching forth from dozens of guns. The only time a Pilot on photo mission will leave his course is when the enemy machines attack him, and then he fights on the defensive. If he can drive off the attacking enemy planes, he sticks to his aerial post and continues his mission. Nothing matters to him except those pictures, which must be taken and delivered to the home station behind the lines.[1767]

The view of the infantry on the ground revealed the danger and fascination of aeronautics. One British officer recalled after observing aerial combat: 'A daring Boche chaser pursues one of our observer planes. Clever diving alone saves our man who gets away within 100 feet of the ground.'[1768] A standard aerial photographic sortie involved the observer sighting the ground track and assisting the pilot in flying to a recognised point. The pilot flew as straight and level a course as possible to achieve a true vertical shot, necessary for the photographic interpreter's analysis or the construction of the mosaic. Maintaining straight and level flight was generally determined by the pilot's own sense of balance or from applications of a mechanical nature, such as improvised instruments showing water levels or plumb lines. Acquiring straight and level coverage was the exception rather than the rule owing to hostile aeroplanes, anti-aircraft, and weather. By war's end the RE 8's observer role was primarily defensive, as an air gunner looking for enemy aeroplanes. The RE 8 pilot took the photographs by pulling the appropriate Bowden cable to expose the plate. The pilot then pulled a second cable to transfer the exposed plate to the plate box. With this action a new plate was placed in position for the next exposure. The observer's aerial reconnaissance function was relegated to changing boxes of plates. The RFC RE 8 pilot, Lieut. Smart, reflected on the reconnaissance mission after one sortie, 'Photography is a good job when you don't get hurt.'[1769]

Aerial observation demanded teamwork. Ideally, pilot and observer worked to achieve synchronised reactions, judgments and decisions. The pilot was to follow the observer's directions promptly. The pilot was said to be 'married' to his observer.[1770] Despite the close relationship, pilots did not consider themselves 'observation pilots'. Amongst American aero squadrons, the crews preferred that the term 'observation' apply to

The exposed photo plates are delivered to the Photographic Section for development and interpretation. (Gorrell Report, NARA)

A poignant photograph revealing the 'do or die' attitude of an aircrew in front of their RE 8. Both crewman and ground crew look exhausted from mission stress. (Nesbit, *Eyes of the RAF*)

the observer alone.[1771] The public portrait of both pilot and observer mirrored the general awe associated with aviation:

> You will be astonished to learn that the average age of R.F.C. pilots doing reconnaissance work is twenty and of observers twenty-two. It requires young blood and muscle to stand the strain, risk and excitement of this branch of the air service. That results so far have more than justified expectations, is a tribute to the skill and bravery of these youngsters.[1772]

ESPRIT DE CORPS

Without doubt the pursuit pilot's exploits were incredible feats of daring, facing the adversary alone in a hostile environment with no safety net other than airmanship skills. The pursuit pilot was in the limelight throughout the war and long afterwards. However, most aviation during the conflict served the ground combatant through aerial observation. It was perhaps the most perilous of all aerial missions because the observer had to accomplish a number of important functions and defend himself at the same time. It was near impossible to do both when tucked inside the cockpit taking photos. The American observer Lt. Percival Hart spoke for his counterparts on coming to terms with the dangers associated with the aerial reconnaissance mission: 'The definite nature of these missions left nothing to the observer's imagination – he had to bring back photographs of exact positions regardless of whether or not he thought it was entirely safe to do so. As time went on this work was entrusted mainly to the most experienced and daring teams.'[1773] Such flying established its own sense of honour. Most recollections suggest the aerial reconnaissance crew were of the 'do or die' kind. The flight was there to acquire 'snaps,' even if it resulted in sacrifice.

The existence of the aerial reconnaissance air crew was tenuous at best. A French Morane-Saulnier T from escadrille C 17 serves as an altar for the mass. (SHD, B80.36)

Similarly, here an RFC FE 2b cockpit serves as pulpit. (IWM)

For the British aviator, such an attitude was part and parcel of whole the military culture. Colonel Squier, the American attaché, observed on 20 April 1915:

> No name is ever made known in connection with any feat however important. I was told that on the day of our arrival Sir John French had asked the commander of the Flying Corps to tell him who had performed a particularly clever bit of work, and he was reminded of the rule of the organisation and withdrew the question. They are so modest that I could not get any of them to admit that they have command of the air, although one hears it everywhere else both in the army and in England.[1774]

The priority in aerial reconnaissance, according to one American flying with the French prior to the US coming into the war, was 'to get back to headquarters with the information you have gained. If you are attacked and see no chance of successfully fighting off the enemy, it is your business to run.'[1775] Confronting the German aviation threat was a subject of debate. One French instructor at the Second Artillery Observation School at Camp de Souge clarified the official line:

> You are young, you are eager; you all want to be aces, shooting down the Boche. But you are not fighters, you are observers. You fight when you must, but your job is to bring back information. That is what we are training you for. We French have had terrible losses. We conserve our men. If possible avoid conflict. Your planes will be slower, less manoeuvrable. Do not hesitate to run.[1776]

Many Americans had their own view on how to take up the fight, and they were critical of this approach. Snyder commented that the 'majority of French aviators would look for enemy aircraft nine-tenths of the time and look at the ground the remainder of the time.'[1777]

COLLECTING INFORMATION

The entire mission revolved around details, be they in the memory of the aircrew or in the photographs. British aerial observers established a reporting format for observations noted on the sortie. Colonel Squier reported in February 1915 that the form contained instructions for observers, and on the back they were to fill out the report in the aeroplane and produce duplicate copies in the field.[1778] This was later changed to an Objective Card describing each target, particularly enemy artillery batteries.[1779] One US Air Service squadron, the 91st, was rigorous in its approach to understanding significant targets. The crews had to be familiar with the key targets of the sector, to include enemy-held towns, supply dumps, and aerodromes. The squadron drilled the members through a close study of aerial photographs posted on the cards, with the names of the various places shown under each. The displayed photographs were studied for several days, after which the captions were removed and the towns, dumps, and airfields memorised. The training resulted in higher-quality reconnaissance. Both pilot and observer knew what to look for. Their routes were covered more accurately and the crew returned with valuable reports.[1780]

Once aloft, crew members described the terrain. One such description portrayed by Lieut. Cecil Lewis in his memoir, *Sagittarius Rising,* contains a hint of ennui. After an initial introduction flying ten hours over the front, Lewis was handed his first real job, to photograph the enemy second-line trenches.

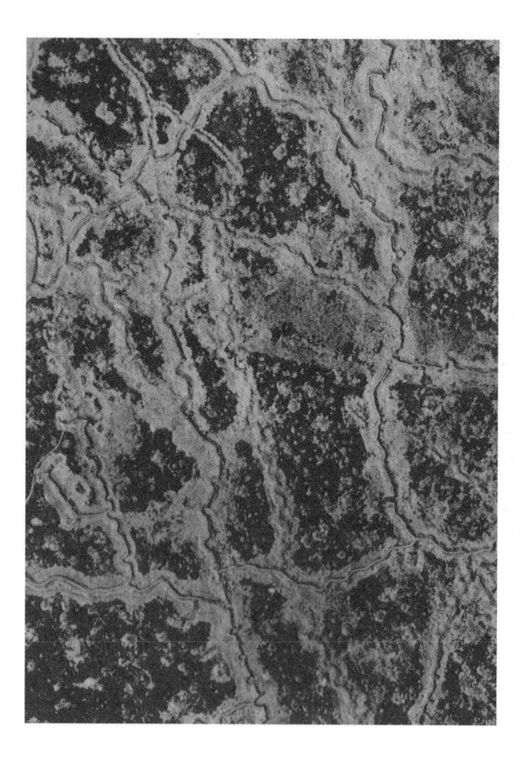

Flying aerial sorties on a regular basis over the same area meant that the aerial observation crew came to know the sector like the palm of their hand. (*Étude et Exploitation des Photographies Aériennes*)

> The lines, from the air, had none of the significance they had from the ground, mainly because all the contours were non-existent. The local undulations, valleys, ravines, ditches, hillsides, which gave advantage to one side or the other, were flattened out. All you saw was two more or less parallel sets of trenches, clearer in some places than in others according to the colour of the earth thrown up in making them. These faced each other across the barren strip of No-Man's Land, and behind them started a complicated network of communication trenches, second-line trenches, and more communication trenches, and then the third-line trenches. The network was more complex at the important positions along the line; but everywhere it was irregular, following the lie of the ground, opening up to a wide mesh at one place, closing up, compact and formidable, at another. As positions were consolidated more trenches were dug, and later, when I came to know my section of the line as well as the palm of my hand, I could tell at a glance what fresh digging had been done since my last patrol.[1781]

An important observation was described by A.J. Insall, a British aerial observer, in his post-war memoir:

> One afternoon, when I was looking down on to a part of the German line below Gommecourt, after spending some twenty minutes photographing a stretch of reserve trench asked for by Third Army, I suddenly became aware of a line of closely grouped whitish dots in a part of the same trench-line a little to the south of the section we had just passed over. The dots were so minute that they were only just distinguishable from the 9,000 feet at which we were flying. I thought for a moment that they might be pit-props, but dismissed the idea as soon as I had made a more careful examination through my field glasses and had noted the typically sound condition of the trench and the bunching of the dots. There was, to my mind, little doubt that these were troops, their helmets catching the light. So I got my pilot to describe a slow S turn, and reloaded my camera. It was the last plate in the box, and I took my time over it. Later that evening our Wing Photographic Officer rang through to congratulate me on this one negative, which, according to him, made the others I had taken that afternoon look rather indifferent. As he was an experienced Fleet Street man and not usually inclined to issue compliments, I was thrilled. The prints he had made from it apparently showed a concentration of troops quite clearly, and a day or two later a note appeared in the official news bulletin that read: 'On the southern part of our line, aerial reconnaissance reported a local concentration of the enemy to the north of Beumont-Hamel, which our artillery dispersed.'[1782]

Even Lewis described a sense of accomplishment upon returning from one flight. He remembers a conversation with his sergeant-gunner after a hair-raising sortie:

> The sergeant came up. 'Are you all right, sir?' 'Fine! And you?' 'Quite, thank you, sir.' 'I thought he'd got us with that second burst.' 'Always turn, sir, as soon as a machine attacks. It can't get its sights on you so easy. And it has to allow for the traverse ... If you'll phone the squadron, sir, and order out a tender and a repair squad, I'll dismount the camera and get a guard put over the machine. You got all the photos, didn't you, sir?' 'Yes, Twenty-two in all.' 'The Corps will be pleased. They wanted them badly.'[1783]

THE PHOTOGRAPHIC LABORATORY

The photographic laboratory was an extension of the aerial observation aeroplane. If the laboratory did not

American photographic section maintenance personnel work on both the French-designed De Ram semi-automatic camera and the British-designed 'L' Type camera. (Gorrell Report, NARA)

have the right chemical mix to produce the necessary prints for a diverse array of customers, the aerial photographic reconnaissance mission was a failure. The Photographic Officer had an impact on the aircrew in many ways. He was ultimately responsible for the camera equipment and supervised installation and extraction of cameras and plates from the aeroplane. Only the Photographic Officer could designate who could touch the camera. The observer was required to operate the camera only over the target area.[1784] The observers were only human. As one veteran commented: 'The worst mistake was to be unaware that the camera was not working.'[1785] The Photographic Officer also governed the laboratory manned by 30 lab specialists – stock controllers, draughtsman, maintenance men, print specialists and motorcycle riders.[1786] Such work was not merely ancillary to the mission. Lieutenant Smart of the 135th Aero Squadron recalled a harrowing mission photographing Nonsard to Thiaucourt. On the return his flight of DH 4s fought off four Fokker DVIIs and two Pfalz D.IIIs. They were lucky and managed to escape and land safely. They then discovered that the Photo Officer, responsible for installing the camera, had neglected to take the protection cover off the lens. 'He had to buy drinks for the squadron that night and from then on, that cap was checked twice before takeoff.'[1787]

Records of the role the photographic laboratory played in the aerial reconnaissance process are scant. One illuminating source came from the diary of Sergeant Eric E. Dixon in his privately published memoir of his photographic laboratory experiences in the US Air Service. His sense of pride in the photographic laboratory's ability to contribute to the fight comes through. Sgt. Dixon described the routine supporting the 1st Aero Squadron: 'Camera and loaded magazines were always ready for instant action. When a mission was scheduled, Perkins took the cameras and magazines to the planes in a truck and installed them as needed.'[1788] The cameras were loaded and downloaded by the photographic lab crew. There was a certain cachet attached to the work:

> The field men were much envied, as they got at first hand all the thrilling reports brought back by the Observers and Pilots, and to see the planes come in bullet riddled and tattered. After a time the cameras were kept in a tent on the field, which also served as the camp barbershop. However, the magazines had to be carried back and forth the mile to the field, and one might occasionally see the field men trudging along with five magazines hung on a pole, suspended between them and carrying one in each hand, thereby making one trip.[1789]

A negative lab mixed the chemicals and processed the negatives. Lab personnel loaded the plates into the camera magazines. Other lab functions included printing, developing, and print drying. Prints were developed on a printing machine. Developers sometimes had 20 prints in the tray at the same time. Immediately after the aeroplane landed, the camera magazines were quickly brought to the negative lab for processing.[1790] The exposed plates were put in a bath of chemical developer. Once the plates had been coated with the chemicals, they were dipped in a 'short stop dip' to neutralise the chemicals before being placed in a fixing bath. The plate negatives were placed on a light box for viewing. Restitution was applied to define the area covered by the photograph. Once that was accomplished the area was outlined on the map. The plate negative was assigned an identification number with pertinent data (number, date, altitude) along the lower edge. A complete set of prints was then printed, developed, fixed, rinsed, and the excess water was 'squeegeed' off. Getting the prints ready for intelligence analysis was the next step. A five-man crew worked one printing machine. The printing, developing, and fixing was accomplished under the haze of an orange light.[1791]

'The photo unit had one redeeming feature; that was we were excused from all drills, guard duty, and K.P. On days we were working we could also go to the head of the chow line, much to the envy of all others.'[1792] Such perks were earned through hard work:

> The print lab was a very busy place. Everyone, including myself, worked like galley slaves and kept at it until the last print was delivered to Intelligence. Sometimes our work would cover only three or four hours of darkroom work, but it usually was much more. Once we worked 36 hours straight, with hardly time to eat. Food was brought in and we grabbed a bite when we could and really turned out the work. We usually ran one or two print crews, but when our personnel was increased, we used three. We used a bromide paper, which was very fast and exposure went from a flash to several seconds. At times a printer could turn out 250 prints an hour ... Despite the long and hard work, there were fond memories of some of the laboratory crew. The use of alcohol to dry the prints had its moments. Not all the alcohol burned, and the men working the rack sometimes got a mild jag on from inhaling the fumes. One man we called 'Gee' who ordinarily was a bit on the sour side would loosen up, get real happy, sing songs, and even dance a jig.[1793]

Based on the operations group records, weather controlled the schedule. 'The ground photographer chooses his weather; the General Staff chooses the aerial photographer's weather. The weather conditions during both the Argonne and St Mihiel operations were exceptionally poor. The most favorable days were crammed with photographic activity, and even days with barely enough light to produce the faintest trace of an image on the photographic plate had to be "photo days".'[1794]

Getting the product out remained a challenge throughout the war. Post-war American staff commentary reflected the ongoing distribution of prints and intelligence needed improvement: 'The current distribution of photos has interfered greatly with the regular running of the office.'[1795] Despite the emphasis on timeliness, aerial photographs from Army-assigned aero squadrons often did not arrive at the Corps until the next day. One post-war assessment groused about staff wasting time awaiting the arrival of the photographs. Once received, it was a 'laborious task' sorting and distributing the materials to other offices and line units. Shift work was not a solution. As a result, staffs worked through the night, caught up on what little sleep they could, and then waited for the next round of aerial photographs and interpretation reports. Solutions focussed on establishing large photographic processing and distribution centers such as the British AP and SS. Had the war continued, the discussions would have become reality.[1796]

CHAPTER 26

The Intrigue of Aerial Photographic Interpreters

In his post-war memoir describing the challenges of the aerial photographic interpreter, James Barnes captured the essence of the work quite brilliantly.

> The reading and interpretation of airplane photographs demands a peculiar mind – the type of mind that would work out chess problems or, nowadays, cross-word puzzles, perhaps. To the uninitiated a photograph of a line of entrenchments and myriad shell holes might mean very little, but to the puzzle solver, working over a clear photograph, with a magnifying glass, those shadows and lines and suggested slopes and rises mean much. They tell a story. Often his imagination is set on fire by some puzzling little, thing, the reason for which he cannot quite discover; and then, all at once, he has it! Those funny, little dots are iron fence posts with strong wires strung along them. The men who went to that big shell hole left no trace of a path, for they reached their hidden machine-gun emplacement by walking along the lower wire as a sailor would use the foot rope on the yard of a sailing vessel. That well-traversed path some hundred yards away, leading to another rounded pit in lower ground is deliberate deception; there are no guns there. The battle between the camera and camouflage was on. It was like a poker game with aces up the sleeve.[1797]

Despite the acknowledged importance of the function, there is little information on individual interpreters, other than in a few romanticised newspaper articles. The lasting legacy is of aviators engaged in combat. The roll call of those interpreting the photographs and performing analysis, so eloquently portrayed by Barnes, was lost as the years passed.

PUBLIC AWARENESS OF AERIAL PHOTOGRAPHIC INTERPRETATION

A captivating sidelight of the aerial reconnaissance and photographic interpretation experience was the public's perception, shaped by popular journals, of a process steeped in the intrigue of intelligence and its mysteries. An air of drama and suspense was portrayed in the 24 November 1917 *Scientific American* article, 'The Camera at the Front'.

> That peaceful-looking camera which we use on Sundays for snapping pictures of our friends and of pretty views, becomes a deadly instrument when it is brought into the military world. It may be the means whereby invaluable fortifications can be quickly destroyed; it may give away important ideas or inventions to the enemy; it may betray the location of important batteries and doom them to early destruction; and it may result in heavy loss of life to troops whose position is divulged by the black-and-white markings of a photographic plate or film. At any rate, the camera in the war zone is no longer a peaceful instrument; it is many times deadlier than its equivalent weight of high explosive.

Discussion of aerial photographic interpretation lent an air of suspense:

> These negatives, or contact prints made from them, by some route or another make their way to spy headquarters, and there they are enlarged and studied by specialists trained in 'reading' photographs. These specialists we are told, can glean a tremendous amount of information from what appeared to the layman to be a most uninteresting and valueless print. And when such informations are properly collated and distributed to the various military departments where they can do the most good – and the most harm to the opponent – the deadliness of the camera begins to dawn on us.[1798]

In July 1918, the *Scientific American* ran a cover article titled, 'The Photographic Eye and Memory of the Military Airman'. Again, the drama was played up:

> It is difficult to conceive of a photographic plate that holds the fate of hundreds of thousands of men, batteries of guns, and possibly the decision of a great battle, in its thin coating of emulsion. Yet there are numerous such plates made every day by intrepid airmen, which, after being developed in less than fifteen minutes after reaching the ground and then flashed on the stereopticon screen in the presence of skilled photograph readers and officers, sign the death warrant of numerous soldiers...[1799]

Such pieces ran for several years. Adding to the intrigue was a series of articles published in *Popular Mechanics* by Douglass Reid, entitled 'The Eye Behind the Lines'.[1800] He wrote of 'The stereoscope, which has even been called by those well qualified to judge, "the deadliest weapon of the war", and the application of the stereoscopic principle to photographs taken with ordinary cameras.'[1801] In 1919 Reid described the process:

> In a picture of a village 15 miles behind the lines, the photographic officer notes buildings that instantly betray themselves as barracks. Definite trails, that only soldiers living in them could make, lead to them. Near by is a villa, further on a town. He sees that the place is a railroad center and a division point. The villa he scans closely. Hard by it is a faint irregular circle on the grass. This was made, he reasons, by horses being led about by orderlies so that they may cool off gradually while their riders are inside. The riders have come for information. The villa is a headquarters. Aside from machine-gun officers, majors are the lowest of foot soldiers that are mounted. Hence the man to whom the majors come for orders must be a colonel or better.
>
> Now behind the villa, the detective spies a larger and more regular circle on the grass. This, he deducts, is made by orderlies exercising horses daily, according to routine. The officer in the villa has a private stable of horses for himself and staff. Now he knows an all-important fact: the German is either a brigadier or major general.
>
> Next he discovers practice trenches and machine-gun pits on the outskirts of the town: an infantry organisation. Then he finds still another circle made by horses, a circle just large enough for one troop of cavalry to use in practising. This tells him that the officer in the villa is a major general, since no brigadier in the German army boasts a troop of cavalry.
>
> He scans this last circle more closely. Its shape informs him that the troop is composed of lancers rather than saber men, since it is round, whereas the circles of saber men riding in practice are in the shape of a figure '8'. Lancers drill, thrusting from the right-hand side only, their horses being trained to step so that weight of both man and beast will be thrown into the blow, thus weaving in and out in a figure eight.
>
> Discovering this, the detective turns over in his mind the names and organisations of the various German major generals. Only one has a troop of lancers attached. Major General von A.
>
> This, then, is he.
>
> What is he planning?
>
> The photographic officer bends over the system of trenches which the enemy infantrymen have created for practice. This is a replica of some little sector of the Allied lines which the Germans are going to attack. They have photographed it, worked it out carefully and duplicated it exactly, in order that their men in the hour of the attack may be perfectly familiar with their objective.
>
> Our detective calls for a map of the Allied lines within a radius of 50 miles. Experience has told him that the Germans will not be planning to attack farther away than that. He goes over this map carefully, and finally finds a plan of trenches in such and such a sector of the French lines, 30 miles south, let us say, that he is safe in concluding is the particular sector the enemy is going to attack.
>
> Again he bends over the picture. The Germans have not completed their work. He knows that they will practise two weeks after their completion and then occupy a day and a night transferring to the place of the battle. So with a final scrutiny of the number of machine-gun practice pits he writes out a report to the army commanding officer informing him that Major General von. A. with so many thousand infantrymen will attack such and such a sector on such and such a day, employing a certain number of machine gunners, and that in the meantime bombers can find Major General von A. at this certain spot. His report is positive – the aerial detective can be relied upon.
>
> The following day the photographic officer sends another camera man to hover over Major General von A. and make another picture, and each day thereafter the same thing, so that when at last the German officer leads his men to the objective, not only is he met and defeated, but his entire record is hanging on the wall where it may be used as a key to his way of preparing, his habits and his weaknesses, against the next time he sallies forth.[1802]

Reality for public consumption was introduced in the 1920s by Capitaine André-H. Carlier and Major Harold E. Porter. Carlier, one of the first French photographic interpreters, instructed the first American photographic interpreters at Cornell University in 1918, and wrote the most detailed assessment of photographic interpretation in *La Photographie Aérienne pendant la Guerre*. Probably the best discussion of photographic interpretation outside of the manuals of the war came from Harold Porter's *Aerial Observation*. Porter's work provided the most extensive look into aerial reconnaissance, covering the challenges of aerial observers during the sortie, the culture of the aviation community at large, and the contribution of the photographic interpreter.

THE FRENCH LEGACY

Perhaps the most passionate and erudite description of the role of intelligence was provided by Pépin in his manual on photographic interpretation: 'The study of an aerial photograph must be confided to specialists and ... these specialists must surround themselves with the largest possible number of means of control and of verifications.'[1803] When Pépin instructed the newly arriving American interpreters, he remarked: 'To study these various documents one must have a very thorough training, a liking for minute research and some aptitude.'[1804] The French approach to aerial photographic interpretation demanded intellectual rigour. 'The reading of the photos may induce certain officers to make deductions. The plain information must however always be transmitted, in order to enable the check to be made by a specialist of photograph interpretation.'

> To summarise, the minute study of aerial photographs is based solely upon a close examination of forms, trails, shadow and relief, and upon a comparison of successive photos. The interpretation requires good sense, method, and a certain knowledge of German theories and practice. I hope that the many and varied photographs will develop the confidence of every one in aerial photography which when thoroughly exploited, enables us to penetrate into the life and even the intentions of the enemy; and in the calmest times as well as the most active, to follow the destructive work of our Artillery and to register the victorious advance of our Infantry.[1805]

Pépin's counterpart, Sous-lieutenant Pouquet, emphasised the need for the interpreter to be engaged with the various elements of aerial observation throughout the cycle of reconnaissance:

French interpreters pore over the battlefield landscape. (SHD, from Christienne and Lissarague, *A History of French Military Aviation*)

> It is obvious that the interpreter should keep in constant liaison with the observers of escadrilles, who may aid him in the study of photographs of their respective sectors, but it appears to me indispensable that the latter should have a room for meetings, study and files, that is absolutely distinct from that of the interpreter, who in moments of extreme activity can only rely on himself...
>
> I shall add a word on the installation as regards material which it appears to me, should be furnished to the person entrusted with the interpretation of the photos. His work must be at the same time accurate and speedy; it requires more reflection than decision; that is to say, order and quiet are necessary. As a consequence, we shall reserve to the interpreter an office where he can shut himself in at his wish; this office may be small, but well lighted (if possible on two sides). His files are absolutely private.[1806]

Such facilities were arranged. An RAF visitor to a reconnaissance escadrille noted the space and light: 'Large light rooms containing every conceivable map, order plan and photograph and separate telephone cabins for the use of artillery observers – equally spacious offices for the Photographic Officer, his 'interpreters' and draughtsmen' ... The Corps Aviation Commander is the source of all wisdom and all orders and information come from him.'[1807]. One checklist summarised what was essential for the successful interpreter:

> Excellent eyesight
> Knowledge of enemy documents, especially those relative to fortification
> Be very attentive and most conscientious
> Know how to interpret and assemble information from all sources
> Have a wide experience in reading photographs
> Finally, have a large supply of patience[1808]

A BRITISH INTERPRETATION

The British view of the aerial photographic interpreter paralleled the French. The ability of a 'Sherlock Holmes' was alluded to in one post-war analysis. 'Minute, almost imperceptible details, overlooked by the amateur, conveyed to the trained eye a wealth of meaning.'[1809] Jones summarised the scope of interpretation: 'Experience in the reading of air photographs had shown that they might be made to reveal considerably more than the positions of trenches and strong points.'[1810] The British camoufleur, Solomon Solomon, further described the function in basic photographic terms: 'The photographer's picture is made by the sun; he makes no personal effort to reproduce or analyse. The sun reproduces for him. His main concerns are purely mechanical, not analytical as are those of the artist, who knows what is meant by the perspective of shadows, and will see where there are other abnormalities besides those for which he has been asked to look, and would see after careful study what escapes the photographer.'[1811]

An illuminating description of the British photographic interpreter experience was provided by the AEF staff member, Captain Stanford, visiting Headquarters British Fifth Army to gain an understanding of its intelligence organisation:

> He must study and get to know thoroughly the enemy lines opposite his Corps front; this is absolutely essential if he is to get through his work in good time and keep his office routine running smoothly. He should read all Intelligence Summaries – and especially reports from the line [patrol reports and reports on raids]; these will be of help to him in interpreting his photographs.[1812]

A competent and trusted photographic interpreter had to understand the needs of the combatant:

> He should have experience of trench life and so get to know what the man in the trench wants to know, and wants to have in the way of photographs. He ought periodically to go round the units of his Corps – at any rate as far as Brigade H.Q. (both Infantry and Artillery) and see that the photographs are getting properly circulated to the right people. In this way he can make himself useful in helping to solve difficulties of interpretation which may have arisen in the lower formations.[1813]

Captain Stanford's observations of British Intelligence illuminated an important feature of the relationship between the intelligence staff and the aviator: 'He should endeavour as far as possible to make his office a place where the pilots and observers drop in for a chat about their work (as well as other things); this will prove of great assistance to his efficiency.'[1814] Interpersonal skills were critical for effective information dissemination to staff personnel: 'It is imperative that he should keep in close touch with the staffs of all the formations in his Corps and with the Intelligence Staff at Army H.Q. He should get to know these officers personally – also the staffs at Wing and Brigade H.Q. R.F.C. He should make a point of being hand in glove with the C.C. of the Squadron to which he is attached – also with all the pilots and observers of that squadron. If he does not do this he will miss a large amount of information which he would be able to collect otherwise.'[1815] The British observer A.J. Insall described that close relationship in his recollection of one memorable intelligence officer serving with the 12th (Army) Wing Headquarters:

> The Wing Photographic Officer was an extraordinary character. I knew him as well as one got to know anyone with whom one had practically daily contact, and he was always welcome in our mess, as incidentally, were the other members of that Command's staff, including the Commander himself, like our own C.O. an Irishman. The Wing Photographic Officer's name was 2nd Lt. E. W. J. Payne. To his friends – and they were legion – he was always known as 'Agony.' He smoked like a chimney, and, no matter who you happened to be, sooner or later when he was around he would lunge forward, a fistful of 'yellow perils' thrust out at you, and you would be bidden uncompromisingly to 'Ave a gasper, cock!'[1816]

AN AMERICAN PERSPECTIVE

The veteran aerial observer, Harold Porter, described the interpreter:

> The interpreter, above all, must have a very lively mental vision. He should be imaginative, not as a writer of fiction is imaginative, but rather as a reader of it ought to be. That is, he ought always to be wondering about the outcome, and trying to guess it before he has finished with the story. He must know just how any given object on the ground will look when seen from above. He must have learned his business from the ground up, and then with equal thoroughness from the air down. He must put himself in the enemy's frame of mind, and know how the enemy applies himself to the arduous duty of thinking. He must know, by constant study, every inch of the country which has been and is to be photographed, so that he won't be misled by the absence of contours from the photographs.[1817]

Edward Steichen provided a more thoughtful perspective. 'There has been a tendency to wrap the technique of aerial photography in a veil of secrecy and mystery. This attitude may have been useful in stimulating curiosity, but there comes a time when such a pose seriously interferes with the free development of a new process.' Steichen remained sensitive to the fact that not all understood aerial photography. 'It is but fair to note the following remark by a colonel of the infantry: "They sent me some of that stuff (referring to aerial photographs), but I couldn't make anything out of them".'[1818]

> The value and meaning of aerial photography has been definitely and conclusively established, and the success with which aerial photographs can be exploited is measured by the natural and trained ability of those concerned with their study and interpretation. The aerial photograph is in itself harmless and valueless. It enters into the category of 'instruments of war' when it has disclosed the information written on the surface of the print. The average vertical aerial photographic print is upon first acquaintance as uninteresting and unimpressive a picture as can be imagined. Without considerable experience and study it is more difficult to read than a map, for it badly represents nature from an angle we do not know.[1819]

One assessment of interpreters concluded, 'He should not be too young.'[1820] The average age for American aviators was 20; for the experienced and more mature interpreters it was 21.

THE ROUTINE

Photographic interpretation was hard work. Intelligence had to be timely and in quantity to have value. Sgt. Dixon's records how intelligence worked with the prints. 'The first set of prints was rushed to the Intelligence Department [G-2] where they were looked over and a decision made on how many prints were wanted from each negative. The number could vary from under ten to 150 or even 200 or more.'[1821] After the war, Major Steichen discussed the photographic interpreter's challenges with his brother-in-law, the author Carl Sandburg. Steichen 'sympathised with the sturdy lads who almost suffocated in the little pup-tent darkrooms, cellars and dugouts used in field service,' while they 'developed the art of reading lines, shadows, blurs, camouflage in the finished prints of the day's duty.'[1822] At the battle of Château-Thierry in June 1918, one American photo section worked six straight days and nights in an improvised laboratory and interpretation room in a tent blacked out with tar paper and thick curtains. They performed 'as bravely and as doggedly as any organisation which ever took the field.'[1823]

In a historical summary of one of the first US Photo Sections (Section No. 1), the hard work throughout the campaign was described:

> At two A.M., one hasn't the 'pep,' the energy, that one had at 8 A.M., the previous morning. But the work must go on, the war cannot wait. In a few hours, men standing in grave conference at headquarters; men standing beside cannon and gun limbers; men looking eagerly at the enemy from the

A tribute to the Photographic Sections of the Great War. The 4th Photographic Section reveals the secret weapon behind victory. 'Photoshop' in more than one sense: a larger-than-life 50cm is revealed with the Section manning the contours. (USAFA McDermott Library Special Collections (MS5))

> front line trenches – will be waiting anxiously for these photographs. For these prints, that disclose and reveal all of Fritz's secrets; that tell where he is strong and ready for attack, where he is weak and demoralised, where he has many cannon, where he has few, and exactly where each and everyone is located. Tomorrow, 'they' will need these prints in order to continue the attack. So tonight, as many another night, we in the lab go on with the work. Other men pore over maps and plans locating the enemy's positions, his batteries, his headquarters, his supply dumps, his telegraph and telephone lines, his innermost, hidden secrets. Still other men walk innumerable miles around and around and around the giant drying frames that dry these prints in several seconds. Still others sort them, number them, count them, wrap them, ready to be delivered to 'them' at the appointed time ... Such is our lab in the long stretches of twenty, thirty hours' work that the war demands.[1824]

The attitude of the photo section mirrored that of Steichen, their staff leader, always thinking the entire time of 'how to spot the enemy's positions, how to knock hell out of the opposition and tear the living guts out of him.'[1825]

THE MATTER OF INTERPRETATION AND EXPLOITATION

The intellectuals of the interpretation discipline such as de Bissy, Pépin, Pouquet, Moore-Brabazon, Campbell and Solomon made it clear in their manuals that fighting a modern war demanded the ability to interpret photographs:

> Aerial photography constitutes at the present time the principal source of information concerning the enemy. It enables us to study his organisation as a whole and in detail and oftentimes even to predict his plans. Nevertheless this study of photographs will be advantageously supplemented, checked and often borne out by other sources of information (terrestrial and aerial observation, reports from

patrols, information obtained by questioning prisoners, observation of flashes from the enemy guns, direction of reports from hostile fire, etc.). But, in case of slightly different information or data, the photograph will almost always settle the discrepancy. Hence, all those who collect data, check them or utilise them, must be able to read and interpret the aerial photographs which are furnished them in various forms by the Aerial Photography Section.[1826]

Aerial photography created a new military intelligentsia. 'Concentrate your whole mind on the particular objects which you are seeking. Do not let your attention wander to subsidiary objectives. Follow every traverse and detail with a pointer in regular and logical order, but be careful not to mark the photograph.'[1827]

Shortly after the Armistice, as the Allied armies settled into the Rhineland to begin the occupation, staffs began a debate on the best way to continue the discipline. In his memorandum 'Aerial Photography, The Matter of Interpretation and Exploitation,' Steichen made it clear that expertise was best established through experience:

> The best French and British interpreters have insisted that the essential requisite for the training of interpreters is that they shall know modern warfare, either from having had experience at the front or from having made extended study of the actual

Members of the 5th Photographic Section and 99th Aero Squadron study the ground of the Meuse-Argonne battlefield. (USAFA McDermott Library Special Collections (SMS57))

ground on reconquered territory. The mistakes that are made by a student in interpretation whose knowledge has been gained entirely from the study of books and pictures are flagrant; a few hours spent on the battlefield studying the ground and organisations in connection with aerial photographs previously made, are of more value than any amount of classroom work.[1828]

Steichen handed the laurels to the French:

> The interpretation of photographs is in no sense an exact science, but is largely a matter of astute deductions, and men picked for this work must have qualifications that will make it possible for them to develop along such lines. The Intelligence Section was not only charged with the interpretation of photographs, but also with ordering them to be made. With the exception of one or two cases the Branch Intelligence Officers know absolutely nothing about photography or its possibilities and consequently were not able to judge as to what could be done or what was desirable. It is believed that the Photographic Interpreter should have a very thorough knowledge of photography. Frequent cases have been noted where an interpreter who knew nothing about photography mistook imperfections in the photographic plate or print for things recorded by the camera. Numerous British officers have stated that the French get more value out of their photographs than the British do. From observation this is also our opinion, and while recognising certain racial characteristics in the French that particularly fit them for meticulous and keen work of this kind, the fact that in the French service the interpretation and all matters pertaining to aerial photography rests in the hands of the French Air Service is chiefly responsible for this superiority.[1829]

Steichen's critiqued the existing order:

> The organisation planned by the Intelligence Section to handle this work was inadequate, and gradually more and more of the work supposed to be done by the Branch Intelligence Officer and his staff of draftsmen fell upon the shoulders of the draftsmen attached to our Photographic Sections, and the better qualified B.I.O.s were quick to realise the value

After the Armistice, three great aerial photographic pioneers, Major Laws, Major Steichen and Lt-Col. Moore-Brabazon share a lighthearted moment. (Moore-Brabazon, *The Brabazon Story*)

The Schweissguth stereoscope was an essential tool for French and American interpretation. The stereoscope acquired the media's hyperbolic description as the 'the Deadliest Weapon of the War'. (Ives, *Airplane Photography*)

of these photographic men and wisely made use of it. The extemporised Photographic Section at Ligny-en-Barrois, which did most of this photographic printing, also, with the help of a French officer, did a major part of the interpretation of the oblique photographs that were sent out, as the Intelligence Personnel was insufficient in number to do the work. Furthermore, this division of the work entails a great deal of unnecessary duplication, as, for instance, in the packing for distribution of prints to their ultimate destination.[1830]

Steichen suggested imitating the French setup: 'Our recommendation would be that the officer charged with interpretation of photographs at an Observation Group be a Photographic Officer belonging to the Photographic Section of the Air Service: this would be along the lines of the French Organisation.'[1831]

Steichen's desire to move the aerial photographic interpretation from the intelligence staff to the aerial photographic expert did not gain the attention of any senior officer within the AEF or Allied counterparts. In post-war thinking, his position met opposition from other members of the new military intelligentsia. Notably, Solomon, the leading British expert on camouflage, dissented. He countered Steichen's position with a descriptive summary. Such advocacy was 'as logical as requesting a compositor to translate and annotate an abstruse work, because he set up the type for its printing.'[1832] Post-war, aerial photographic technology continued improving through the use of film cameras. However, the skill of aerial photographic

The US Air Service's fabled 94th Aero Squadron acquired a Spad XIII configured with an aerial camera compartment labelled distinctly with 'Photo'; a portent of the future. Faster aircraft dominated tactical aerial reconnaissance for the rest of the 20th century. (USAFA McDermott Library Special Collections (SMS5))

interpretation waned throughout the Allied forces. By the mid-1920s it was a lost art.

The human experience behind aerial reconnaissance and photographic interpretation clearly showed how a cadre of courageous, hard-working and intelligent people could directly influence a well-established military culture. The aviators had accomplished aerial observation, one of the most dangerous assignments of the war, on a daily basis in all kinds of harsh and lethal conditions. The mission did not stop at the landing ground. It required a new breed of expertise in production, dissemination and interpretation. The latter determined the ultimate value of aerial reconnaissance. All efforts by aviation to accomplish the reconnaissance were as nothing until the interpretation was successfully accomplished. In the immediate post-war review, Steichen wrote that greater expertise in the photographic medium could have improved the process of interpretation. Post-war, the emphasis was on improving the technology of the aircraft, with little interest in refining and developing the intelligence aspects.

With the decline in the 1920s, and further neglect in the Depression era of the 1930s, aerial photography did not advance save in the greater use of film. Not until the start of another war would the need resurface for skilled professionals in the art of aerial photographic interpretation.

Ansaldo SVA 5 in flight. The fastest aerial reconnaissance aeroplane in the First World War – the precursor of twentieth-century aerial reconnaissance. (Gregory Alegi, *Ansaldo SVA 5*)

CONCLUSION

Intelligence and Modern Warfare

THE MILITARY INTELLIGENCE RESPONSE TO THE STATIC BATTLEFIELD

Rapid troop movement in the first weeks of the First World War quickly degenerated into static trench warfare. In this strategic siege spanning an entire continent, combatants became entrapped in a stationary battle of interminable duration, continuously monitored by new technologies, and forced to suffer from a vast inventory of powerful new weapons. Contemporary accounts paid tribute to the effective use of aeroplanes to direct artillery fire. One eloquent and fatalistic German soldier fighting at Chauffour Ravine during the battle at Verdun in late June 1916 put it best: 'A silent enemy flyer pursued the individual wanderer, and French artillery shells accompanied his path. To run, stand still or lay down, it was all the same.'[1833]

Aerial reconnaissance quickly emerged during the early phases of the First World War as a significant information-gathering device which tapped into new sources of combat knowledge about the enemy, giving field commanders and strategic planners unprecedented insights into enemy dispositions, capabilities, and, by extrapolation, battlefield intentions. However reluctantly, commanders, faced with the conundrum that traditional military doctrine failed to achieve breakthroughs and only yielded large casualty lists, turned to new means to achieve tactical and strategic advantages over their enemies, especially aerial reconnaissance and signals intelligence. The increasing accuracy and timeliness of intelligence from these sources gave commanders new confidence in their ability to plan effectively. 'Under the conditions of modern warfare, no army can long exist without using every possible means of gathering information; and of all these means aerial photographs present probably the best medium.'[1834]

Command-level willingness to adapt to the emergence of new technology-driven information sources altered conventional battle planning by making intelligence a major consideration. This new emphasis on intelligence as a planning factor drove the evolution of intelligence to achieve unprecedented levels of precision and timeliness. Artillery targeting rapidly matured to take advantage of the overhead perspective, both to identify targets and to assess damage inflicted on the enemy. The French Plan Directeur and British Firing Maps provided an extremely detailed analysis of enemy positions. Intelligence reports were transmitted by every means to speed their dissemination to the command element for planning purposes and to the front lines for execution of those plans. Dissemination devices ranged from the traditional use of carrier pigeons and fireworks to electronic communications networks throughout the trench networks. This ensured effective liaison, coordination, and information sharing among the Allies. In addition, the needs of the three largest Allied armies, spread from the English Channel to the Alps, drove the first-ever attempt to mass-produce intelligence. All of these transformations created a new way of thinking about and applying intelligence to campaign planning. As a result, the First World War became the first information and intelligence war.

CHALLENGES TO MILITARY INTELLIGENCE EFFECTIVENESS

Intelligence cannot win battles or wars on its own but only contributes to the operational planning and execution process. Allied commanders had not initially embraced either photographic or signals intelligence with confidence. That confidence only grew with the ability of these disciplines to provide usable, accurate, and timely insights about the enemy. However, as significant battlefield success never materialised, military seniors expressed their exasperation that even aerial photography did not result in substantial progress. As the battle of the Somme raged, the BEF's senior intelligence officer, Brigadier-General John Charteris, wrote to the senior British intelligence officer at the War Ministry, Major-General Macdonogh,

'Airplanes as a battle means of information have failed us.'[1835] This opinion reflected the pervading illusion that intelligence could be a panacea for wartime success. It shows a misunderstanding of the limited role of intelligence in the command decision-making process, one which could not consistently overcome traditionalist military doctrine such as the value of frontal assaults. First World War intelligence often fixated on assessing enemy order of battle to extrapolate (often inaccurately) the depth of enemy strength.[1836] Allied commanders, however, maintained their confidence in aerial photographic intelligence until the autumn of 1918 when the war returned to a mobile battleground. It was then that aerial reconnaissance and photographic interpretation was left behind by the rapidly evolving front line.

Aerial photography held sway on what constituted reality in the front line. It overcame many enemy deception efforts, especially the extensive use of camouflage, through the use of rapidly evolving and creative photographic interpretation skills. The war of illusion was waged by warriors and the intelligentsia of the art world. Both sides conducted psychological war based on the perceived reality of the photograph: paint, canvas, and netting were the weapons of the camoufleur and stereoscopes the weapon of the imagery interpreters. What reconnaissance could not penetrate was any enemy retreat underground and any night-time movement, except through traces left as tell-tale indicators visible in daylight. Nevertheless, aerial reconnaissance and photographic interpretation remained the most important way of perceiving the enemy and making sense of his operations.

MILITARY INTELLIGENCE INTERACTION WITH THE OUTSIDE WORLD

The entire landscape of the war was brought to the home front of all combatants through the lens of the aerial camera. Magazines such as *Scientific American*, *Popular Science Monthly*, and the *Army and Navy Register* brought the mysteries of aerial reconnaissance and camouflage to public notice. Aerial photography conveyed the extent of damage to population centres hes with vivid clarity. The public accepted the hyperbole about the photographic interpreter penetrating the image, seeing through the camouflage and detecting targets for the infantry and artillery. The label 'the deadliest weapon of the war' applied to the simple stereoscope tool brought an overwrought, popular slant to an otherwise obscure craft. Intense public interest in the tools of intelligence continues to the present day.

Information acquisition came at a price. The growing demand for ever larger volumes of aerial photographs pressured Allied air services and industry to produce cameras with optical lenses with sufficient resolution for interpretation and cartographic purposes. The war effort made the military increasingly dependent on civilian production. At one point, the shortage of high quality glass for camera optics created a crisis. The French managed to meet their objectives thanks to a limited, higher quality pre-war optics industrial base. The British optics industry, however, could not meet the demand from its own industry, which prompted a plea to the public to turn in whatever personal camera lenses they had available. Moreover, the Allies began forcing German reconnaissance aeroplanes to land behind Allied lines in order to capture their lenses for use in their own camera systems. The optics crisis also directly affected British aeroplane design. By 1917, wider fuselage frames were necessary to mount the new internal camera bays. In the last year of the war, the British conducted ministry-level negotiations with the French to acquire additional lenses as well as commenced production of their own higher quality optics for a new series of state-of-the-art aerial cameras. At the time of the Armistice, such military-industry partnerships had alerted the nations to new vulnerabilities arising from lack of an industrial infrastructure – vulnerabilities that created additional dependency on allies.

ALLIED INTELLIGENCE RELATIONSHIPS

The First World War defined how modern states shared intelligence. The Anglo-French relationship was unique. Each nation supported and cooperated with the other while continually suspecting each other's intentions. From the beginning, both shared technical lessons that advanced knowledge of the principles, practice, and potential of aerial reconnaissance, which led to an intelligence revolution. When the Americans arrived in 1917, not only did they benefit from the joint experience of three years of warfare, they adopted the entire process for generating intelligence on the battlefield. French-American relations were especially critical, and French techniques, handbooks, manuals, and technical information provided the standard by which American aviation operated. American intelligence seniors opted for British intelligence methods based on staff visits to Allied headquarters, and adopted the British position of BIO at each squadron. However, the close relationship of Americans with their French aviator and photographic interpreter counterparts at the unit level overcame the preferences of those senior officers. Even at the headquarters of the American Expeditionary

Forces, General Pershing's senior intelligence advisor was Eugène Pépin, the foremost French aerial photographic interpretation officer of the entire war, who had been personally selected for the position by Général Pétain. Consequently, the aerial reconnaissance and photographic techniques used in the St Mihiel and Meuse-Argonne campaigns had a distinct French flavour. Interestingly, the extent of the French-American intelligence connection has been almost universally overlooked since the end of the Great War. The Anglo-American bias likely predominated in accounts because of language and cultural ties.

WARTIME ACCOMPLISHMENTS OF AERIAL RECONNAISSANCE AND PHOTOGRAPHIC INTERPRETATION

The accomplishments and ultimate shortcomings of aerial reconnaissance and photographic interpretation in the First World War were eloquently captured by the legendary British photographic interpreter, Constance Babington-Smith, in her Second World War memoir:

> By the time of the Armistice, photographic intelligence had proved itself, and was recognised as the indispensable eye of a modern army. But largely because of the technical limitations of the day – the performance and range of the aircraft and the scope of the cameras – it had become essentially regarded as tactical in value. After the war, the concept became 'frozen stiff' in the thinking of the staff colleges of the world.[1837]

Aerial reconnaissance and photographic interpretation was supplemented by the traditional but still important collection of intelligence from prisoners, deserters, spies, reconnaissance patrols, captured documents, and the direct observation of the battlefield by field commanders. The urgency of the early campaigns prompted critical thinking from a cadre of artillery-trained experts to help define how information could best be applied to targeting. A military intelligentsia met that need, a pantheon of visionaries such as Henderson, De Bissy, Grout, Pépin, Weiller, Sykes, Moore-Brabazon, Campbell, Lewis, Laws, and Steichen. All made significant breakthroughs that advanced information's role in achieving successful military operations, defined aerial platforms as reconnaissance vehicles, or improved the aerial photograph to become the definitive intelligence source. Their leadership made an indelible mark on all sectors of the front. Their achievements are largely forgotten. The final resting place of Lieutenant-General Sir David Henderson makes the point. He rests under a modest tombstone in a public cemetery.

POSTWAR DECLINE IN THE STATUS OF AERIAL PHOTOGRAPHIC RECONNAISSANCE

Change came slowly to labour-intensive aspects of aerial reconnaissance and photographic interpretation techniques after the Armistice. Aircraft technology, on the other hand, advanced steadily over the next 20 years. Camera and film technology also advanced, with film replacing photographic plates as the storage media and automatic film cameras becoming the mainstays for the remainder of the century. The weak link in post-war aerial reconnaissance was the human in the loop, the aerial observer. With the downgrading of the aerial observer position, single-seat aerial reconnaissance platforms capable of flying at higher altitudes and at faster speeds became the preferred mode of aerial photographic collection.

Despite the intelligence lessons offered by the war and especially the final campaigns, most Allied ground commanders continued to insist that aviation's role was limited to cooperating with ground troops by providing tactical reconnaissance. American commanders in particular held to the traditional position that battlefield success relied solely on the ability of the infantry to drive back the enemy and conquer territory. The veteran aerial observer, Harold Porter, echoed this attitude in his comprehensive 1921 analysis of aerial observation by quoting Pershing, who had declared during the war that it was necessary to 'utilise all the auxiliary fighting arms in helping the infantry to get forward to the objective of his attack, which is the enemy's line, and in that aviation plays a very large part that requires a very careful and close training in time of peace with the infantryman who carries a gun. You must protect him from attack by the enemy's aviation and from observation in order to save him from being the object of the enemy's artillery destruction.'[1838]

The debate over the role of aerial reconnaissance and photographic interpretation continued during the postwar era. The controversy concerning Brigadier General Mitchell did not alter the US establishment's dogma, although recognition of aerial reconnaissance and photographic interpretation's potential in modern battle did not disappear from the thinking of senior commanders. They clearly treasured what aviation had provided. When Air Service and later Air Corps officers argued for more control of aviation resources, Army commanders were not tolerant of such innovation, especially coupled with Air Corps demand for strategic bombing. Despite the recognition that American aerial reconnaissance and photographic interpretation equated to modern battlefield success, little progress was made in the interwar period defining their role for the next war.[1839]

The French quickly hid their reconnaissance and photographic interpretation techniques and processes under a veil of secrecy. The strategic advances made by the Weiller Group and associated photographic interpretation had strongly influenced French military leadership. On 11 November 1918, Maréchal Foch concluded his presentation of the Legion of Honour to Capitaine Paul-Louis Weiller for his aerial reconnaissance accomplishments over the entire war with the admonition to keep this method of reconnaissance secret in order 'to preserve for France the considerable lead it had then attained over the rest of the world in the conduct of the war.'[1840] French postwar military doctrine espoused a greater role for aerial reconnaissance, with emphasis on technological advances in film cameras. Unfortunately, national-level endorsement did not translate into effective action. Aerial reconnaissance and analysis failed during the German Sitzkrieg of 1939, followed by the Blitzkrieg and the rapid capitulation of France in 1940.

British attention to aerial reconnaissance reached a plateau in the interwar years despite the ringing endorsement of Field-Marshal Haig in his postwar writings. For example, as early as 1923, the RAF Director of Training, Air Commodore T. C. R. Higgins, expressed concern because 'The question has arisen concerning the interpreting or extracting information from aeroplane photographs. It appears to run some risk of becoming a lost art, and one that requires very considerable practice.'[1841] His correspondent, Air Commodore H.R.M. Brooke-Popham, noted increasing concern about the status of the Branch Intelligence Sections at flying squadrons, and he concluded, 'I am very glad that this question is being taken up again and I think it is most important. Otherwise, the Army will quite forget that we can take photographs.' The issue had developed over the commitment of resources, especially what was required to train the next generation of aerial photographic interpreters. As a shared responsibility between the Royal Army and the RAF, joint training remained in jeopardy because of competing priorities.

UNCHANGING PRINCIPLES

The First World War was the first information and intelligence war, and it established abiding principles and expectations that underlie all subsequent intelligence operations. Aerial reconnaissance and photographic interpretation have held the rapt attention of national and combat leaders since. These information sources demonstrated the extent to which intelligence could influence command-level planning and operations. Interestingly, despite the great efforts put into developing cameras for employment on aerial platforms, not one new photographic principle was discovered and applied during the war. Aerial photography made a difference in providing accurate, current, and insightful intelligence – in describing battle scenes, assessing effects from artillery barrages, facilitating terrain analysis, uncovering camouflage and deception, and enabling the ultimate post-battle assessment of the 'period of destruction'.

The initiative of the aerial reconnaissance pioneers achieved more than simply marrying contemporary intelligence needs to available technologies. They established the foundations of modern intelligence. Since the Great War, many of the practices of aerial reconnaissance and much of the architecture supporting photographic interpretation have undergone rapid change and adaptation to new technologies and new wartime requirements. Tactical reconnaissance by today's unmanned aerial vehicles abides by many of the standards employed by early twentieth century manned aeroplanes. Modern strategic reconnaissance continues the First World War strategical goal of examining the defence and industrial infrastructure within enemy borders. Modern spacecraft-borne imagery is directly linked to the pioneering efforts of the early aviators, interpreters, and all their colleagues.

Nowhere has this entire evolutionary development based on First World War principles been more successfully demonstrated than in the first Gulf War in 1991 when space-borne imagery provided Allied commanders with the intelligence that enabled the famous 'end run' around the flank of the Iraqi positions. The seemingly unchanging principles behind these fields of endeavour underlie the operational processes to this day. Ironically, experience demonstrates that military intelligence and the information that is shared remains a perishable commodity.

Brigadier-General (later Lieutenant-General) Sir David Henderson. (Courtesy Andrew Renwick, RAF Museum)

An extraordinary testament to humility. The man who laid the groundwork for the RFC and gave purpose to aerial reconnaissance, Lieutenant-General Sir David Henderson, shares the same gravesite and modest gravestone with his son Captain Ian Henderson at Girvan (Doune) Cemetery. Captain Henderson was killed in an aeroplane accident in 1918. (Courtesy of Steve Erskine, The Scottish War Graves Project)

APPENDIX A

Aerial Reconnaissance Platforms

YEAR	AEROPLANE	TYPE	CAMERA
1914	**FRENCH DESIGN**		
1914	Blériot XI (FR)	Reconnaissance	Hand Held
1914	Blériot XI (GB)	Reconnaissance	Pan-Ross Hand Held
1914	Blériot XI (BE)	Reconnaissance	
1914	Blériot XI Artillerie (FR)	Reconnaissance	Hand Held
1914	Blériot Parasol (GB)	Reconnaissance	Pan-Ross Hand Held
1914	Breguet AG 4 (FR)	Reconnaissance	Hand Held
1914	Caudron G.3 (FR)	Reconnaissance	Hand Held
1914	Deperdussin (FR)	Reconnaissance	Hand Held
1914	Deperdussin (BE)	Reconnaissance	
1914	Henri Farman (HF) 20 (FR)	Reconnaissance	Hand Held
1914	Henri Farman (HF) 20 (BE)	Reconnaissance	Hand Held
1914	JERO HF 20 (BE)	Reconnaissance	
1914	Henri Farman F.20 'Longhorn' (GB)	Reconnaissance	Hand Held
1914	Maurice Farman (MF) 7 (FR)	Reconnaissance	Hand Held
1914	Maurice Farman MF 11 (FR)	Reconnaissance	Hand Held
1914	Maurice Farman MF 11 (BE)	Reconnaissance	Hand Held
1914	Maurice Farman 'Shorthorn' (GB)	Reconnaissance	Hand Held
1914	Maurice Farman S.7 (GB)	Reconnaissance	Hand Held
1914	Morane-Saulnier (MS) Type L (FR)	Reconnaissance	Hand Held
1914	Morane-Saulnier Type L (GB)	Reconnaissance	Pan-Ross Hand Held
1914	Morane-Saulnier (MS) Type L (BE)	Reconnaissance	
1914	Nieuport IV (RU)	Reconnaissance	Ulyanin, Potte
1914	REP N (FR)	Reconnaissance	Hand Held
1914	Voisin 3 (FR)	Reconnaissance	Hand Held
BRITISH DESIGN			
1914	Avro 504 (GB)	Reconnaissance	Pan-Ross Hand Held
1914	Blériot Experimental (BE) 2a; BE 2b (GB)	Reconnaissance	Pan-Ross Hand Held, Aeroplex Hand Held
1914	BE 8 (GB)	Reconnaissance	Hand Held
1914	Reconnaissance Experimental (RE) 1 (GB)	Reconnaissance	Hand Held
1914	RE 5 (GB)	Reconnaissance	Hand Held

YEAR	AEROPLANE	TYPE	CAMERA
1914	Wight Pusher Seaplane (GB)	Reconnaissance	Hand Held
1915	**FRENCH DESIGN**		
1915	Blériot XI (GB)	Reconnaissance	Aeroplex Hand Held
1915	Blériot XI (BE)	Reconnaissance	
1915	Caudron G.3 (FR)	Reconnaissance; Army Cooperation	26 cm
1915	Caudron G.3 (GB)	Corps Machine for Artillery Work	'A' Type
1915	Caudron G.3 (BE)	Reconnaissance; Army Cooperation	26 cm
1915	Caudron G.3 (IT)	Reconnaissance	Lamperti
1915	Caudron G.4 (FR)	Reconnaissance; Army Cooperation	26 cm; 50 cm; 120 cm
1915	Caudron G 4 (GB)	Reconnaissance	'A' Type, Hand Held
1915	Caudron G.4 (BE)	Reconnaissance; Army Cooperation	26 cm; 50 cm; 120 cm
1915	Farman F.40 (FR)	Reconnaissance; Army Cooperation	26 cm; 50 cm; 120 cm
1915	Henri Farman HF 20 (FR)	Reconnaissance; Army Cooperation	26 cm
1915	Henri Farman HF 20 (BE)	Reconnaissance; Army Cooperation	26 cm
1915	JERO HF 20 (BE)	Reconnaissance; Army Cooperation	26 cm
1915	Henri Farman 'Longhorn' (GB)	Corps Machine for Artillery Work	'A' Type
1915	Maurice Farman MF 7 (FR)	Reconnaissance; Army Cooperation	26 cm
1915	Maurice Farman MF 11 (FR)	Reconnaissance; Army Cooperation	26 cm; 50 cm; 120 cm
1915	Maurice Farman MF 11 (BE)	Reconnaissance; Army Cooperation	26 cm; 50 cm
1915	Maurice Farman MF 11 (IT)	Reconnaissance	Lamperti
1915	JERO MF 11 (BE)	Reconnaissance; Army Cooperation	26 cm; 50 cm
1915	Maurice Farman 'Shorthorn' (GB)	Corps Machine for Artillery Work	'A' Type; 'C' Type
1915	Morane-Saulnier Type L Parasol (GB)	Reconnaissance; Army Cooperation	'A' Type; 'C' Type
1915	Morane-Saulnier Type LA Parasol (GB)	Reconnaissance; Army Cooperation	'A' Type; 'C' Type
1915	Nieuport IV (RU)	Reconnaissance	Ulyanin, Potte
1915	Nieuport 10 (FR)	Reconnaissance; Army Cooperation	26 cm; 50 cm
1915	Nieuport 12 (FR) (GB)	Two Seater Fighter Reconnaissance	'A' Type; 'C' Type

YEAR	AEROPLANE	TYPE	CAMERA
1915	Voisin 5 (FR)	Reconnaissance, Artillery Registration	Douhet-Zollinger; 26 cm
1915	**BRITISH DESIGN**		
1915	BE 2a; BE 2b; BE 2c (GB)	Corps Machine for Artillery Work	Aeroplex Hand Held; "A' Type; 'C' Type
1915	Fighting Experimental (FE) 2a (GB)	Two Seater Fighter Reconnaissance	'A' Type; 'C' Type
1915	RE 1 (GB)	Corps Machine for Artillery Work	'A' Type
1915	RE 5 (GB)	Corps Machine for Artillery Work	'A' Type, 'C' Type
1915	RE 7 (GB)	Corps Machine for Artillery Work	'A' Type, 'C' Type
1915	Sopwith 'Baby' Floatplane (GB)	Float Seaplane	'A' Type; Hand Held
1915	Vickers Fighting Biplane (FB) 5 (GB)	Two Seater Fighter Reconnaissance	'A' Type; 'C' Type
1915	**RUSSIAN DESIGN**		
1915	Il'ya Muromets	Strategical Reconnaissance	Ulyanin, Potte
1915	Anatra D Anade (RU)	Reconnaissance	Ulyanin, Potte
1915	Lebed (7, 11, 12) (RU)	Reconnaissance	Ulyanin, Potte
1915	Mosca MB (RU)	Reconnaissance	Ulyanin, Potte
1915	Sikorsky S-16 (RU)	Reconnaissance	Ulyanin, Potte
1915	**FRANCO-BRITISH DESIGN**		
1915	Franco-British Aviation Flying Boat (FR)	Float Seaplane	26 cm
1915	Franco-British Aviation Flying Boat (GB)	Float Seaplane	'A' Type
1916	**FRENCH DESIGN**		
1916	Caudron G.3 (FR)	Reconnaissance; Army Cooperation	26 cm; 50 cm; 120 cm
1916	Caudron G.3 (GB)	Corps Machine for Artillery Work	'A' Type
1916	Caudron G.3 (IT)	Reconnaissance; Artillery Spotting	Lamperti
1916	Caudron G.4 (FR)	Reconnaissance; Army Cooperation	26 cm; 50 cm; 120 cm
1916	Caudron G.4 (GB)	Reconnaissance	'A' Type
1916	Caudron G.6 (FR)	Long-Range Reconnaissance	26 cm; 50 cm; 120 cm
1916	Farman F.40 (FR)	Reconnaissance; Army Cooperation	26 cm; 50 cm; 120 cm
1916	Farman F.40-A2 (FR)	Reconnaissance; Army Cooperation	26 cm; 50 cm; 120 cm
1916	Farman F.40 (BE)	Reconnaissance; Army Cooperation	26 cm; 50 cm; 120 cm
1916	Farman XXVII (RU)	Reconnaissance	Ulyanin, Potte
1916	Farman XXX (RU)	Reconnaissance	Ulyanin, Potte

YEAR	AEROPLANE	TYPE	CAMERA
1916	JERO F.40 (BE)	Reconnaissance; Army Cooperation	26 cm; 50 cm; 120 cm
1916	Farman-Nélis GN1 (BE)	Reconnaissance, Army Cooperation	26 cm; 50 cm
1916	Letord 1 A3 (FR)	Reconnaissance; Army Cooperation	26 cm; 50 cm; 120 cm
1916	Maurice Farman MF 11 (FR)	Reconnaissance; Army Cooperation	26 cm; 50 cm; 120 cm
1916	Maurice Farman MF 11 (BE)	Reconnaissance; Army Cooperation	26 cm; 50 cm
1916	Maurice Farman 'Shorthorn' (GB)	Corps Machine for Artillery Work	'A' Type, 'C' Type, 'E' Type
1916	Morane-Saulnier Type LA Parasol (GB)	Corps Machine for Artillery Work	'A' Type, 'C' Type
1916	Morane-Saulnier P (GB)	Corps Machine for Artillery Work	'A' Type, 'C' Type
1916	Morane-Saulnier BB (GB)	Corps Machine for Artillery Work	'A' Type, 'C' Type
1916	Morane-Saulnier T (FR)	Long-Range Reconnaissance	26 cm; 50 cm; 120 cm
1916	Nieuport 12 (FR)	Reconnaissance	26 cm
1916	Nieuport 12 (GB)	Two Seater Fighter Reconnaissance	'A' Type, 'C' Type
1916	Nieuport 17 (GB)	Single Seat Fighter Reconnaissance	C' Type, 'E' Type
1916	R.4 (FR)	Long-Range Reconnaissance	26 cm; 50 cm; 120 cm
1916	Salmson-Moineau (SM) 1 (FR)	Long-Range Reconnaissance	26 cm; 50 cm; 120 cm
1916	Tellier T.3 (FR)	Flying Boat	Hand Held; 26 cm 'Grand Champ'
1916	Voisin 5 (FR)	Reconnaissance, Artillery Registration	26 cm; 50 cm; 120 cm
1916	**BRITISH DESIGN**		
1916	BE 2c; BE 2d; BE 2e (GB)	Corps Machine for Artillery Work	'A' Type, 'C' Type; 'E' Type
1916	BE 2c (BE)	Reconnaissance; Army Cooperation	26 cm; 50 cm
1916	FE 2b; FE 2c; FE 2d (GB)	Two Seater Fighter Reconnaissance	'A' Type, 'C' Type; 'E' Type
1916	Martinsyde G.100 (Elephant) (GB)	Corps Machine for Artillery Work	'A' Type, 'C' Type; 'E' Type
1916	RE 7 (GB)	Corps Machine for Artillery Work	'A' Type, 'C' Type; 'WA' Type
1916	RE 8 (GB)	Corps Machine for Artillery Work	'A' Type, 'C' Type
1916	Short 184 (GB)	Float Seaplane	'A' Type; Hand Held
1916	Sopwith 1 1/2 Strutter (GB)	Two Seater Fighter Reconnaissance	'A' Type; 'C' Type; 'E' Type

YEAR	AEROPLANE	TYPE	CAMERA
1916	Sopwith 1 A2 (FR)	Reconnaissance; Army Cooperation	26 cm; 50 cm; 120 cm
1916	**RUSSIAN DESIGN**		
1916	Il'ya Muromets (RU)	Strategical Reconnaissance	Ulyanin, Potte
1916	Anatra DS Anasal (RU)	Reconnaissance	Ulyanin, Potte
1916	Lebed 10 (RU)	Reconnaissance	Ulyanin, Potte
1916	**FRANCO-BRITISH DESIGN**		
1916	Franco-British Aviation Flying Boat (GB)	Float Seaplane	'A' Type; Hand Held
1916	Franco-British Aviation Flying Boat (FR)	Float Seaplane	26 cm
1916	**UNITED STATES DESIGN**		
1916	Curtiss H 12 (GB)	Flying Boat	Houghton-Butcher Naval Camera; 'A' Type
1917	**FRENCH DESIGN**		
1917	AR 1 A2 (FR)	Reconnaissance; Army Cooperation	26 cm 'Grand Champ'; 50 cm; 120 cm
1917	AR 2 A2 (FR)	Reconnaissance; Army Cooperation	26 cm 'Grand Champ'; 50 cm; 120 cm
1917	Breguet 14 A2 (FR)	Reconnaissance; Army Cooperation	26 cm 'Grand Champ'; 50 cm 120 cm
1917	Caudron G.4 (FR)	Corps Machine for Artillery Work	26 cm 'Grand Champ'; 50 cm; 120 cm
1917	Caudron G.4 (IT)	Reconnaissance; Artillery Spotting	Lamperti
1917	Caudron G.6 (FR)	Long-Range Reconnaissance	26 cm 'Grand Champ'; 50 cm; 120 cm
1917	F.40 series (FR)	Reconnaissance; Army Cooperation	26 cm 'Grand Champ'; 50 cm; 120 cm
1917	Farman F.40 (BE)	Reconnaissance; Army Cooperation	26 cm; 50 cm; 120 cm
1917	Farman-Nélis GN2; GN4; GN5 (BE)	Reconnaissance, Army Cooperation	26 cm; 50 cm
1917	Letord 1/2/4/5 (FR)	Reconnaissance; Army Cooperation	26 cm 'Grand Champ'; 50 cm; 120 cm
1917	Maurice Farman MF 11 (FR)	Reconnaissance; Army Cooperation	26 cm 'Grand Champ'; 50 cm; 120 cm
1917	Maurice Farman MF 11 (BE)	Reconnaissance; Army Cooperation	26 cm 'Grand Champ'; 50 cm; 120 cm
1917	Morane-Saulnier P (FR)	Reconnaissance; Army Cooperation	26 cm 'Grand Champ'
1917	Morane-Saulnier T (FR)	Reconnaissance; Army Cooperation	26 cm 'Grand Champ'; 50 cm; 120 cm
1917	Caudron R.4 (FR)	Long-Range Reconnaissance	26 cm 'Grand Champ'; 50 cm; 120 cm
1917	Salmson S.M.1 (FR)	Reconnaissance; Army Cooperation	26 cm 'Grand Champ'; 50 cm; 120 cm
1917	Spad VII (FR)	Single Seater Fighter	26 cm 'Grand Champ'; 50 cm; Piazza
1917	Spad VII (US)	Single Seater Fighter	'C' Type

YEAR	AEROPLANE	TYPE	CAMERA
1917	Spad XI A2 (FR)	Reconnaissance; Army Cooperation	26 cm 'Grand Champ'; 50 cm
1917	Voisin 7 (FR)	Army Cooperation	26 cm 'Grand Champ'; 50 cm
1917	**BRITISH DESIGN**		
1917	Airco De Havilland (DH 4) (GB)	Two Seater Fighter Reconnaissance;	'L' Type; 'P' Type
		Short Distance Day Bomber	'L' Type; 'P' Type
1917	Armstrong Whitworth F.K.8 (GB)	Corps Machine for Artillery Work	'L' Type; 'P' Type
1917	BE 2d; BE 2e, BE 2f; BE 2g (GB)	Corps Machine for Artillery Work	'A' Type; 'E' Type; 'L' Type; 'P' Type
1917	BE 2c (BE)	Reconnaissance; Army Cooperation	
1917	BE 12a (GB)	Corps Machine for Artillery Work	'L' Type; 'LB' Type
1917	Bristol 'Fighter' F2B (GB)	Two Seater Fighter Reconnaissance; Corps Machine for Artillery Work	'L' Type; 'P' Type
1917	FE 2d (GB)	Corps Machine for Artillery Work	'E' Type; 'L' Type; 'P' Type
1917	Felixstowe F.2A (GB)	Large Flying Boat	Houghton-Butcher Naval Camera; 'A' Type, 'P' Type
1917	Felixstowe F.3 (GB)	Large Flying Boat	Houghton-Butcher Naval Camera; 'A' Type, 'P' Type
1917	RE 8 (GB)	Corps Machine for Artillery Work	'L' Type; 'LB' Type
1917	RE 8 (BE)	Reconnaissance; Army Cooperation	
1917	Short 184 (GB)	Float Seaplane	'A' Type, 'P' Type
1917	Sopwith 1 1/2 Strutter (GB)	Two Seater Fighter Reconnaissance	'C' Type;'E' Type; 'P' Type
1917	Sopwith 1 A2 (FR)	Reconnaissance; Army Cooperation	26 cm 'Grand Champ'; 50 cm; 120 cm
1917	Sopwith 1 A2 (BE)	Reconnaissance; Army Cooperation	26 cm 'Grand Champ'; 50 cm; 120 cm
1917	Sopwith Torpedo Plane (GB)	Torpedo Aeroplane	'A' Type, 'P' Type
1917	Wight (GB)	Float Seaplane	'A' Type, 'P' Type
1917	**ITALIAN DESIGN**		
1917	Ansaldo SVA 5 (IT)	Reconnaissance	Lamperti
1917	Pomilio D (IT)	Reconnaissance	Lamperti
1917	Savoia-Pomilio S.P. 3 (IT)	Reconnaissance	Lamperti
1917	SAML S.1; S.2 (IT)	Reconnaissance; Artillery Spotting	Lamperti
1917	**RUSSIAN DESIGN**		
1917	Il'ya Muromets	Strategical Reconnaissance	Ulyanin, Potte
1917	Anatra DSS (RU)	Reconnaissance	Ulyanin, Potte

YEAR	AEROPLANE	TYPE	CAMERA
1917	**UNITED STATES DESIGN**		
1917	Curtiss H 12 (GB)	Flying Boat	Houghton-Butcher Naval Camera; 'A' Type, 'P' Type
1918	**FRENCH DESIGN**		
1918	AR 1 A2 (FR)	Reconnaissance; Army Cooperation	26 cm 'Grand Champ'; 50 cm; 120 cm
1918	AR 2 (US)	Reconnaissance/ Photographic	26 cm; 50 cm; 'Short Focus Cameras'
1918	AR 2 A2 (FR)	Reconnaissance; Army Cooperation	26 cm 'Grand Champ'; 50 cm; 120 cm
1918	AR 1 (US)	Reconnaissance/ Photographic	26 cm; 50 cm; 'Short Focus Cameras'
1918	Breguet 14 A2 (FR)	Reconnaissance; Army Cooperation	26 cm 'Grand Champ'; 50 cm; 120 cm; De Ram
1918	Breguet 14 A2 (BE)	Reconnaissance; Army Cooperation	26 cm; 50 cm; 120 cm
1918	Breguet 14 A2 (US)	Reconnaissance/ Photographic	26 cm 'Grand Champ; 50 cm; 120 cm; 'Short Focus Cameras'
1918	Caudron G.6 (FR)	Long-Range Reconnaissance	26 cm 'Grand Champ'; 50 cm; 120 cm
1918	Caudron R.11 (FR)	Reconnaissance; Army Cooperation	26 cm 'Grand Champ'; 50 cm; 120 cm
		Long Range Reconnaissance	26 cm 'Grand Champ'; 50 cm; 120 cm
1918	Georges-Levy 40 HB2 (FR)	Flying Boat	26 cm 'Grand Champ'; Hand Held
1918	Salmson 2 A2 (FR)	Reconnaissance; Army Cooperation	26 cm; 50 cm; 120 cm; De Ram; Automatique
1918	Salmson 2 A2 (US)	Reconnaissance/ Photographic	26 cm; 50 cm; 120 cm; De Ram
1918	Spad VII (FR)	Single Seater Fighter	26 cm 'Grand Champ', 50 cm; Piazza
1918	Spad XI A2 (FR)	Reconnaissance; Army Cooperation	26 cm 'Grand Champ'; 50 cm; 120 cm
1918	Spad XI A2 (BE)	Reconnaissance; Army Cooperation	26 cm; 50 cm; 120 cm
1918	Spad XI (US)	Reconnaissance/ Photographic	26 cm 'Grand Champ'; 50 cm
1918	Spad XIII (FR)	Single Seater Fighter	26 cm 'Grand Champ'; 50 cm; Piazza
1918	Spad XVI A2 (FR)	Reconnaissance; Army Cooperation	26 cm 'Grand Champ'; 50 cm
1918	Spad XVI (US)	Reconnaissance/ Photographic	26 cm 'Grand Champ'; 50 cm
1918	Tellier T.3 (FR)	Flying Boat	Hand Held; 26 cm 'Grand Champ'
1918	Voisin 10 (FR)	Night Reconnaissance	26 cm 'Grand Champ'
1918	**BRITISH DESIGN**		
1918	Airco De Havilland DH 9 (GB)	Two Seater Fighter Reconnaissance;	'L' Type; 'LB' Type; 'BM' Type; 'P' Type
1918	Airco De Havilland DH 9a (GB)	Two Seater Fighter Reconnaissance	'L' Type; 'P' Type

1918	Armstrong Whitworth FK 8 (GB)	Corps Machine for Artillery Work	'L' Type; 'LB' Type; 'P' Type
1918	BE 2d; BE 2e; BE 2f; BE 2g (GB)	Corps Machine for Artillery Work	'L' Type; 'LB' Type; 'P' Type
1918	Bristol 'Fighter' F2B (GB)	Two Seater Fighter Reconnaissance; Corps Machine for Artillery Work	'L' Type; 'LB' Type; 'BM' Type; 'P' Type
1918	DH 4 (GB)	Two Seater Fighter Reconnaissance	'L' Type; 'LB' Type; 'BM' Type, 'P' Type
1918	DH 4 (US)	Two Seater Fighter Reconnaissance	'L' Type; 26 cm; 50 cm; 120 cm; De Ram
1918	H.12 (GB)	Large Flying Boat	Houghton-Butcher Naval Camera; 'P' Type
1918	Felixstowe F.2A (GB)	Large Flying Boat	Houghton-Butcher Naval Camera; 'P' Type
1918	Felixstowe F.3 (GB)	Large Flying Boat	Houghton-Butcher Naval Camera; 'P' Type
1918	RE 8 (GB)	Corps Machine for Artillery Work	'L' Type; 'LB' Type; 'BM' Type; 'P' Type
1918	RE 8 (BE)	Reconnaissance; Army Cooperation	26 cm; 50 cm
1918	Sopwith 1 A2 (FR)	Reconnaissance; Army Cooperation	26 cm 'Grand Champ'; 50 cm; 120 cm
1918	Sopwith 1 A2 (BE)	Reconnaissance; Army Cooperation	26 cm; 50 cm; 120 cm
1918	Sopwith 1 1/2 Strutter (US)	Reconnaissance/ Photographic	26 cm 'Grand Champ'; 50 cm
1918	Sopwith Torpedo Plane (GB)	Torpedo Aeroplane	Houghton-Butcher Naval Camera; 'P' Type
1918	Short 184 (GB)	Float Seaplane	Houghton-Butcher Naval Camera; 'P' Type
1918	Wight (GB)	Float Seaplane	Houghton-Butcher Naval Camera; 'P' Type
1918	**ITALIAN DESIGN**		
1918	Ansaldo SVA 5 (IT)	Reconnaissance	Lamperti
1918	Savoia-Pomilio S.P. 3 (IT)	Reconnaissance	Lamperti
1918	Savoia-Pomilio S.P. 4 (IT)	Reconnaissance	Lamperti
1918	SAML S.1; S.2 (IT)	Reconnaissance; Artillery Spotting	Lamperti
1918	SIA 7B1 (IT)	Reconnaissance	Lamperti
1918	Pomilio E (IT)	Reconnaissance	Lamperti

Abbreviations listed are the more common reference. i.e., Blériot Experimental (BE)

BE – Belgium FR – France GB - Great Britain IT – Italy RU – Russia US – United States

'Re-Organisation of Photographic Activities', TNA: PRO AIR 2/163
Barker, Ralph, *The Royal Flying Corps in World War I*
Carlier, André H., *La Photographie Aérienne Pendant la Guerre*
Bruce, J.M., *The Aeroplanes of the Royal Flying Corps (Military Wing)*
Davilla, James, *French Aircraft of the First World War*
Di Martino, Basilo, *Ali Sulle Trincee*
Gorrell Report, National Archives, RG 120
Pieters, Walter, *The Belgian Air Service in the First World War*
Nesbit, Roy, *Eyes of the RAF: A History of Photo-Reconnaissance*
Ulyanin, Yuriy Alekseyevich, *Pioneer of Russian Aviation*

APPENDIX B

Allied Aerial Cameras

Camera	Nationality	Timeframe	Manufacturer	Lens Focal Length	Magazine	Plate Size
Pan-Ross'	GB	1914	Pan-Ross	6"	12 plates	5" x 4"
Douhet-Zollinger	F	1915	Douhet-Zollinger	10 cm	film	6 cm x 20 m
A' Type	GB	1915-1918	Thornton-Pickard	8 1/4", 10"	plate (Mackenzie-Wishart)	5" x 4"
26 cm	F	1915-1918	De Maria, Gaumont	26 cm	Gaumont (12 plate magazine)	13 x 18 cm
Model F-1 Camera	US	1918		26 cm		
C' Type	GB	1915-1917	Thornton-Pickard	8 1/2"; 12", 17"	18 plates	5" x 4"
WA' Type	GB	1916-1917	British RFC Repair Depot	7", 8", 10", 14"; (Wide Angle)	plates	8 1/2" x 6 1/2"
B' Type	GB	1916	British RFC Repair Depot	Captured German 10", 14" 20", 27" Lenses	plates	8 1/2" x 6 1/2"
				British 20"; 47"	plates	
E' Type	GB	1916-1917	Thornton-Pickard	8 1/2", 10"	36 plate magazine	5" x 4"
EB' Type	GB	1916-1917	Thornton-Pickard	20"	36 plate magazine	
53'	GB	1916-1917		53 "	plate	5" x 4"
50 cm	F	1916-1918	De Maria, Gaumont	50 cm; 52 cm	Gaumont (12 plate magazine)	18 x 24 cm
Model F-2 Camera	US	1918		50 cm; 52 cm	Gaumont (12 plate magazine)	18 x 24 cm
26 cm 'Grand Champ'	F	1917-1918	De Maria, Gaumont	26 cm (Wide Angle)	Gaumont (12 plate magazine)	10 x 24 cm
Model F-3 Camera	US	1918		26 cm		
120 cm	F	1916-1918	De Maria, Gaumont	120 cm	Gaumont (12 plate magazine)	18 x 24 cm
Model F-4 Camera	US	1918		120 cm		
'F' Type	GB	1916	Thornton-Pickard	8 "	5" x 4" on 25 or 50 foot spools	50 feet provided 120 exposures

Camera	Nationality	Timeframe	Manufacturer	Lens Focal Length	Magazine	Plate Size
'L' Type	GB	1917-1918	Thornton-Pickard	6", 8 1/2", 10"; 20"	18 plate magazine	5" x 4"
	US	1918	Eastman Kodak Co.			
'LB' Type	GB	1917-1918	Thornton-Pickard	4", 5", 8", 12", 20"	18 plate magazine	5" x 4"
'BM' Type	GB	1917-1918	Thornton-Pickard	7", 8", 10" 14" 20", 36"	12 or 18 plate magazine	18 x 24 cm; 8 1/2" x 6 1/2"
Piazza	IT	1917-1918	Piazza	24 cm	36 plates	13 x 18 cm
Model DR-1 Camera	FR	1918	De Ram	50 cm	50 plate magazine	18 x 24 cm
Model DR-2 Camera	FR	1918	De Ram	26 cm	50 plate magazine	18 x 24 cm
Model DR-3 Camera	FR	1918	De Ram	50 cm	50 plate magazine	18 x 24 cm
Model DR-4 Camera	US	1918	De Ram; Burke and James Instrument Co.	50 cm	50 plate magazine	18 x 24 cm
'P' Type; Type P14	GB	1917-1918	H. Munro	10"	plate	5" x 4"
Type 18	GB	1918	Houghton-Bucher	6", 8", 10"	plate	18 x 24 cm
Private Cameras'	US	1918	Eastman Kodak and others	short focus cameras	film	
'K' Camera	US	1918	Eastman Kodak	12", 20", 40"	film	18 x 24 cm
Model G.E.M. Camera	US	1918	Groom, Elliot & Mottlau Co.	8 1/2", 12"	36 exposures	6" by 6" film
'LA' Camera	US	1918	Eastman Kodak	8 1/2", 10"	24 plate magazine	5" x 4"
Automatique	FR	1918	Duchatellier	24 cm	film	18 cm film spool
Ulyanin	RU	1914–17	Ulyanin	21 cm	6 plates	13 x 18 cm
Potte	RU	1914–17	Potte	21 cm	[film magazine] 50 frames	13 x 18 cm
Lamperti	IT	1915–18	Lamperti e Garbagnati	18 cm; 24 cm; 43 cm	12, 24, 48 plates	13 x 18 cm

FR – France
GB – Great Britain
IT – Italy
RU – Russia
US – United States

de Brunoff, Maurice, *1914-1918, L'Aeronautique pendant la Guerre Mondiale*
Carlier, André H., *La Photographie Aérienne Pendant la Guerre*
Di Martino, Basilo, *Ali Sulle Trincee*
Gorrell Report, Series G, Roll 24 'Cession of Optical Instruments to the American and British Government,' 4
Parry, David, *Descriptions of Aerial Reconnaissance Cameras Used, or Evaluated for use by the Royal Flying Corps and Royal Air Force*
Photographic Museum at RAF Cosford – Circa 2003 – POC Dave Humphrey
Pyner, Alf, *Air Cameras R.A.F. and U.S.A.A.F. 1915-1945*
Toye, *History of Photography*
Ulyanin, Yuriy Alekseyevich, *Pioneer of Russian Aviation*

APPENDIX C

Notes on Air Reconnaissance

(1) Now that squadrons of the Royal Flying Corps have been organised in wings, it is desirable that some indication should be given of the methods of reconnaissance which have been employed in the first period of the war, in order that advantage may be taken of the experience gained, and that the reconnaissance of the enemy may be as complete in the future as it has been in the past.

(2) Wings which are allotted to armies are army troops, and the aeroplanes of each wing are at the disposal of the Wing Commander, under instructions of the Army Commander, for such temporary allotment or detachment to Army Corps or Divisions as may be, from time to time, advisable.

(3) The missions on which aeroplanes may be dispatched are of four main types:

(a) *Destructive.* It is a general rule that hostile aircraft should be attacked wherever met with, unless, for such reasons as the distance to be traversed, or the importance of the duty on which he is already engaged, the flier should consider it undesirable to engage. For this reason every British aeroplane carries a suitable equipment. Some are armed with machine-guns. In all other two-seater aeroplanes the observer carries a rifle or carbine and the pilot a pistol, with sufficient ammunition. In a single-seated aeroplane a pistol is the weapon which has been found suitable. For the attack of troops or material on the ground, bombs or steel arrows are carried as ordered.

For the attack of aircraft no orders are required to be issued, except such as will ensure the presence of a sufficient force of aeroplanes to destroy or drive away the enemy's aircraft.

The attack of troops on the ground may be undertaken at the discretion of the flier, provided always that sufficient care is exercised in avoiding the attack of localities where any considerable injury is likely to be inflicted on the persons of peaceful inhabitants. The destruction of material, and especially the dispatch of expeditions against any particular objective, should not usually be undertaken without reference to General Headquarters, in order that the latest information as to the value of the objective may be obtained, so as to ensure that it is worth the risk incurred.

(b) *Artillery Observation and Reconnaissance in the Immediate Front of the Troops.* This can be very well undertaken by aeroplanes allotted to, or acting under the orders of, Army Corps or Divisional Commanders.

The particular uses of such reconnaissance are: The location of the enemy's defences, gun positions, or reserves; and the observation of artillery fire.

If these duties be delegated by armies to Army Corps it is desirable that the area to be covered by such reconnaissance as distinct from artillery observation should be defined.

It will usually be sufficient if it extends to eight or nine miles in front of the line. The Wing-Commander should ensure that no part of the front is left unobserved, by indicating the dividing line between the areas to be reconnoitred by the different corps or divisions. A slight overlapping should be allowed for, more especially on the flanks of the armies.

(c) *Tactical Reconnaissance for Armies.* These have usually been called 'Tactical Reconnaissances'; but the use of the words is likely to lead to misunderstanding, as the information obtained may be either tactical or strategical. The area to be covered by such reconnaissances may extend to some 20 miles from the front, and is usually somewhat extended to the flanks, overlapping to some extent the reconnaissance on either side. The information obtained by such reconnaissance should be brought in the first place to the Wing-Commander at Army Headquarters, and by him passed to the Army Headquarters Staff.

Aeroplane patrols to drive off hostile aeroplanes should also be ordered by Wing-Commanders.

(d) *Strategical Reconnaissance undertaken by Order of General Headq*uarters. Reconnaissance of this kind may be carried out either by aeroplanes sent direct from General Headquarters or by the aeroplanes of armies, under instructions from General Headquarters, according to the conditions of weather, distance, and aeroplanes available. Those sent out by General Headquarters return there and report direct; those sent out by wings, on the instructions of General Headquarters, return to the Wing-Commander, and the information is sent to General Headquarters through the army staff, the original report being forwarded as soon as possible by the Wing-Commander to Headquarters, Royal Flying Corps.

No strategical reconnaissance will be carried out by armies other than allotted to them by General Headquarters. In the event of Army Commanders requiring a strategical reconnaissance in addition to those already so ordered, they will ask General Headquarters to issue the necessary instructions.

A.J. MURRAY, Lieutenant-General
of the General Staff

GENERAL HEADQUARTERS
15th January 1915

APPENDIX D

GENERAL HEADQUARTERS

THE ARMIES OF THE NORTH AND NORTHEAST
GENERAL STAFF
STUDY AND EXPLOITATION OF AERIAL PHOTOGRAPHS

15 January 1918.

I. Text.

FIRST PART.—PRINCIPLES.

CHAPTER I.

PRINCIPLES ON WHICH REST THE ORGANIZATION AND DEFENSE OF THE GERMAN POSITIONS.

Aerial photography originated with trench warfare. It made rapid progress and has become one of the most important sources of information at the commander's disposal. In fact, it alone makes possible the exact location of the enemy's defensive works and their detailed study.

The enemy, realizing its importance, tries to render this study difficult. Skillful camouflages, a large number of defenses and imitation works are some of the means employed.

As a result, the study of aerial photographs must be entrusted to specialists, who should be provided with all possible means of verification.

Of the indispensable kinds of information, the fundamental one is that of the defensive plans of the enemy. (The officers who will have to study, from the photos, the French defenses, must also be thoroughly conversant with our instruction of August 22, 1917, on field defenses, and the employment of all kinds of troops.) These principles are contained in the various rules and instructions emanating from the high command, especially the following:

Lessons drawn from the war, relating to field fortifications. (June 1915).

Lessons drawn from the battle of the Somme (June 24 to November 26, 1916), by the first German Army (General F.von Below).

German regulations relating to stationary [positional] *warfare for all arms* (First part, 13 Nov. 13 and Dec. 15, 1916; new edition, Aug. 17, 1917; eighth part, Dec. 1, 1916; appendix 10 June 1917).

Lessons taught by the last battles at Verdun (G.H.Q. Dec. 25, 1916).

Orders of Gen. Sixt von Armin, commander of the Fourth Army, under date of June 30, 1917 (appendix to the B.R. of the G.H.Q. of the armies of the N. and N.E. of Sept. 13, 1917).

The reading of these documents one to understand the German operations since the beginning of stationary [positional] warfare, by following its evolutions. This simplifies its study and explains its reason.

It seems helpful to give a historical sketch of the principles of defensive fighting and of the working methods of our adversaries since the end of the year 1914.

PERIOD I.—END OF 1914 TO OCTOBER, 1915.

Originally, trenches were dug to protect the fighting line, after it had become stationary. Therefore they contained the larger part of the forces engaged. The supporting troops, smaller in numbers, sheltered themselves in other trenches a little farther back.

Plate I. (Captions not in original document.) The German front line in 1915 usually consisted of two continuous trenches.

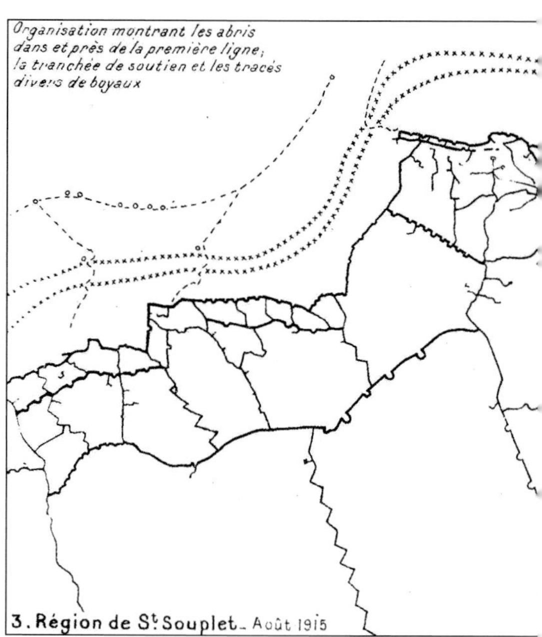

Barbed wire entanglements covered an area 5 to 10 metres wide. The trenches were irregular to better protect against flank attacks.

This state of affairs continued for a year, *the fundamental principle* being that the first line trench should be held under all circumstances or retaken immediately, in cased it should be penetrated by the enemy. (Instructions of Gen. von Below, commander of the Second Army, under date of August 1, 1915.) The period from the end of 1914 to October, 1915, was characterized by *the small depth of the German lines.*

By reference to photos (Plate I) before November 1915, it may be seen that the first line usually consisted of two continuous trenches. In front of these were barbed wire entanglements from 5 to 10 meters wide, preceded by numerous sentry posts. The trenches themselves were quite irregular, for protection against flank attacks. Sharpshooters were stationed at the firing notches. The communication trenches were from 2 to 4 meters wide. The regular trenches were narrow and not very deep (1.4 m., 1.8 m., occasionally 2 m., rarely 2.5 m.). Bomb-proof dugouts for the garrison were in the rear, but in immediate proximity to the first line trenches. They usually had two entrances opening from a circulation trench or boyau. The second trench was from 50 to 100 meters back of the first. It was also protected by barbed wire entanglements, and had dugouts for the supporting troops, whose number was about one-third as many as were in the first line.

Boyaus connected the first and second trenches, two for each company sector. Other boyaus communicated between the companies in line and the reserve troops, two or three boyaus for each battalion sector. The reserve troops lived in villages or camps, protected by trenches, and were often assigned the special role of "artillery protection."

Battery emplacements almost always consisted of plain earth breastworks connected with a boyau. Several casemates had, however, already appeared. The trenches were habitually protected against flanking by machine guns in the trenches themselves, with dugouts and protected firing platforms, and by several casemated guns in the first line.

Concrete first appeared in the spring of 1915, but it was generally used only for observation posts and flanking guns.

At least one or two kilometers back of the first line was the supporting line, usually with rather temporary defenses—one or two lines of trenches without dugouts or boyaus, but protected by barbed-wire entanglements.

PERIOD II.—OCTOBER, 1915, TO OCTOBER 1916.

Photographs of the German defenses taken during the winter of 1915-1916 (Plate II) show that important fortifications had been build along nearly the whole front. The experience acquired as a result of our September offensive had borne fruit. The lines had been reorganized, with reference to the three following points of view:

Economy of forces.
Reduction of losses.
Logical utilization of terrain.

Economy of forces was obtained by depth arrangement. The first line no longer consisted of two trenches, but of three, all provided with dugouts. Back of this was an intermediate line, the line of artillery protection; farther back, a second line, consisting of two trenches with dugouts, one in front of the crest, the other on the opposite slope. (Redistribution of first troops according to the Stein arrangement, in case of attack—Report Ia No. 1797 of Sept. 8, 1916—first line trenches, defense troops; second line trenches, defense and counteroffense troops; third line trenches, counteroffense troops.)

Reduction of losses was sought in:

(a) Deepening and widening the trenches—whence the disappearance of the circulation trenches and of the firing notches in unprotected terrains.
(b) Widening the communication trenches.
(c) Reenforcing the covered dugouts and especially in replacing them by underground dugouts.
(d) Building narrow-gauge railways (0.6 and 0.4 meters) to serve the first lines.
(e) Constructing casements to replace the old breastworks.

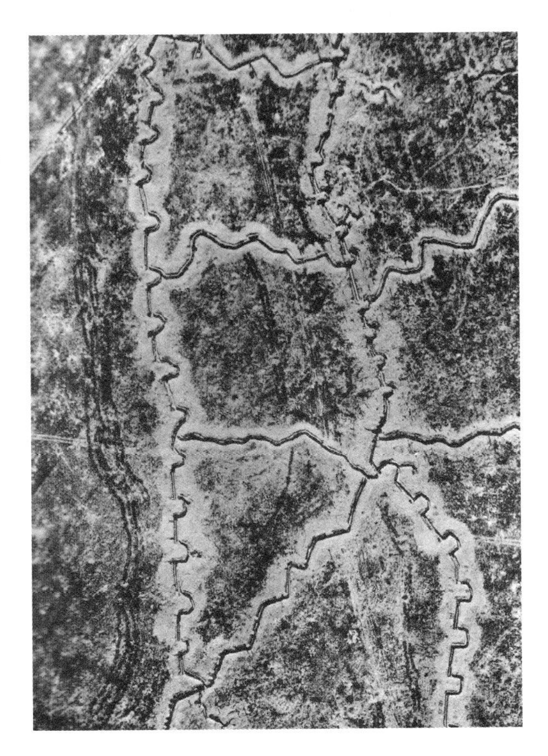

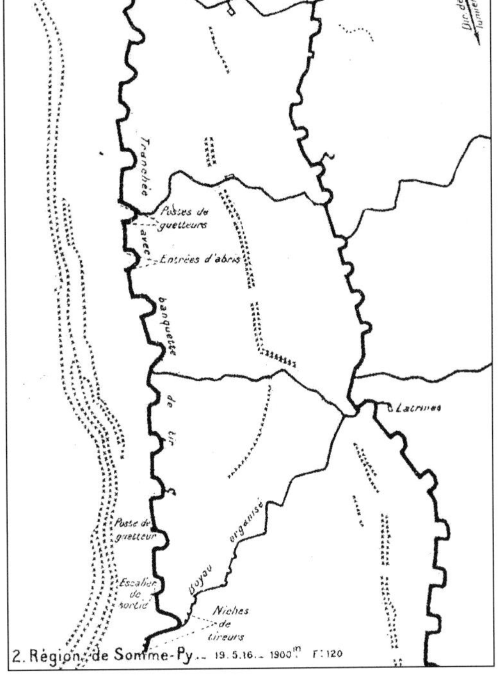

Plate II. Photographic aerial reconnaissance of the German front line in late 1915-early 1916 revealed substantial fortifications constructed along the entire Western Front.

Utilization of the terrain was carefully studied for the purpose of an obstinate defense. They endeavored to render impossible the breaking of their lines, or at least to limit our progress beyond their first line by successive supporting lines, generally continuous oblique organized trenches or boyaus ("suspenders") connecting the first and second lines.

PERIOD III.—OCTOBER, 1916, TO JUNE, 1917.

After the hard battles of 1916 at Verdun and on the Somme, the German staff published, in November and December, 1916, "New Regulations for Stationary Warfare." The first part (General instructions on the organization of positions and the details of organization) and the eighth part of these regulations (Principles for conducting defensive battles) are of great interest.

A The following are the *salient points:*
- (a) Maximum utilization of the reverse slope.
- (b) Organization of extensive fortified areas, several lines deep and thoroughly organized.
- (c) Distribution on the terrain of flanking machine-guns (checkerboard arrangement) and dugouts forming the *skeleton of all the infantry fighting lines.*
- (d) Successive lines of troops in depth.—Small garrison in the front line, with shallow concrete shelters so far as possible, the underground dugouts being reserved for the rear lines. (One about every 150 meters for a group of 8 men, on the Hindenburg line.)
- (e) Extensive use of diagonal defensive trenches, especially near the sector limits.
- (f) Execution of urgency works, picketing the lines, narrow gauge railways, flanking sections and dugouts (position framework), observation posts, headquarters, auxiliary defenses, ammunition depots, drainage, etc.

B. Regarding *construction and organization details, the above-mentioned regulations recommend* (Plate IV):
- (a) Giving to the auxiliary defenses an irregular outline and arranging them in sections separated by spaces, in such fashion that they may be effectually flanked by machine guns.
- (b) Still further widening the trenches and boyaus. (This width attains 3.2 and 3.4 meters on the Hindenburg line in front of the British.)
- (c) To dispose the mortars outside of the firing trenches.
- (d) To establish a large number of observation posts at successive intervals from the front.

It was in accordance with these principles that the Hindenburg position was constructed, during the winter of 1916-17.

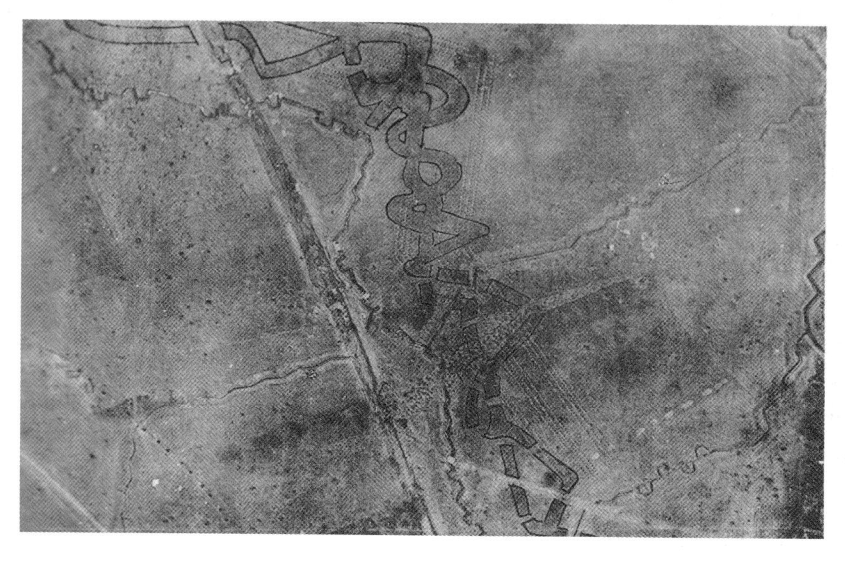

Plate IV. (There was no plate III.) The interpretation opposite above of this photograph in the region of St Quentin in April 1917 distnguishes between 'réseau [networks] réel' and 'réseau simulé'.

The German front line reflects the lessons learned at Verdun and on the Somme, resulting in the Hindenburg Line. Auxiliary defences are irregular in outline. The trenches are widened to at least 3.2 metres opposite the British sectors

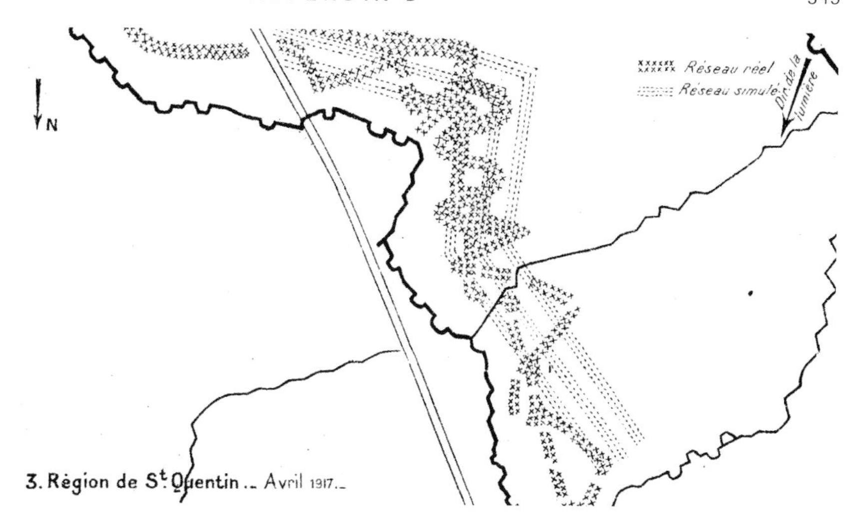

PERIOD IV.—JUNE, 1917.

The French-English 1917 spring offensive led the German General Staff to publish, June 10, 1917, a Supplement to the Regulations for Stationary Warfare, in which were summarized the most important principles in these regulations.

This supplement was annotated and enlarged by Gen. Sixt von Armin (commander of the Fourth Army) in an army order under the date of June 30, 1917. This order is certainly one of the most important documents that has fallen into our hands. Here are some essential extracts:

Necessity of camouflage (Plate V, 2).—"Strength in a defensive battle depends essentially on the precautions taken to conceal from the eyes of the enemy all our means of fighting. These fighting means (trenches, dugouts, machine-gun and battery emplacements) will *surely* be destroyed, if they appear on the enemy's *aerial photographs*."

Depth organization.—"There must be substituted for the old system of position, which may be plotted and destroyed by the enemy, a defensive zone organized in depth."

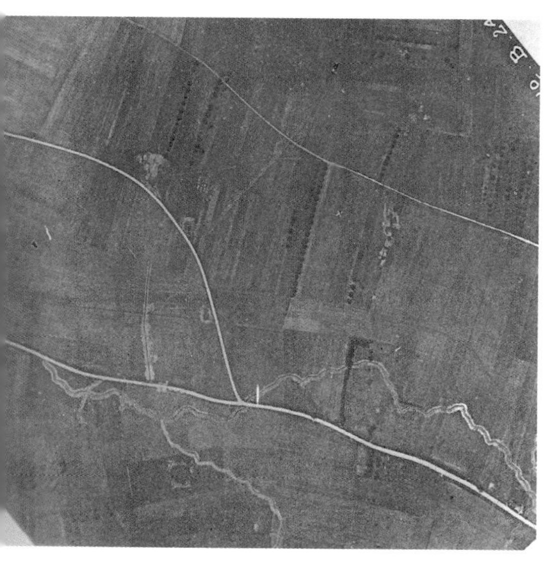

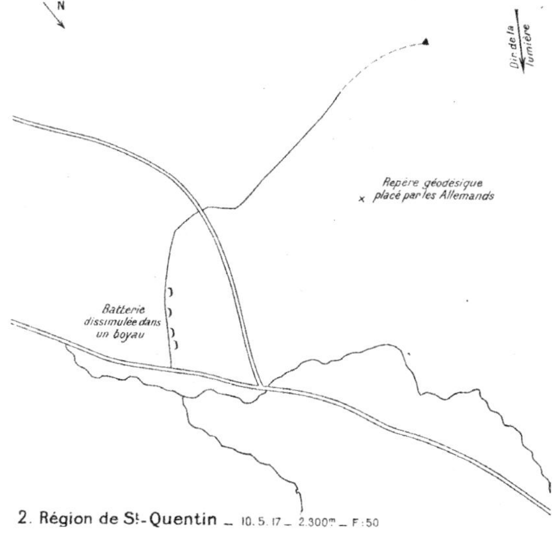

Plate V, 2. The Germans have learned that positional warfare required the use of camouflage.

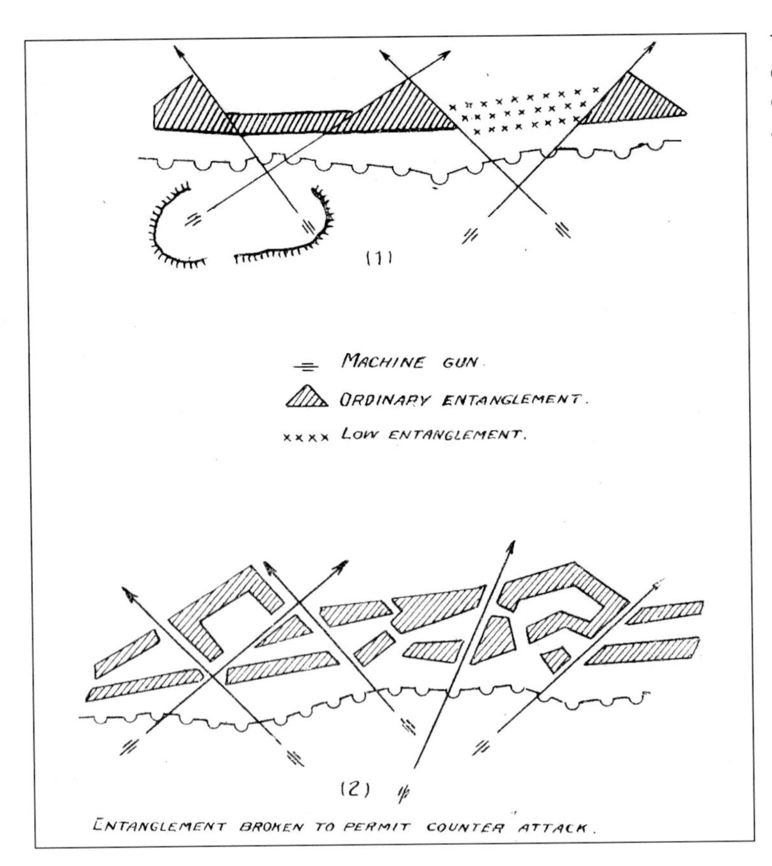

The defensive zone is in a checkerboard arrangement of isolated groups of soldiers and machine guns.

Utilization of shell holes by the fighting line.—"During a battle it is no longer necessary to have continuous trenches in the front line. They may be replaced by shell holes held by bunches of men and isolated machine guns, *arranged in checkerboard.*

"In front of the first row of shell holes will be placed an irregular barbed-wire entanglement, as continuous as possible. It is also advisable to protect with barbed wire the holes in front of the first line, to prevent their occupation by the enemy.

"Farther back, it is preferable to organize isolated defensive works, by surrounding the shell holes with auxiliary defenses in such a way as to leave passages for counterattacking troops."

Placing supporting troops and reserves.—"A large part of the supporting troops and of the reserves will be sheltered in the open field, in shell holes, woods, ravines, anywhere to avoid aerial observation. (Villages must be avoided, as they draw the enemy's fire.) These troops assist in the formation of a continuous line, avoiding so far as possible the view of the enemy. Thus a supporting line is established for the defense troops placed at successive intervals in *front* of it."

First-line defenses.—"These must include several lines of trenches, protected by a strong barbed-wire entanglement, with passages for assaulting troops. Deep dugouts will be constructed only on the second and third lines. The first line will have only small dugouts, capable of holding about one-sixth of the occupants of this line."

New-line defenses.—"The organization of new lines back of the first will be governed by the same principles as above. If possible, the construction of dugouts for the garrison must be undertaken in succession for the whole depth of the zone. (For one-sixth of the men in the first trench, two-sixths in the second trench, three-sixths back of the second trench as far as the line of artillery defense.)

"By the creation of numerous shelters in the rear of the lines, preparations are made for the opening of the defensive battle, the orderly retreat of the troops in the trenches, and their redistribution on the intermediate terrain.

Order of urgency of the works (Plate V, 1).—"Staking out the works, placing auxiliary defenses, bomb-proof shelters, digging trenches."

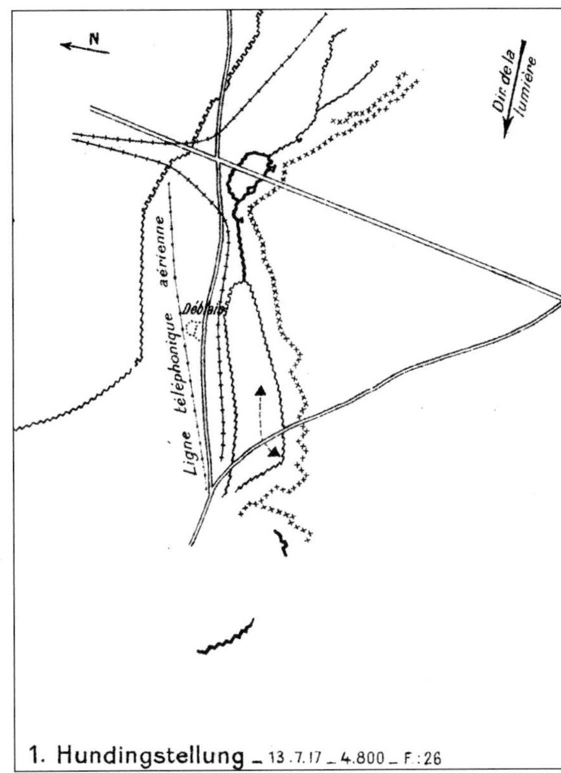

Plate V, 1. The 1917 polices resulted in more shelters being built towards the rear echelon areas in preparation for a defensive battle.

Machine-gun emplacements.—"Machine-gun locations should not be chosen in front positions, but on the sides of hills or in depressions, at points convenient for flanking. The enemy should come upon machine guns unexpectedly. Therefore, it is not desirable to install them in salients, but to the right or left or in the rear. In the salients only fake emplacements should be constructed."

Laying out boyaus.—"In order to prevent the enemy from forming an exact idea of our system of defense, the boyaus connecting the different lines may be given an oblique direction. This arrangement gives the appearance of the meshes of a net, in which the boyaus, provided with auxiliary defenses and organized as firing trenches, have at the same time defensive functions. (Riegel.)

"Everywhere must be kept in mind the necessity of escaping aerial observation."

PERIOD V.—AUGUST, 1917.

The supplement of June 10, 1917, appeared a little in advance of a new edition of the First Part (Section A) of the Regulations for Stationary Warfare, which reaffirmed the necessity, for all defensive works, of organization in depth for a mobile and active defense.

Fighting zones.—But where, in the appendix of June 10, 1917, it was written: "The fight in or for the first line has become the fight for the first position," the new regulation reads: "*The fight must be conducted not about the lines, but in the fighting zones.*"

The fighting zone comprises a number of organizations (trenches, boyaus, supporting points, etc.) designated to provide for the *defense* by infantry and artillery, *lines of communication, replenishment and good conditions of living.* Its depth could reach several kilometers. Its defense was to be organized by sectors.

Several fighting zones will be constructed, at least 3 km. from each other so that artillery preparation can not be made on two zones at the same time. A fighting zone should include:

(a) *An ordinary fighting zone*, organized to repel any surprise attack and considered as the covering zone (Vorfeld-zone).

(b) *A grand fighting zone* (Grosskampfzone). It is in this that enemy attacks, even the most violent must be stopped. It may coincide with the ordinary fighting zone. If, however, the latter zone is unfavorably located, the grand fighting zone will be located and organized further back. The grand fighting zone should be constructed in the ordinary fighting zone but with much more care and thoroughness. Still further back, at least one rear fighting zone (Rueckwaertige Kampfzone).
The principles already stated for selecting trench locations also apply to selecting zone locations:
Need of good rear communications,
Need of good artillery observation,
For trenches, the use of opposite slope, wherever compatible with good artillery observation.

Location of batteries.—The regulations of August 17 insist upon the necessity of making shelters for the men, munitions and artillery. Mobility being the best protection against enemy fire, it prescribes the organization, in advance, of extra emplacements with dugouts. It goes even further, and this is new in saying that "during a battle or for special objects, the use of unprepared locations is often advisable."

Order of Urgency of Defensive Works.—The order of urgency of the works has not been changed: flanking positions and dugouts, observation stations and headquarters, barbed wire entanglements, communications (liaisons). Earthworks (terrassements) are only made as a last extremity.

The historical sketch, just given, shows the evolution of the German theory of defense. The fighting *in or for the first line* has become the fight *in the fighting zone*. This doctrine may be modified in the future.

Whatever changes may be made, it is probable that aerial photos and information from prisoners, observation stations, etc., will enable us to keep pace with them. Perhaps the difficulties encountered by the officers who have to interpret the photos will become greater. They will always be less than those of our adversary to conceal his system of defense.

APPENDIX E

Aerial Photography and Interpretation

HEADQUARTERS AIR SERVICE, S.O.S., A.E.F.,
OFFICE OF C.A.S.
PHOTOGRAPHIC SECTION

December 26th 1918.

PART I

AERIAL PHOTOGRAPHY

The Matter of Interpretation and Exploitation

The importance and value of aerial photography in warfare has been clearly demonstrated to our army during the last few months of military activities.

This has been made clear in spite of the fact that photography was certainly not used as fully or as intelligently as it could have been. The following reasons are given:

(a) Lack of knowledge in the army at large as to the value of photography and its uses.

(b) Lack of competent and trained personnel for the interpretation and exploitation of photographs.

(c) Division of responsibility for photography between the Air Service and Intelligence Section.

In reference to (a) it would seem advisable that all field officers have a certain knowledge of aerial photography and the role it plays in modern warfare. They should be given illustrated lectures on what photography can do in the way of aiding them, and they should all have a summary knowledge of the interpretation of photographs. It is further recommended that popular lectures be given to field troops, this chiefly to acquaint them as to how the enemy can see, by means of aerial photography, just what they are doing, even miles behind the lines. By illustrated lectures or pamphlets they can be shown that carelessness in making paths at the front, or of allowing camouflage to deteriorate etc., they are endangering their lives as well as the success of their operations. It would be well for every soldier to know that the enemy observation planes flying high overhead are a much more dangerous enemy to them than those that come with bombs or harass them by machine-gun fire.

In reference to (b) and (c) the Photographic Section in America gave considerable attention to the interpretation of photographs in training its photographic officers and photographic laboratory men. As a matter of fact there were a number of men in the Photographic Section who, as a result of this training, were better qualified to be entrusted with the responsibility of interpreting photographs than were some of the Branch Intelligence Officers working with the observation groups. It is quite obvious that the interpretation of photographs is, from a military point of view, the most important phase of the work, and too much stress cannot be laid on choosing and training the right men to do the same. The best French and British interpreters have insisted that the essential requisite the training of interpreters is that they shall know modern warfare, either from having had experience at the front or from having made extended study of the actual ground on reconquered territory. The mistakes that are made by a student in interpretation whose knowledge has been gained entirely from the study of books and pictures are flagrant, a few hours spent on the battlefield studying the ground and organizations in connection with aerial photographs previously made, are of more value than any amount of class room work. The interpretation of photographs is in no sense an exact science, but is largely a matter of astute deductions, and men picked for this work must have qualifications that will make it possible for them to develop along such lines. The Intelligence Section was not only charged with the interpretation of photographs, but also with ordering them to be made. With the exception of one or two cases the Branch Intelligence Officers know absolutely nothing about photography or its possibilities and consequently were not able to judge as to what could be done or what was desirable. It is believed that the Photographic Interpreter should have a very thorough knowledge of photography. Frequent cases have been noted and where an interpreter who knew nothing about photography mistook imperfections in the photographic plate or print for things recorded by the camera. Numerous British officers have stated that the French get more value out of their photographs than the British do. From observation this is also our opinion, and while recognizing certain racial characteristics in the French that particularly fit them for meticulous and keen work of this kind, the fact that in the French service the interpretation and all matters pertaining to aerial photography rests in the hands of the French Air Service is chiefly responsible for this superiority. The organization planned by the Intelligence Section to handle this work was inadequate, and gradually more and more of the work supposed to be done by the Branch Intelligence Officer and his staff of draftsmen fell upon the shoulders of the draftsmen attached to our Photographic Sections, and the better qualified B.I.O's were quick to realize the value of these photographic men and wisely made use of it. In a similar way the Intelligence Section planned to take over the mass reproduction of all photographs, the making of assemblages, and the exploitation of the information contained in the photographs for training purposes. As a matter of fact very little work in exploiting photographs for field training was done by the Intelligence Section. It is reported that the mass photographic printing they planned to do for the St. Mihiel offensive was very unsatisfactory because of slow production, and some of the pictures did not arrive until two days after the attack. In the first Argonne offensive this work of mass printing was turned over to the Photographic Section, all the printing required was done promptly and in due time under very difficult conditions. The extemporized Photographic Section at Ligny-en-Barrois, which did most of this photographic printing, also, with the help of a French officer, did a major part of the interpretation of the oblique photographs that were sent out, as the Intelligence Personnel was insufficient in number to do the work. Furthermore, this division of the work entails a great deal of unnecessary duplication, as, for instance, in the packing for distribution of prints to their ultimate destination. Our recommendation would be that the officer charged with interpretation of photographs at an Observation Group, be a Photographic Officer belonging to the Photographic Section of the Air Service: this would be along the lines of the French Organization.

With aerial photography clearly understood and an organization adequate to exploit same, this Photographic Officer becomes a very important figure in the Air Service. He is chosen first for his special and natural aptitude, such as keenness, imagination, tact, astuteness and personality. It is desirable that he shall have had previous photographic experience and under any conditions that he take the full course in photo laboratory work. He should also receive the training currently given to an observer and be on a flying status, as it is desirable that this officer make occasional flights over the enemy lines. In this way he will become personally acquainted with his sector and have a better understanding of conditions. He will also in this manner develop the Air Service point of view and will more fully understand his observers and be able to intelligently advise them, as well as secure information from them. He must have the photographic point of view so that he may know just how much he can get out of the laboratory. This will further make the whole Photo Section a more vital working force as it will stimulate the whole section and keep up their interest which is always important where such quantities of work must be speedily turned out. With the aid of the observers and the Photo Section he will interpret the photographs and properly record them. With the aid of the observers and the Photo Section he will interpret the photographs and properly record them. From this point onward the photographic material properly belongs within the domain of the Intelligence Section, so that this Photographic Officer in reality acts as a liaison between the Air Service and the Intelligence Section. This Photographic Officer should not command or administer the Photo Section proper, but simply direct this work. The administration and laboratory work should be entrusted to a subordinate officer. This Photographic Information Officer should be responsible for the photographs but not for the photography. He should be in a position to know both what is wanted and what can be accomplished. Up to the present photography has been a commodity without enough customers, and each army needs officers qualified to drum up trade and to show the army what they can get and how they can use it.

The Photographic Section was rarely asked to make assemblages except for office ornaments. Stereoscopic prints were only occasionally called for. Not once was a large assemblage requested of an entire sector. Initiative in these direction (sic) on the part of the Photographic Sections was discouraged and there was a general atmosphere concerning these things which suggested that it was none of our affair and did not conform to Army Intelligence Regulations.

The fact an adequate supply of proper material and sufficient personnel was at last beginning to arrive from America placed us in a position to undertake new things. The following is cited as an instance: A number of specialists, stereoscopic printers had been trained at the Photo School, 2nd A.I.C., with a view to assigning them to the various Photo Sections at the front. All photo missions could then have been sent out in stereo-prints as well as the usual single prints. This would have materially increased the work of the Photo Section in the field. The experiment was successful, and in a short time this would have become a general practice, and the value of the photographs to those studying them would have been very materially increased.

In this connection it is noted that we recommended the adoption of our Richard stereoscope to the Intelligence Section. This was done and the instruments were secured by their Supply Service, but, to the best of our information they were never issued to the Branch Intelligence Officers and the Photographic Sections were consequently never called upon to make the stereopositives required for use in these instruments. It was on the strength of this failure that the Photographic Section took the initiative and decided to force the stereographs on the market instead of waiting for the request.

In addition to having a Photographic Information Officer with each Photo Section at a Corps Observation Group, a Chief Photographic Information Officer should similarly be attached to the Air Service Army Headquarters Staff. It would be the duty of this officer to co-ordinate the work of the Corps Group officers and hold the army negative and print files. All negatives from Corps Sections to be sent to the Army Section at the end of two weeks so as to relieve the Corps Section of the task of making the reprints required. This reprinting in a Corps Group during periods of great activity unnecessarily complicates their work. This Army Photo Officer would also receive first copies of all prints from the Corps Sections. He will therefore be acquainted with the work of the army and can furnish adjoining armies with any data of interest to them.

In this case he also forms the liaison for photographic information matters between the Air Service and G-2. His office would also furnish a center from which photographic material could be drawn for training purpose at the flying schools. He would also have at his disposal an army base laboratory plant for making reproductions, assemblies, copies and enlargements for general distribution throughout the army. This base laboratory plant to be operated by a regular Photographic Section.

Not having had a Photographic Officer at Army Headquarters, except during the last few weeks of the war has often made the co-ordination very difficult, and it has necessitated the writer's personally taking charge of photographic operations during active periods.

E. STEICHEN
Major, A.S.

APPENDIX F

The Importance of Aerial Photographs in Intelligence

LECTURE GIVEN AT 2ND A.I.C.
BY THOMAS E. HIBBON
1ST LT, 49TH INF. ATT. A.S.

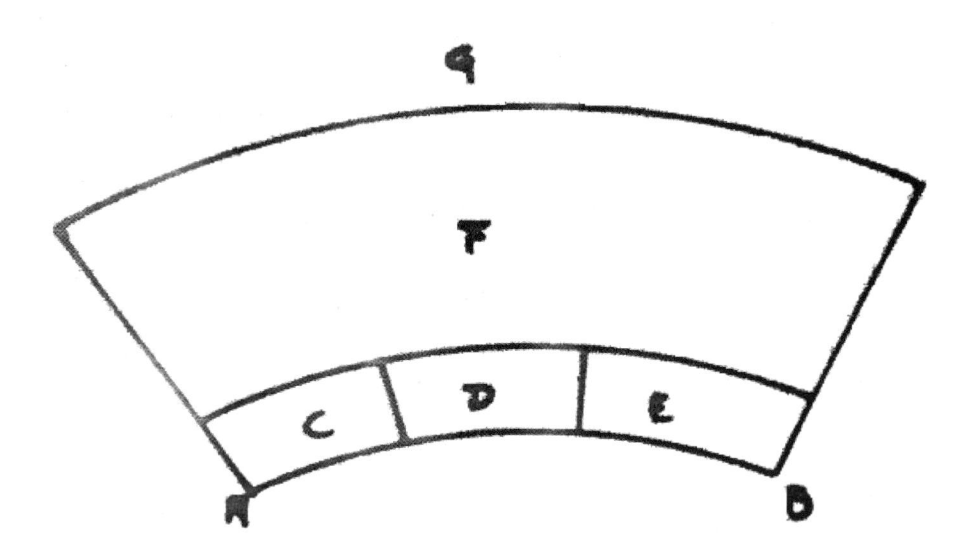

In order that a given sector such as AB may be kept under constant surveillance both photographic and visual it is divided into subsectors each assigned to a squadron according to the nature of the squadron (Divisional, Corps, Army).

Thus a particular squadron would be assigned to cover each of the areas C, D, E, F, and G. In the regular course of its work each squadron would take photographs as frequently as weather and tactical times permitted. In addition as the headquarters received information from other sources, special missions would be assigned. In this way every point of the units' front being accounted for there would be a constant and uniform flow of information to the general headquarters covering the entire sector. The fact that some sectors are larger and more distant than others as in the case of G above, is compensated for in that the squadron working over them has less artillery work and stronger and faster planes.

Key to Chart (author)
A–B – Enemy Front Line of the Western Front
C, D, E – Squadrons or escadrilles assigned to a particular division cover the enemy area facing the division (i.e. US 1st Division, 2nd Division, 42nd Division)
F – Squadrons or escadrilles assigned to Corps (Corps d'Armée) area of responsibility
G – Squadrons or escadrilles assigned to Army (strategical) area of responsibility

Notes

1. GHQ, [French] Armies of the North and Northeast, *Study and Use of Aerial Photographs* (transl.), 15 January 1918, sec. I, 10, NARA, RG 120, Box 819.
2. Jonathan Crary. *Techniques of the Observer: On Vision and Modernity in the Nineteenth Century* (Cambridge, MA: MIT Press, 1991), 13.
3. Theodore M. Knappen, *Wings of War: An Account of the Important Contribution of the United States to Aircraft Invention, Engineering, Development and Production during the World War* (New York: G. P. Putnam's Sons, 1920), 197–198.
4. Capitaine Jean de Bissy, *Illustrations to Accompany Captain De Bissy's Notes Regarding the Interpretation of Aeroplane Photographs*, October 1915 (transl., n.p., n.d.). In this instance the title Capitaine, as a French word, is italicised. In this work the first time a non-English word is used it is italicised, but later uses are not, since the use is quite frequent and italicizing all is distracting to the reader.
5. Sir Walter A. Raleigh and H. A. Jones, *The War in the Air, Being the Story of the Part Played in the First World War by the Royal Air Force* (6 vols., Oxford: The Clarendon Press, 1922-37; reprint, London: The Imperial War Museum, and Nashville, TN: The Battery Press, 1998), I, 75. Raleigh authored the first volume and Jones the others. Hereafter cited as Raleigh for references to vol. 1, and as Jones for vols. 2-6.
6. Raleigh, I, 177–178.
7. Thomas G. Fergusson, *British Military Intelligence, 1870–1914* (Frederick, MD: University Publications of America, Inc., 1984), 159.
8. Fergusson, 174.
9. *Field Intelligence, Its Principles and Practices* in Fergusson, 174.
10. Fergusson, 179–180.
11. Fergusson, 182.
12. Brigadier-General Sir David Henderson, *The Art of Reconnaissance*, 3rd ed., 3rd Printing (London: John Murray, 1916), iii.
13. 'Aeronautics,' June 1921, 1, Part XXXVII, sec. 3, in American Mission with the Commanding General, Allied Forces of Occupation, Mainz, Germany, 'A Study of the Organisation of the French Army Showing its Development as a Result of the Lessons of the World War and Comprising Notes on Equipment and Tactical Doctrine Developed in the French Army, 1914–11921,' NARA, RG 120, Box 819.
14. Gilbert T. Woglom, *Parakites* (New York: G. P. Putnam's Sons, 1896), 41–44.
15. James W. Bagley, *The Use of the Panoramic Camera in Topographic Surveying* (Washington, DC: GPO, 1917), 12.
16. Roy C. Nesbit, *Eyes of the RAF: A History of Photo-Reconnaissance* (Phoenix Mill, UK: Alan Sutton Publishing, Ltd., 1996), 8.
17. Wireless and radio were interchangeable terms. Radio will be the primary term used in this book. Raleigh, I, 170.
18. 'Lecture on Aerial Photography,' 25 April 1923, NASM, A30.2/36.
19. *Rapport du Capitaine du Genie Saconney, au sujet des experiences de Cerfs-Volants Effectuées Avec Nacelle Lestrée à bout de Cable*, 21 February 1910, *Service Historique de la Défense* (SHD); Marcellin Hodeir, 'La Photographie Aérienne: 'de la Marne à la Somme,' 1914–1916,' *Révue Historique des Armées*, no. 2 (1996): 108 (translation by Commandant Marc Rivière, French Air Force).
20. Foch, quoted in Georges Blond, *The Marne* (trans., London: Macdonald and Company, 1965), 163.
21. Haig, quoted in Major-General Sir Frederick Sykes, *From Many Angles, An Autobiography* (London, Toronto, Bombay, Sidney: George G. Harrap & Company Ltd., 1942), 105.
22. 'The Origins of the Italian Air Force,' http://www.finn.it/regia/html/origini.htm, accessed on 19 April 2010.
23. 'Air Reconnaissance of Turks' Position,' *New York Times*, October 23, 1911, 4, ProQuest Historical Newspapers *The New York Times*.
24. 'The Origins of the Italian Air Force.'
25. *Manoeuvres du 6ème Corps d'Armée*, Flight Log, 18 September 1911, *Service Historique de la Défense* (SHD).

26 James J. Davilla and Arthur M. Soltan, *French Aircraft of the First World War* (Stratford, CT: Flying Machines Press, 1997), 55.
27 'Report on Aeronautical Matters in Foreign Countries for 1913,' 16, The National Archive, (TNA): Public Record Office (PRO): AIR 1/7/6/98/20.
28 Davilla and Soltan, 1.
29 Davilla and Soltan, 1–2.
30 'Report on Aeronautical Matters,' 16, TNA, PRO: AIR 1/7/6/98/20.
31 'Report on Aeronautical Matters,' 16, TNA, PRO: AIR 1/7/6/98/20.
32 'Report on Aeronautical Matters,' 16, TNA, PRO: AIR 1/7/6/98/20.
33 Raleigh, I, 178.
34 'The Paris Aero Salon 1913,' *FLIGHT*, January 10, 1914.
35 Gaston Labussière, 'La Photographie Aérienne,' in Maurice de Brunoff, ed., *1914–1918: L'Aéronautique pendant la Guerre Mondiale* (Paris: M. de Brunoff, 1919), 183 (translation provided by Commandant Marc Rivière).
36 John H. Morrow, Jr., *The First World War in the Air: Military Aviation from 1909 to 1921* (Washington, DC, and London: Smithsonian Institution Press, 1993), 34.
37 Hodeir, 108.
38 *Lieut* Maurice Marie Eugène Grout, Personnel Records, *Service Historique de la Défense* (SHD).
39 Hodeir, 108.
40 'Report on Aeronautical Matters,' 12, TNA, PRO: AIR 1/7/6/98/20.
41 Charles Christienne and Pierre Lissarague, *A History of French Military Aviation* (Washington, DC: Smithsonian Institution Press), 59.
42 'Aeronautics,' June 1921, NARA, RG 120, Box 819, 1.
43 Christienne and Lissarague, 59.
44 Christienne and Lissarague, 58.
45 Douglas Porch, *The French Secret Services: From the Dreyfus Affair to the Gulf War* (New York: Farrar, Straus, and Giroux, 1995), 56.
46 Porch, 56.
47 Wilhelm F. Flicke, *War Secrets in the Ether*, ed. by Sheila Carlisle (2 vols., Laguna Hills, CA: Aegean Park Press, 1977), I, 23.
48 Eric Ash, *Sir Frederick Sykes and the Air Revolution 1912–1918* (London: Frank Cass, 1999), 11–14; Malcom Cooper, *The Birth of Independent Air Power: British Air Policy in the First World War* (London: Allen & Unwin, 1986), 23–25.
49 Brigadier-General Sir David Henderson, 'Design of a Scouting Aeroplane,' *FLIGHT*, 18 May 1912.
50 Henderson, 'Design of a Scouting Aeroplane'.
51 Lieut Charles W. Gamble, 'The Technical Aspects of British Aerial Photography during the War 1914–1918,' TNA, PRO: AIR 1/2397/267/7; Nesbit, 11.
52 Nesbit, 10.
53 Nesbit, 11.
54 The Royal Naval Air Service was the naval counterpart of the RFC until the RAF was established in 1918; Nesbit, 11.
55 Raleigh, I, 250.
56 'Royal Flying Corps (Military Wing) at Netheravon. The Concentration Camp,' *FLIGHT*, June 26, 1914.
57 'Royal Flying Corps (Military Wing at Netheravon'.
58 Nesbit, 13.
59 Unknown source quoted in Gamble, 'Technical Aspects of British Aerial Photography,' TNA, PRO: AIR 1/2397/267/7.
60 TNA, PRO: AIR 1/2397/267/7.
61 'Royal Flying Corps (Military Wing) at Netheravon. The Concentration Camp,' *FLIGHT*, June 26, 1914.
62 Davilla and Soltan, 3.
63 Davilla and Soltan, 219.
64 Raleigh, I, 260.
65 J. M. Bruce, *The Aeroplanes of the Royal Flying Corps (Military Wing)* (London: Putnam, 1982), xvi.
66 Gamble, 'The Technical Aspects of British Aerial Photography during the War 1914–1918,' TNA, PRO: AIR 1/2397/267/7.
67 Paul-Louis Weiller, 'The French Aviation of Recognition,' in de Brunoff, 63 (translation provided by Commandant Marc Rivière).
68 Bruce, *Aeroplanes of the Royal Flying Corps (Military Wing)*, 353.
69 John Terraine, *Mons: The Retreat to Victory* (New York: The Macmillan Company, 1960), 31.
70 Dennis E. Showalter, *Tannenberg: Clash of Empires* (Hamden, CT: Archon Books, 1991), 152–153.
71 W. M. Lamberton, comp., *Reconnaissance & Bomber Aircraft of the 1914–1918 War*, ed. by E. F. Cheesman (Letchworth, UK: Harleyford Publications, Ltd., and Los Angeles, CA: Aero Publishers, Inc., 1962), 9.
72 Flicke, I, 3.
73 Flicke, I, 6-7.
74 Christienne and Lissarague, 61; André-H. Carlier, *La Photographie Aérienne pendant la Guerre* (Paris: Libraire Delagrave, 1921), 16 (translation provided by Commandant Marc Rivière).
75 Basil Collier, *Heavenly Adventurer: Sefton Brancker and the Dawn of British Aviation* (London; Secker & Warburg, 1959), 36.
76 Terraine, 31.
77 Bruce, *Aeroplanes of the Royal Flying Corps (Military Wing)*, xvi.
78 Boulogne-la-Grasse is located southeast of Amiens; original text says only Boulogne; Terraine, 31; Ralph Barker, *The Royal Flying Corps in World War I* (London: Constable & Robinson Ltd, 1995), 32.
79 Lyn Macdonald, *1914: The Days of Hope* (New York: Atheneum, 1987), 74.
80 Bruce, *Aeroplanes of the Royal Flying Corps (Military Wing)*, 131–132; Terraine, 61-62.

81 Capt. Philip Joubert de la Ferté quoted in Terraine, 61-62.
82 Terraine, 61-62.
83 Nesbit, 16.
84 Account related by J. T. C. Moore-Brabazon to Frederic Coleman, quoted in Frederic Coleman, *With Cavalry in 1915: The British Trooper in the Trench Line* (Toronto: William Briggs, 1916), 130–132.
85 Christienne and Lissarague, 62.
86 Christienne and Lissarague, 63.
87 Christienne and Lissarague, 64.
88 Macdonald, *1914*, 75; Raleigh, I, 298–300; Constance Babington-Smith, *Air Spy: The Story of Photo Intelligence in World War II* (New York: Ballantine Books, 1957), 9; Brig-Gen Sir J.E. Edmonds (general editor), *Military Operations: France and Belgium, 1914*, 3rd ed. (2 vols., London: Macmillan and Co., Ltd., 1922–25; reprint, London: The Imperial War Museum and Nashville, TN: The Battery Press, 1996), I, 50. Hereafter, titles in this series, 1914–1918, will be referred to as Edmonds, followed by the war year; see bibliography for complete citations for each title.
89 Field-Marshal Viscount French of Ypres, *1914* (London: Constable and Company, Ltd., and Boston: Houghton Mifflin Company, 1919), 47.
90 Edmonds, *1914*, I, 52.
91 Terraine, 77-78.
92 Macdonogh letter in Air Vice Marshal Brooke-Popham, 'RAF Staff College Commandant's Lectures The War of 1914 - 1918,' TNA, PRO: AIR 69/1.
93 Major-General Sir Edward Spears, *Liaison 1914*, 2nd ed. (London: Cassell & Co., 1968), 137–138.
94 Terraine, 77-78.
95 Sir Edward Spears, 137–138.
96 Raleigh, I, 304.
97 Bruce, *Aeroplanes of the Royal Flying Corps (Military Wing)*, 112.
98 Gen Alexander von Kluck, *The March on Paris and the Battle of the Marne 1914* (trans., London: Edward Arnold, 1920), 40.
99 Lieut. Strange flew a French-designed, British-built aeroplane called the Henry Farman F.20; Bruce, *Aeroplanes of the Royal Flying Corps (Military Wing)*, 233.
100 Raleigh, I, 327–328.
101 Field-Marshal French, 43–44.
102 Field-Marshal French, 44.
103 Edmonds, *1914*, I, 76.
104 Raleigh, 304.
105 Holger H. Herwig, *The Marne, 1914* (New York: Random House, 2009), 153–154.
106 Edmonds, *1914*, I, 66.
107 Terraine, 124–1125.
108 Raleigh, I, 304, 314; Babington-Smith, 9.
109 Raleigh, I, 294–1295.
110 W. S. Douglas, 'Notes on Equipment of Units of RFC with BEF,' 27 September 1914, TNA, PRO: AIR 1/836/204/5/268.
111 Sykes, 130.
112 Norman Macmillan, *Sir Sefton Brancker* (London: William Heinemann Ltd., 1935), 61.
113 Macdonald, *1914*, 244.
114 M. L. Skelton, 'Major H. D. Harvey-Kelly, Commanding Officer, No. 19 Squadron,' *Cross and Cockade: The Journal of the Society of World War I Aero Historians* 16, No. 4 (Winter 1975): 365.
115 Raleigh, I, 313; Terraine, 150.
116 Field-Marshal French, 91.
117 Robert Esnault-Pelterie (REP) was an early design French aéroplane. Christienne and Lissarague, 78-79.
118 Edmonds, *1914*, I, 241.
119 Von Kluck, 73.
120 Barbara Tuchman, *The Guns of August* (New York: Ballantine, 1962), 443–444.
121 Macdonald, *1914*, 245.
122 Michael Occleshaw, *Armour against Fate: British Military Intelligence in the First World War* (London: Columbus Books, 1986), 56.
123 Robert B. Asprey, *The First Battle of the Marne* (Philadelphia: J.B. Lippincott Company, 1962), 85; Edmonds, *1914*, I, 248.
124 Occleshaw, 56.
125 Lt Col Georges Bellenger, letter to François Baban, 6 February 1954, *Service Historique de la Défense* (SHD) (translation provided by Commandant Marc Rivière).
126 Asprey, 86.
127 Asprey.
128 Raleigh, I, 327–328.
129 Edmonds, *1914*, I, 270; Macdonald, *1914*, 245.
130 Bellenger letter, *Service Historique de la Défense* (SHD).
131 Tuchman, 445.
132 Davilla and Soltan, 3.
133 Christienne and Lissarague, 79.
134 Bellenger.
135 von Kluck, 93.
136 Bellenger; Hodeir, 108.
137 Christienne and Lissarague, 78
138 Porch, 73.
139 Occleshaw, 57.
140 Porch, 75.
141 Tuchman, 429; Porch, 74.
142 Douglas W. Johnson, *Topography and Strategy in the War* (New York: Henry Holt and Company, 1917), 31.
143 Brindejonc des Moulinais, 'Journal de Bord, Guerre Aérienne,' in Charles Henri Le Goffic, *Général Foch at the Marne: An Account of the Fighting in and near the Marshes of Saint-Gond* (trans., New York: E. P. Dutton and Company, 1918), 199–200.
144 Flicke, I, 23–24.

145 Terraine, 205.
146 Des Moulinais, 199–200.
147 The Blériot XI flew at 60mph. Armament then consisted of one carbine. Blond, 163–164; Davilla and Soltan, 56.
148 Other sources state escadrille 10 deployed to Mery-sur-Seine, south of Mailley. Blond, 163–164; Service historique de l'armée de l'air France (SHAA), *Les escadrilles de l'aéronautique militaire française: Symbolique et histoire 1912–1920* (Paris: Sepeg International, 2004), 42.
149 Blond, 164.
150 Davilla and Soltan, 56; Weiller, in de Brunoff, 63; Blond, 164.
151 Sir John Slesser, 'Air Reconnaissance in Open Warfare,' TNA, PRO: AIR 75/131; Edmonds, *1914*, I, 299.
152 Des Moulinais, 200.
153 Blond, 170–172.
154 Blond, 182.
155 Flicke, I, 23–24.
156 Edmonds, *1914*, I, 309–311.
157 Edmonds, *1914*, I, 319.
158 Des Moulinais, 200–201
159 Edmonds, *1914*, I, 333.
160 Edmonds, *1914*, I.
161 Slesser, TNA, PRO: AIR 75/131; Edmonds, *1914*, I, 340.
162 Unknown source quoted in Slesser, TNA, PRO: AIR 75/131.
163 Slesser, 'Air Reconnaissance,' TNA, PRO: AIR 75/131.
164 Von Kluck, 136.
165 Edmonds, *1914*, I, 359.
166 Des Moulinais, 201–202.
167 Field-Marshal French, 140.
168 Edmonds, *1914*, I, 341–342.
169 Général Joffre quoted in Raleigh, I, 334–1335; Spears, 414.
170 Raleigh, I, 324.
171 Nikolas Gardner, *Trial by Fire: Command and the British Expeditionary Force in 1914* (Westport, CT: Praeger, 2003), 86.
172 No. 3 Squadron diary, quoted in Macdonald, *1914*, 300.
173 Gardner, 87.
174 Gardner, 92-93.
175 Macdonald, *1914*, 310.
176 Gardner, 88.
177 Gardner.
178 Field-Marshal French quoted in Gardner, 88-89.
179 Gardner, 88-89.
180 Gardner, 89.
181 'Papers Collected by Air Vice Marshal Sir W. G. H. Salmond,' TNA, PRO: AIR 20/705.
182 Col George O. Squier, 'Intelligence Services of the British Army in the Field,' 16 February 1915, 5, Squier Papers, USAMHI; Jones, II, 85-87.
183 Gardner, 93.
184 Anthony Farrar-Hockley, *Death of an Army* (London: Arthur Barker, Ltd., 1967), 110.
185 Labussière, in de Brunoff, 183 (translation provided by James Davilla).
186 Alain Dégardin, 'Un Observateur de la Guerre 1914–1918: Eugène Pépin,' in *Vues d'en Haut: La Photographie Aérienne pendant la Guerre de 1914–1918* (Paris: Musée de l'Armée, 1989), 17 (translation provided by Commandant Marc Rivière).
187 Weiller, 68.
188 Christienne and Lissarague, 80.
189 Nicholas Watkis, *The Western Front from the Air* (Stroud, UK: Sutton Publishing, Ltd., 1999), 8-9; Jones, II, 88-89; Edmonds, *1914*, I, 420; A. J. Insall, *Observer: Memoirs of the R.F.C., 1915–1918* (London: William Kimber, 1970), 152.
190 Jones, II, 88-89.
191 Insall, 151–152.
192 Capitaine Edouard Barès quoted in Hodeir, 109.
193 Hodeir, 109.
194 Hodeir, 8.
195 Dégardin, in *Vues d'en Haut*, 18.
196 Labussière, in de Brunoff, 183; Carlier, 16.
197 Von Kluck, 166.
198 Macdonald, *1914*, 310.
199 Gardner, 89
200 Field-Marshal French, 189.
201 J.B.A. Bailey, *Field Artillery and Firepower* (Oxford: The Military Press, 1989), 128–129.
202 Johnson, *Topography and Strategy*, 7.
203 Johnson, *Topography and Strategy*, 11.
204 Johnson, *Topography and Strategy*, 23–24.
205 Count A. E. W. Edward Gleichen, ed., *Chronology of the Great War, 1914–1918* (3 vols., London: Ministry of Information, 1918–20; reprint, 3 vols. in 1, London: Greenhill Books, 2000), 38.
206 'Aeronautics,' June 1921, NARA, RG 120, Box 819, 1–2.
207 Christienne and Lissarague, 70.
208 Christienne and Lissarague, 69.
209 The captive balloon was acquired from the Royal Navy. Note in Edmonds, *1914*, II, 303.
210 Christienne and Lissarague, 69.
211 Christienne and Lissarague, 82.
212 Christienne and Lissarague, 82.
213 Sykes, 147–148.
214 Unknown British officer quoted in Harold Porter, *Aerial Observation: The Airplane Observer, the Balloon Observer, and the Army Corps Pilot* (New York: Harper & Brother Publishers, 1921), 167.
215 'Papers Collected by Salmond,' TNA, PRO: AIR 20/705.
216 Jones, II, 82; Lamberton, 10.
217 J. T. C. Moore-Brabazon, Baron Brabazon, *The Brabazon Story* (London: William Heinemann, Ltd., 1956), 88.

218 Occleshaw, 28.
219 Squier, 'Intelligence Services,' Squier Papers, USAMHI.
220 Squier, 'Intelligence Services.'
221 Farrar-Hockley, 54–156.
222 Ash, 61.
223 Insall, 18; 'Memorandum on New Organisation of the Royal Flying Corps,' 29 November 1914, TNA, PRO: AIR 1/751/204/4/8.
224 Ash, 61.
225 'Memorandum of New Organisation of RFC,' TNA, PRO: AIR 1/751/204/4/8.
226 'Orders for Reconnaissance from 24th Oct. 1914 to 28th Nov. 1914,' TNA, PRO: AIR 1/751/204/4/8.
227 'Organisation of the Royal Flying Corps, British Expeditionary Force, December 1914,' TNA, PRO: AIR 1/751/204/4/8.
228 'Organisation of the Royal Flying Corps, BEF,' TNA, PRO: AIR 1/751/204/4/8.
229 Ash, 61.
230 'Notes on Aerial Reconnaissance' is presented in Appendix C. Sykes, 146.
231 Ash, 61.
232 'Aeronautics,' June 1921, NARA, RG 120, Box 819, 3.
233 Davilla and Soltan, 3
234 'Aeronautics,' June 1921, NARA, RG 120, Box 819, 3.
235 Davilla and Soltan, 3.
236 Davilla and Soltan, 4.
237 Davilla and Soltan.
238 Dégardin, in *Vues d'en Haut*, 17.
239 Grand Quartier Général Défenses et Différents Points du Territoire ... Dossier 7, 'Service Aéronautique, No. 5034,' SHAA, A19.
240 Carlier, 17, 19; Hodeir, 110; Dégardin, in *Vues d'en Haut*, 17.
241 Labussière, in de Brunoff, 183.
242 Carlier, 110; Birger Stichelbaut, *First World War aerial photography. An archeological perspective.* Ph.D. Dissertation (Ghent, BE: University of Ghent, 2009), 12.
243 Peter Chasseaud, '1:10,000 Regular Series Trench Maps,' in *British Western Front Mapping 1914–118: Summary of Geodetic Situation with Reference to the Operations Maps Actually in Use in 1918*, in *The Imperial War Museum Trench Map Archives CD-ROM* (London: The Naval & Military Press and The Imperial War Museum, 2000).
244 'Papers Collected by Salmond,' TNA, PRO: AIR 20/705. Peter Mead, *The Eye in the Air: History of Air Observation and Reconnaissance for the Army, 1785–1945* (London: Her Majesty's Stationery Office, 1986), 66.
245 Jones, II, 88.
246 Weiller, in de Brunoff, 64–168.
247 'Aeronautics,' June 1921, NARA, RG 120, Box 819, 3.
248 Weiller, in de Brunoff, 63, 68; Davilla and Soltan, 220.
249 Carlier, 18.
250 Davilla and Soltan, 220.
251 Davilla and Soltan, 211.
252 Davilla and Soltan, 207.
253 Davilla and Soltan, 220.
254 Raleigh, I, 170; Maj E. G. Toye, RFC, 'History of Photography,' TNA, PRO: AIR 1/724/91/1.
255 Nesbit, 16–18.
256 Lt Col J. T. C. Moore-Brabazon, 'Photography in the RAF,' TNA, PRO: AIR 1/724/91/1.
257 Moore-Brabazon, *The Brabazon Story*, 92-93.
258 Malcom Brown, ed., *The Imperial War Museum Book of the First World War: A Great Conflict Recalled in Previously Unpublished Letters, Diaries, Documents, and Memoirs* (London: Sidgwick & Jackson; and Norman, OK: University of Oklahoma Press, 1991), 134.
259 Jones, II, 88-89.
260 1st Lieut C. C. Darley, 'Notes on Photography in No. 3 Squadron, Royal Flying Corps, December 1914 - End of February, 1915,' TNA, PRO: AIR 1/2395/255/1; Insall, 154.
261 Darley, 'Notes on Photography in No. 3 Squadron,' TNA, PRO: AIR 1/2395/255/1
262 Moore-Brabazon, 'Photography in the RAF,' TNA, PRO: AIR 1/724/91/1.
263 Toye, 'History of Photography,' TNA, PRO: AIR 1/724/91/1.
264 Moore-Brabazon recalled him possessing a 'quite terrifying' personality. Trenchard was a large man, possessing an enormous voice, by which he earned his nickname, 'Boom.' Moore-Brabazon, *The Brabazon Story*, 94.
265 'Capt. [C. D. M.] Campbell's Notes on Aerial Photography,' TNA, PRO: AIR 1/7/6/98/20; Nesbit, 22
266 Darley, 'Notes on Photography in No. 3 Squadron,' TNA, PRO: AIR 1/2395/255/1; 'Papers Collected by Salmond,' TNA, PRO: AIR 20/705.
267 Toye, 'History of Photography,' TNA, PRO: AIR 1/724/91/1.
268 Darley, 'Notes on Photography in No. 3 Squadron,' TNA, PRO: AIR 1/2395/255/1.
269 Darley, 'Notes on Photography in No. 3 Squadron,' TNA, PRO: AIR 1/2395/255/1; Jones, II, 88-89; Insall, 152–153.
270 Toye, 'History of Photography,' TNA, PRO: AIR 1/724/91/1.
271 Jones, II, 88; Richard T. Bickers, *The First Great Air War* (London: Hodder & Stoughton, 1988), 72; Mead, 66; Watkis, 9; Babington-Smith, 10.
272 Darley, 'Notes on Photography in No. 3 Squadron' TNA, PRO: AIR 1/2395/255/1; Jones, II, 88-89; Insall, 152–153.
273 Insall, 152–153; Jones, II, 89.
274 John Charteris, *At G.H.Q.* (London: Cassell and Company, Ltd., 1931), 77.

275 Charteris, *At G.H.Q.*, 77.
276 Andrew Boyle, *Trenchard* (New York: W.W. Norton & Company Inc., 1962), 130.
277 Jones, II, 90–91.
278 Babington-Smith, 9; Lamberton, 10; Watkis, 2.
279 Moore-Brabazon, 'Photography in the RAF,' TNA, PRO: AIR 1/724/91/1.
280 Charteris, *At G.H.Q.*, 77.
281 Jones, II, 90–91; Edmonds, *1915*, I, 85-86.
282 Charteris, *At G.H.Q.*, 82.
283 Lamberton, 10.
284 Bailey, 131.
285 Charteris, *Haig*, 82.
286 Maj J. E. Hahn, *The Intelligence Service Within the Canadian Corps, 1914–1918* (Toronto: The Macmillan Company of Canada, 1930), xv-xvi.
287 Col George O. Squier, 'The Map Room and Printing Room,' 20 April 1915, Squier Papers, USAMHI.
288 Hahn, xvii.
289 Jones, II, 102; Nesbit, 27.
290 Hahn, xvi-xvii.
291 Charteris, *At G.H.Q.*, 88-89.
292 Peter Chasseaud, *Topography of Armageddon: A British Trench Map Atlas of the Western Front, 1914–1918* (Lewes, UK: Mapbooks, 1991), 8; John R. Innes, *Flash Spotters and Sound Rangers: How They Lived, Worked and Fought in the First World War* (London: George Allen & Unwin, Ltd., 1935; reprint, Trowbridge, UK: Redwood Books, Ltd., 1997), 36.
293 Moore-Brabazon, *The Brabazon* Story, 95
294 Jones, II, 80.
295 German report quoted in Porter, 117.
296 Gamble, 'Technical Aspects of British Aerial Photography,' TNA, PRO: AIR 1/2397/267/7; Jones, IV, 107.
297 Jones, IV, 102; Edgar Charles Middleton, *The First World War in the Air* (4 vols., London: The Waverly Book Company, Ltd., 1920), III, 114–1115.
298 Lieut Charles Gamble, 'Photographic Section of the French Flying Corps,' TNA, PRO: AIR 1/404/97/1917.
299 Toye, 'History of Photography,' TNA, PRO: AIR 1/724/91/1.
300 Squadron Commander Letter, 4 September 1915, in Gamble, 'Photographic Section of the French Flying Corps,' TNA, PRO: AIR 1/404/97/1917.
301 Weiller, in de Brunoff, 69.
302 Christienne and Lissarague, 71.
303 Dégardin, in *Vues d'en Haut*, 20–21; Eugène Pépin Papers, MA&E, Paris (translation provided by Commandant Marc Rivière); Carlier, 18.
304 Pépin Papers, 3, MA&E, Paris; Carlier, 18.
305 Watkis, 9–10.
306 Charteris, *At G.H.Q.*, 82.
307 Edmonds, *1915*, II, 142–143.
308 Squier, 'Intelligence Services,' Squier Papers, 5, USAMHI; Jones, II, 85-87; Capt. D.S. Lewis, 'Procedure of Ranging,' 25 January 1915, TNA, PRO: AIR 1/834/204/5/240.
309 Lamberton, 9–10; Johnson, *Topography and Strategy*, 150–151; Lewis, 'Procedure of Ranging,' TNA, PRO: AIR 1/834/204/5/240.
310 Insall, 171–172.
311 Insall.
312 Baring, 306–307.
313 Maj J. P. C. Sewell, 'Notes ... on French Corps Aviation,' TNA, PRO: AIR 1/475/15/312/201.
314 Note in Jones, II, 123–124; Sykes, 155.
315 Ash, 67.
316 Edmonds, *1915*, II, 143; Insall, 63. Lieut-Col. Jack Salmond assumed command of 2nd Wing and Lieut.-Col. W. Sefton Brancker took command of 3rd Wing. Barker, 88.
317 Insall, 63.
318 Lieut-Col. Marindin, 'Conference Notes, 23 July 1915,' TNA, PRO: AIR 1/127/15/40/147.
319 Insall, 155.
320 Hahn, xvii-xviii.
321 Hahn, xviii.
322 Nesbit, 30.
323 Weiller, in de Brunoff, 71.
324 Bailey, 132.
325 Barker, 111.
326 Davilla and Soltan, 5.
327 Christienne and Lissarague, 91.
328 Christienne and Lissarague, 91; Weiller, in de Brunoff, 75.
329 Christienne and Lissarague, 181; Jacques Mousseau, *Le siècle de Paul-Louis Weiller, 1893–1993*, (Paris: Éditions Stock, 1998), 129; *Biographie P.L. Weiller*, Service Historique de la Défense (SHD)
330 Christienne and Lissarague, 181; Mousseau, 134.
331 Edmonds, *1915*, II, 142–143.
332 The aerial photograph pioneer, Lt. C. C. Darley, was shot down by the famous Eindecker ace, Max Immelmann. Barker, 111; Insall, 73.
333 Henri Pétain, *Verdun* (New York: The Dial Press, 1930), 192.
334 Christienne and Lissarague, 97.
335 Christienne and Lissarague, 90.
336 Lamberton, 11.
337 Johnson, *Topography and Strategy*, 44–147.
338 Porch, 96-97, 103.
339 Porch, 102–103.
340 Christienne and Lissarague, 72.
341 Christienne and Lissarague, 95; Lamberton, 11.
342 'Aeronautics,' June 1921, NARA, RG 120, Box 819, 4.
343 'Aeronautics,' June 1921, NARA, RG 120, Box 819, 4–15.
344 'Aeronautics,' June 1921, NARA, RG 120, Box 819, 4–15; Jones, II, 165.
345 Christienne and Lissarague, 72; Davilla and Soltan, 7.
346 Hodeir, 115. This was also known as *Section Artillerie Lourde* (SAL).

347 Alan D. Toelle, *Breguet 14* (Berkhamsted, UK: Albatros Productions, Ltd., 2003), 2.
348 Christienne and Lissarague, 72; Davilla and Soltan, 7.
349 Christienne and Lissarague, 106.
350 Christienne and Lissarague, 77.
351 Lieut Paul-Louis Weiller quoted in Hodeir, 116.
352 Hodeir, 115–116.
353 Hodeir, 115–116; Jones, II, 176–177.
354 'Notes on the French System of Intelligence during the Battle of the Somme,' 23 September 1916, 4–15, TNA, PRO: WO 198/983.
355 Hodeir, 115; Christienne, 98.
356 Christienne and Lissarague, 72.
357 Hodeir, 116.
358 Weiller, in de Brunoff, 79; Jones, II, 177.
359 David Mason, *Verdun* (Gloucestershire, UK: The Windrush Press, 2000), 144.
360 *The History of the 6th Garde Infantry Regiment* (n.p., 1931) (translation provided by Tom Gudmestad).
361 Colonel Höfl, *The History of the Royal Bavarian 20th Infantry Regiment, 'Prinz Franz'* (Munich: n.p., 1929) (translation by Tom Gudmestad).
362 Pétain, *Verdun*, 191.
363 Francis Martel, *Pétain: Verdun to Vichy* (New York: E.P. Dutton & Company, Inc. 1943), 37.
364 Basil H. Liddell Hart, *Foch: The Man of Orleans* (Boston: Little, Brown, and Company, 1932), 181.
365 Dégardin, in *Vues d'en Haut*, 22.
366 Pépin Papers, MA&E, Paris.
367 Edmonds, *1916*, I, 86-87.
368 'Proposals for a Long Distance Machine,' 9 March 1916, encl. 1A to 'Letter from Col R.F.C. E. F. D. 28/2,' TNA: PRO AIR 2/16
369 Jones, II, 146–147; Edmonds, *1916*, I, 85.
370 'Proposals for a Fighting Machine,' TNA, PRO: AIR 2/16.
371 Edmonds, *1916*, I, 86-87.
372 Bailey, 133.
373 Bailey, 134.
374 Bailey, 135.
375 Jones, II, 147; Lamberton, 11; Edmonds, *1916*, I, 85.
376 Edmonds, *1916*, I, 87-88; Gerald Gliddon, *The Battle of the Somme* (Phoenix Mill, UK: Sutton Publishing, 1987), 427.
377 Moore-Brabazon, 'Photography in the RAF,' TNA, PRO: AIR 1/724/91/1.
378 Jones, II, 177; Edmonds, *1916*, I, 88; Mead, 76.
379 Charteris, *At G.H.Q.*, 155.
380 Middleton, III, 120.
381 Campbell, 'Notes on Aerial Photography,' TNA, PRO: AIR 1/7/6/98/20.
382 Capt C. D. M. Campbell, 'Progress of Photography in Royal Flying Corps at Home, to 31st December, 1915,' 22 January 1916, TNA, PRO: AIR 1/724/91/1.
383 'The General Officer Commanding, 6th Brigade,' 3 February 1916, TNA, PRO: AIR 1/724/91/1, 87/7298.
384 Maj Edward Steichen, 'Aerial Photography: The Matter of Interpretation and Exploitation,' Gorrell Report, Ser. G, Roll 24, NARA.
385 Edmonds, *1916*, I, 268.
386 Edmonds, *1916*, I, 286–287.
387 Campbell, 'Notes on Aerial Photography,' TNA, PRO: AIR 1/7/6/98/20; 2nd Lt V. F. Bryce, Photographic Officer, 1st Brigade, BEF, Report, 1 June 1916, TNA, PRO: AIR 1/7/6/98/20.
388 Edmonds, *1916*, I, 300.
389 Jones, II, 197.
390 Edmonds, *1916*, I, 303.
391 Jones, II, 209.
392 Edmonds, *1916*, I, 304.
393 Charteris, *At G.H.Q.*, 77.
394 The RFC at the Somme comprised approximately 3 per cent of the BEF. On 1 July 1916, the RFC suffered the death of just five airmen. Ash, 207.
395 Edmonds, *1916*, I, 477.
396 Edmonds, *1916*, II, 62.
397 Watkis, 11.
398 'Brig-Gen John Charteris to Maj-Gen Macdonogh, War Ministry,' 28 July 1916, TNA: PRO WO 158/897; Charteris, *At G.H.Q.*,159.
399 Friedrich Steinbrecher quoted in Robert B. Asbrey, *The German High Command at War: Hindenburg and Ludendorff Conduct World War I* (New York: Willam Morrow and Company, 1991), 245-246.
400 Ludendorff quoted in Porter, 17.
401 Edmonds, *1916*, II, 304.
402 Edmonds, *1916*, II, 347.
403 Barker, 173–174.
404 Great Britain, General Staff (Intelligence), General Headquarters, *Notes on the Interpretation of Aeroplane Photographs* (France: Army Printing and Stationery Services, November 1916). Hereafter, the three editions of this work will be distinguished by their Stationery Services numbers: S.S. 445 (1916), S.S. 550 (1917), and S.S. 631 (1918).
405 Nesbit, 42.
406 Toye, 'History of Photography,' TNA, PRO: AIR 1/724/91/1.
407 TNA, PRO: AIR 1/724/91/1.
408 Mousseau, 139–142; *Biographie P.L. Weiller.*
409 J. M. Bruce and Jean Noel, *The Breguet 14* (Leatherhead, UK: Profile Publications, 1967), 7.
410 Christienne and Lissarague, 181. 'Boche' was a derogatory French term roughly translated as cabbage or blockhead and applied by the French to German soldiers. Note in Herwig, 62.
411 'Do your best, come what may.' Translation by Mary Elizabeth Ruwell. Mousseau, 143–144.
412 *Study and Exploitation of Aerial Photographs*, 7.
413 'Military Aviation,' in Maurer Maurer, ed., *The US Air Service in World War I* (4 vols., Washington, DC: The Office of Air Force History, 1978), II, 41.

414 'Military Aviation,' in Maurer, ed., *US Air Service*, II, 41.
415 'Military Aviation,' in Maurer, ed., *US Air Service*, II, 49.
416 'Military Aviation', in Maurer II.
417 Brig Gen George W. Goddard, *Overview: A Life-Long Adventure in Aerial Photography* (Garden City, NY: Doubleday & Company, 1969), 10.
418 Benedict Crowell and Robert F. Wilson, *The Armies of Industry: Our Nation's Manufacture of Munitions for a World in Arms, 1917–1918* (2 vols., New Haven, CT: Yale University Press, 1921), II, 419.
419 General von Below quoted in Baring, 200.
420 General von Below quoted in Baring, 201.
421 Davilla and Soltan, 10.
422 A.R. was the primary designation of the A.R.1. *Aviation Militaire* referred to it as '*Avion de Reconnaissance*.' The Renault firm called it '*Avion Renault*' or '*Avant Renault*.' Finally, the French production directorate *STAé* just called it A.R. 1. Davilla and Soltan, 9–10.
423 J. M. Bruce, *The de Havilland D.H. 4* (Leatherhead, UK: Profile Publications, Ltd., 1965), 5.
424 Maj Charles C. Turner, *The Struggle in the Air 1914–1918* (New York: Longmans, Green & Co. and London: Edward Arnold, 1919), 203.
425 'Notes on the French System of Intelligence,' 23 September 1916, TNA: PRO WO 158/983.
426 Jones, III, 314–315.
427 Jones, III, 315.
428 Edmonds, *1917*, I, 11.
429 Edmonds, *1917*, I, 15.
430 Jones, III, 307.
431 Jones, III, 308.
432 Bailey, 139.
433 Charteris, *Haig*, 131.
434 Analysis of the changes observed in the Siegfried Stellung are found in *Étude et Exploitation des Photographies Aériennes* (see Appendix E). David French in Dockrill and French, 76-77; Jones, III, 305–306.
435 David French in Dockrill and French, 77.
436 Edmonds, *1917*, I, 87-88; Jones, II, 318.
437 Edmonds, *1917*, I, 88-89.
438 Général Nivelle had replaced Général Joffre as Général Commandant en Chef des Armées. Edmonds, *1917*, I, 88-89.
439 David French in Dockrill and French, 77-78.
440 Edmonds, *1917*, I, 88-89.
441 Edmonds, *1917*, I.
442 Von Below quoted in Baring, 199.
443 Reported by Lieut Peters and Lieut Balfour of RFC No. 70 Squadron, in Oldham, 9.
444 Edmonds, *1917*, I, 92.
445 Jones, III, 305–306.
446 *Study and Use of Aerial Photographs*, 15 January 1918, sec. I, 10, NARA, RG 120, Box 819.
447 Maj Joseph M. Hanson, ed., *The World War through the Telebinocular* (Meadville, PA: Keystone View Company, 1928), 24.
448 *Study and Use of Aerial Photographs*, 15 January 1918, sec. I, 10n, NARA, RG 120, Box 819.
449 Chasseaud, *Topography of Armageddon*, 11.
450 Ironically, the Germans recaptured the same ground they had effectively scorched during the following 1918 Michael Offensive. David T. Zabecki, *The German 1918 Offensives* (New York: Routledge, 2006), 28; Christienne and Lissarague, 111.
451 David French in Dockrill and French, 68.
452 General Neville quoted in Doughty, 343.
453 Doughty, 349–350.
454 Davilla and Soltan, 12.
455 Doughty, 353–354.
456 Pépin Papers, MA&E, Paris.
457 Davilla and Soltan, 12.
458 Dégardin, in *Vues d'en Haut*, 22.
459 David French in Dockrill and French, 69.
460 Edmonds, *1917*, I, 305.
461 Edmonds, *1917*, I, 312–313.
462 Edmonds, *1917*, I, 311.
463 Edmonds, *1917*, I, 314.
464 Hahn, xix-xx.
465 Interestingly, Bailey entitles this segment '1917: Destruction Reaches Its Zenith,' Bailey, 140.
466 *Artillery Operations of the Ninth British Corps at Messines, June, 1917*, 115–116.
467 *Artillery Operations of the Ninth British Corps at Messines, June, 1917*, 117.
468 John Ferris, 'The British Army and Signals Intelligence in the Field during the First World War,' *Intelligence and National Security* 3, No. 4 (October 1988): 39.
469 A 'black-list' was compiled of the various sources, Jones, IV, 117.
470 Jones, IV, 115–116.
471 Brig Gen Dennis Nolan, Headquarters American Expeditionary Force (HQ AEF) G-2, later concluded that the Air Service's goniometric role was useful only while the army engaged in trench warfare. He expressed the remarkable assumption that 'the service lost much of its value, and it is doubtful that in any future war that its duties will again fall on the intelligence,' and he amazingly concluded that 'We fell heir to it, as we did to many other things, because it was new and the Intelligence Section seemed as good a place as any for it.' Nolan, 'Codes & Ciphers,' 15, Nolan Papers, USAMHI.
472 *Artillery Operations of the Ninth British Corps at Messines, June, 1917*, 115–116.
473 Mead, 85-86; *Artillery Operations of the Ninth British Corps at Messines*, 116.
474 E. P. F. Rose and M. S. Rosenbaum, 'British Military Geologists through War and Peace in the 19th and 20th Centuries,' in James R. Underwood and Peter

L. Guth, eds., *Military Geology in War and Peace* (Boulder, CO: The Geological Society of America, Inc., 1998), 32–33.

475 *Artillery Operations of the Ninth British Corps at Messines, June, 1917*, 1.
476 *Artillery Operations of the Ninth British Corps at Messines, June, 1917*, 35.
477 *Artillery Operations of the Ninth British Corps at Messines, June, 1917*, 95.
478 Edmonds, *1917*, II, 42–43.
479 *Artillery Operations of the Ninth British Corps at Messines, June, 1917*, 52.
480 *Artillery Operations of the Ninth British Corps at Messines, June, 1917*, 118.
481 Jones, IV, 118–119.
482 Jones, IV, 122.
483 Jones, IV, 135.
484 Mead, 85-86; *Artillery Operations of the Ninth British Corps at Messines, June, 1917* (Washington, DC: GPO, 1917), 116.
485 Jones, IV, 135–136.
486 *Study and Exploitation of Aerial Photographs*, 9–10.
487 Occleshaw, 66-67.
488 Jones, IV, 156–157.
489 Jones, IV.
490 Jones, VI, 23.
491 Jones, V, 437.
492 Christienne and Lissarague, 110; Davilla and Soltan, 9.
493 Lamberton, 62.
494 Christienne and Lissarague, 181; Mousseau, 144–1145.
495 Mousseau, 153–154.
496 *Study and Exploitation of Aerial Photographs*, 7-8.
497 *Study and Exploitation of Aerial Photographs*, 11–12.
498 British armour attacked with bales of timber serving as self-contained trench fillers. Upon arriving at the trench obstacle, the tanks offloaded the bale and continued the attack. Johnson, *Battlefields of the World War*, 166–167; Bailey, 142.
499 Chasseaud, '1914–118 Trench Maps.'
500 Edmonds, *1917*, III, 230.
501 Mead, 92-94.
502 Lyn Macdonald, *To the Last Man, Spring 1918* (New York: Carroll & Graf Publishers, Inc.), xxxii-xxxiv.
503 Edmonds, *1918*, I, 108.
504 Salmond's illustrious career from the first days as an aerial observer to his appointment as RFC commander on 18 January 1918 was an important statement on the role that reconnaissance played in British aviation. His counterpart, Trenchard, had returned to England to serve with the newly created Air Ministry Trenchard's assignment became Chief of Air Staff (CAS) for the RAF. Jones, IV, 266; Edmonds, *1918*, I, 108–109.
505 Lieut Thomas M. Hughes quoted in Brown, 136.
506 Jones, IV, 266–268.
507 'Protection of Artillery Positions from Air Observation,' captured German document, TNA, PRO: AIR 1/2124/207/74/3.
508 'Intelligence Summary,' 19 August 1918, TNA, PRO: AIR 1/2124/207/74/3.
509 TNA, PRO: AIR 1/2124/207/74/3.
510 Rudolf Stark quoted in William E. Fischer, Jr., *The Development of Military Night Aviation to 1919* (Maxwell AFB, AL: Air University Press, 1998), 104.
511 Porch, 107; David French in Dockrill and French, 92.
512 Edmonds, *1918*, I, 111.
513 Zabrecki, *The German 1918 Offensives*, 130.
514 David French in Dockrill and French, 92.
515 Bailey, 144.
516 David T. Zabecki, *Steel Wind, Colonel Georg Bruchmüller and the Birth of Modern Artillery* (Westport, CT: Praeger, 1994), 48.
517 Stephen Westman quoted in Malcom Brown, *The Imperial War Museum Book of 1918, Year of Victory* (London: Sidgwick & Jackson, 1998), 38.
518 Charteris, *At G.H.Q.*, 289–290.
519 Zabecki, *The German 1918 Offensives*, 139; Edmonds, *1918*, I, 162.
520 Mead, 128.
521 'Official History,' in Mead, 129.
522 Mead, 130.
523 Unknown GQG intelligence senior quoted in David Kahn, *The Codebreakers, The Story of Secret Writing* (New York: The Macmillan Company, 1967), 341.
524 Hahn, xx-xxi.
525 Trenchard quoted in Edmonds, *1918*, I, 167–168.
526 Edmonds, *1918*, I, 167–168.
527 Author discussions with Dr Michael Neiberg; Jones VI, 405–407.
528 Jones, VI, 405–407.
529 'Conference at Dury at 11.0 p.m., Sunday, March 24th, 1918,' TNA: PRO WO 158/48.
530 René Martel, *L'Aviation Francaise de Bombardment (Des Origines au 11 Novembre 1918)* (Paris: Paul Hartmann, 1939), 317 (translation provided by Allen Suddaby).
531 Jones, IV, 456–457.
532 Jones, IV, 457–458.
533 Mead, 127, Nesbit, 43.
534 Sykes, 215.
535 Cooper, 129.
536 Lieut Charles Gamble, 'Memorandum ... to Major Moore-Brabazon,' 4 February 1918, in 'Re-Organisation of Photographic Activities,' TNA, PRO: AIR 2/163.
537 Toye, 'History of Photography,' TNA, PRO: AIR 1/724/91/1.
538 'Summary of Indications of Enemy Preparations for an Offensive in Flanders,' Conger Papers, Box 2, USAMHI.

539 Jones, IV, 368.
540 Mead, 131.
541 Edmonds, *1918*, II, 273–274.
542 Mead, 131; Jones, IV, xxx.
543 Edmonds, *1918*, II, 374.
544 'Summary of Indications of Enemy Preparations for an Offensive in Flanders,' Conger Papers, Box 2, USAMHI.
545 Corporal Martin E. Friedmann, 'History of Photographic Section #1, Air Service, U.S.A., 1918,' 1, Gorrell Report, ser. G, r. 24, NARA.
546 Eric E. Dixon, *Diary of Eric E. Dixon, Whippany, N.J., U.S.A.: An American Citizen* (Whippany, NJ: n.p., n.d.), 49, Dixon Papers, USAMHI.
547 Friedmann, 'History of Photographic Section #1,' 2, Gorrell Report, ser. G, r. 24, NARA.
548 'Final Report of Chief of Air Service,' in Maurer, ed., *US Air Service*, I, 182.
549 Friedmann, 'History of Photographic Section #1,' Gorrell Report, ser. G, r. 24, NARA.
550 HQ First Army Corps, A.E.F., Office of Chief of Air Service, 'Report of Operations, For April 20, 1918,'Gorrell Report, ser. C, r. 12, NARA.
551 'Final Report of Chief of Air Service,' in Maurer, ed., *US Air Service*, I, 210
552 'Final Report of Chief of Air Service,' in Maurer, ed., *US Air Service*, I, 182.
553 Sam H. Frank, *American Air Service Observation in World War I*, Ph.D. Dissertation (Gainesville, FL: University of Florida, 1961), 216.
554 Friedmann, 'History of Photographic Section #1,' HQ, 1st Army Corps, Office of Chief of Air Service, 'Report of Operations, for April 20, 1918,' Gorrell Report, ser. C, r. 12, NARA.
555 Maj Edward Steichen, 'American Aerial Photography at the Front,' 2, NASM, A30.2/24.
556 Brig Gen Dennis E. Nolan, 'G-2-C Topography,' 5, Nolan Papers, USAMHI.
557 Christienne and Lissarague, 73.
558 Knappen, 199–202.
559 Christienne and Lissarague, 73.
560 'Aeronautics,' June 1921, 8-9, NARA, RG 120, Box 819.
561 Author conversation with Dr James Davilla.
562 J. T. C. Moore-Brabazon, 'Photography's Part in the War,' TNA, PRO: AIR 1/724/91/2.
563 John W. Stuart Gilchrist. *An Aerial Observer in World War I* (n.p.: privately printed, 1966), 127.
564 'Final Report of Chief of Air Service,' in Maurer, ed., *US Air Service*, I, 207.
565 'Aeronautics,' June 1921, 8-9, NARA, RG 120, Box 819.
566 'Intelligence Summary,' 9 August 1918, TNA, PRO: AIR 1/2124/207/74/3.
567 'Intelligence Summary,' 19 September 1918, TNA, PRO: AIR 1/2124/207/74/3.
568 Author conversation with Dr James Davilla.
569 Christienne and Lissarague, 181; *Biographie P.L. Weiller.*
570 Mousseau, 160; David Jean, Georges-Didier Rohrbacher, Bernard Palmieri, and Service historique de l'armée de l'air France, Les escadrilles de l'aéronautique militaire française:Symbolique et histoire 1912–1920 (Paris: Sepeg International, 2004), 420.
571 Christienne and Lissarague, 183; Mousseau, 162.
572 G-2, 1st Army Corps, 'Distribution of Photographs made by the Aviation Section of the Corps,' 11 August 1918, Gorrell Report, ser. C, r. 12.
573 Jones, VI, 415.
574 Frank, *American Air Service Observation*, 218; 'Letter to the Director of Air Service, French Sixth Army from the Chief of Air Service, I Army Corps, American Expeditionary Forces, July 9, 1918,' in US Army, Office of Military History, *United States Army in the World War, 1917–1919* (17 volumes, Washington, DC: GPO, 1948; reprint, Washington, DC: US Army Center for Military History, 1988-92), III, 397–398.
575 General Rawlinson quoted in Zabecki, *The German 1918 Offensives*, 317.
576 Sykes, 202.
577 Edmonds, *1918*, I, 12.
578 Erich von Ludendorff, *Ludendorff's Own Story, August 1914–1November 1918: The First World War from the Siege of Liege to the Signing of the Armistice as Viewed from the Grand Headquarters of the German Army* (2 vols., New York: Harper & Brothers Publishers, 1919), II, 326.
579 Bailey, 149.
580 Hahn, xxi.
581 'Squadron Leader F.C.V. Laws O.B.E. to G.O.C., Advanced Operations,' 18 November 1918, TNA, PRO: AIR 1/724/91/2.
582 'Final Report of Chief of Air Service,' in Maurer, ed., *US Air Service*, I, 221–223.
583 Frank, *American Air Service Observation*, 266–267; 'Final Report of Chief of Air Service,' in Maurer, ed., *US Air Service*, I, 226–228.
584 G-2 to Chief of Staff, 'Account of the Operation,' 7, Drum Papers, Box 14, USAMHI.
585 'Final Report of Chief of Air Service,' in Maurer, ed., *US Air Service*, I, 239.
586 Nolan, 'G-2-C Topography,' 12, Nolan Papers, USAMHI.
587 Maj James Barnes, 'History of Photographic Section #4, Air Service, U.S.A. 1918,' Gorrell Report, ser. G, r. 24, NARA.
588 Frank, *American Air Service Observation*, 312.
589 G-2 to Chief of Staff, 'Account of the Operation,' 51, Drum Papers, Box 14, USAMHI.
590 1st Lt Lawrence L. Smart, *The Hawks that Guided the Guns* (n.p.: privately Printed, 1968), 44.
591 Smart, *The Hawks that Guided the Guns*, 45.
592 G. T. Lindstrom, 'Photographic Missions,' Gorrell Report, ser. N, r. 53, NARA.

593 Nolan, 'Codes and Ciphers,' 14, Nolan Papers, USAMHI.
594 G-2 to Chief of Staff, 'Account of the Operation,' 7, Drum Papers, Box 14, USAMHI; 'Final Report of Chief of Air Service,' in Maurer, ed., *US Air Service*, I, 239.
595 'Historical Account of Organisation and Functioning of Subdivision,' 62, Drum Papers, Box 14, USAMHI.
596 Barnes, 'History of Photographic Section #4,' Gorrell Report, ser. G, r. 24, NARA.
597 G-2 to Chief of Staff, 'Account of the Operation,' 56-57, Drum Papers, Box 14, USAMHI.
598 Steichen, 'American Aerial Photography at the Front,' 1, NASM, A30.2/24.
599 'Historical Account,' 61, Drum Papers, Box 14, USAMHI.
600 'Final Report of Chief of Air Service,' in Maurer, ed., *US Air Service*, I, 249–250; Frank, *American Air Service Observation*, 353; G-2 to Chief of Staff, 'Account of the Operation,' 56-57, Drum Papers, Box 14, USAMHI.
601 G-2 to Chief of Staff, 'Account of the Operation,' 48, Drum Papers, Box 14, USAMHI.
602 G-2, 1st Army, 'Account of the Operation,' Drum Papers, Box 14, USAMHI.
603 'Account of the Operation.'
604 'Final Report of Chief of Air Service,' in Maurer, ed., *US Air Service*, I, 213; Frank, *American Air Service Observation*, 292.
605 'Final Report of Chief of Air Service,' in Maurer, ed., *US Air Service*, I, 152.
606 'Final Report of Chief of Air Service,' in Maurer, ed., *US Air Service*, I, 154.
607 'Final Report of Chief of Air Service', in Maurer I.
608 G-2, 1st Army, 'Account of the Operation,' 56-57, Drum Papers, Box 14, USAMHI.
609 G-2, 1st Army, 'Account of the Operation,' 14, Drum Papers, Box 14, USAMHI.
610 Porter, 158.
611 Steichen, 'American Aerial Photography at the Front,' 1, NASM, A30.2/24.
612 Pershing quoted in Knappen, 212–213; Powell, 293.
613 'Operations Report for November 5, 1918,' HQ 3rd Army Corps, Office of Chief of Air Service, Gorrell Report, ser. C, r. 12, NARA.
614 Frank, *American Air Service Observation*, 363–364.
615 Frank, *American Air Service Observation*, 353.
616 Edmonds, *1918*, V, 576-577.
617 Moore-Brabazon, 'Photography's Part in the War,' TNA, PRO: AIR 1/724/91/2.
618 TNA, PRO: AIR 1/724/91/2.
619 TNA, PRO: AIR 1/724/91/2.
620 French term for photographic plate is *cliché*. Lt G.W. Gill and 1st Lt Thomas E. Hibben, 'Preparation of Photographic Missions,' Lectures given at 2nd Aviation Instruction Center, Gorrell Report, Ser. J, Roll 34, NARA; Solomon J. Solomon, *Strategic Camouflage* (New York: E.P. Dutton and Company), 1.
621 'Lecture on Aerial Photography, Centre d'Etudes de l'Aeronautique,' 25 April 1923. NASM A30.2/36.
622 *Study and Use of Aerial Photographs*, 15 January 1918, sec. I, 34–135, NARA, RG 120, Box 819; 'Aerial Photography for Pilots,' 21 August 1918, 20, NASM, A30.2/2; 'Lecture on Aerial Photography, Centre d'Etudes de l'Aeronautique,' 25 April 1923. NASM A30.2/36.; *Notes on the Interpretation of Aeroplane Photographs*, March 1917 (S.S. 550), 5, and February 1918 (S.S. 631), 3; Solomon, 4; Hahn, 21.
623 Hahn, 21.
624 Porter, 184.
625 Porter,181.
626 'Characteristics of the Ground and Landmarks in the Enemy Lines Opposite the British Front from the Sea to St Quentin' TNA, PRO: AIR 10/61.
627 *Notes on the Interpretation of Aeroplane Photographs*, March 1917 (S.S. 550), 5, and February 1918 (S.S. 631), 3; Solomon, 4; Hahn, 21.
628 1st Lt Thomas E. Hibben, 'Interpretation of Aerial Photography,' 49th Inf. Att. A.S. Gorrell Report, ser. J, r. 34, NARA.
629 'Table of Information for Taking Aerial Photographs,' Gorrell Report, ser. G, r. 24, NARA.
630 Gill and Hibben, 'Preparation of Photographic Missions,' Gorrell Report, ser. J, r. 34, NARA.
631 Ives, 40.
632 Christienne and Lissarague, 70.
633 *Notes on the Interpretation of Aeroplane Photographs*, March 1917 (S.S. 550), 10, and February 1918 (S.S. 631), 11; Gill and Hibben, 'Preparation of Photographic Missions,' Gorrell Report, ser. J, r. 34, NARA; *Study and Use of Aerial Photographs*, 15 January 1918, sec. I, 33, NARA, RG 120, Box 819; Pépin, 'Lecture on the Study and Interpretation of Aerial Photographs,' 1, NASM.
634 Gill and Hibben, 'Preparation of Photographic Missions,' Gorrell Report, ser. J, r. 34, NARA; Porter, 176–177; Hart, 37; 1st Army Historical Files, NARA, RG 120, Box 22; Ives, 39–40.
635 Hahn, 22.
636 Ives, 39–40.
637 G-2, 1st Army Corps, 'Summary of Intelligence,' Nolan Papers, USAMHI.
638 Gill and Hibben, 'Preparation of Photographic Missions,' Gorrell Report, ser. J, r. 34, NARA; Hahn, 36; Hart, 36–37.
639 Ives, 39–40.
640 Lieut Michel quoted in Porter, 173.
641 'Lecture on Aerial Photography, Centre d'Etudes de l'Aeronautique,' 25 April 1923. NASM A30.2/36.
642 Hart, 36–37.
643 Hahn, 36–37.
644 1st Army Historical Files, NARA, RG 120, Box 22.
645 Gamble, 'Photographic Section of the French Flying Corps,' TNA, PRO: AIR 1/404/97/1917.
646 *Notes on the Interpretation of Aeroplane Photographs*, March 1917 (S.S. 550), 10, and February 1918 (S.S. 631), 11; Porter, 163; Hahn, 37–38.

647 Jonathan Crary, *Techniques of the Observer: On Vision and Modernity in the Nineteenth Century* (Cambridge, MA: MIT Press, 1991), 117–118; Arthur W. Judge, *Stereoscopic Photography* (London: Chapman & Hall, Ltd., 1950), 14–115.
648 Herman von Helmholtz, *Treatise on Physiological Optics* (trans.; n.p., n.d.), quoted in Crary, 122–124.
649 Moore-Brabazon, *The Brabazon* Story, 104.
650 'Notes on Photography from Aeroplanes,' 20 December 1914, TNA, PRO: AIR 1/836/204/5/268.
651 Carlier, 76.
652 Carlier.
653 Carlier.
654 Author discussion with Nicholas Watkis.
655 Annotation example provided by Nicholas Watkis.
656 'Interpretation of Aeroplane Photographs in Mesopotamia,' TNA, PRO: AIR 10/1001.
657 TNA, PRO: AIR 10/1001.
658 1st Army Historical Files, NARA, RG 120, Box 22.
659 Gill and Hibben, 'Preparation of Photographic Missions,' Gorrell Report, ser. J, r. 34, NARA.
660 1st Army Historical Files, NARA, RG 120, Box 22.
661 'Aerial Photography for Pilots,' 21, NASM, A30.2/2.
662 'Aeronautics,' June 1921, 24–125, NARA, RG 120, Box 819.
663 'Lecture on Aerial Photography, Centre d'Etudes de l'Aeronautique,' 1923, NASM, A30.2/36.
664 G-2, 1st Army Corps, 'Summary of Intelligence,' Nolan Papers, USAMHI.
665 *Notes on the Interpretation of Aeroplane Photographs*, March 1917 (S.S. 550), 5, and February 1918 (S.S. 631), 3; Hahn, 21.
666 *Study and Use of Aerial Photographs*, 15 January 1918, sec. II, 22, NARA, RG 120, Box 819; 'Depots de Munitions Ennemis, Woevre et Hauts de Meuse,' Mitchell Papers, LC.
667 'Historical Account,' 57, Drum Papers, USAMHI.
668 1st Lt J. L. Stewart, 'Report on the Organisation and Operation of Intelligence 'A' at the British II Army (with Special Attention to the Photographic Section, Enemy Defences and Rear Organisations),' 18–29 September 1918, Conger Papers, Box 2, USAMHI.
669 'Organisation of Intelligence,' HQ, British 5th Army, Conger Papers, Box 2, USAMHI.
670 S. T. Hubbard, Jr., 'Memorandum for Col Nolan, Subject: 'Report on Trip to British Front',' 3 January 1918, 6, Conger Papers, Box 2, USAMHI.
671 Stewart, 'Report on the Organisation and Operation of Intelligence 'A',' Conger Papers, Box 2, USAMHI.
672 'Aerial Photography. Statement Made by Squadron Leader F. C. V. Laws O.B.E. to Mr. A. J. Insall on 3rd June, 1924,' TNA, PRO: AIR 1/724/91/2.
673 Stewart, 'Report on the Organisation and Operation of Intelligence 'A',' Conger Papers, Box 2, USAMHI.
674 'Historical Account,' 63, Drum Papers, USAMHI.
675 'Notes on Branch Intelligence,' 91st Aero Squadron, 1 November 1918, Gorrell Report, ser. M, r. 40, NARA.
676 'Historical Account of Organisation and Functioning of Subdivision,' 62, Drum Papers, Box 14, USAMHI.
677 'Provisional Manual of Operations,' in Maurer, ed., *US Air Service*, II, 272.
678 The location of this postwar archive is unknown. Some files of divisional photographs exist, albeit incomplete. It has been suggested that the WW I glass negatives were recycled in the 1950s for silver recovery; see 'Final Report of Chief of Air Service,' in Maurer, ed., *US Air Service*, I, 154–1155.
679 'Conference du SLT [Sous-Lieutenant] Pouquet sur le Fonctionnement du Service d'Interpretation des Photographies dans un Secteur Offensif faite au G.D.E., le 27 December 1916,' SHAA; transl. as 'Working of the Service of Interpretation of Photographs in an Offensive Sector (Conference by Sub/Lieutenant Pouquet),' 1, NARA, RG 120, Box 819; Coleman, 'Interpretation of Aerial Photographs,' 1921, NARA, RG 120, Box 819.
680 Pouquet, 'Working of the Service of Interpretation of Photographs,' 1, NARA, RG 120, Box 819.
681 *Illustrations to Accompany Captain De Bissy's Notes*, 1.
682 *Illustrations to Accompany Captain De Bissy's Notes.*
683 Steichen, 'American Aerial Photography at the Front,' 2, NASM, A30.2/24.
684 Capt Jean de Bissy, 'Note Concernant l'Interpretation Méthodiqie de Photographies Aériennes,' 19 November 1915, SHAA.
685 *Bulletin du Service de la Photographie Aérienne aux Armées*, 7 January 1916 (Paris: Imprimerie Nationale, 1916), 7; *Bulletin of the Aerial Photography Department in the Field* (Washington, DC: GPO, 1918).
686 Occleshaw, 66.
687 Edmonds, *1916*, I, 132.
688 Moore-Brabazon, *The Brabazon* Story, 95-96.
689 'Scale of Issue of Periodical Publications,' Conger Papers, Box 2, USAMHI.
690 'Report on Day Bombardment Training,' Gorrell Report, ser. J, r. 35, NARA.
691 *Notes on the Interpretation of Aeroplane Photographs* (Washington, DC: GPO, 1918), reprint of February 1918 (S.S. 631).
692 *Study and Utilization of Aerial Photographs* (transl., [France]: Headquarters American Expeditionary Forces, France July, 1918), NARA, RG 120, Box 6202.
693 GHQ, [French] Armies of the North and North East, *Study and Exploitation of Aerial Photographs*, Translated and Issued by the Division of Military Aeronautics, US Army (Washington, DC: GPO, July 1918).
694 GHQ, [French] Armies of the North and North East, *Study and Use of Aerial Photography* (Washington,

DC: Government Printing Office), November 1921, copy in Part XLIII, sec. 5, in American Mission with the Commanding General, Allied Forces of Occupation, Mainz, Germany, 'A Study of the Organisation of the French Army ...,' NARA, RG 120, Box 819.

695 G-2, 1st Army, 'Account of the Operation,' 67, Drum Papers, Box 14, USAMHI.

696 Steichen, 'Aerial Photography,' Gorrell Report, ser. G, r. 24, NARA.

697 Pouquet, 'Working of the Service of Interpretation of Photographs,' 3, NARA, RG 120, Box 819.

698 Gill and Hibben, 'Preparation of Photographic Missions,' Gorrell Report, ser. J, r. 34, NARA.

699 'Balloon Training,' slide 59, Gorrell Report, ser. F, r. 23, NARA.

700 Squier, 'Map Room and Printing Room,' Squier Papers, USAMHI.

701 1st Lt Thomas E. Hibben, 'Topography, Lecture delivered at the Second Aviation Instruction Center,' Gorrell Report, ser. J, r. 34, NARA; 'Balloon Training,' Gorrell Report, ser. F, r. 23, NARA.

702 Chasseaud, *Topography of Armageddon*, 8-9.

703 Chasseaud, '1:10,000 Regular Series Trench Maps,' in *British Western Front Mapping.*

704 Draft notes show that 'and the nation' was subsequently deleted from the text. Nolan, 'G-2-C Topography,' Nolan Papers, USAMHI.

705 Maxel, 'Instructions on the Intelligence Service,' Conger Papers, Box 2, USAMHI.

706 Porter,121–123.

707 Hahn, 96.

708 Général Henri Pétain, French General Staff and 3rd Bureau No. 1077/3, 'Notice about Maps and Special Plans to be Drawn by the Armies,' Conger Papers, Box 2, USAMHI.

709 Hibben, 'Topography,' Gorrell Report, ser. J, r. 34, NARA.

710 Pouquet, 'Working of the Service of Interpretation of Photographs,' 1, NARA, RG 120. Box 819; 'Balloon Training,' Gorrell Report, ser. F, r. 23, NARA.

711 Chasseaud, *British Western Front Mapping.*

712 Pétain, 'Notice about Maps and Special Plans,' Conger Papers, Box 2, USAMHI.

713 Pétain, 'Notice about Maps and Special Plans,' Conger Papers, Box 2, USAMHI; 'Notes on the French System of Intelligence,' 6, TNA: PRO WO 158/983; Marie-Anne de Villèle, Agnès Beylot, and Alain Morgat, eds., *Du Paysage à La Carte: Trois Siècles de Cartographie Militaire de la France* (Vincennes, FR: Ministère de la Défense, SHAA, 2002), 129–130 (translation by Commandant Marc Rivière).

714 Hibben, 'Topography,' Gorrell Report, ser. J, r. 34, NARA; 'Balloon Training,' Gorrell Report, ser. F, r. 23, NARA.

715 Chasseaud, '1:10,000 Regular Series Trench Maps,' in *British Western Front Mapping.*

716 'Balloon Training,' Gorrell Report, ser. F, r. 23, NARA; Chasseaud, '1:10,000 Regular Series Trench Maps,' in *British Western Front Mapping.*

717 Chasseaud, *Topography of Armageddon*, 11.

718 Pétain, 'Notice about Maps and Special Plans,' Conger Papers, Box 2, USAMHI; 'Notes on the French System of Intelligence,' 6, TNA, PRO: WO 158/983; Villèle, 129–130 (translation by Commandant Marc Rivière).

719 Maj Kerr T. Riggs, 'Memorandum to Chief of Intelligence 'A,' 14 February 1918, Subject: 'Memorandum Report of Visit to HQ 1st French Army,' Conger Papers, USAMHI.

720 Author discussion with Peter Chasseaud.

721 'Balloon Training,' Gorrell Report, ser. F, r. 23, NARA.

722 Stewart, 'Report on the Organisation and Operation of Intelligence 'A',' Conger Papers, Box 2, USAMHI.

723 Chasseaud, *British Western Front Mapping.*

724 Hubbard, 'Report on Trip to British Front,' Conger Papers, Box 2, USAMHI.

725 Hibben, 'Topography,' Gorrell Report, ser. J, r. 34, NARA.

726 Pétain, 'Notice about Maps and Special Plans,' Conger Papers, Box 2, USAMHI.

727 G-2, 1st Army, to Chief of Staff, 'Account of the Operation,'47, Drum Papers, Box 14, USAMHI.

728 Pétain, 'Notice about Maps and Special Plans,' Conger Papers, Box 2, USAMHI.

729 'Aerial Photography for Pilots,' 21, NASM, A30.2/2.

730 Hahn, 126.

731 'Photographs and the Map, Appendix D,' in Campbell, 'Notes on Aerial Photography,' TNA, PRO: AIR 1/7/6/98/20.

732 Stewart, 'Report on the Organisation and Operation of intelligence 'A',' Conger Papers, Box 2, USAMHI.

733 G-2, 1st Army, to Chief of Staff, 'Account of the Operation,'48, Drum Papers, Box 14, USAMHI.

734 'Aeronautics,' June 1921, 22, NARA, RG 120, Box 819.

735 Chasseaud, *British Western Front Mapping.*

736 *Illustrations to Accompany Captain De Bissy's Notes*, 3.

737 Post of Command (PC) was translated from the French term for command post; the AEF adopted that term for operations at the front. *Illustrations to Accompany Captain De Bissy's Notes*, 3; Pouquet, 'Working of the Service of Interpretation of Photographs,' NARA, RG 120, Box 819.

738 'Historical Account of Organisation and Functioning of Subdivision,' 57, Drum Papers, Box 14, USAMHI.

739 'Aerial Photography for Pilots,' 22, NASM, A30.2/2.

740 1st Lt Thomas E. Hibben, 'The Importance of Aerial Photographs in Intelligence,' Lecture given at 2nd A.I.C., Gorrell Report, ser. J, r. 34, NARA.

741 Brooks, 93; G. A. Kiersch and J. R. Underwood, Jr., 'Geology and Military Operations, 1800–1960: An Overview,' in Underwood and Guth, 12; Alfred H.

Brooks, 'The Use of Geology on the Western Front,' in D. White, ed., *Shorter Contributions to General Geology, 1920* (Washington, DC: GPO, 1921), 93.
742 Brooks, in White, 93; Kiersch and Underwood, in Underwood and Guth, 12.
743 de Villèle, 129–130 (translation by Commandant Marc Rivière).
744 Chasseaud states it best on the challenge of mobile operations: 'The faster the movement, the smaller the scale - otherwise units and formations would be off the sheet before they received it!' Chasseaud, '1:10,000 Regular Series Trench Maps,' in *British Western Front Mapping*.
745 de Villèle, 129–130 (translation by Commandant Marc Rivière).
746 Hibben, 'Importance of Aerial Photographs,' 1, Gorrell Report, ser. J, r. 34, NARA.
747 'Aeronautics,' June 1921, 24–125, NARA, RG 120, Box 819.
748 Hahn, 263
749 *Study and Use of Aerial Photographs*, 15 January 1918, sec. II, 1, NARA, RG 120, Box 819; *Bulletin du Service de la Photographie Aérienne*; *Bulletin of the Aerial Photography Department*, 7; *Illustrations to Accompany Captain De Bissy's Notes*, 1.
750 Hahn, 263.
751 Hibben, 'Importance of Aerial Photographs,' Gorrell Report, ser. J, r. 34, NARA.
752 Hardy, 'Interpretation of Aerial Photography,' 2, NASM, A30.2/28.
753 Charteris quoted by Occleshaw, 100–101.
754 2nd Lt Arthur A. Zimmerman, 'Report on Visit to VIIIth French Army, February 22 to March 2, 1918,' 7 March 1918, 4–16, Nolan Papers, USAMHI.
755 *Study and Use of Aerial Photographs*, 15 January 1918, sec. I, 38, NARA, RG 120, Box 819.
756 Hubbard, 'Report on Trip to British Front,' Conger Papers, Box 2, USAMHI.
757 1st Army Corps, 'Summary of Intelligence,' Nolan Papers, USAMHI.
758 Hahn, 188.
759 *Study and Use of Aerial Photographs*, 15 January 1918, sec. III, 15, NARA, RG 120, Box 819.
760 Terrence J. Finnegan, 'Military Intelligence at the Front 1914–1918,' *Studies in Intelligence* 53, No. 4 (December 2009), 32.
761 Innes, 67.
762 William A. Morgan, 'Invasion on the Ether: Radio Intelligence at the Battle of St Mihiel, September 1918,' *Military Affairs* 51, No. 12 (April 1987): 57-61.
763 *Illustrations to Accompany Captain De Bissy's Notes*, 1.
764 Pépin, 'Lecture,' 1917, 1, NARA.
765 Dixon, *Diary*, 52; Maj David M. Henry, 'Report on my Visit to the British Army for the Purpose of Observing the Work of the Intelligence Section of the Royal Flying Corps,' 7 March 1918, 8, Conger Papers, USAMHI.
766 Riggs, 'Memorandum Report of Visit to Headquarters 1st French Army.'
767 *Study and Use of Aerial Photographs*, 15 January 1918, sec. I, 24, NARA, RG 120, Box 819.
768 *Study and Use of Aerial Photographs*, 15 January 1918, sec. I, 24, NARA, RG 120, Box 819.
769 'Lecture on Aerial Photography, Centre d'Etudes de l'Aeronautique,' 25 April 1923. NASM A30.2/36.
770 Hahn, 260.
771 'Aerial Photography for Pilots,' 21, NASM, A30.2/2.
772 *Illustrations to Accompany Captain De Bissy's Notes*, 3.
773 Christienne and Lissarague, 181–183.
774 Mousseau, 162.
775 Christienne and Lissarague, 181–183.
776 Kirke, senior GHQ intelligence officer, quoted in David French in Dockrill and French, 76.
777 Lieut Estienne, 'Army Photographic Reconnaissance as it Should Be,' 29 March 1923, 2, NASM, A30.2/34.
778 Pépin, 'Lecture on the Study and Interpretation of Aerial Photographs,' 10 December 1917, 2–3, NASM.
779 'Aeronautics,' June 1921, 7-8, NARA, RG 120, Box 819.
780 *Aviation Militaire* was also referred to as either *Director of Service Aéronautique* or the Aeronautical Command at GQG; 'Aeronautics,' June 1921, 20, NARA, RG 120, Box 819.
781 Maxel, 'Instructions on the Intelligence Service,' Conger Papers, Box 2, USAMHI.
782 *Study and Use of Aerial Photographs*, 15 January 1918, sec. I, 29, NARA, RG 120, Box 819.
783 'Instructions on the Employment of Aerial Observation in Liaison with Artillery,' in Maurer, ed., *US Air Service*, II, 181.
784 Martel, 19–20.
785 Martel, 20.
786 Martel , 250–252.
787 'Notes on the French System of Intelligence during the Battle of the Somme,' 6-7.
788 Maxel, 'Instructions on the Intelligence Service,' Conger Papers, Box 2, USAMHI.
789 Maxel, 'Instructions on the Intelligence Service.'
790 'Notes on the French System of Intelligence during the Battle of the Somme,' 9, TNA, PRO: WO 198/983.
791 1st Lt Colman D. Frank, 'Report of a Visit to VIIIth Army Headquarters to Attend the Interrogatory of Prisoners Taken in the Coup de Main on the Lorraine Sector,' 20 February 1918, Conger Papers, USAMHI.
792 Zimmerman, 'Report on Visit to VIIIth French Army,' 2–3, Nolan Papers, USAMHI.
793 Pétain, 'Notice about Maps and Special Plans,' Conger Papers, Box 2, USAMHI.
794 Pétain, 'Notice about Maps and Special Plans.'
795 Zimmerman, 'Report on Visit to VIIIth French Army,' 2–3, Nolan Papers, USAMHI.

796 'Notes on the French System of Intelligence,' 6-7, TNA, PRO: WO 198/983.
797 'Notes on the French System of Intelligence,' 6-7, TNA, PRO: WO 198/983.
798 'Notes on the French System of Intelligence,' 9, TNA, PRO: WO 198/983.
799 The U.S. report referred to the *Chef d'Cartographie* as Head of the Cartographie. The more correct French term is used hereafter. Zimmerman, 'Report on Visit to VIIIth French Army,' 2–3, Nolan Papers, USAMHI.
800 Zimmerman, 'Report on Visit to VIIIth French Army,' 2–3, Nolan Papers, USAMHI.
801 Canevas de Tir means 'triangulation for the direction of artillery fire.'
802 Pétain, 'Notice about Maps and Special Plans,' Conger Papers, Box 2, USAMHI.
803 'Instructions on the Functions of the Corps Intelligence Service,' Conger Papers, USAMHI.
804 'Lecture on Aerial Photography, Centre d'Etudes de l'Aeronautique,' 25 April 1923. NASM A30.2/36.
805 'Notes on the French System of Intelligence,' 6-7, TNA, PRO: WO 198/983.
806 Gamble, 'Photographic Section of the French Flying Corps,' TNA, PRO: AIR 1/404/97/1917.
807 *Study and Use of Aerial Photographs*, 15 January 1918, sec. I, 26–27, NARA, RG 120, Box 819.
808 *Study and Use of Aerial Photographs*, 15 January 1918, sec. I, 24–125, NARA, RG 120, Box 819.
809 'Notes on the French System of Intelligence,' 5-6, TNA, PRO: WO 198/983.
810 Pétain, 'Notice about Maps and Special Plans,' 5, Conger Papers, Box 2, USAMHI; *Aerial Observation for Artillery, Based on the French Edition of December 29, 1917* ([Washington, DC: GPO?], May 1918), 8.
811 Pétain, 'Notice about Maps and Special Plans,' 1–2, Conger Papers, Box 2, USAMHI.
812 Pétain, 'Notice about Maps and Special Plans,' 5, Conger Papers, Box 2, USAMHI; *Aerial Observation for Artillery*, 8.
813 'Notes on the French System of Intelligence,' 10, TNA, PRO: WO 198/983.
814 Maxel, 'Instructions on the Intelligence Service,' Conger Papers, Box 2, USAMHI.
815 'Aeronautics,' June 1921, 23, NARA, RG 120, Box 819.
816 Sewell, 'Notes ... on French Corps Aviation,' TNA, PRO: AIR 1/475/15/312/201.
817 Author discussion with Dr James Davilla.
818 Squadron strength reflects an average count; no definitive number of aéroplanes was assigned to French *escadrilles* despite those indicated in the report. Sewell, 'Notes ... on French Corps Aviation,' TNA, PRO: AIR 1/475/15/312/201.
819 'Notes on the French System of Intelligence,' 7, TNA, PRO: WO 198/983.
820 Sewell, 'Notes ... on French Corps Aviation,' TNA, PRO: AIR 1/475/15/312/201.
821 'Lecture on Aerial Photography, Centre d'Etudes de l'Aeronautique,' 25 April 1923. NASM A30.2/36.
822 'Notes on the French System of Intelligence,' 7, TNA, PRO: WO 198/983.
823 *Study and Use of Aerial Photographs*, 15 January 1918, sec. I, 26, NARA, RG 120, Box 819.
824 'Memorandum Report of Visit to HQ 1st French Army,' Conger Papers, USAMHI.
825 Zimmerman, 'Report on Visit to VIIIth French Army,' 1–2, Nolan Papers, USAMHI.
826 Lieut Michel quoted in Porter, 174.
827 Lieut Michel quoted in Porter, 175.
828 Sewell, 'Notes ... on French Corps Aviation,' TNA, PRO: AIR 1/475/15/312/201.
829 1st Lt Edward Steichen, 'Memorandum for Major Decker,' 21 December 1917, NASM.
830 'Lecture on Aerial Photography, Centre d'Etudes de l'Aeronautique,' 25 April 1923. NASM A30.2/36.
831 'Lecture on Aerial Photography, Centre d'Etudes de l'Aeronautique,' 25 April 1923. NASM A30.2/36.
832 Carlier, 60–61.
833 'Lecture on Aerial Photography, Centre d'Etudes de l'Aeronautique,' 25 April 1923. NASM A30.2/36.
834 Lieut Michel quoted in Porter, 173–174.
835 Pouquet, 'Working of the Service of Interpretation of Photographs,' NARA, RG 120, Box 819.
836 Gamble, 'Photographic Section of the French Flying Corps,' TNA, PRO: AIR 1/404/97/1917.
837 Lieut Michel quoted in Porter, 174.
838 'Lecture on Aerial Photography, Centre d'Etudes de l'Aeronautique,' 25 April 1923. NASM A30.2/36.
839 'Lecture on the Study and Interpretation of Aerial Photographs,' 10 December 1917, 1, NARA.
840 'Notes on the French System of Intelligence,' 11, TNA, PRO: WO 198/983.
841 'Lecture on Aerial Photography, Centre d'Etudes de l'Aeronautique,' 25 April 1923. NASM A30.2/36.
842 'Lecture on Aerial Photography, Centre d'Etudes de l'Aeronautique,' 25 April 1923. NASM A30.2/36.
843 Capt Charles W. Lawrence, 'Organisation and Assignment of Observation Aviation Observation Airplanes and their Equipment,' Air Corps Tactical School, Maxwell Field, Alabama, 1937–38, Air Force Historical Office Archives, IRIS ref. A2810, Maxwell AFB, AL.
844 Anthony C. Cain, *The Forgotten Air Force: French Air Doctrine in the 1930s* (Washington, DC: Smithsonian Institution Press, 2002), 11.
845 Pépin Papers, MA&E, Paris.
846 Dégardin, 22.
847 'Study and Use of Aerial Photographs,' Section II, 1.
848 G–2, 1st Army, AEF to Chief of Staff, 1st Army, AEF, 'Report,' 53.
849 Pépin Papers, MA&E, Paris.

850 Eugène Pépin, *Chinon*, (Paris: Henri Laurens); Eugène Pépin, *The Glory of the Loire*, (New York: The Viking Press, 1971).
851 Pépin biography in Pépin Papers, MA&E, Paris.
852 Dr Eugène Pépin, *The Legal Status of the Airspace in the Light of Progress in Aviation and Astronautics*, (Montreal: Institute of International Air Law, McGill University, 1957), 1.
853 Air Comm H. R. M. Brooke-Popham, 'Personalities of Du Peuty and Trenchard, The War of 1914–1918,' RAF Staff College, TNA, PRO: AIR 69/1.
854 Thomas G. Fergusson, *British Military Intelligence, 1870–1914*, (Frederick, MD: University Publications of America, Inc., 1984), 159.
855 Squier, 'Intelligence Services,' 16 February 1915, 2, Squier Papers, USAMHI.
856 Maj David M. Henry, 'Report on my Visit to the British Army for the Purpose of Observing the Work of the Intelligence Section of the Royal Flying Corps,' 7 March 1918, Conger Papers, USAMHI.
857 Field-Marshal French, 92.
858 Occleshaw, 246–247.
859 Occleshaw, 347–348.
860 Marshall-Cornwall quoted in Occleshaw, 325, 326.
861 Lloyd George quoted in Occleshaw, 331.
862 'Intelligence Organisation and Establishment,' 1 November 1917, TNA: PRO WO 152/262.
863 Watkis, 10.
864 Henry, 'Report on my Visit to the British Army,' Conger Papers, Box 2, USAMHI.
865 Hahn, 129.
866 Henry, 'Report on my Visit to the British Army,' Conger Papers, Box 2, USAMHI.
867 Hubbard, 'Report on Trip to British Front,' Conger Papers, Box 2, USAMHI.
868 Stewart, 'Report on the Organisation and Operation of Intelligence 'A',' Conger Papers, Box 2, USAMHI.
869 Gliddon, 18–19.
870 'Aerial Photography for Pilots,' 5, NASM, A30.2/2.
871 Henry, 'Report on my Visit to the British Army,' Conger Papers, Box 2, USAMHI.
872 'Organisation of Intelligence,' HQ, British 5th Army, Conger Papers, Box 2, USAMHI.
873 Henry, 'Report on my Visit to the British Army,' Conger Papers, Box 2, USAMHI.
874 Henry, 'Report on my Visit to the British Army,' Conger Papers, Box 2, USAMHI.
875 Moore-Brabazon, 'Photography's Part in the War,' TNA, PRO: AIR 1/724/91/2.
876 Hahn, 87.
877 Hubbard, 'Report on Trip to British Front,' 16, Conger Papers, Box 2, USAMHI.
878 Hubbard, 'Report on Trip to British Front,' 17, Conger Papers, Box 2, USAMHI.
879 Stewart, 'Report on the Organisation and Operation of Intelligence 'A',' Conger Papers, Box 2, USAMHI.
880 Henry, 'Report on my Visit to the British Army,' Conger Papers, Box 2, USAMHI.
881 Stewart, 'Report on the Organisation and Operation of Intelligence 'A',' Conger Papers, Box 2, USAMHI.
882 Hubbard, 'Report on Trip to British Front,' Conger Papers, Box 2, USAMHI.
883 Hahn, 119.
884 Edwin C. Vaughan, *Some Desperate Glory: The World War I Diary of a British Officer* (New York: Simon and Schuster Inc., and Touchstone Books, 1981), 154.
885 Hahn, 258–259.
886 'Organisation of Intelligence,' HQ, British 5th Army, Conger Papers, Box 2, USAMHI.
887 Henry, 'Report on my Visit to the British Army,' Conger Papers, Box 2, USAMHI.
888 Henry, 'Report on my Visit to the British Army.'
889 Henry, 'Report on my Visit to the British Army'; Hahn, 259–260.
890 'Organisation of Intelligence,' HQ, British 5th Army, Conger Papers, Box 2, USAMHI; 'Summary of Indications of Enemy Preparations for an Offensive in Flanders,' Conger Papers, Box 2, USAMHI.
891 'Organisation of Intelligence,' HQ, British 5th Army, Conger Papers, Box 2, USAMHI.
892 Henry, 'Report on my Visit to the British Army,' Conger Papers, Box 2, USAMHI.
893 Henry, 'Report on my Visit to the British Army,' Conger Papers, Box 2, USAMHI.
894 Henry, 'Report on my Visit to the British Army,' Conger Papers, Box 2, USAMHI'; 'Summary of Indications of Enemy Preparations for an Offensive in Flanders,' Conger Papers, Box 2, USAMHI.
895 Henry, 'Report on my Visit to the British Army,' Conger Papers, Box 2, USAMHI.
896 'Organisation of Intelligence,' HQ, British 5th Army, Conger Papers, Box 2, USAMHI.
897 'Organisation of Intelligence.'
898 'Telephone call between Colonel Moore-Brabazon and Capt Wall,' 13 November 1918, TNA, PRO: AIR 1/724/91/2; Laws, 'Statement made ... to Mr. A. J. Insall,' TNA, PRO: AIR 1/724/91/2.
899 Hahn, 70.
900 Hahn, 72; 83.
901 Hahn, 82.
902 Laws, 'Statement made ... to Mr. A. J. Insall,' TNA, PRO: AIR 1/724/91/2.
903 TNA, PRO: AIR 1/724/91/2.
904 Toye, 'History of Photography,' TNA, PRO: AIR 1/724/91/1.
905 TNA, PRO: AIR 1/724/91/1.
906 Hahn, 258–259.
907 Hahn, 261.
908 Stewart, 'Report on the Organisation and Operation of Intelligence 'A',' Conger Papers, Box 2, USAMHI.
909 Henry, 'Report on my Visit to the British Army,' Conger Papers, Box 2, USAMHI.

910 Stewart, 'Report on the Organisation and Operation of Intelligence 'A',' Conger Papers, Box 2, USAMHI.
911 Campbell, 'Notes on Aerial Photography,' TNA, PRO: AIR 1/7/6/98/20.
912 Jones, II, 177.
913 'Provision and Supply,' in 'Re-Organisation of Photographic Activities,' TNA, PRO: AIR 2/163.
914 Toye, 'History of Photography,' TNA, PRO: AIR 1/724/91/1.
915 Edmonds, *1916*, I, 131–132.
916 Occleshaw, 63.
917 2Lt J.T.C. Moore-Brabazon, 'For Promotion to Lt. and Refund of £ 75 paid for R.A.C. Certificate,' TNA, PRO: WO 339/23903.
918 Moore-Brabazon, 48.
919 Coleman, 130.
920 Moore-Brabazon, 84–86.
921 Moore-Brabazon, 89.
922 Moore-Brabazon, 92.
923 Occleshaw, 64.
924 Moore-Brabazon, 92.
925 Moore-Brabazon, 'Photography in the RAF.'
926 Occleshaw, 64.
927 Moore-Brabazon, 96.
928 'Parliamentary Secretary, Ministry of Transport,' *Modern Transport*, January 12, 1924. Moore-Brabazon Papers, RAF Museum, London.
929 'Lord Brabazon of Tara Dies; First Briton to Fly in England; Aviator had Been Early Auto Racer and Minister of Transport for Churchill,' Special to the *New York Times*, May 18, 1964. ProQuest Historical Newspapers *The New York Times* (1851- 2001), 29.
930 2nd Lt Prentiss M. Terry, 91st Aero Squadron, 1st Army, 'Notes on Branch Intelligence,' 19 August 1918, Gorrell Report, ser. M, r. 44, NARA.
931 Wing Cdr Seddon, 'Report ... on Co-Operation between Britain and American Air Services,' 13 September 1917, TNA, PRO: AIR 2/35.
932 Seddon, 'Report ... on Co-Operation between Britain and American Air Services,' TNA, PRO: AIR 2/35.
933 Barnes, 'History of Photographic Section #4,' 1, Gorrell Report, ser. G, r. 24, NARA.
934 Roy Talbert, *Negative Intelligence: The Army and the American Left, 1917–1941* (Jackson, MS: University Press of Mississippi, 1991), 20–21.
935 Maj E. Alexander Powell, *The Army Behind the Army* (New York: Charles Scribner's Sons, 1919), 334.
936 U.S. War Department, 'Report of the Director of Military Aeronautics,' in *War Department Annual Reports, 1918* (3 vol., Washington DC: GPO, 1919), I, 1382.
937 Crowell and Wilson, II, 419–420.
938 Nolan, 'G-2-C, Topography,' 3–4, Nolan Papers, USAMHI.
939 'Report of the Director of Military Aeronautics,' in *War Department Annual Reports, 1918*, I, 1382-83.
940 G-2, GHQ, AEF, to Chief of Staff, 'Report,' 15 June 1919, NASM.
941 G-2, GHQ, AEF, to Chief of Staff, 'Report,' 15 June 1919, NASM.
942 Alfred E. Bellinger, 'Relations of the Air Service with G-2,' Gorrell Report, ser. M, r. 41, NARA.
943 Brig Gen Dennis E. Nolan, 'Intelligence,' 21, in Nolan Papers, USAMHI.
944 G-2, GHQ, AEF, to Chief of Staff, 'Report,' 15 June 1919, NASM.
945 Nolan, 'G-2-C, Topography,' 4, Nolan Papers, USAMHI.
946 'History of the Air Service,' I, Gorrell Report ser. C, r. 12, NARA; Frank, *American Air Service Observation*, 126.
947 'History of the Air Service,' I, Gorrell Report, ser. C, r. 12, NARA; Frank, *American Air Service Observation*, 127.
948 James Cooke, *The US Air Service in the Great War, 1917–1919* (Westport, CT: Praeger, 1996), 59-60.
949 G-2, 1st Army, to Chief of Staff, 'Account of the Operation,' 50, Drum Papers, Box 14, USAMHI.
950 Nolan, 'G-2-C Topography,' 7-8, Nolan Papers, USAMHI.
951 'Historical Account,' 57, Drum Papers, USAMHI.
952 'The Development of Aerial Photography,' 12 July 1918, Gorrell Report, ser. G, r. 24, NARA.
953 G-2, 1st Army, to Chief of Staff, 'Account of the Operation,' 60, 48, Drum Papers, Box 14, USAMHI.
954 1st Lt John Snyder, 'Notes on Reconnaissance,' Lecture given at 2nd A.I.C.,' Gorrell Report, ser. J, r. 34, NARA.
955 G-2, 1st Army, to Chief of Staff, 'Account of the Operation,'45, Drum Papers, Box 14, USAMHI.
956 'Tactical History of Corps Observation: The First Corps Observation Group in the Toul Sector,' in Maurer, ed.,*US Air Service*, I, 173.
957 Snyder, 'Notes on Reconnaissance,' Gorrell Report, ser J, r. 34, NARA.
958 This quotation is the only documented format for a WW I aerial photographic interpretation report that the author was able to uncover; to what extent this form was actually used is unknown. Terry, 'Notes on Branch Intelligence,' Gorrell Report, ser. M, r. 44, NARA.
959 Pépin Papers, MA&E, Paris.
960 Nolan, 'G-2-C Topography,' 8, Nolan Papers, USAMHI.
961 'Photography,' Gorrell Report, ser. G, r. 24, NARA.
962 Dixon, *Diary*, 51.
963 G-2, 1st Army, to Chief of Staff, 'Account of the Operation,' 52, Drum Papers, Box 14, USAMHI.
964 Frank, *American Air Service Observation*, 219.
965 G-2, 1st Army, to Chief of Staff, 'Account of the Operation,' 36–37, Drum Papers, Box 14, USAMHI.
966 Barnes, *From Then Till Now*, 477; Barnes, 'History Photographic Section #4,' 2–3, Gorrell Report, ser. G, r. 24, NARA; Penelope Niven, *Steichen* (New York: Clarkson Potter Publishers, 1997), 449–450.
967 Barnes, 'History Photographic Section #4,' 4.
968 Niven, 450.

969 Niven, 450–451.
970 Barnes, 'History Photographic Section #4,' 11.
971 Barnes, *From Then Till Now*, 487.
972 Barnes, 'History Photographic Section #4,' 13–14.
973 Barnes, 'History Photographic Section #4,' 13–14.
974 Barnes, 'History Photographic Section #4,' 15–17.
975 Barnes, 'History Photographic Section #4,' 21–22.
976 Barnes, 'History Photographic Section #4,' 21–22.
977 Barnes, 'History Photographic Section #4,' 18.
978 Steichen, 'Aerial Photography,' Gorrell Report, ser. G, r. 24, NARA.
979 G-2, 1st Army, to Chief of Staff, 'Account of the Operation,'56-57, Drum Papers, Box 14, USAMHI.
980 'Provisional Manual of Operations,' Maurer, ed., *US Air Service*, II, 275.
981 'Tentative Manual for the Employment of Air Service,' Maurer, ed., *US Air Service*, II 326.
982 'Tentative Manual for the Employment of Air Service,' Maurer, ed., *US Air Service*, II 342.
983 'Tentative Manual for the Employment of Air Service,' Maurer, ed., *US Air Service*, II 342.
984 Steichen, 'Aerial Photography,' Gorrell Report, ser. G, r. 24, NARA.
985 Maurer, *Aviation in the US Army 1919–1938* (Washington, DC: Office of Air Force History, 1987), 430.
986 Steichen's daughter, Dr Mary Steichen Calderone, was a noted physician and co-founder of the Sex Information and Education Council of the United States. Dr Mary Steichen Calderone 5/29/86 Telephone Interview, NYC, National Air and Space Museum files.
987 Goddard, p. 20.
988 Robert E. Hood, *12 at War, Great Photographers Under Fire* (New York: G.P. Putnam's Sons, 1967), 71–72.
989 The book is paginated by photographs rather than text. Edward Steichen, *A Life in Photography* (Garden City, NY: Doubleday & Company, Inc. 1981), chap. 5.
990 Niven, 448.
991 Niven.
992 Steichen, xx.
993 Steichen, chap. 5.
994 Steichen, chap. 5; Barnes, 477.
995 Niven, 457.
996 'Major Campbell Letter to Major Barnes,' 30 January 1918, NASM.
997 Steichen, chap. 5; Hood, 71–72; NASM, Notes on Steichen's Military Personnel File.
998 Steichen, chap. 5; Hood, 74–75.
999 Christopher Phillips, *Steichen at War* (New York: H.N. Abrams, 1981) 1.
1000 Steichen, chap. 5; Niven, 463.
1001 Steichen, chap. 5.
1002 Steichen, chap. 12.
1003 Hood, 76–77.
1004 Hood, 71–72.
1005 'Aerial Photography for Pilots,' NASM, A30.2/2.
1006 *Illustrations to Accompany Captain De Bissy's Notes*, xx.
1007 *Bulletin of the Aerial Photography Department in the Field*, 8.
1008 *Bulletin of the Aerial Photography Department in the Field.*
1009 Porter,195–196.
1010 *Bulletin of the Aerial Photography Department in the Field*, 8.
1011 *Illustrations to Accompany Captain De Bissy's Notes*, 3.
1012 *Notes on the Interpretation of Aeroplane Photographs*, March 1917 (S.S. 550), 9; 'Aerial Photography for Pilots,' 24, NASM, A30.2/2.
1013 *Study and Use of Aerial Photographs*, 15 January 1918, sec. II, 8, NARA, RG 120, Box 819.
1014 *Bulletin of the Aerial Photography Department in the Field*, 16.
1015 *Illustrations to Accompany Captain De Bissy's Notes*, 3.
1016 *Study and Use of Aerial Photographs*, 15 January 1918, sec. II, 6, NARA, RG 120, Box 819; Hardy, 'Interpretation of Aerial Photography,' 3, NASM, A30.2/28; Porter, 197; Hahn, 22.
1017 *Study and Use of Aerial Photographs*, 15 January 1918, sec. II, 6, NARA, RG 120, Box 819.
1018 *Bulletin of the Aerial Photography Department in the Field*, 7; Porter, 83-84.
1019 Hardy, 'Interpretation of Aerial Photography,' 3, NASM, A30.2/28.
1020 Pépin, 'Lecture on the Study and Interpretation of Aerial Photographs,' 1, 10 December 1917, NASM; *Study and Use of Aerial Photographs*, 15 January 1918, NARA, RG 120, Box 819; Hahn, 190.
1021 *Study and Use of Aerial Photographs*, 15 January 1918, sec. II, 20, NARA, RG 120, Box 819.
1022 *Notes on the Interpretation of Aeroplane Photographs*, March 1917 (S.S. 550), 9, and February 1918 (S.S. 631), 4–15.
1023 Porter, 83–84.
1024 *Bulletin of the Aerial Photography Department in the Field*, 11–16; *Illustrations to Accompany Captain De Bissy's Notes*, 3; 'Aerial Photography for Pilots,' 24, NASM, A30.2/2.
1025 Pépin, 'Lecture on the Study and Interpretation of Aerial Photographs,' 10 December 1917, 4–16, NASM.
1026 *Notes on the Interpretation of Aeroplane Photographs*, March 1917 (S.S. 550), 9, and February 1918 (S.S. 631), 4–15.
1027 *Study and Use of Aerial Photographs*, 15 January 1918, sec. II, 20, NARA, RG 120, Box 819.
1028 Each sector was compared to a human hand and arm, with the main arteries running down the arm and feeding the spread fingers through the smaller veins

which lead to the finger tips and return. *Notes on the Interpretation of Aeroplane Photographs*, March 1917 (S.S. 550), 9.

1029 Edmonds, *1916*, I, 157–158.

1030 G-2, 1st Army, 'Summary of Intelligence,' 2 October 1918, NARA, RG 120, Box 3370.

1031 G-2, 1st Army Corps, 'Summary of Intelligence,' 9 July 1918, Drum Papers, Box 14, USAMHI.

1032 Other sap descriptions varied with the source, to include winding, sawtooth, and 'cronolated.' 'Aerial Photography for Pilots,' 26, NASM, A30.2/2; *Bulletin of the Aerial Photography Department in the Field*, 25–26; Hahn, 188.

1033 Haig's 28 August 1917 diary quoted in Occleshaw, 67.

1034 *Study and Use of Aerial Photographs*, 15 January 1918, sec. II, 8, NARA, RG 120, Box 819; Hardy, 'Interpretation of Aerial Photography,' 4, NASM, A30.2/28; *Notes on the Interpretation of Aeroplane Photographs*, February 1918 (S.S. 631), 8; Hahn, 30–31.

1035 G-2, 1st Army Corps, 'Summary of Intelligence,' 9 July 1918, Drum Papers, USAMHI.

1036 Pépin, 'Lecture on the Study and Interpretation of Aerial Photographs,' 10 December 1917, 5, NASM.

1037 *Illustrations to Accompany Captain De Bissy's Notes.*

1038 *Illustrations to Accompany Captain De Bissy's Notes.*

1039 *Bulletin of the Aerial Photography Department in the Field*, 27–35; Hardy, 'Interpretation of Aerial Photography,' 3, NASM, A30.2/28; Brooks, in White, 104; 'Aerial Photography for Pilots,' 25–26, NASM, A30.2/2; Pépin, 'Lecture on the Study and Interpretation of Aerial Photographs,' 10 December 1917, 5, NASM; 'Aerial Photography for Pilots' 25–26, NASM, A30.2/2.

1040 Porter, 202.

1041 Pepin Papers, A&E, Paris.

1042 Pouquet, 'Working of the Service of Interpretation of Photographs,' NARA, RG 120, Box 819.

1043 Hahn, 73; Porter, 192–193.

1044 Porter, 209.

1045 *Notes on the Interpretation of Aeroplane Photographs*, March 1917 (S.S. 550), 7, and February 1918 (S.S. 631), 7; Pépin, 'Lecture on the Study and Interpretation of Aerial Photographs,' 10 December 1917, 10, NASM.

1046 G-2, 1st Army Corps, 'Summary of Intelligence,' 9 July 1918, Drum Papers, USAMHI.

1047 *Notes on the Interpretation of Aeroplane Photographs*, March 1917 (S.S. 550), 8, and February 1918 (S.S. 631), 6.

1048 Hardy, 'Interpretation of Aerial Photography,' NASM, A30.2/28; Porter, 190.

1049 *Notes on the Interpretation of Aeroplane Photographs*, March 1917 (S.S. 550), 8, and February 1918 (S.S. 631), 6-7; Hardy, 'Interpretation of Aerial Photography,' NASM, A30.2/28.

1050 *Illustrations to Accompany Captain De Bissy's Notes*, 2.

1051 *Illustrations to Accompany Captain De Bissy's Notes*, 1–2; Hahn, 27–28.

1052 *Notes on the Interpretation of Aeroplane Photographs* March 1917 (S.S. 550), 9, and February 1918 (S.S. 631), 4.

1053 *Illustrations to Accompany Captain De Bissy's Notes*, 1; *Study and Use of Aerial Photographs*, 15 January 1918, NARA, RG 120, Box 819; Hahn, 22; 72-73.

1054 Hahn, 23.

1055 Porter, 201.

1056 Hahn, 73; *Notes on the Interpretation of Aeroplane Photographs*, March 1917 (S.S. 550), 9, and February 1918 (S.S. 631), 5.

1057 *Study and Use of Aerial Photographs*, 15 January 1918, sec. II, 11, NARA, RG 120, Box 819.

1058 Brooks, in White, 98-99; Hahn, 73; *Notes on the Interpretation of Aeroplane Photographs*, March 1917 (S.S. 550), 9.

1059 *Notes on the Interpretation of Aeroplane Photographs*, February 1918 (S.S. 631), 9.

1060 Pépin, 'Lecture on the Study and Interpretation of Aerial Photographs,' 10 December 1917, NASM; *Notes on the Interpretation of Aeroplane Photographs*, February 1918 (S.S. 631), 10.

1061 G-2, 2nd Army, 'Effectiveness of Different Weapons,' 25 October. 1918, NARA, RG 120, Box 10.

1062 'Aerial Photography for Pilots,' 27, NASM, A30.2/2.

1063 *Illustrations to Accompany Captain De Bissy's Notes*, 3.

1064 *Illustrations to Accompany Captain De Bissy's Notes*, 3–4.

1065 *Illustrations to Accompany Captain De Bissy's Notes*, 3.

1066 Pépin, 'Lecture on the Study and Interpretation of Aerial Photographs,' 10 December 1917, 6, NASM.

1067 *Notes on the Interpretation of Aeroplane Photographs*, March 1917 (S.S. 550), 5.

1068 *Notes on the Interpretation of Aeroplane Photographs*, February 1918 (S.S. 631), 5.

1069 *Illustrations to Accompany Captain De Bissy's Notes*, 3–4.

1070 *Study and Use of Aerial Photographs*, 15 January 1918, sec. II, 5, NARA, RG 120, Box 819.

1071 *Study and Use of Aerial Photographs*, 15 January 1918, sec. II, 5, NARA, RG 120, Box 819.

1072 Pépin, 'Lecture on the Study and Interpretation of Aerial Photographs,' 10 December 1917, 1, NASM; *Notes on the Interpretation of Aeroplane Photographs*, March 1917 (S.S. 550), 5.

1073 'Aerial Photography for Pilots,' 27, NASM, A30.2/2.

1074 Hahn, 25; Hardy, 'Interpretation of Aerial Photography,' NASM, A30.2/28.

1075 Porter, 203.
1076 *Illustrations to Accompany Captain De Bissy's Notes*, 4.
1077 G-2, 1st Army Corps, 'Summary of Intelligence,' 2 September 1918, Drum Papers, USAMHI.
1078 'G-2, 2nd Army, 'Effectiveness of Different Weapons,' NARA, RG 120, Box 10.
1079 *Bulletin of the Aerial Photography Department in the Field*, 48-51.
1080 Pépin, 'Lecture on the Study and Interpretation of Aerial Photographs,' 10 December 1917, 5, NASM; Porter, 205; Hardy, 'Interpretation of Aerial Photography,' 6, NASM, A30.2/28; Hahn, 26.
1081 *Notes on the Interpretation of Aeroplane Photographs*, March 1917 (S.S. 550), 6, and February 1918 (S.S. 631), 6.
1082 'Aerial Photography for Pilots,' 27-28, NASM, A30.2/2; Hahn, 26.
1083 Supreme War Council, Inter-Allied Aviation Committee, 'Draft Minutes of the 1st Session ... held in the Council Chamber, Versailles,' 9 May 1918, TNA: PRO AIR 1/2299/209/77/29.
1084 Capt. Henry A. Wilsdon, 'Photography's Notable Part in the War,' *Vanity Fair*, December 1918, 52.
1085 Christienne and Lissarague, 181.
1086 Estienne, 'Army Photographic Reconnaissance as it Should Be,' 5, NASM, A30.2/34.
1087 G-2, 1st Army, to Chief of Staff, 'Account of the Operation,' 45-47, Drum Papers, Box 14, USAMHI; Henry, 'Report on my Visit to the British Army,' Conger Papers, Box 2, USAMHI.
1088 Snyder, 'Notes on Aerial Reconnaissance,' 2, Gorrell Report, ser. J, r. 34, NARA.
1089 *Notes on the Interpretation of Aeroplane Photographs*, March 1917 (S.S. 550), 7-8, and February 1918 (S.S. 631), 7; Pépin, 'Lecture on the Study and Interpretation of Aerial Photographs,' 10 December 1917, 9, NASM; *Study and Use of Aerial Photographs*, 15 January 1918, sec. II, 37, NARA, RG 120, Box 819; 'Aerial Photography for Pilots,' 30, NASM, A30.2/2.
1090 Stewart, 'Report on the Organisation and Operation of intelligence 'A',' Conger Papers, Box 2, USAMHI.
1091 G-2, 1st Army, 'Summary of Intelligence,' 14 September 1918, Drum Papers, USAMHI.
1092 *Study and Use of Aerial Photographs*, 15 January 1918, sec. I, 24, NARA, RG 120, Box 819; Hahn, 35-36.
1093 *Study and Use of Aerial Photographs*, 15 January 1918, sec. II, 6-7, NARA, RG 120, Box 819.
1094 'Aerial Photography for Pilots,' 30, NASM, A30.2/2; Hahn, 35-36.
1095 5th Group, RAF, 'Result of Aerial Photography by the R.N.A.S. in 1917,' 13 February 1918, TNA PRO: AIR 1/93/15/9/240.
1096 G-2, 1st Army, 'Summary of Intelligence,' 2 September 1918, Nolan Papers, USAMHI.
1097 Ian Hogg, *The Guns 1914-118* (New York: Ballantine Books, Inc., 1971), 141.
1098 Hogg, 137.
1099 Henry W. Miller, *The Paris Gun: The Bombardment of Paris by the German Long Range Guns and the Great German Offensives of 1918* (New York: Jonathan Cape & Harrison Smith. 1930), 200-203.
1100 *Bulletin of the Aerial Photography Department in the Field*, 63-64; Pepin, 'Lecture on the Study and Interpretation of Aerial Photographs,' 10 December 1917, 8, NARA; 'Aerial Photography for Pilots,' 29, NASM, A30.2/2.
1101 Hogg, 141.
1102 Miller, *The Paris Gun*, 200-203.
1103 Snyder, 'Notes on Reconnaissance,' 3, Gorrell Report, ser. J, r. 34, NARA.
1104 *Notes on the Interpretation of Aeroplane Photographs*, February 1918 (S.S. 631), 10.
1105 Hahn, 36; 'Aerial Photography for Pilots,' 31, NASM, A30.2/2.
1106 Snyder, 'Notes on Reconnaissance,' 3, Gorrell Report, ser. J, r. 34, NARA.
1107 'Summary of Indications of Enemy Preparations for an Offensive in Flanders,' Conger Papers, Box 2, USAMHI.
1108 *Aerial Observation for Artillery*, 49.
1109 *Notes on the Interpretation of Aeroplane Photographs*, February 1918 (S.S. 631), 10.
1110 *Notes on the Interpretation of Aeroplane Photographs*, February 1918 (S.S. 631), 10; Pepin, 'Lecture on the Study and Interpretation of Aerial Photographs,' 10 December 1917, 11, NARA.
1111 'Index to Periodical Summaries of Aeronautical Information,' TNA, PRO: AIR 1/2124/207/74/3.
1112 Henry, 'Report on my Visit to the British Army,' Conger Papers, Box 2, USAMHI.
1113 Author discussion with Steven Suddaby.
1114 Martel, 269.
1115 G-2, 1st Army Corps, 'Summary of Intelligence,' 15 September 1918, Drum Papers, USAMHI.
1116 5th Group, RAF, 'Result of Aerial Photography,' 13 February 1918, TNA PRO: AIR 1/93/15/9/240.
1117 TNA PRO: AIR 1/93/15/9/240.
1118 'Study by Corporal Mallet of the R.N.A.S. Photographs Nos. 527 to 540, Reconnaissance of November 21st 1916,' in 'Photographic Studies' III (January-Sept 1918), TNA, PRO: AIR 1/82/15/9/204.
1119 5th Group, RAF, 'Result of Aerial Photography,' 13 February 1918, TNA PRO: AIR 1/93/15/9/240.
1120 TNA PRO: AIR 1/93/15/9/240.
1121 'Study by Corporal Mallet of the R.N.A.S. Photographs,'TNA, PRO: AIR 1/82/15/9/204.
1122 5th Group, RAF, 'Result of Aerial Photography,' 13 February 1918, TNA PRO: AIR 1/93/15/9/240.
1123 The 'big camera' was most likely the 47-in (119.38-cm) lens aircraft camera. The newly arrived DH 4 had

an operational ceiling of 20,000 feet and most likely flew the sorties for this historic coverage. 5th Group, RAF, 'Result of Aerial Photography,' 13 February 1918, TNA PRO: AIR 1/93/15/9/240.

1124 A. J. Jackson, *De Havilland Aircraft Since 1909* (London: Putnam & Company, Ltd., 1962), 56.

1125 *Illustrations to Accompany Captain De Bissy's Notes*, 1.

1126 *Study and Use of Aerial Photographs*, 15 January 1918, sec. III, 3–4, NARA, RG 120, Box 819.

1127 Pépin, 'Lecture on the Study and Interpretation of Aerial Photographs,' 10 December 1917, 8, NASM.

1128 'Aerial Photography for Pilots,' 28, A 30.2/2.

1129 5th Group, RAF, 'Result of Aerial Photography,' 13 February 1918, TNA PRO: AIR 1/93/15/9/240.

1130 Williams, 4.

1131 Author discussion with Dr James Davilla.

1132 Williams, 28.

1133 Williams, 56.

1134 Pépin Papers, MA&E, Paris.

1135 The editor's note in Martel, 238, describes the planning for bombing German and Austro-Hungarian cities.

1136 Hahn, 187; Porter, 191.

1137 Porter, 209.

1138 G-2, 2nd Army, 'Effectiveness of Different Weapons,' NARA, RG 120, Box 10.

1139 'Artillery Operations of the Ninth British Corps at Messines,' 23–24.

1140 'Artillery Operations of the Ninth British Corps at Messines,' 32–33.

1141 'Artillery Operations of the Ninth British Corps at Messines,' 28.

1142 *Illustrations to Accompany Captain De Bissy's Notes*, 1; 'Aerial Photography for Pilots,' 28, NASM, A30.2/2; *Bulletin of the Aerial Photography Department in the Field*, 58-62.

1143 *Study and Use of Aerial Photographs*, 15 January 1918, sec. I, 16, NARA, RG 120, Box 819.

1144 Hahn, 33–34.

1145 *Illustrations to Accompany Captain De Bissy's Notes*, 2.

1146 *Notes on the Interpretation of Aeroplane Photographs*, March 1917 (S.S. 550), 6-7, and February 1918 (S.S. 631), 8-9.

1147 *Bulletin of the Aerial Photography Department in the Field*, 66.

1148 Hardy, 'Interpretation of Aerial Photography,' 6, NASM, A30.2/28.

1149 Bailey, 59.

1150 Hardy, 'Interpretation of Aerial Photography,' 2, NASM, A30.2/28.

1151 *Aerial Observation for Artillery*, 7; *Bulletin of the Aerial Photography Department in the Field*, 65.

1152 'Aerial Photography for Pilots,' 22, NASM, A30.2/2.

1153 *Study and Use of Aerial Photographs*, 15 January 1918, sec. III, 24, NARA, RG 120, Box 819.

1154 *Study and Use of Aerial Photographs*, 15 January 1918, sec. III, 3–4, NARA, RG 120, Box 819.

1155 Innes, 49.

1156 Mead, 100–101; Innes, 69.

1157 Innes, 69.

1158 Hahn, 262–263.

1159 *Study and Use of Aerial Photographs*, 15 January 1918, sec. III, 17, NARA, RG 120, Box 819.

1160 'A Study of the Organisation of the French Army,' 91, NARA, RG 120, Box 819.

1161 *Study and Use of Aerial Photographs*, 15 January 1918, sec. II, 16–17, NARA, RG 120, Box 819.

1162 'A Study of the Organisation of the French Army,' 91, NARA, RG 120, Box 819.

1163 'A Study of the Organisation of the French Army,' 94–95, NARA, RG 120, Box 819.

1164 Hardy, 'Interpretation of Aerial Photography,' 5, NASM, A30.2/28.

1165 Hardy, 'Interpretation of Aerial Photography,' 5, NASM, A30.2/28.

1166 'A Study of the Organisation of the French Army,' 92–93, NARA, RG 120, Box 819.

1167 'Aerial Photography for Pilots,' 31, NASM, A30.2/2; Hardy, 'Interpretation of Aerial Photography,' 4, NASM, A30.2/28.

1168 Hahn, 26–27.

1169 Hahn, 8.

1170 Pépin, 'Lecture on the Study and Interpretation of Aerial Photographs,' 10 December 1917, 11, NASM.

1171 Hardy, 'Interpretation of Aerial Photography,' 4, NASM, A30.2/28; Pépin, 'Lecture on the Study and Interpretation of Aerial Photographs,' 10 December 1917, 11, NASM.

1172 Hahn, 8.

1173 Hardy, 'Interpretation of Aerial Photography,' 5, NASM, A30.2/28.

1174 Georges Blanchon quoted in Porter, 209–210.

1175 Pépin, 'Lecture on the Study and Interpretation of Aerial Photographs,' 10 December 1917, 11, NASM; Jones, II, 177–178.

1176 'A Study of the Organisation of the French Army,' 4–15, NARA, RG 120, Box 819.

1177 Hardy, 'Interpretation of Aerial Photography,' 5, NASM, A30.2/28.

1178 Hardy, 'Interpretation of Aerial Photography,' 5, NASM, A30.2/28.

1179 'Aerial Photography for Pilots,' 30, NASM, A30.2/2.

1180 Hardy, 'Interpretation of Aerial Photography,' 5, NASM, A30.2/28.

1181 Hahn, 29–30; *Study and Use of Aerial Photographs*, 15 January 1918, NARA, RG 120, Box 819; Hardy, 'Interpretation of Aerial Photography,' 5, NASM, A30.2/28; 'Aerial Photography for Pilots,' 30, NASM, A30.2/2; *Notes on the Interpretation of Aeroplane*

Photographs, March 1917 (S.S. 550), 8, and February 1918 (S.S. 631), 7.
1182 'Aerial Photography for Pilots,' 30, NASM, A30.2/2.
1183 W. E. Pitman, 'American Geologists at War: World War I,' in Underwood and Guth, 43.
1184 'Aerial Photography for Pilots,' 31, NASM, A30.2/2.
1185 Hardy, 'Interpretation of Aerial Photography,' 5, NASM, A30.2/28.
1186 *Study and Use of Aerial Photographs*, 15 January 1918, sec. II, 17, NARA, RG 120, Box 819.
1187 'A Study of the Organisation of the French Army,' 97, NARA, RG 120, Box 819.
1188 'A Study of the Organisation of the French Army,' 97, NARA, RG 120, Box 819.
1189 'A Study of the Organisation of the French Army,' 99, NARA, RG 120, Box 819.
1190 Moore-Brabazon, 'Photography's Part in the War,' TNA, PRO: AIR 1/724/91/2.
1191 'Aeronautics,' June 1921, 24–125, NARA, RG 120, Box 819.
1192 Pouquet, 'Working of the Service of Interpretation of Photographs,' NARA, RG 120, Box 819.
1193 *Study and Use of Aerial Photographs*, 15 January 1918, sec. I, 21–22, NARA, RG 120, Box 819.
1194 'Summary of Indications of Enemy Preparations for an Offensive in Flanders,' Conger Papers, Box 2, USAMHI.
1195 *Study and Use of Aerial Photographs*, 15 January 1918, sec. II, 12, NARA,.RG 120, Box 819.
1196 G-2, 1st Army Corps, 'Summary of Intelligence,' Drum Papers, USAMHI.
1197 Hahn, 188.
1198 'Summary of Indications of Enemy Preparations for an Offensive in Flanders,' Conger Papers, Box 2, USAMHI.
1199 Cooke, 24–125.
1200 Snyder, 'Notes on Reconnaissance,' 5, Gorrell Report, ser. r. 34, NARA.
1201 *Notes on the Interpretation of Aeroplane Photographs*, March 1917 (S.S. 550), 9–10, and February 1918 (S.S. 631), 11.
1202 Snyder, 'Notes on Reconnaissance,' 2, Gorrell Report, ser. J, r. 34, NARA.
1203 *Study and Use of Aerial Photographs*, 15 January 1918, sec. III, 23, NARA, RG 120, Box 819.
1204 *Study and Use of Aerial Photographs*, 15 January 1918, sec. II, 22–23, NARA, RG 120, Box 819.
1205 Pépin, 'Lecture on the Study and Interpretation of Aerial Photographs,' 10 December 1917, 12, NASM; Porter, 193–194.
1206 Pépin, 'Lecture on the Study and Interpretation of Aerial Photographs,' 10 December 1917, 11, NASM.
1207 Pépin, 'Lecture on the Study and Interpretation of Aerial Photographs,' 10 December 1917, 12, NASM.
1208 Porter, 122.
1209 'Instructions on the Employment of Aerial Observation in Liaison with Artillery,' in Maurer, ed., *US Air Service*, II, 182.
1210 *Notes on the Interpretation of Aeroplane Photographs*, March 1917 (S.S. 550), 9–10, February 1918 (S.S. 631), 11.
1211 Pépin, 'Lecture on the Study and Interpretation of Aerial Photographs,' 10 December 1917, 12–13, NASM.
1212 Coleman, 'Interpretation of Aerial Photographs,' 1921, 4, NARA, RG 120, Box 819.
1213 *Aerial Observation for Artillery*, 10–11.
1214 Pépin, 'Lecture on the Study and Interpretation of Aerial Photographs,' 10 December 1917, 13, NASM; 'Aerial Photography for Pilots,' 22, NASM, A30.2/2.
1215 *Study and Use of Aerial Photographs*, 15 January 1918, sec. III, 23, NARA, RG 120, Box 819.
1216 *Study and Use of Aerial Photographs*, 15 January 1918, sec. III, 23, NARA, RG 120, Box 819.
1217 'Notes on the French System of Intelligence,' 5, TNA, PRO: WO 198/983.
1218 *Aerial Observation for Artillery*, 11.
1219 'Aeronautics,' June 1921, NARA, RG 120, Box 819, 24–125.
1220 'Notes on the French System of Intelligence during the Battle of the Somme,' 9.
1221 'Notes on the French System of Intelligence during the Battle of the Somme.'
1222 *Aerial Observation for Artillery*, 10–11.
1223 *Aerial Observation for Artillery*, 11.
1224 'Notes on the French System of Intelligence,' 4, TNA, PRO: WO 198/983.
1225 'Notes on the French System of Intelligence,' 4–15, TNA, PRO: WO 198/983.
1226 'Notes on the French System of Intelligence,' 11, TNA, PRO: WO 198/983.
1227 Pépin, 'Lecture on the Study and Interpretation of Aerial Photographs,' 10 December 1917, 14, NASM.
1228 Pouquet, 'Working of the Service of Interpretation of Photographs,' NARA, RG 120, Box 819.
1229 Pouquet, 'Working of the Service of Interpretation of Photographs,' NARA, RG 120, Box 819.
1230 Porter, 192.
1231 Pouquet, 'Working of the Service of Interpretation of Photographs,' NARA, RG 120, Box 819.
1232 Porter, 192.
1233 Pouquet, 'Working of the Service of Interpretation of Photographs,' NARA, RG 120, Box 819.
1234 Pouquet, 'Working of the Service of Interpretation of Photographs,' NARA, RG 120, Box 819.
1235 *Study and Use of Aerial Photographs*, 15 January 1918, sec. III, 25, NARA, RG 120, Box 819.
1236 Porter, 192.
1237 Pouquet, 'Working of the Service of Interpretation of Photographs,' NARA, RG 120, Box 819.
1238 Porter, 192.

1239 Pouquet, 'Working of the Service of Interpretation of Photographs,' NARA, RG 120, Box 819.
1240 *Study and Use of Aerial Photographs*, 15 January 1918, sec. II, 12, NARA, RG 120, Box 819.
1241 Hahn, 28.
1242 'Summary of Indications of Enemy Preparations for an Offensive in Flanders,' Conger Papers, Box 2, USAMHI.
1243 Porter, 192.
1244 'Aeronautics,' June 1921, 22–23, NARA, RG 120, Box 819.
1245 Pouquet, 'Working of the Service of Interpretation of Photographs,' NARA, RG 120, Box 819.
1246 Steichen, 'American Aerial Photography at the Front,' 1, NASM, A30.2/24.
1247 Capt I. W. Learmont, R.F.C., 'Report ... of a Visit to the French Bombardment Group at Malzeville,' TNA, PRO: AIR 1/475/15/312/201.
1248 Martel, 161.
1249 'Memorandum Report of Visit to HQ 1st French Army,' Conger Papers, USAMHI.
1250 Learmont, 'Report ... of a Visit to the French Bombardment Group,' TNA, PRO: AIR 1/475/15/312/201.
1251 Martel, 161.
1252 Martel.
1253 Henry, 'Report on My Visit to the British Army,' 10, Conger Papers, Box 2, USAMHI.
1254 Pépin Papers, MA&E, Paris.
1255 'Detailed Raid Report, Capt. A. Gray, commanding RFC No. 55 Squadron, 11 December 1917,' in Williams, 105.
1256 Morris, 75.
1257 The type of camera was not identified; it probably referred to a simple hand-held camera available on the market. 'Development of Aerial Photography,' Gorrell Report, ser. J, r. 34, NARA.
1258 'Report on Day Bombardment Training,'Gorrell Report, ser. J, r. 35, NARA.
1259 George K. Williams, *Biplanes and Bombsights: British Bombing in World War I* (Maxwell AFB, AL: Air University Press), 4
1260 Estienne, 'Army Photographic Reconnaissance as it Should Be,' 5, NASM, A30.2/34.
1261 Martel, 161.
1262 'Report on Day Bombardment Training,' Gorrell Report, ser. J, r. 35, NARA.
1263 'Report on Day Bombardment Training.'
1264 'Report on Day Bombardment Training.'
1265 'Report on Day Bombardment Training.'
1266 'Report on Day Bombardment Training.'
1267 Pouquet, 'Working of the Service of Interpretation of Photographs,' NARA, RG 120, Box 819.
1268 'A Study of the Organisation of the French Army,' 26, NARA, RG 120, Box 819.
1269 Barnes, *From Then Till Now*, 495.
1270 Carlier, 200. Translation by Mary Elizabeth Ruwell.
1271 A.T. Fuller, 'Mysteries of Camouflage,' 16 August 1919, *Army and Navy Register*, 222–223.
1272 *Study and Use of Aerial Photographs*, 15 January 1918, sec. I, 3, NARA, RG 120, Box 819.
1273 Sir Joshua Reynolds in Solomon, 53.
1274 Solomon, vii; Carlier, 144–1161.
1275 American Mission with the Commanding General, Allied Forces of Occupation, Mainz, Germany, Part XXXVIII, 'Camouflage,' Sec 2, July 1921, 9–11, NARA, RG 120, Box 819.
1276 Edmonds, *1916*, I, 82-83.
1277 Edmonds, *1916*, I.
1278 Stephen Kern, *The Culture of Time and Space 1880–1918* (Cambridge, MA: Harvard University Press, 1983), 303; André Mare and Laurence Graffin, *Carnets de Guerre, 1914–1918* (Paris: Herscher, 1996), 46.
1279 Kern, 303.
1280 Anne Distel, *Dunoyer de Segonzac* (New York: Crown Publishers, 1980), 93.
1281 André Mare was another known artist who was a camoufleur. He later served as an instructor of camouflage with the British and Italians. Mare and Graffin, 46; Kern, 303.
1282 Edmonds, *1916*, I, 82-83.
1283 Solomon, vii.
1284 'Camouflage,' Sec 2, July 1921, 25, NARA, RG 120, Box 819.
1285 'Camouflage,' Sec 2, July 1921, 17–19, NARA, RG 120, Box 819.
1286 'Camouflage,' Sec 2, July 1921, 23, NARA, RG 120, Box 819.
1287 'Camouflage,' Sec 2, July 1921, 20, NARA, RG 120, Box 819.
1288 Solomon, 51.
1289 'Camouflage,' Sec 2, July 1921, 27, NARA, RG 120, Box 819.
1290 Porter, 175–176; Barnes, *From Then Till Now*, 495; Ives, 244.
1291 'Camouflage,' Sec 2, July 1921, 27, NARA, RG 120, Box 819.
1292 Hardy, 'Interpretation of Aerial Photography,' 1, NASM, A30.2/28.
1293 Hanson, 15.
1294 Fuller, 222–223.
1295 Solomon, 48.
1296 Supreme War Council, Draft Minutes, 9 May 1918, TNA, PRO: AIR 1/2299/209/77/29.
1297 TNA, PRO: AIR 1/2299/209/77/29.
1298 'Camouflage,' Sec 2, July 1921, 76, NARA, RG 120, Box 819.
1299 Fuller, 222–223; Supreme War, Draft Minutes, 9 May 1918, TNA, PRO: AIR 1/2299/209/77/29.
1300 Fuller; TNA, PRO: AIR 1/2299/209/77/29.
1301 Captain F. T. Colby quoted in *Camouflage for the Troops of the Line*, Ed. by Army War College, January

1918, (Washington: Government Printing Office, 1920), 8.

1302 'Notes on Field Fortifications for the Use of Troops of all Arms,' in *Trench Fortifications, 1914–1918: A Reference Manual* (London: The Imperial War Museum, and Nashville, TN: The Battery Press, Inc. 1998), 16–17.

1303 Hardy, 'Interpretation of Aerial Photography,' 3, NASM, A30.2/28.

1304 Hahn, 10.

1305 Solomon, 8.

1306 *Study and Use of Aerial Photographs*, Section II, 20; Hardy, 'Interpretation of Aerial Photography,' 1, NASM, A30.2/28.

1307 'Camouflage,' Sec 2, July 1921, 50–51, NARA, RG 120, Box 819.

1308 *Notes on the Interpretation of Aeroplane Photographs*, February 1918 (S.S. 631), 3.

1309 Edmonds, *1916*, I, 84.

1310 Fuller, 222–223.

1311 Edmonds, *1916*, I, 84.

1312 Barnes, *From Then Till Now*, 495.

1313 Supreme War Council File, Draft Minutes, 9 May 1918, TNA, PRO: AIR 1/2299/209/77/29.

1314 Edmonds, *1916*, I, 84.

1315 Martin Blumenson, *The Patton Papers, 1885–1940* (2 vols., Boston: Houghton Mifflin Company, 1972), II, 582.

1316 Guidance provided to U.S. aviators through Captain F. T. Colby's 'Notes of Artillery Camouflage and its Employment,' HQ AEF, October 1917, Jefferson Hayes Davis Papers, McDermott Library, USAFA, SMS 57.

1317 Colby, 'Notes of Artillery Camouflage and its Employment.'

1318 'Camouflage for Troops of the Line,' 12.

1319 Hanson, 17–18.

1320 Pépin, 'Lecture on the Study and Interpretation of Aerial Photographs,' 10 December 1917, 7, NASM; *Notes on the Interpretation of Aeroplane Photographs*, February 1918 (S.S. 631), 9.

1321 'Camouflage,' Sec 2, July 1921, 62-63, NARA, RG 120, Box 819.

1322 Général Vincent Memorandum to Artillery Commander of US 3rd Division, Divisional Artillery Commander of French 39th Division, and Heavy Artillery Commander of the 38th Army Corps, 'Necessity for Battery Camouflage,' in *United States Army in the World War 1917–1919*, III, 590.

1323 'Camouflage,' Sec 2, July 1921, 54–157, NARA, RG 120, Box 819.

1324 'Camouflage,' Sec 2, July 1921, 76, NARA, RG 120, Box 819.

1325 Hardy, 'Interpretation of Aerial Photography,' 1923, 6, NASM, A30.2/28; 'Aerial Photography for Pilots,' 29, NASM, A30.2/2.

1326 Supreme War Council, Draft Minutes, 9 May 1918, TNA, PRO: AIR 1/2299/209/77/29.

1327 Frank, *American Air Service Observation*, 394; 'Instructions on the Employment of Aerial Observation,' in Maurer, ed., *US Air Service*, II, 182.

1328 Col E. S. Gorrell, 26 October 1917 Memorandum, quoted in Toelle, 67.

1329 *Notes on the Interpretation of Aeroplane Photographs*, March 1917 (S.S. 550), 6, and February 1918 (S.S. 631), 8.

1330 Lieut I. M. Bonham-Carter, 'Notes,' 1 November 1914, TNA, PRO: AIR 1/834/204/5/240.

1331 Edmonds, *1916*, I, 82.

1332 Weiller in de Brunoff, 64.

1333 Solomon, 22.

1334 Ernst Jünger, *Storm of Steel: From the Diary of a German Storm-Troop Officer on the Western Front* (trans., London: Chatto & Windus, 1929; reprint New York: Howard Fertig, 1996), 16–17.

1335 Von Kathen, 'Memorandum for Encampments,' in HQ 26th Division, 'Intelligence Bulletin,' 26 July 1918,' NARA, RG 120, Box 1.

1336 Supreme War Council, Draft Minutes, 9 May 1918, TNA, PRO: AIR 1/2299/209/77/29.

1337 'Intelligence Summary,' 19 August 1918, TNA, PRO: AIR 1/2124/207/74/3.

1338 Porch, 113.

1339 Radio intercept units had to be on guard when discovering an absence of enemy communications reflecting either a change in the command and control of a particular sector or a deceptive cover for an impending operation. Général Ribadeau Dumas quoted in Porch, 113, n. 84 and n. 543.

1340 David French in Dockrill and French, 79-83.

1341 The RFC also maintained an aggressive offensive strategy that would have overlooked signs for the buildup while engaging tactical targets. Author discussion with James Streckfuss; David French in Dockrill and French, 79-83.

1342 'Supreme War Council File,' Draft Minutes, 9 May 1918, TNA, PRO: AIR 1/2299/209/77/29.

1343 'Camouflage,' Sec 2, July 1921, 7, NARA, RG 120, Box 819.

1344 Brig Gen William Mitchell, *Memoirs of World War I: From Start to Finish of Our Greatest War* (New York: Random House, 1960), 242–243.

1345 German report quoted in 'Camouflage for Troops of the Line,' 14.

1346 Fuller, 222–223; Porter, 178.

1347 Capt O. O. Ellis and Capt E. B. Garey, *The Plattsburg Manual: A Manual for Military Training* (New York: The Century Co., 1917), 213.

1348 *French Manual on Trench Warfare, 1917–1918: A Reference Manual, French General Staff*, reprint of *Manual of the Chief of Platoon of Infantry*, 1918 (London: The Imperial War Museum and Nashville, TN: The Battery Press, Inc. 2002), 234.

1349 Office of the Chief of Air Service, 'Recommendations, Command, Equipment, Organisation,' Gorrell Report, ser. C, r. 12, NARA.
1350 'Aeronautics,' June 1921, 1, NARA, RG 120, Box 819.
1351 Lt Col Brereton quoted in Knappen, 199–200.
1352 Capt Eddie Rickenbacker, *Fighting the Flying Circus* (New York, Fredrick A. Stokes Co., [c1919]; reprint, Chicago: R.R. Donnelley & Sons Company, 1997), 158.
1353 Philip wrote to Quentin's brother Theodore, Jr., 16 July 1918, quoted in Bert Frandsen, *Hat in the Ring: The Birth of American Air Power in the First World War* (Washington, DC: Smithsonian Books, 2003), 176–177. In WW II, Theodore, Jr., led a unit at Utah beach on D-Day and was awarded the Medal of Honor.
1354 Squier to War Department, 26 February 1915, quoted by James J. Cooke, *The US Air Service in the Great War, 1917–1919* (Westport, Connecticut: Praeger, 1996), 15.
1355 Brigadier-General Henderson had temporarily left the RFC to take command of the 1st Division. Colonel Sykes assumed command of the RFC until 22 December 1914. 'Memorandum on New Organisation of the Royal Flying Corps,' 29 November 1914, TNA, PRO: AIR 1/751/204/4/8; Sykes, 146.
1356 'Notes on Air Reconnaissance,' Appendix C.
1357 'Memorandum on New Organisation of the Royal Flying Corps,' 29 November 1914, TNA, PRO: AIR 1/751/204/4/8.
1358 'Organisation of RFC BEF December 1914,' TNA, PRO: AIR 1/751/204/4/8.
1359 HQ, RFC, 'Orders for Reconnaissance,' 24 October to 28 November 1914,' TNA, PRO: AIR 1/751/204/4/8.
1360 French doctrine quoted in Porter, 126.
1361 *Aerial Observation for Artillery*, 9.
1362 Christienne and Lissarague, 70.
1363 *Aerial Observation for Artillery*, 6-7.
1364 *Aerial Observation for Artillery*, 9.
1365 'A Study of the Organisation of the French Army,' 105, NARA, RG 120, Box 819.
1366 Cooke, 24–125.
1367 Christienne and Lissarague, 73.
1368 'Commanding Officers of Balloon Groups and Balloon Companies,' 4 October 1918, 1st Army, Gorrell Report, ser. F, r. 22, NARA.
1369 'Balloon Section Report,' 31 December 1918, Historical, 36, Gorrell Report, ser. F, r. 23, NARA.
1370 Snyder, 'Notes on Reconnaissance,' Gorrell Report, ser. J, r. 34, NARA.
1371 'Notes on the French System of Intelligence,' 7, TNA, PRO: WO 198/983.
1372 'Aeronautics,' June 1921, 21, NARA, RG 120, Box 819; Snyder, 'Notes on Reconnaissance,' Gorrell Report, ser J, r. 34, NARA.
1373 Gliddon, 18–19.
1374 *Aerial Photography*, Air Publication 838. TNA, PRO: AIR 10/853.
1375 'Aeronautics,' June 1921, 17, NARA, RG 120, Box 819.
1376 'Aeronautics,' NARA, RG 120, Box 819.
1377 Christienne and Lissarague, 183.
1378 Nesbit, 42–43.
1379 Carlier, 30–31; 'Final Tactical Report of Chief of Air Service,' Maurer, ed., *US Air Service*, I, 226–227.
1380 Davilla and Soltan, 503.
1381 Gliddon, 18–19.
1382 Gliddon.
1383 *Aerial Photography*. Air Publication 838. TNA, PRO: AIR 10/853.
1384 'Aeronautics,' June 1921, 19, NARA, RG 120, Box 819.
1385 'Aeronautics,' June 1921, 10–11, NARA, RG 120, Box 819.
1386 This was also known as *Section Artillerie Lourde* (SAL); Toelle, 2; Hodeir, 115.
1387 'Aeronautics,' June 1921, 10–11, NARA, RG 120, Box 819.
1388 The actual *escadrille* aéroplane inventory varied throughout the war. Author discussion with Dr James Davilla; 'Aeronautics,' June 1921, 10–11, NARA, RG 120, Box 819.
1389 'Proposals for a Long Distance Machine,' TNA, PRO: AIR 2/16.
1390 Elmer R. Haslett, 1st Lieutenant, 'Notes on the Duties of the Operations Officer of an Observation Squadron or Group,' USAF Academy McDermitt Library Special Collections, SMS 57.
1391 Gliddon, 18–19; Riggs, 'Report on Visit to British Army for Purpose of Studying Branch Intelligence,' Conger Papers, USAMHI; Lamberton, 11.
1392 'Central Flying School,' TNA, PRO: AIR 1/1625/204/89/7.
1393 'Proposals for a Fighting Machine,' TNA, PRO: AIR 2/16.
1394 British GHQ, France, 'Fighting in the Air,' February 1918, in Jones, 102.
1395 'Notes on Air Fighting from Royal Flying Corps,' September-December 1916, TNA, PRO: AIR 1/1625/204/89/7.
1396 'Final Report of Chief of Air Service,' Maurer, ed., *US Air Service*, I, 181; Frank, *American Air Service Observation*, 222–223.
1397 'Notes on Air Fighting from Royal Flying Corps,' TNA, PRO: AIR 1/1625/204/89/7.
1398 TNA, PRO: AIR 1/1625/204/89/7.
1399 'Aerial Photography for Pilots,' 1, NASM, A30.2/2.
1400 Gilchrist, 127.
1401 Hart, 37.
1402 'Fighting in the Air,' Jones, 102.
1403 Lieut. John Brophy, Diary, in Brereton Greenhous, ed., *A Rattle of Pebbles: The First World War Diaries*

of *Two Canadian Airmen* (Ottawa: Canadian Government Publishing Centre, 1987), 35.

1404 Lindstrom, 'Photographic Missions,' Gorrell Report, ser. N, r. 53, NARA.

1405 'Development of Aerial Photography,' Gorrell Report, ser. J, r. 34, NARA.

1406 Author discussion with Dr James Davilla.

1407 Crowell and Wilson, II, 418; Powell, 294.

1408 Hart, 36–37; 1st Lt John Snyder, 'Notes on Army Observation,' Gorrell Report, ser. J, r. 34, NARA; Watkis, 12–13.

1409 Porter, 96-97.

1410 Lindstrom, 'Photographic Missions,' Gorrell Report, ser. N, r. 53, NARA.

1411 Lamberton, 9–11; Barnes, *From Then Till Now*, 426.

1412 Lamberton, 193.

1413 Edmonds, *1915*, II, 142–143.

1414 Edmonds, *1916*, I, 89; *Aerial Observation for Artillery*, 2; 'A Study of the Organisation of the French Army,' 101–102, NARA, RG 120, Box 819.

1415 Sewell, 'Notes ... on French Corps Aviation,' TNA, PRO: AIR 1/475/15/312/201.

1416 'A Study of the Organisation of the French Army,' 103, NARA, RG 120, Box 819.

1417 Neil Fraser-Tytler, *With Lancashire Lads and Field Guns in France, 1915–1918* (Manchester, UK: J. Heywood, 1922), 65-66.

1418 'Final Report of Chief of Air Service,' Maurer, ed., *US Air Service*, I, 178.

1419 'Aerial Photography for Pilots,' 13, NASM, A30.2/2.

1420 Porter, 169.

1421 Neil L. Smith, 'History of 3 Squadron,' AFC, RAAF, URL: <http://www.3squadron.org.au/history.htm>, accessed 10 January 2005.

1422 'Aerial Photography for Pilots,' 13, NASM, A30.2/2.

1423 *Aerial Observation for Artillery*, 11.

1424 'A Study of the Organisation of the French Army,' 101, NARA, RG 120, Box 819.

1425 Edmonds, *1918*, I, 169.

1426 Philip Katcher, 'Bulletin of Information Air Service American Expeditionary Forces, Notes for Divisional Staffs, Commanding Officers and Aerial Observers on Work of Infantry Contact Airplanes,' 24 September 1918, Katcher Papers, USAMHI.

1427 2nd Lieutenant Joseph E. Eaton, 'Methods in Observation Practiced with Fifth Corps First American Army on the Fronts, St Mihiel Sector, September 12 to 16, 1918 Argonne-Meuse Sector, September 12 to November 11, 1918,' *Air Service Information Circular, II*, 20 September 1920, No. 115., 4–15.

1428 'Final Tactical Report of Chief of Air Service,' Maurer, ed., *US Air Service*, I, 254.

1429 'Combat Activities of Army Group Air Service during November 1–2, 1918,' Gorrell Report, ser. C, r. 12, NARA.

1430 Cooke, 59-60.

1431 'Tentative Manual for the Employment of Air Service,' Maurer, ed., *US Air Service*, II, 342.

1432 Frank, *American Air Service Observation*, 220.

1433 Lieut Michel quoted in Porter, 173.

1434 Snyder, 'Notes on the Relative Importance of Information to Be Obtained by Observers when on a Visual Reconnaissance,' Gorrell Report, ser. J, r. 34, NARA.

1435 Lieut Michel in Porter, 175.

1436 Squier, 'Map Room and Printing Room,' Squier Papers, USAMHI.

1437 Eaton, 'Methods in Observation Practiced with Fifth Corps First American Army on the Fronts, St Mihiel Sector, September 12 to 16, 1918 Argonne-Meuse Sector, September 12 to November 11, 1918,' 15.

1438 Major-General Sir William Sefton Brancker quoted in Porter, 175.

1439 Frank, *American Air Service Observation*, 234; Snyder, 'Notes on the Relative Importance of Information,' Gorrell Report, ser. J, r. 34, NARA.

1440 Frank, 234; Snyder, 'Notes on the Relative Importance of Information To be Obtained by Observers when a Visual Reconnaissance.'

1441 Eaton,'Methods in Observation Practiced with Fifth Corps First American Army on the Fronts, St Mihiel Sector, September 12 to 16, 1918 Argonne-Meuse Sector, September 12 to November 11, 1918,' 15.

1442 'Notes on Night Reconnaissance and Bombing,' July 1918, Gorrell Report, ser. J, r. 35, NARA.

1443 Lt Col J. T. C. Moore-Brabazon, 'Night Photography,' 1 November 1918, TNA, PRO: AIR 1/1118/204/5/2041.

1444 Martel, 282.

1445 Capitaine Faure, 'Instruction Relative aux Bombardments par Avions,' quoted in Fischer, 103.

1446 Brig Gen von Sauberzweig, 'Summary of Air Transportation,' quoted in Fischer, 105.

1447 Fischer, 106–107.

1448 'Aerial Photography for Pilots,' 2–3, NASM, A30.2/2.

1449 'Aeronautics,' June 1921, 5-6, NARA, RG 120, Box 819.

1450 'Aeronautics,' NARA, RG 120, Box 819.

1451 'Notes on Observation from Aeroplanes,' TNA, PRO: AIR 10/193.

1452 Ives, 61; Barnes, *From Then Till Now*, 485.

1453 'Final Report of Chief of Air Service,' Maurer, ed., *US Air Service*, I, 183.

1454 Mead, 62-63; Lamberton, 10.

1455 Gamble, 'Technical Aspects of British Aerial Photography,' TNA, PRO: AIR 1/2397/267/7.

1456 Moore-Brabazon, 'Photography's Part in the War,' TNA, PRO: AIR 1/724/91/2.

1457 Ives, 45.

1458 Moore-Brabazon, *The Brabazon Story*, 101.

1459 Ives, 44; Barnes, *From Then Till Now*, 485.

1460 Gamble, 'Photographic Section of the French Flying Corps,' TNA, PRO: AIR 1/404/97/1917.

1461 Moore-Brabazon, *The Brabazon Story*, 100–101.
1462 Gamble, 'Photographic Section of the French Flying Corps,' TNA, PRO: AIR 1/404/97/1917.
1463 TNA, PRO: AIR 1/404/97/1917.
1464 Moore-Brabazon, 'Photography's Part in the War,' TNA, PRO: AIR 1/724/91/2; Toye, 'History of Photography,' TNA, PRO: AIR 1/724/91/1.
1465 Ives, 45.
1466 Moore-Brabazon, 'Photography's Part in the War,' TNA, PRO: AIR 1/724/91/2.
1467 Toye, 'History of Photography,' TNA, PRO: AIR 1/724/91/1.
1468 Moore-Brabazon, *The Brabazon Story*, 101.
1469 Toye, 'History of Photography,' TNA, PRO: AIR 1/724/91/1.
1470 'Final Report of Chief of Air Service,' Maurer, ed., *US Air Service*, I, 151.
1471 Moore-Brabazon, 'Photography in the RAF,' TNA, PRO: AIR 1/724/91/1.
1472 Ives, 61; Barnes, *From Then Till Now*, 485.
1473 Steichen, 'Memorandum for Major Decker,' NASM.
1474 Col Raynal C. Bolling, 'Memorandum: 'Situation Regarding Glass for Air Service Photographic Lenses',' 22 January 1918, Gorrell Report, ser. G, r. 24, NARA.
1475 Maj James Barnes, 'Conference on the Optical Glass and Aviation Lens Situation,' 25 January 1918, NASM.
1476 Barnes, 'Conference on the Optical Glass and Aviation Lens Situation,' NASM
1477 Campbell, Letter to Maj James Barnes, 30 January 1918, NASM.
1478 Maj James Barnes, 'Photographic Matters,' 17 February 1918, NASM.
1479 Maj E. G. Toye, Letter to Maj James Barnes, 23 February 1918, NASM.
1480 'Appointment of Officer to Represent Air Service and U.S.Gov't. in Glass and Lens Negotiation,' 26 March 1918, NASM.
1481 'Appointment of Officer to Represent Air Service and U.S.Gov't. in Glass and Lens Negotiation.'
1482 Maj Edmond Gros, 'Notes ... on Conference on Glass Situation Held at the Bureau of Monsieur Ganne,' 25 April 1918, NASM; 'Official Report of the Conference Held at the Office of the Under Secretary of State of the Presidency of the Council,' 25 April 1918, Gorrell Report, ser. G, r. 24, NARA.
1483 Gros, 'Conference on Glass Situation,' NASM; 'Official Report of the Conference,' 25 April 1918,' Gorrell Report, ser. G, r. 24, NARA.
1484 Gros, 'Conference on Glass Situation,' NASM.
1485 Gros, 'Conference on Glass Situation.'
1486 Barnes, *From Then Till Now*, 487.
1487 Barnes, *From Then Till Now*, 487–488; 'Official Report; Cession of Optical Instruments to the American and English Governments,' Gorrell Report, ser. G, r. 22, NARA.
1488 Lt Col J. T. C. Moore-Brabazon, 'Report on ... Visit to the Conference Held in Paris,' 16 June 1918, NASM.
1489 Moore-Brabazon, 'Photography's Part in the War,' TNA, PRO: AIR 1/724/91/2.
1490 'Plates and Development of Plates,' Steere Papers, USAMHI.
1491 Dégardin, in *Vues d'en Haut*, 20–21; Pépin Papers, MA&E, Paris; Carlier,18
1492 Toye, 'History of Photography, TNA, PRO: AIR 1/724/91/1.
1493 Carl W. Ackerman, *George Eastman* (Boston: Houghton Mifflin Company, 1930), 296; Crowell and Wilson, II, 422–423; Powell, 294–295.
1494 Ackerman, 289.
1495 Ackerman, 299.
1496 Crowell and Wilson, II, 422–423; Powell, 294–295; Ives, 144; Carlier, 36.
1497 Ackerman, 296.
1498 Crowell and Wilson, II, 422–423; Powell, 294–295.
1499 Crowell and Wilson, II, 420; Powell, 294.
1500 George Eastman quoted in Ackerman, 302.
1501 Gill and Hibben, 'Preparation of Photographic Missions,' Gorrell Report, ser. J, r. 34, NARA.
1502 Porter, 171.
1503 Lieutenant-Colonel J.T.C. Moore-Brabazon, 'Inter-Allied Photographic Conference, August 19, 1918,' Group Captain Frederick Laws Papers, RAF Museum, London.
1504 Moore-Brabazon, 'Inter-Allied Photographic Conference, August 19, 1918.'
1505 Crowell and Wilson, II, 425; Porter, 175.
1506 *Aerial Photography*, Air Publication 838, TNA, PRO: AIR 10/853.
1507 Porter, 175–176.
1508 Insall, 158.
1509 Porter, 175–176; Barnes, *From Then Till Now*, 495; Ives, 244.
1510 Crowell and Wilson, II, 419–420.
1511 Lt Col E. S. Gorrell, 'Memorandum for Colonel Dunwoody,' 21 December 1917, Gorrell Report, ser. G, r. 24, NARA.
1512 Crowell and Wilson, II, 419; Ives, 45–46.
1513 Barnes, *From Then Till Now*, 487.
1514 Steichen, 'Memorandum for Major Decker,' NASM.
1515 Steichen, 'Aerial Photography,' Gorrell Report, ser. G, r. 24; Toulmin, 317–319.
1516 Steichen, 'Aerial Photography'; Toulmin.
1517 Moore-Brabazon, *The Brabazon* Story, 96-97.
1518 Lt Col Edward Steichen, 'Report of Inter-Allied Photographic Meeting,' 17 October 1918, Gorrell Report, ser. G, r. 24, NARA.
1519 Lt Col J. F. C. Moore-Brabazon, 'R.A.F. Night Photography,' TNA, PRO: AIR 1/2395/255/1.
1520 Steichen, 'Report of Inter-Allied Photographic Meeting,' Gorrell Report, ser. G, r. 24, NARA.
1521 Hanson, 15.

1522 Gamble, 'Photographic Section of the French Flying Corps,' TNA, PRO: AIR 1/404/97/1917.
1523 Toye, 'History of Photography,' TNA, PRO: AIR 1/724/91/1.
1524 Campbell, 'Notes on Aerial Photography,' TNA, PRO: AIR 1/7/6/98/20.
1525 '2nd Lt Kippel, OIC Photographic Section, 1st Wing, R.F.C.,' 14 June 1915, TNA, PRO: AIR 1/2151/209/3/244.
1526 Dixon, *Diary*, 50.
1527 Lieut.-Col J. T. C. Moore-Brabazon, 'OIC Photos in the Field,' 21 June 1915, TNA, PRO: AIR 1/2151/209/3/244.
1528 Carlier, 65.
1529 Lieutenant Michel quoted in Porter, 174.
1530 Crowell and Wilson, II, 425–426; Powell, 296–297.
1531 'General Lecture on Cameras,' Steers Papers, USAMHI; Campbell, 'Notes on Aerial Photography,' TNA, PRO: AIR 1/7/6/98/20.
1532 Author discussion with Alf Pyner.
1533 Barnes, *From Then Till Now*, 424–1425.
1534 The name of the camera varied with the source; Penros, Pan-Rose, and Pan-Ross were mentioned. See Appendix B for a list of all cameras of this type as Pan-Ross. Gamble, 'Technical Aspects of British Aerial Photography,' TNA, PRO: AIR 1/2397/267/7; Author discussion with Alf Pyner.
1535 6 in. equals 15.24cm. Gamble, 'Technical Aspects of British Aerial Photography,' TNA, PRO: AIR 1/2397/267/7.
1536 'Aerial Photography for Pilots,' 21 August 1918, NASM, A30.2/2.
1537 'Aerial Photography for Pilots,' NASM, A30.2/2.
1538 'Notes on Aerial Photography,' Part II, Baghdad, 1918, TNA, PRO: AIR 10/1001.
1539 Nesbit, 51.
1540 'Aeronautics,' June 1921, 3, NARA, RG 120, Box 819.
1541 'Aerial Photography for Pilots,' 21 August 1918, NASM, A30.2/2.
1542 Carlier, 26–28.
1543 Moore-Brabazon, 'Photography's Part in the War,' TNA, PRO: AIR 1/724/91/2.
1544 Metric equivalents are provided to understand the relationship between the French cameras. Campbell indicates 'a 12 in (30.48cm) equivalent focus, and has a back focal length of 6 1/4 in (15.875cm) ... Assuming 6,000 feet (1828.8 metres) to be the lowest an aeroplane can fly over the enemy lines with comparative safety ... The lens gave a scale of 1 mile to 10.6 inches (1,609 metres to 26.92cm) ... Pictures were made to have a scale of one mile to 21.2 inches (1,609 metres to 53.848cm), or 1/3,000 roughly ... The team showed the images of varying size taken from one standpoint, concluding that a focal length of 8 in (20.32cm).' Campbell, 'Notes on Aerial Photography,' TNA, PRO: AIR 1/7/6/98/20.
1545 TNA, PRO: AIR 1/7/6/98/20.
1546 Moore-Brabazon, 'Photography in the RAF,' TNA, PRO: AIR 1/724/91/1.
1547 Moore-Brabazon, *The Brabazon Story*, 94.
1548 Moore-Brabazon, 'Photography's Part in the War,' TNA, PRO: AIR 1/724/91/2.
1549 Insall, 73.
1550 Insall, 153–154.
1551 Camera lenses set for infinity allow for photography at a distance; Campbell, 'Notes on Aerial Photography,' TNA, PRO: AIR 1/7/6/98/20.
1552 Jones, II, 89-90; Watkis, 9; Babington-Smith, 10; Mead, 66; Gamble, 'Technical Aspects of British Aerial Photography,' TNA, PRO: AIR 1/2397/267/7.
1553 Toye, 'History of Photography,' 27, TNA, PRO: AIR 1/724/91/1.
1554 Jones, II, 89-90; Watkis, 9; Gamble, 'Technical Aspects of British Aerial Photography,' TNA, PRO: AIR 1/2397/267/7.
1555 Lieut.-Col. Trenchard commented, 'The three squadrons under my command are delighted with the camera, and the only crab that we have yet against it is that the slides are apt to break rather easily, a defect which is easily remedied by replacing the broken slides with new ones.'; Lieut.-Col. Trenchard memo to O.C. No. 2 Wing, 28 March 1915 in J.T.C. Moore-Brabazon Papers, RAF Museum, London; Capt Cuddell, Asst Dir of Mil Aeronautics, 3 March 1915, War Office, TNA, PRO: AIR 1/2151/209/3/244
1556 'Photography in the RAF,' TNA, PRO: AIR 1/724/91/1.
1557 Insall, 155.
1558 Jones, II, 89-90; Gamble, 'Technical Aspects of British Aerial Photography,' TNA, PRO: AIR 1/2397/267/7.
1559 Toye, 'History of Photography,' 17–18, TNA, PRO: AIR 1/724/91/1.
1560 Toye, 'History of Photography,' TNA, PRO: AIR 1/724/91/1.
1561 Air Vice Marshal H. R. M. Brooke-Popham, 'The War of 1914–118,' TNA, PRO: AIR 69/1.
1562 Gamble, 'Photographic Section of the French Flying Corps'; Labussière, in de Brunoff, 178; 'Aerial Photography for Pilots,' 21 August 1918, NASM, A30.2/2.
1563 Ralph Pulitzer, *Over the Front in an Aeroplane and Scenes inside the French and Flemish Trenches* (New York and London: Harper & Brothers Publishers, 1915), 4.
1564 Nesbit, 35.
1565 Moore-Brabazon, 'Photography's Part in the War,' TNA, PRO: AIR 1/724/91/2.
1566 Toye, 'History of Photography,' TNA, PRO: AIR 1/724/91/1.
1567 Moore-Brabazon, 'Photography's Part in the War,' TNA, PRO: AIR 1/724/91/2; Toye, 'History of Photography,' TNA, PRO: AIR 1/724/91/1; Ives, 109–110.
1568 Author discussion with Alf Pyner.

1569 'Report on Day Bombardment Training,' Gorrell Report ser. J, r. 35, NARA.
1570 Ives, 43; Crowell and Wilson, II, 420.
1571 '1916: 47' Lens Aircraft Cameras,' and '1918: Hand Held Oblique Aircraft Camera Type P 14,' RAF Museum, London.
1572 Toye, 'History of Photography,' TNA, PRO: AIR 1/724/91/1; Ives, 109–110; Moore-Brabazon, 'Photography's Part in the War,' TNA, PRO: AIR 1/724/91/2; Nesbit, 35.
1573 Ives, 43–45.
1574 'B' Type camera labels covered several variants. One was originally a cinematic movie camera. The variant developed at Repair Depots was sometimes referred to as the 'EB' Type camera. Toye, 'History of Photography,' TNA, PRO: AIR 1/724/91/1; Author discussion with Alf Pyner.
1575 Nesbit, 35.
1576 Gen H. R. M. Brooke-Popham, 'General Notes of German Aeroplanes,' TNA, PRO: AIR 1/7/6/98/20.
1577 Toye, 'History of Photography,' TNA, PRO: AIR 1/724/91/1.
1578 Brooke-Popham, 'General Notes of German Aeroplanes,' TNA, PRO: AIR 1/7/6/98/20.
1579 Photo of a 47-in lens camera shows a moniker of 'B.E.3' on the housing, RAF Museum, London.
1580 Moore-Brabazon, *The Brabazon* Story, 104.
1581 Laws later in life stated that the 'L' stood for Laws. Transcripts from 1972 taped interview with Group Captain Laws, F.C.V. Laws Papers, RAF Museum, London.; Nesbit, 40.
1582 Moore-Brabazon, 'Photography in the RAF,' TNA, PRO: AIR 1/724/91/1.
1583 Toye, 'History of Photography,' TNA, PRO: AIR 1/724/91/1; Ives, 109–110.
1584 Nesbit, 42.
1585 Ives, 117–120.
1586 Dixon, *Diary*, 50.
1587 The American version of the 'L' Type camera is sometimes referred to as the 'LA' Type camera. Steichen, 'Aerial Photography,' Gorrell Report, ser. G, r. 24, NARA.
1588 Ives, 87, 102; Crowell and Wilson, II, 421.
1589 W. O. Farnsworth, 'Photographic Officer Zone of Advance,' 20 December 1918, Gorrell Report, ser. A, r. 2, NARA.
1590 Toye, 'History of Photography,' TNA, PRO: AIR 1/724/91/1; Nesbit, 42.
1591 Moore-Brabazon, 'Photography's Part in the War,' TNA, PRO: AIR 1/724/91/2.
1592 Clarence Winchester and F. L. Wills, *Aerial Photography, A Comprehensive Survey of Its Practice & Development* (Boston: American Photographic Publishing, Co., 1925), 84.
1593 Nesbit, 51.
1594 Steichen, 'Memorandum for Major Decker,' NASM.
1595 Toye, 'History of Photography,' TNA, PRO: AIR 1/724/91/1; Moore-Brabazon, 'Inter-Allied Photographic Conference, August 19, 1918.'
1596 Author discussions with Nicholas Watkis and Michael Mockford.
1597 The 'P' Type was also designated as the P14. Winchester and Wills, 78.
1598 Author discussion with Alf Pyner.
1599 Toye, 'History of Photography,' TNA, PRO: AIR 1/724/91/1.
1600 Author discussions with Nicholas Watkis and Michael Mockford.
1601 Gill and Hibben, 'Preparation of Photographic Missions,' Gorrell Report, ser. J, r. 34, NARA; Carlier, 23.
1602 'Lecture on Aerial Photography, Centre d'Etudes de l'Aeronautique,' 25 April 1923. NASM A30.2/36.
1603 Ives, 43; 89.
1604 W. O. Farnsworth, 'Photographic Officer Zone of Advance,' 20 December 1918, Gorrell Report, ser. A, r. 2, NARA.
1605 'Lecture on Aerial Photography, Centre d'Etudes de l'Aeronautique,' 25 April 1923. NASM A30.2/36.
1606 Labussière, in de Brunoff, 184; Ives, 43; Carlier, 22–25; Gill and Hibben, 'Preparation of Photographic Missions,' Gorrell Report, ser. J, r. 34, NARA.
1607 Hart, 36–37.
1608 Gill and Hibben, 'Preparation of Photographic Missions,' Gorrell Report, ser. J, r. 34, NARA; Carlier, 22–25
1609 'Lecture on Aerial Photography, Centre d'Etudes de l'Aeronautique,' 25 April 1923. NASM A30.2/36.
1610 'Lecture on Aerial Photography, Centre d'Etudes de l'Aeronautique,' 25 April 1923. NASM A30.2/36; 'Nomenclature Chart,' Gorrell Report, ser. F, r. 22, NARA.
1611 'The Camera at the Front: Why the Military Authorities Have Use for Their Own Cameras but None for the Enemy's,' *Scientific American* (24 November 1917): 380, 390.
1612 Gamble, 'Photographic Section of the French Flying Corps'; Carlier, 25.
1613 Steichen, 'Memorandum for Major Decker,' NASM.
1614 Lieut Michel quoted in Porter, 172.
1615 Steichen, 'American Aerial Photography at the Front,' 3, NASM, A30.2/24.
1616 Moore-Brabazon in his 'Inter-Allied Photographic Conference' notes stated the French De Ram weighed 75 pounds while the American version weighed up to 100 pounds. Capt Proctor, 'Training Section, Photographic Branch Memorandum for Major Sullivan, Subject: 'Situation De Ram Camera'', Gorrell Report, ser. G, r. 24, NARA.
1617 'Final Report of Chief of Air Service,' Maurer, ed., *US Air Service*, I, 402; Crowell and Wilson, II, 422; Ives, 116, Carlier, 30–31.

1618 Ives, 123.
1619 Maj James Barnes, 'Memo to Lieut-Colonel Dunwoody, Subject: 'De Ram Camera'', 29 November 1917, Gorrell Report, ser. G, r. 24, NARA; Toulmin, 320.
1620 'Lecture on Aerial Photography, Centre d'Etudes de l'Aeronautique', 25 April 1923. NASM A30.2/36.; Carlier, 31–32.
1621 The Americans tested the De Ram in the DH 4. Steichen, 'American Aerial Photography at the Front', 3, NASM, A30.2/24.
1622 Steichen, 'Memorandum for Major Decker', NASM.
1623 Steichen, 'American Aerial Photography at the Front', 3, NASM, A30.2/24.
1624 'Final Report of Chief of Air Service', Maurer, ed., *US Air Service*, I, 402; Crowell and Wilson, II, 422.
1625 Steichen, 'Aerial Photography', Gorrell Report, ser. G, r. 24, NARA; Crowell and Wilson, II, 422.
1626 Lt Col J. T. C. Moore-Brabazon, 'Report on ... Visit to the Conference Held in Paris', 16 June 1918, NASM.
1627 Steichen, 'Aerial Photography', Gorrell Report, ser. G, r. 24, NARA; Crowell, II, 422.
1628 Steichen, 'Aerial Photography', Gorrell Report, ser. G, r. 24, NARA; 'Final Report of Chief of Air Service', Maurer, ed., *US Air Service*, I, 402; Crowell and Wilson, II, 422.
1629 Steichen, 'American Aerial Photography at the Front', 3, NASM, A30.2/24; 'Final Report of Chief of Air Service', Maurer, ed., *US Air Service*, I, 154.
1630 Eaton, 'Methods in Observation Practiced with Fifth Corps First American Army on the Fronts, St Mihiel Sector, September 12 to 16, 1918 Argonne-Meuse Sector, September 12 to November 11, 1918', 16.
1631 'Final Report of Chief of Air Service', Maurer, ed., *US Air Service*, I, 278–279.
1632 Dixon, *Diary*, 50.
1633 Carlier, 30.
1634 Moore-Brabazon, 'Inter-Allied Photographic Conference, August 19, 1918.'
1635 Carlier, 17.
1636 Moore-Brabazon, 'Inter-Allied Photographic Conference, August 19, 1918;' Carlier, 34–135.
1637 Ives, 46.
1638 Davilla and Soltan, 54–161; Bruce, *Aeroplanes of the Royal Flying Corps (Military Wing)*,128–132; Lamberton, 70–71, 216–217; Kenneth Munson, *Bombers: Patrol and Reconnaissance Aircraft, 1914–1919* (New York: Macmillan Publishing Co., Inc., 1968), 97.
1639 Munson, 14–115; Lamberton, 48.
1640 Douglas, 'Notes on Equipment of Units of RFC with BEF', TNA, PRO: AIR 1/836/204/5/268.
1641 Insall, 80.
1642 Bruce, *Aeroplanes of the Royal Flying Corps (Military Wing)*, 344–1370; Lamberton, 46-51, 214–215; Munson, 129–132.
1643 Nesbit, 12.
1644 Collection S.H.D., 377/094, SHAA.
1645 Martel, 77-78.
1646 Bruce, *Aeroplanes of the Royal Flying Corps (Military Wing)*, 242.
1647 Davilla and Soltan, 222–232; Bruce, *Aeroplanes of the Royal Flying Corps (Military Wing)*, 241–246; Lamberton, 86-87, 218–219.
1648 Peter Kilduff, 'Kenneth P. Littauer: From the Lafayette Flying Corps to the 88th Aero Squadron', *Journal of the Society of World War I Aero Historians*, 13, No. 1, 25.
1649 Guttman, 9–10.
1650 Munson, 147.
1651 Davilla and Soltan, 149–156; Bruce, *Aeroplanes of the Royal Flying Corps (Military Wing)*, 194–196; Lamberton, 80–81, 216; Guttman, 35; Munson, 147–148.
1652 Bruce, *Aeroplanes of the Royal Flying Corps (Military Wing)*, 406–407.
1653 'Notes on Air Fighting from Royal Flying Corps', TNA, PRO: AIR 1/1625/204/89/7.
1654 TNA, PRO: AIR 1/1625/204/89/7.
1655 Bruce, *Aeroplanes of the Royal Flying Corps (Military Wing)*, 403–418; Lamberton, 63, 214–1215.
1656 Peter Cooksley, *Sopwith Fighters in Action* (Carrollton, TX: Squadron/Signal Publications, Inc., 1991), 14.
1657 J. M. Bruce, *War Planes of the First World War, Fighters, Great Britain* (2 vols., Garden City, NY: Doubleday and Company, Inc., 1968), II, 115–117; Munson, 135.
1658 Bruce, *War Planes of the First World War, Fighters*, II, 115–117.
1659 'Price-List of Various British Airframes and Engines', in Jones, Appendices, 155–156.
1660 J. M. Bruce, *The Sopwith 1½ Strutter* (London: Profile Publications, 1967); Bruce, *Aeroplanes of the Royal Flying Corps (Military Wing)*, 499-509; Davilla and Soltan, 465–473; Christienne and Lissarague, 107; Munson, 58; Lamberton, 58-59, 216–217.
1661 C.D. Smart, 'The Front Cockpit: The Diary of Lt. C.D. Smart, No. 5 Squadron, RFC', *Cross and Cockade, Journal of the Society of World War I Aero Historians* 10, No. 1 (Spring 1969), 17.
1662 Lamberton, 56.
1663 Lamberton.
1664 Bruce, *Aeroplanes of the Royal Flying Corps (Military Wing)*, 458–464; Munson, 57; Lamberton, 56-58, 214–1215.
1665 Davilla and Soltan, 101–125; Munson, 65, 144; Lamberton, 72, 216–217.
1666 Steichen, 'American Aerial Photography at the Front', 3, NASM, A30.2/24.
1667 Smart, 21.
1668 Colin Owers, *De Havilland Aircraft of World War I*, Vol. I, *D.H.1 -D.H.4* (Boulder, CO: Flying Machine

Press, 2001); J. M. Bruce, *The de Havilland D.H.4* (London: Profile Publications, 1965); J. M. Bruce, *The American D.H.4* (London: Profile Publications, 1966); Bruce, *Aeroplanes of the Royal Flying Corps (Military Wing)*, 276–280; Lamberton, 36; Munson, 138.
1669 Owers, 17.
1670 Bruce, *Aeroplanes of the Royal Flying Corps (Military Wing)*, 49-61; Smart, 29; Lamberton, 214–215; Munson, 137–139.
1671 Colin A. Owers, John S. Guttman, and James J. Davilla, *Salmson Aircraft of World War I* (Boulder, CO: Flying Machine Press, 2001), 15; Munson 142–143.
1672 Munson, 143.
1673 Davilla and Soltan, 439–446; Munson, 63; Lamberton, 218–219.
1674 Goddard, 7-8.
1675 'Aeronautics,' June 1921, 1, NARA, RG 120, Box 819.
1676 Maurice Marie Eugène Grout personnel records, SHAA.
1677 Hodeir, 116; Labussière in de Brunoff, 184; Carlier, 19.
1678 Watkis, 11; Toye, 'History of Photography,' TNA, PRO: AIR 1/724/91/1.
1679 Nesbit, 40–42.
1680 Moore-Brabazon, 103.
1681 Campbell, 'Progress of Photography,' TNA, PRO: AIR 1/724/91/2.
1682 Toye, 'History of Photography,' TNA, PRO: AIR 1/724/91/1.
1683 Middleton, III, 119–120.
1684 Middleton, III.
1685 Royall Tyler, G-2, HQ AEF, 'Letter to Col Nolan,' 7 March 1918, Conger Papers, Box 2, USAMHI.
1686 1st Lt John Stewart, 'Report,' 4 October 1918, Conger Papers, Box 2, USAMHI.
1687 'Correspondence,' 23 March 1918, Conger Papers, Box 2, USAMHI.
1688 *British Trench Warfare*, 64.
1689 *British Trench Warfare*.
1690 Brown, 136.
1691 Brown, 153.
1692 Barnes, *From Then Till Now*, 488.
1693 Frank, *American Air Service Observation*, 160.
1694 'History of the Training Section, Air Service, A.E.F.,' Gorrell Report, ser. J, r. 34, NARA.
1695 'History of the Training Section, Air Service, A.E.F.'
1696 Hubbard, 'Memorandum for Col Nolan,' 29 December 1917, Conger Papers, Box 2, USAMHI.
1697 Cooke, 22–23.
1698 'History of the Training Section,' Gorrell Report, ser. J, r. 34, NARA.
1699 Harry A. Toulmin, Jr., *Air Service, American Expeditionary Force, 1918* (New York: D. Van Nostrand Company, 1927), 320.
1700 'Final Report of Chief of Air Service,' Maurer, ed., *US Air Service*, I, 154.
1701 Barnes, *From Then Till Now*, 483.
1702 1st Lt Theordore J.Lindoff, 'Photographic Mission,' Lindoff Papers, USAMHI.
1703 Gilchrist, 48.
1704 Frank, *American Air Service Observation*, 154; Arthur Sweetser, *The American Air Service: A Record of Its Problems, Its Difficulties, Its Failures, and Its Final Achievements* (New York: D. Appleton and Co., 1919), 116–117.
1705 Snyder, 'Notes on Army Observation,' Gorrell Report, ser. J, r. 34, NARA.
1706 'History of the Training Section,' Gorrell Report, ser. J, r. 34.
1707 Barnes, *From Then Till Now*, 475–476.
1708 Frank, 204.
1709 Barnes, *From Then Till Now*, 488.
1710 Barnes, *From Then Till Now*, 488.
1711 Barnes, *From Then Till Now*, 495.
1712 'History of Photographic Section #5, Air Service, U.S.A.,' 1918, Gorrell Report, ser. G, r. 24, NARA.
1713 Porter, 166
1714 'Report of the Director of Military Aeronautics,' I, 1388.
1715 'History of Photographic Section #5.
1716 George Eastman quoted in Ackerman, 296.
1717 George Eastman quoted in Ackerman, 298.
1718 George Eastman quoted in Ackerman, 304–1305.
1719 Porter, 171.
1720 Barnes, *From Then Till Now*, 500.
1721 'Final Report of Chief of Air Service,' Maurer, ed., *US Air Service*, I, 151.
1722 Porter, 171; Ackerman, 306.
1723 *Scientific Monthly*, April 1918, quoted in Ackerman, 306.
1724 Barnes, *From Then Till Now*, 479.
1725 Barnes, *From Then Till Now*, 482.
1726 Niven, 461.
1727 Frank, *American Air Service Observation*, 160.
1728 Goddard, 8.
1729 Barnes, *From Then Till Now*, 482.
1730 Goddard, 10.
1731 Goddard, 8.
1732 Goddard, 9.
1733 Sgt-Maj Haslitt quoted in Goddard, 26.
1734 Barnes, *From Then Till Now*, 501.
1735 *The Airscout's Snapshot*, 28 August 1918, Steere Papers, USAMHI.
1736 *The Airscout's Snapshot*.
1737 Jones, II, 82.
1738 'Instructions on the Employment of Aerial Observation,' in Maurer, ed., *US Air Service*, II, 182.
1739 Jones, II, 82.
1740 C.D. Smart, 'The Front Cockpit,' 23.
1741 Gliddon, 44.

1742 Insall, 155–156.
1743 Morris, 102–103.
1744 Gamble, 'Technical Aspects of British Aerial Photography,' TNA, PRO: AIR 1/2397/267/7.
1745 'Military Aviation,' Maurer, ed., *US Air Service*, II, 49.
1746 Turner, 198–199.
1747 Porter, 165.
1748 W. 'Sholto' Douglas, 'Notes on First Steps in Air Photography on the Western Front, January to March 1915,' TNA, PRO: AIR 69/2.
1749 TNA, PRO: AIR 69/2.
1750 Hart, 43.
1751 Douglas, 'First Steps in Air Photography,' TNA, PRO: AIR 69/2; Jones, II, 89.
1752 Douglas, 'Notes on Equipment of Units of RFC with BEF,' 27 September 1914, TNA, PRO: AIR 1/836/204/5/268.
1753 Dixon, *Diary*, 50.
1754 Snyder, 'Notes on Army Observation,' Gorrell Report, ser. J, r. 34, NARA.
1755 'Aerial Photography for Pilots,' 12, NASM, A30.2/2
1756 Lindstrom, 'Photographic Missions,' Gorrell Report, ser. N, r. 53, NARA.
1757 Frank, *American Air Service Observation*, 220–221.
1758 'Tactical History of Corps Observation,' Maurer, ed., *US Air Service*, I, 181; 'Report on Day Bombardment Training,' Gorrell Report, ser. J, r. 35, NARA.
1759 Snyder, 'Notes on Reconnaissance,' Gorrell Report, ser. J, r. 34, NARA.
1760 'Report on Day Bombardment Training,' Gorrell Report, ser. J, r. 35, NARA.
1761 Snyder, 'Notes on Army Observation,' Gorrell Report, ser. J, r. 34, NARA.
1762 'Notes on Army Observation.'
1763 Sewell, 'Notes ... on French Corps Aviation,' TNA, PRO: AIR 1/475/15/312/201.
1764 Hart, 43.
1765 'Aerial Photography for Pilots,' 9, NASM, A30.2/2.
1766 Lieut Michel quoted in Porter, 174
1767 Unknown source quoted in Porter, 225.
1768 2nd Lieut Frank Warren quoted in Brown, *Imperial War Museum Book of 1918*, 17.
1769 P.D. Townshend, *Eye in the Sky 1918: Reflections of a World War One Pilot on Artillery and Infantry Co-operation Duties*, 29-31; C.D. Smart, 'The Front Cockpit,' 20.
1770 'Aerial Photography for Pilots,' 9, NASM, A30.2/2; Porter, 221.
1771 Porter, 218.
1772 Henry Bruno, 'The Eyes in the Air,' *Popular Science Monthly* (1918): 511.
1773 Hart, 37.
1774 Squier, 'Map Room and Printing Room,' Squier Papers, USAMHI.
1775 Carroll D. Winslow, *With the French Flying Corps* (New York: Charles Scribner's Sons, 1917), 183.
1776 Unknown source quoted in Gilchrist, 48–49.
1777 Snyder, 'Notes on Army Observation,' Gorrell Report, ser. J, r. 34, NARA.
1778 Squier, 'Intelligence Services,' Squier Papers, USAMHI.
1779 *Aerial Observation for Artillery*, 8.
1780 'History of the 91st Aero Squadron,' in Frank, *American Air Service Observation*, 233.
1781 Lewis, 'Procedures of Ranging,' 55, TNA, PRO: AIR 1/834/204/5/240.
1782 Insall, 155–156.
1783 Lewis, 'Procedures of Ranging,' 61, TNA, PRO: AIR 1/834/204/5/240.
1784 'Notes on Branch Intelligence,' 91st Aero Squadron, 1 November 1918, 21, Gorrell Report, ser. M, r. 40, NARA; 'Final Report of Chief of Air Service,' Maurer, ed., *US Air Service*, I, 173; Porter, 170.
1785 Porter, 165.
1786 Terry, 'Notes on Branch Intelligence,' Gorrell Report, ser. M, r. 40, NARA; 'Final Report of Chief of Air Service,' Maurer, ed., *US Air Service*, I, 173; Porter, 170.
1787 Smart, *The Hawks that Guided the Guns*, 43.
1788 Dixon, *Diary*, 51.
1789 'History of Photographic Section #2, Air Service, U.S.A.,' 1918, Gorrell Report, ser. G, r. 24, NARA.
1790 Dixon, *Diary*, 49-50.
1791 Dixon, *Diary*, 51.
1792 Dixon, *Diary*, 53.
1793 Dixon, *Diary*, 52.
1794 Steichen, 'American Aerial Photography at the Front,' 2, NASM, A30.2/24.
1795 G-2, 1st Army, to Chief of Staff, 'Account of the Operation,' 67, Drum Papers, Box 14, USAMHI.
1796 'Account of the Operation.'
1797 Barnes, 494–495.
1798 'Camera at the Front,' *Scientific American* (24 November 1917): 380.
1799 'The Photographic Eye and Memory of the Military Airman,' in *Scientific American* (27 July 1918), 61.
1800 Douglass Reid, 'The Eye Behind the Lines,' *Popular Mechanics* (March-May 1919).
1801 Hanson, 15.
1802 Reid quoted in Hanson, 8–10.
1803 *Study and Use of Aerial Photographs*, Section I, 4.
1804 Pépin, 'Lecture on the Study and Interpretation of Aerial Photographs,' 10 December 1917, 2, NASM.
1805 Pépin, 'Lecture on the Study and Interpretation of Aerial Photographs,' 10 December 1917, 14, NASM.
1806 Pouquet, 'Working of the Service of Interpretation of Photographs,' NARA, RG 120, Box 819.
1807 Sewell, 'Notes ... on French Corps Aviation,' TNA, PRO: AIR 1/475/15/312/201.
1808 'Lecture on Aerial Photography, Centre d'Etudes de l'Aeronautique,' 25 April 1923. NASM A30.2/36.
1809 Middleton, III, 122.
1810 Jones, II, 177.

1811 Solomon, 53.
1812 Capt Stanford, 'Summary of the Qualifications of a Branch Intelligence Officer,' 15 February 1918, in 'Organisation of Intelligence,' HQ, British 5th Army, Conger Papers, Box 2, USAMHI.
1813 Stanford, 'Summary of the Qualifications,' Conger Papers, Box 2, USAMHI.
1814 Stanford, 'Summary of the Qualifications.'
1815 Stanford, 'Summary of the Qualifications.'
1816 Insall, 156–157.
1817 Porter, 183.
1818 Steichen, 'American Aerial Photography at the Front,' 1, NASM, A30.2/24.
1819 Steichen, 'American Aerial Photography at the Front,' 1, NASM, A30.2/24.
1820 Stanford, 'Summary of the Qualifications,' Conger Papers, Box 2, USAMHI.
1821 Dixon, *Diary*, 51.
1822 Niven, 460.
1823 Porter, 174.
1824 Friedmann, 'History of Photographic Section #1, Air Service,' U.S.A. 1918.
1825 The noted author, Carl Sandburg, was a close friend of Edward Steichen and remembered Steichen's descriptions of his wartime experiences; Sandberg's wife, Lilian Steichen Sandburg, was Steichen's younger sister; Niven, 461.
1826 Pépin, 'Lecture on the Study and Interpretation of Aerial Photographs,' 10 December 1917, 7, NASM.
1827 *Notes on the Interpretation of Aeroplane Photographs*, March 1917 (S.S. 550), 5, and February 1918 (S.S. 631), 3.
1828 Steichen, 'Aerial Photography,' Gorrell Report, ser. G, r. 24, NARA.
1829 Steichen, 'Aerial Photography,' Gorrell Report, ser. G, r. 24, NARA.
1830 The French officer was Capt. Pépin. Steichen, 'Aerial Photography,' Gorrell Report, ser. G, r. 24, NARA.
1831 Steichen, 'Aerial Photography,' Gorrell Report, ser. G, r. 24, NARA.
1832 Solomon, 53; Solomon was not appreciated by everyone. Major-General Brooke-Popham after the war commented, 'Solomon is the painter and as you no doubt know he was given a commission as a Colonel and sent out to France to give expert advice on camouflage. This paper [authored by Solomon] is rather a good proof of the futility of people without commonsense posing as experts.' Brooke-Popham memo in 'Notes on Col. Soloman's Ideas on German Camouflage of 30 Mar 1918,' TNA, PRO: AIR 1/1/4/21.
1833 Unidentified source, quoted by Höfl (translation by Thomas Gudmestad).
1834 Hibben, 'Interpretation of Aerial Photography,' Gorrell Report, ser. J, r. 34, NARA.
1835 Charteris to Gen Macdonogh, 28 July 1916, TNA: PRO: WO 158/897.
1836 David French makes this point in 'Failures of Intelligence' in Dockerill and French. David French, 79-83.
1837 Babington-Smith, 9–10.
1838 Acceptance of new weapons such as the tank remained equally problematic. The prevailing postwar thinking of the American military showed cavalry still had an important role to play in twentieth century warfare. Armour would have to wait. It was indicative that the leading practitioner of American armoured combat in the Second World War, George Patton, saw the handwriting on the wall and returned to postings in the cavalry in the early 1920s. Like the tank, information war had its time in the limelight at a later date in the twentieth century. General Pershing quoted in Porter, 18–19.
1839 Author's discussions with Dr David Mets.
1840 Foch quoted in Christienne, 184.
1841 Air Commodore T. C. R. Higgins to Air Commodore H. R. M. Brooke-Popham, 28 February 1923, in 'Instruction Regarding Interpretation of Aerial Photographs,' TNA: PRO: AIR 2/277.
1842 Roeckel quotations in the colour section are from a Capitaine Roeckel letter to Lieutenant Farré, in *Sky Fighters of France*.

Bibliography

ARCHIVAL AND MANUSCRIPT SOURCES

AIR UNIVERSITY LIBRARY, MAXWELL AIR FORCE BASE, ALABAMA

Chester, Einar W., Captain. 'The Development and Use of Air Forces from January 1, 1916, to the End of the War.' The Infantry School, Fort Benning, GA. Company Officer's Course, 1925-26

'Fundamental Principles for the Employment of the Air Services.' Army War College. Command Course No. 3, 1925-26

Lawrence, Charles W., Captain. 'Organization and Assignment of Observation Aviation Observation Airplanes and their Equipment.' Air Corps Tactical School, Maxwell Field, Alabama, 1937-38. Air Force Historical Office Archives, IRIS ref. A2810

LIBRARY OF CONGRESS, MANUSCRIPTS DIVISION, WASHINGTON DC

Brigadier General William Mitchell Papers

'Depots de Munitions Ennemis, Woevre et Hauts de Meuse.' Diary, 21 April 1917

'Gares Ennemies, en Face du Front de la II-Armée.'

George Goddard Papers, Manuscripts Division

W.B. Prince Collection, Diary, Map Division

Musée de l'Air et de l'Espace, Le Bourget, France
Eugène Pépin Papers

NATIONAL AIR AND SPACE MUSEUM, WASHINGTON DC

'Aerial Photography for Pilots,' 21 August 1918. A30.2/2

'Appointment of Officer to Represent Air Service and U.S. Gov't. in Glass and Lens Negotiation,' 26 March 1918

Assistant Chief of Staff for Intelligence, General Headquarters, American Expeditionary Forces 'Report' to Chief of Staff, American Expeditionary Forces, 15 June 1919

Barnes, James, Major. 'Conference on the Optical Glass and Aviation Lens Situation,' 25 January 1918

—— Letter to Colonel Van Horn. Subject: 'Conference on the Optical Glass and Aviation Lens Situation,' 25 January 1918

—— Letter to Major J. T. C. Moore-Brabazon. Subject: 'Photographic Matter,' 17 February 1918

—— 'Photographic Matters,' 17 February 1918

Campbell, C. D. M., Major. 'Letter to [Major James] Barnes,' 30 January 1918

Estienne, Lieutenant. 'Army Photographic Reconnaissance as it Should Be,' 29 March 1923 A30.2/34

Gros, Edmond, Major. 'Notes by Major Gros on Conference on Glass Situation Held at the Bureau of Monsieur Ganne,' 25 April 1918

—— 'Interpretation of Aerial Photography,' I.S. 282 (EDA). NASM A30.2/28

—— 'Lecture on Aerial Photography, Centre d'Etudes de l'Aeronautique,' 25 April 1923 NASM A30.2/36

Letter to Colonel J.G. Morrow, O.I.C. Photographic Section. Subject: 'Photographic Cameras,' 9 May 1918

Moore-Brabazon, J. T. C., Major. 'Letter to [Major] Barnes,' 12 February 1918

—— Lieutenant-Colonel. Report of 'Moore-Brabazon's Visit to the Conference, Held in Paris, on the 16th June, 1918.'
'On the Study of German Organizations by Means of Aerial Photographs. Sapping Works.' A30.2/17
Pépin, Eugène, Lieutenant. 'A Lecture on the Study and Interpretation of Aerial Photographs,' 10 December 1917. War Department, Office of the Chief Signal Officer, Information Section – Air Division, Washington, DC, Stencil No. 682
Steichen, Edward J. 'American Aerial Photography at the Front.'
—— Army Personnel File
—— Captain. Letter to Major Barnes. Subject: 'Optical Glass Situation,' 6 March 1918
—— First Lieutenant. 'Memorandum for Major Decker, Photographic Division, Signals Corps Washington D.C.,' 21 December 1917
'Tactical History of Corps Observation A.E.F.' *Air Service Information Circular* 1, No. 75 (12 June 1920)
Telephone Interview with Dr Mary Steichen Calderone, Daughter of Edward Steichen, 29 May 1986, New York City
Toye, E. G., Major. 'Letter to [Major James] Barnes,' 23 February 1918 United States Army. Air Service, Chief Photographic Officer. Letter to Chief of the Air Service
Subject: 'Appointment of Officer to Represent Air Service and U.S. Gov't in Glass and Lens Negotiation,' 16 March 1918
United States. War Department, Office of the Director of Military Aeronautics. 'Aerial Photography for Pilots, Study in Detail of the Part that Photography Plays in the Present War, Lecture on Aerial Photography for Pilots,' 21 August 1918. A30.2/2
—— *Aerial Photography in Modern Warfare.* 19 April 1920. NASM Archives A30.2

NATIONAL ARCHIVES AND RECORDS ADMINISTRATION, WASHINGTON DC

Gorrell Report

Microfilm citation: Gorrell's History of the American Expeditionary Forces Air Service, 1917-1919. Washington, DC: National Archives and Records Service, 1974. 58 microfilm reels
'Balloon Section Report,' 31 December 1918. Gorrell Report, Series F, Roll 23
'Balloon Training.' Gorrell Report, Series F, Roll 23
Barnes, Major James. 'History of Photographic Section #4, Air Service, U.S.A., 1918.' Gorrell Report, Series G, Roll 24
—— 'Memo to Lieut-Colonel Dunwoody, Subject: 'De Ram Camera.' 29 November 1917. Gorrell Report, Series G, Roll 24
Bellinger, Alfred E. 'Relations of the Air Service with G-2.' Gorrell Report, Series M, Roll 41
Bolling, Raynal C., Colonel. Memorandum, Subject: 'Situation regarding Glass for Air Service Photographic Lenses,' 22 January 1918. Gorrell Report, Series G, Roll 24
'Combat Activities of Army Group Air Service during November 1-2, 1918.' Gorrell Report, Series C, Roll 12
'Commanding Officers of [1st Army] Balloon Groups and Balloon Companies,' 4 October 1918 Gorrell Report, Series F, Roll 22
'The Development of Aerial Photography. Treatise by Photo Department, 12 Jul 1918, HQ, A.S., S.O.S.' Gorrell Report, Series G, Roll 24, NARA
Farnsworth, W. O. 'Photographic Officer Zone of Advance,' 20 December 1918. Gorrell Report, Series A, Roll 2
First Army Corps. General Staff, Second Section. 'Distribution of Photographs Made by the Aviation Section of the Corps,' 11 August 1918. Gorrell Report, Series C, Roll 12
—— Headquarters, Office of Chief of Air Service. 'Report of Operations, for April 20, 1918.' Gorrell Report, Series C, Roll 12
Friedmann, Martin E., Corporal. 'History of Photographic Section #1, Air Service, U.S.A., 1918.' Gorrell Report, Series G, Roll 24
Gill, G. W., Lieutenant, and First Lieutenant Thomas E. Hibben. 'The Preparation of Photographic Missions. Notes from Lectures of the Photographic Department,' Observers School, 2nd Aviation Instruction Center. Gorrell Report, Series J, Roll 34

Gorrell, E.S., Lieutenant Colonel. 'Memorandum for Colonel Dunwoody,' 21 December 1917 Gorrell Report, Series G, Roll 24
Hibben, Thomas E., First Lieutenant. 'The Importance of Aerial Photographs in Intelligence.' Lecture given at 2nd A.I.C. Gorrell Report, Series J, Roll 34
—— 49th Infantry Att. A. S. 'Interpretation of Aerial Photography.' Gorrell Report, Series J, Roll 34
—— 'Topography. Lecture delivered at the Second Aviation Instruction Center.' Gorrell Report, Series J, Roll 34
'History of Photographic Section #2, Air Service, U.S.A.,' 1918. Gorrell Report, Series G, Roll 24
'History of Photographic Section #5, Air Service, U.S.A.,' 1918. Gorrell Report, Series G, Roll 24
'History of the Air Service.' Gorrell Report, Series C, Roll 12
'History of the Training Section, Air Service, A.E.F.' Gorrell Report, Series J, Roll 34
Lindstrom, G. T. 'Photographic Missions.' Gorrell Report, Series N, Roll 53
'Nomenclature Chart.' Gorrell Report, Series F, Roll 22
'Notes on Branch Intelligence,' 91st Aero Squadron. Issued by Second Section, General Staff, American Expeditionary Forces, 1 November 1918. Gorrell Report, Series M, Roll 40
Office of the Chief of Air Service. 'Recommendations, Command, Equipment, Organization.' Gorrell Report, Series C, Roll 12
'Official Report; Cession of Optical Instruments to the American and English Governments.' Gorrell Report, Series G, Roll 24
'Official Report of the Conference held at the Office of the Under Secretary of State of the Presidency of the Council.' 25 April 1918. Gorrell Report, Series G, Roll 24
'Photography.' Gorrell Report, Series G, Roll 24
Proctor, Captain. 'Training Section. Photographic Branch Memorandum for Major Sullivan, Subject: 'Situation on De Ram Camera.' Gorrell Report, Series G, Roll 24
'Report on Day Bombardment Training.' Flying Officers School, St. Maixent. Gorrell Report Series J, Roll 35
Snyder, John, First Lieutenant. 'Notes on Army Observation.' Gorrell Report, Series J, Roll 34
——. 'Notes on Reconnaissance,' Lecture given at 2nd A.I.C.' Gorrell Report, Series J, Roll 34
——. 'Notes on the Relative Importance of Information to be Obtained by Observers When On a Visual Reconnaissance.' Gorrell Report, Series J, Roll 34
Steichen, Edward, Lieutenant Colonel. 'Aerial Photography: The Matter of Interpretation and Exploitation,' 26 December 1918. Gorrell Report, Series G, Roll 24
—— 'Report of Inter-Allied Photographic Meeting,' 17 October 1918. Gorrell Report, Series G, Roll 24
'Table of Information for Taking Aerial Photographs.' Gorrell Report, Series G, Roll 24
Terry, Prentiss M., Second Lieutenant, 91st Aero Squadron. 'Notes on Branch Intelligence,' 19 August 1918. Gorrell Report, Series M, Roll 44
Third Army Corps. Headquarters, Office of Chief of Air Service. 'Operations Report for November 5, 1918.' Gorrell Report, Series C, Roll 12

Record Group 120

Air Service. American Expeditionary Forces. *History of the Air Services*
Series C. *Tactical Units.* 14 Volumes
Series E. *Squadron Histories.* 26 Volumes
Series G. *Photographic Section.* 4 Volumes
Series N. *Duplicates of all First Army Material.* 23 Volumes

American Mission with the Commanding General, Allied Forces of Occupation, Mainz

Germany. 'A Study of the Organisation of the French Army Showing its Development as a Result of the Lessons of the World War and Comprising Notes on Equipment and Tactical Doctrine Developed in the French Army, 1914-1921.' Prepared by Colonel Le Vert Coleman. Box 819
Part XXXVII. 'Aeronautics.' Section 3. June 1921
Part XXXVIII. 'Camouflage.' Section 2. July 1921

Part XL. 'Infantry: Its Organization, Its Equipment, and Its Present and Future Use Based on the Lessons of the War.' 1921
Section 5. Publication on its Employment, Liaisons. 1921
Part XLIII. 'Interpretation of Aerial Photographs.' November 1921
Assistant Chief of Staff for Intelligence, First Army. 'Summary of Intelligence.' 2 October 1918 Box 3370
First Army Historical Files. Box 22
'Interpretation of Aerial Photographs.' Prepared by Colonel LeVert Coleman, November 1921 Box 819
Pouquet, Sous Lieutenant. 'Working of the Service of Interpretation of Photographs in an Offensive Sector (Conference by Sub/Lieutenant Pouquet).' Box 819
Second Army, Headquarters, General Staff, Second Section. 'Effectiveness of Different Weapons against Each Other.' Supplement to Summary of Intelligence, No. 14, 25 October 1918. Translation of German Document. Box 10
Study and Utilization of Aerial Photographs. Translation from the French edition of 15 January 1918. Headquarters American Expeditionary Forces, France. July, 1918. Box 6202
26th Division. 'Intelligence Bulletin,' 26 July 1918.' Box 1
Von Kathen. 'Memorandum for Encampments.' In Headquarters, 26th Division, 'Intelligence Bulletin,' 26 July 1918.' Box 1

THE NATIONAL ARCHIVES – PUBLIC RECORD OFFICE, LONDON

Air Ministry Record (AIR)

AIR 1/7/6/98/20. Captain C. D. M. Campbell. 'Notes on Aerial Photography.'
AIR 1/7/6/98/20. Captain C. D. M. Campbell. 'Report of Photographic Section, R.F.C.'
AIR 1/7/6/98/20. General H. R. M. Brooke-Popham. 'General Notes on German Aeroplanes.'
AIR 1/7/6/98/20. 'Packet of Photographic Reproductions (RNAS),' 12 December 1918
AIR 1/7/6/98/20. 'Report on Aeronautical Matters in Foreign Countries for 1913.'
AIR 1/7/6/98/20. 2nd Lieut V. F. Bryce, Photographic Officer, 1st Brigade, BEF. 'Report,' 1 June 1916
AIR 1/82/15/9/204. Corporal Mallet. 'Study ... of the R.N.A.S. Photographs No. 579 to 584 Reconnaissance of 17 December 1916.' In Photographic Studies III (January – September 1918)
AIR 1/93/15/9/240. 5th Group, R.A.F. 'Result of Aerial Photography by the R.N.A.S. in 1917,' 13 February 1918
AIR 1/123/15/40/144. 'Notes on Captive Balloons for Use in Land Warfare.'
AIR 1/127/15/40/147. Lieutenant Colonel C. C. Marindin. 'Conference Notes,' 23 July 1915
AIR 1/138/15/40/287. 'Camera Vibration.'
AIR 1/404/97/1917. Lieutenant Charles W. Gamble. 'Photographic Section of the French Flying Corps.'
AIR 1/404/97/1917. Lieutenant Charles W. Gamble. 'Visits to Manufacturers.'
AIR 1/475/15/312/201. Major J. P. C. Sewell, R.A.F. 'Notes ... French Corps Aviation,' August 1918
AIR 1/475/15/312/201. Captain I. W. Learmont, R.F.C. 'Report ... of a Visit to the French Bombardment Group at Malzeville.'
AIR 1/724/91/1. Captain C. D. M. Campbell, R.F.C. 'Progress of Photography in Royal Flying Corps at Home, to 31st December, 1915,' 22 January 1916
AIR 1/724/91/1, 87/7298. 'The General Officer Commanding, 6th Brigade, Royal Flying Corps Memorandum. Signed by C. C. Marindin, Lieut-Colonel,' 3 February 1916
AIR 1/724/91/1. Major E. G. Toye. 'History of Photography (being a General History of Air Photography, its Cameras, Lens, etc.) ... , 1914–1918.'
AIR 1/724/91/2. 'Aerial Photography. Statement made by Squadron Leader F. C. V. Laws to Mr. A. J. Insall,' 3 June 1924
AIR 1/724/91/2. Lieutenant Colonel C. J. Burch. 'Notes on Report Dealing with Camera Vibration,' 8 February 1916
AIR 1/724/91/2. Lieutenant Colonel J. T. C. Moore-Brabazon. 'Photography in the R.A.F 1914–1918.'
AIR 1/724/91/2. 'Squadron Leader F. C. V. Laws, O.B.E., to G.O.C., Advanced Operations,' 18 November 1918
AIR 1/724/91/2. Lieutenant Colonel J. T. C. Moore-Brabazon. 'Photography's Part in the War.'
AIR 1/724/91/2. 'Telephone Call between Colonel [J. T. C.] Moore-Brabazon and Capt Wall,' 13 November 1918
AIR 1/751/204/4/8. Headquarters, R.F.C. 'Orders for Reconnaissance from 24th Oct. 1914 to 28th Nov. 1914.'
AIR 1/751/204/4/8. 'Memorandum of New Organization of the Royal Flying Corps' 29 November 1914
AIR 1/751/204/4/8. 'Organization of the Royal Flying Corps, British Expeditionary Force,' December 1914.'
AIR 1/ 834/204/5/240. Captain D. S. Lewis. 'Procedure of Ranging.' 25 January 1915

AIR 1/834/204/5/240. 'H.Q. R.F.C., G 165/1. Officer Commanding No. 9 Squadron,' 15 January 1915
AIR 1/834/204/5/240. Lieutenant I. M. Bonham-Carter, R.F.C. 'Notes,' 1 November 1914
AIR 1/836/204/5/268. 'Notes on Equipment of Units of RFC with BEF,' 27 September 1914
AIR 1/836/204/5/268. 'Photography from Aeroplanes,' 20 December 1914
AIR 1/1118/204/5/2041. Colonel J. T. C. Moore-Brabazon. 'Night Photography,' 1 November 1918
AIR 1/1139/204/5/2305. Report of 'Visits to Paris of Gen. [H. R. M.] Brooke-Popham & Gen [Hugh] Trenchard Feb 1916 - July 1917.'
AIR 1/1211/204/5/2627. 'Instructions Regarding the Co-Operation of Aeroplanes with Other Arms.'
AIR 1/1625/204/9/7. 'Central Flying School.'
AIR 1/1625/204/89/7. 'Notes on Air Fighting from Royal Flying Corps, Sept 1916–December 1916.'
AIR 1/2124/207/74/3. 'Index to Periodical Summaries of Aeronautical Information.'
AIR 1/2124/207/74/3. 'Intelligence Summary,' 19 August 1918
AIR 1/2124/207/74/3. 'Protection of Artillery Positions from Air Observation.' Captured German document
AIR 1/2151/209/3/240. Brigadier General W.S. Brancker, R.A. and RFC. 'Aerial Defence of Paris. Notes.'
AIR 1/2151/209/3/244. 'Capt. Cuddell, Assistant Director of Military Aeronautics, 3 Mar 1915, War Office, London, S.W.'
AIR 1/2151/209/3/244. Lieutenant J. T. C. Moore-Brabazon. 'OIC Photos in the Field,' 21 June 1915
AIR 1/2151/209/3/244. 'Photographic Lorries.'
AIR 1/2299/209/77/29. Supreme War Council. 'Draft Minutes of the 1st Session of the Versailles Inter-Allied Aviation Committee held in the Council Chamber, Versailles on 9th May, 1918, at 10.0 a.m.'
AIR 1/2395/255/1. Lieutenant Colonel J. F. C. Moore-Brabazon. 'R.A.F. Night Photography.'
AIR 1/2395/255/1. Squadron Leader C. C. Darley. 'Notes on Photography in No. 3 Squadron, Royal Flying Corps. December 1914 – End of February 1915.'
AIR 1/2397/267/7. Charles W. Gamble. 'The Technical Aspects of British Aerial Photography during the War 1914-1918.'
AIR 2/16. 'Letter from Col. E. F. D., R.F.C., 28/2. Enclo. 1A., 'Proposals for a Long Distance Machine.'
AIR 2/35. Wing Commander Seddon. 'Report ... on Co-Operation between Britain and American Air Services,' 13 September 1917
AIR 2/36. Flight Commander Michell. 'Report on Photography and Recommendations,' 31 January 1917
AIR 2/39. 'Extracts from Official French Manual on Photographic Reconnaissance,' 20 August 1925. Translated
AIR 2/163. 'Re-Organisation of Photographic Activities,' 24 January 1917
AIR 2/277. Air Commodore T. C. R. Higgins. Memorandum to Air Officer Commanding, Inland Area, R.A.F., Course: 'Interpretation of Air Photographs.'
AIR 2/277. 'Instruction Regarding Interpretation of Aerial Photographs,' 2 February 1923
AIR 2/277. War Office Memorandum to Secretary of the Air Ministry, 13 May 1925
AIR 10/61. 'Characteristics of the Ground and Landmarks in the Enemy Lines Opposite the British Front from the Sea to St Quentin.'
AIR 10/853. 'Aerial Photography.' Air Publication 838
AIR 10/1120. 'Notes on the Interpretation of Air Photographs.' February 1925.
AIR 10/5521. 'Illustrated Hand-Book for R.A.F. Intelligence Officers Employed on Interpretation of Air Photographs. December 1940.'
AIR 10/1001. 'Notes on Aerial Photography. Part II. The Interpretation of Aeroplane Photographs in Mesopotamia.'
AIR 1/2151/209/3/244. '2nd Lt Kippel, OIC Photographic Section 1st Wing, R.F.C.,' 14 June 1915
AIR 20/705. 'Papers Collected by Air Vice Marshal Sir W. G. H.Salmond. Period 1916-1922.'
AIR 69/1. Air Vice Marshal H. R. M. Brooke-Popham. 'The War of 1914-1918.' RAF Staff College
AIR 69/1. 'Personalities of [Colonel] Du Peuty [French Air Service] and [General Hugh] Trenchard,' in 'The War of 1914-1918.' RAF Staff College
AIR 69/2. W. 'Sholto' Douglas. 'Notes on First Steps in Air Photography on the Western Front, January to March 1915.'
AIR 75/131. 'Air Reconnaissance in Open Warfare. The Story of Two Incidents in the Advance to the Aisne. Sept. 1914.' Official Papers of Sir John Slesser

War Office Papers (WO)

WO 158. 'Intelligence Corps Organization and Establishment.' 1 November 1917

WO 158/48. 'Conference at Dury at 11.0 p.m., Sunday, March 24th, 1918.'
WO 158/897. 'Brig-Gen John Charteris to Gen Macdonough, War Ministry (London), General Headquarters,' 28 July 1916
WO 158/983. 'Notes on the French System of Intelligence during the Battle of the Somme,' 23 September 1916
WO 158/898. 'General Macdonogh, February 1917.'
WO 297/61. 'German System of Squaring Maps.'
WO 339/23903. '2Lt J. T. C. Moore-Brabazon, R.F.C. (S.R.), 'For Promotion to Lt. and Refund of £75 for R.A.C. Certificate.'

ROYAL AIR FORCE MUSEUM, LONDON
Year 1916. 47' Lens Aircraft Camera
Year 1918. Hand Held Oblique Aircraft Camera Type P 14

J.T.C. Moore-Brabazon Papers
'Parliamentary Secretary, Ministry of Transport,' *Modern Transport*, January 12, 1924

SERVICE HISTORIQUE DE L'ARMÉE DE L'AIR, VINCENNES, FRANCE
Bellenger, Lieutenant Colonel Georges. Letter to François Baban, 6 February 1954
Collection S.H.D., 377/094
'Cours de Perfectionnement (Décembre 1916 – Janvier 1917).' A 278. d. 4
de Bissy, Jean, Capitaine. 'Note Concernant l'Interpretation Methodiqie de Photographies Aériennes,' 19 November 1915
Grand Quartier Général Défenses et Différents Points du Territoire, Artillerie Lourde, Services (Septembre 1914 – Novembre 1919). Dossier 7, Service de Photographie Aérienne (23 Octobre 1914-14 Avril 1919). A.19
Grout, Maurice Marie Eugène, Lieutenant. Personnel Records
Manouevres du 6ème Corps d'Armée. Flight Log, 18 September 1911
'Organisation et Administration des Unités.' A.20
Pouchet, Sous Lieutenant. 'Conference du SLT Pouquet sur le Fonctionnement du Service d'Interpretation des Photographies dans un Secteur Offensif faite au G.D.E., le 27 December 1916.' A 278. d. 4
Rapport du Capitaine du Genis Saconney, au Sujet des Experiences de Cerfs-Volants Effectuées Avec Nacelle Lestrée à bout de Cable. 21 February 1910

SERVICE HISTORIQUE DE L'ARMÉE DE TERRE, VINCENNES, FRANCE
Extrait du Plan Directeur, S.T.C.A., 6ème Armée. 'Carte d'activité de l'Artillerie ennemie du 1er au 15 Juin 1917.'
—— 'Carte des Renseignements Recueillis sur l'Ennemie dans la Zône du C.A.' [1917]
S.595. Bois Alberick. SP. III Armée. C. 28D 052/653

US AIR FORCE ACADEMY, MCDERMOTT LIBRARY, COLORADO SPRINGS, COLORADO
Jefferson Hayes Davis Papers
Colby, F. T., Captain and Captain Cushing-Donnell. 'Notes on Artillery Camouflage and Its Employment.' Headquarters, American Expeditionary Forces, France, October, 1917. SMS 57
Eaton, Joseph E., 2nd Lieutenant. 'Methods in Observation Practiced with Fifth Corps First American Army on the Fronts, St. Mihiel Sector, September 12 to 16, 1918 Argonne-Meuse Sector, September 12 to November 11, 1918,' *Air Service Information Circular, II*, Washington: Government Printing Office, 20 September 1920, No. 115. SMS 57
Haslett, Elmer R., First Lieutenant. 'Bulletin of Information Air Service American Expeditionary Forces. Notes on the Duties of the Operations Officer of an Observation Squadron or Group.' SMS 57
'Liaison for All Arms.' General Headquarters, American Expeditionary Forces, France. June 1918. SMS 57
'Notes for Observers and Pilots in Army Corps Aviation Squadrons.' SMS 57

US ARMY MILITARY HISTORY INSTITUTE, CARLISLE, PENNSYLVANIA
Arthur L. Conger Papers
'Channels through which Air Photographs are Ordered and Distributed.'
'Correspondence,' 23 March 1918. Box 2

'Example of Summary Indications of Enemy Preparations for an Offensive in Flanders (from Air Photographs),' September 1918
Frank, Coleman D., First Lieutenant. 'Report of a Visit to VIIIth Army Headquarters to Attend the Interrogatory of Prisoners Taken in the Coup de Main on the Lorraine Sector,' 20 February 1918
Henry, Daniel M., Major. 'Report on My Visit to the British Army for the Purpose of Observing the Work of the Intelligence Section of the Royal Flying Corps,' 18-29 September 1918
Hubbard, S. T., Jr., Captain. 'Memorandum for Col. Nolan,' 29 December 1917. Box 2
—— 'Memorandum for Col. Nolan, Subject: 'Report on Trip to British Front,' 3 January 1917 [i.e., 1918].'
'Instructions on the Functions of the Corps Intelligence Service.'
Maxel, Général, Vth Army Staff, 2nd Bureau. 'Instructions on the Intelligence Service in the Divisions,' 16 January 1917
'Organization of Intelligence, H.Q. British 5th Army.'
Pétain, Henri, Général. 'French General Staff and 3rd Bureau No. 1077/3. Notice about Maps and Special Plans to be Drawn by the Armies.' Box 2
Riggs, Kerr T., Major. 'Memorandum to Chief Intelligence 'A,' 14 February 1918, Subject: 'Memorandum Report of Visit to Headquarters 1st French Army.'
—— 'Report on Visit to British Army for Purpose of Studying Branch Intelligence.'
'Scale of Issue of Periodical Publications.'
Stanford, Captain. 'Capt Stanford's Summary of the Qualifications of a Branch Intelligence Officer,' 15 February 1918, in 'Organization of Intelligence,' H.Q. British 5th Army Box 2
Stewart, John L., First Lieutenant. 'The Organization and Operation of Intelligence 'A' at the British II Army.' September 1918
—— '1st Lt John Stewart Report.' 4 October 1918. Box 2
'Summary of Indications of Enemy Preparations for an Offensive in Flanders (from Air Photographs) together with a Note on the System of Collecting and Disseminating the Information.' Prepared by Air Photo Officer, II British Army, July 1918
'To Show Stream of Intelligence Up and Down (Corps Heavy Artillery Divisions).'
Tyler, Royall. Headquarters, American Expeditionary Forces, Office of the Assistant Chief of Staff for Intelligence. 'Letter to Col Nolan.' 7 March 1918. Box 2

Dennis E. Nolan Papers

Assistant Chief of Staff for Intelligence, 1st Army Corps. 'Summary of Intelligence.'
'Codes & Ciphers.'
'G-2-C, Topography.'
'Intelligence.'
Zimmerman, Arthur A. 'Report on Visit to VIIIth French Army, February 22 to March 2, 1918,' 7 March 1918

George O. Squier Papers

Squier, George O., Colonel. 'The Flying Corps, April 20, 1915.'
—— 'Intelligence Services of the British Army in the Field, February 16, 1915.'
—— 'The Map Room and Printing Room, April 20, 1915.'

Hugh Drum Papers

Assistant Chief of Staff for Intelligence to Chief of Staff, 1st Army, American Expeditionary Forces. 'Account of the Operation from the Standpoint of Enemy Works.' Box 14
'Historical Account.'

Philip Katcher Papers

Katcher, Philip. 'Bulletin of Information Air Service American Expeditionary Forces, Notes for Divisional Staffs, Commanding Officers and Aerial Observers on Work of Infantry Contact Airplanes,' 24 September 1918

Theodore J. Lindoff Papers

Lindoff, Theodore J., First Lieutenant. 'Photographic Missions.'

William J. Steere Papers

The Airscout's Snapshot, 28 August 1918
'General Lecture on Cameras.'
'Plates and Development of Plates.'

CONTEMPORARY SOURCES

Ackerman, Carl W. *George Eastman*. Boston: Houghton Mifflin Company, 1930

L'Aéronautique pendant la Guerre Mondiale, 1914-1918. Edited by Maurice de Brunoff. Paris: M. de Brunoff, 1919

Aerial Observation for Artillery, Based on the French Edition of December 29, 1917 [Washington, DC: Government Printing Office?], May 1918

'Air Reconnaissance of Turks' Position,' *The New York Times*. October 23, 1911 ProQuest Historical Newspapers, *The New York Times* (1851-2001)

Artillery Operations of the Ninth British Corps at Messines, June, 1917. Edited at the U.S. Army War College. Washington, DC: Government Printing Office, 1917

Bagley, James W. *The Use of the Panoramic Camera in Topographic Surveying*. Washington, DC: Government Printing Office, 1917

Baring, Maurice. *Flying Corps Headquarters: 1914-1918*. London: William Heinemann, 1930; reprint, London: Buchan & Enright, 1985. Reprint of *R.F.C. H.Q., 1914-1918*. London: G. Bell and Sons, 1920

Barnes, James. *From Then Till Now: Anecdotal Portraits and Transcript Pages from Memory's Tablets*. New York: D. Appleton-Century Company, 1934

Bethell, H. A., Colonel. *Modern Artillery in the Field*. London: Macmillan and Co., Ltd., 1911

Blumenson, Martin, editor. *The Patton Papers 1885-1940*. 2 volumes. Boston: Houghton Mifflin Company, 1972

Brooks, Alfred H. 'The Use of Geology on the Western Front.' In United States Geological Survey. Professional Papers. *Shorter Contributions to General Geology, 1920*. Washington, DC: Government Printing Office, 1921

Brown, Malcom, editor. *The Imperial War Museum Book of the First World War: A Great Conflict Recalled in Previously Unpublished Letters, Diaries, Documents, and Memoirs*. London: Sidgwick & Jackson, 1991

Bruno, Henry. 'The Eyes in the Air.' *Popular Science Monthly* (1918): 508-511

Bulletin du Service de la Photographie Aérienne Aux Armées, 7 January 1916. Paris: Imprimerie Nationale, 1916

Bulletin of the Aerial Photography Department in the Field. Washington, DC: Government Printing Office, 1918

'The Camera at the Front: Why the Military Authorities Have Use for their Own Cameras but None for the Enemy's.' *Scientific American* (24 November 1917): 380, 390

Camouflage for the Troops of the Line. Edited by the Army War College. January 1918. Washington: Government Printing Office, 1920

Carlier, Andre H. *La Photographie Aérienne Pendant la Guerre*. Paris: Librairie Delagrave, 1921

Charteris, John. *At G.H.Q.* London: Cassell and Company, 1931

—— *Haig*. London: Cassell and Company, 1930

Coleman, Frederic. *With Cavalry in 1915: The British Trooper in the Trench Line*. Toronto: William Briggs, 1916

Crowell, Benedict and Robert Forrest Wilson. *The Armies of Industry: Our Nation's Manufacture of Munitions for a World in Arms, 1917-1918. Illustrated with Photographs from the Collections of the War and Navy Departments*. 2 volumes. New Haven, Yale University Press, 1921

De Bissy, J. de Lannoy, Captain. *Illustrations to Accompany Captain De Bissy's Notes Regarding the Interpretation of Aeroplane Photographs (Ia/1311)*. Translated. N.p., n.d

—— *Notes Regarding the Interpretation of Aeroplane Photographs (Ia/1311)*. Translated N.p., n.d

—— Commandant. 'Les Photographies Aériennes et leur Étude au Point de Vue Militaire.'

Revue Militaire Francaise 12 (April-June 1924): 257-277

Dixon, Eric E. *Diary of Eric E. Dixon, Whippany, N.J., U.S.A.: An American Citizen* Whippany, NJ: n.p., n.d. Copy in Eric E. Dixon Papers, U.S. Army Military History Institute, Carlisle, Pennsylvania

Douglas, Sholto, Marshal of the Royal Air Force, Lord Douglas of Kirtleside, *Years of Combat* [Autobiography]. London: Collins, 1963

Duval, Jacques. *L'Armée de L'Air*. Paris: L'Édition Française Illustrée, 1918

Edmonds, James E., Brigadier-General, [General Editor]. *Military Operations, France and Belgium, 1914*. Compiled by Brigadier-General James E. Edmonds. 3rd revised edition. 2 volumes. London: Macmillan and Co., Ltd., 1933; reprint, London: Imperial War Museum and Nashville, TN: Battery Press, 1995-96

—— *Military Operations, France and Belgium, 1915*. Compiled by Brigadier-General James E. Edmonds and Captain

G. C. Wynne. London: Macmillan and Co., Ltd., 1927-28; reprint, London: Imperial War Museum and Nashville, TN: Battery Press, 1995

—— *Military Operations, France and Belgium, 1916*. Compiled by Brigadier-General James E. Edmonds and Captain Wilfrid Miles. 2 volumes. London: Macmillan and Co., Ltd., 1932-38; reprint, London: Imperial War Museum and Nashville, TN: Battery Press, 1993-95

—— *Military Operations, France and Belgium, 1917*. Compiled by Brigadier-General James E. Edmonds, Captain Cyril Falls, and Captain Wilfrid Miles. 3 volumes. London: Macmillan and Co., Ltd., 1940-48; reprint, London: Imperial War Museum and Nashville, TN: Battery Press, 1991

—— *Military Operations, France and Belgium, 1918*. Compiled by Brigadier-General James E. Edmonds and Lieutenant-Colonel R. Maxwell-Hyslop. 5 volumes. London: Macmillan and Co., Ltd., 1935-47; reprint, London: Imperial War Museum and Nashville, TN: Battery Press, 1993-94

Ellis, O. O., Captain, and Captain E. B. Garey. *The Plattsburg Manual: A Handbook for Military Training*. New York: The Century Co., 1917

Farré, Lieutenant Henry. *Sky Fighters of France: Aerial Warfare 1914-1918*. Boston, NY: Houghton Mifflin, 1919

Finne, K.N. Edited by Carl J. Bobrow and Von Hardesty. *Igor Sikorsky, The Russian Years*. Washington DC: Smithsonian Institution Press, 1987. Tr. from *Russkiye vozdushnyye bogatyri I.I. Sikorskogo*, Belgrade, 1930

Flicke, Wilhelm F. *War Secrets in the Ether*. Edited by Sheila Carlisle. 2 volumes. Laguna Hills, CA: Aegean Park Press, 1977. Original typescript translation of *Kriegsgehemeinisse im Aether* by the National Security Agency, 1953, is located at National Archives and Records Administration, Record Group 38, Records of the Office of the Chief of Naval Operations (Records of the Naval Security Group), Box 16

France, General Headquarters, Armies of the North and Northeast. *Bulletin of the Aerial Photography Department in the Field*. Washington, DC: Government Printing Office, 1918

—— General Headquarters. Armies of the North and Northeast. *Bulletin of the Aerial Photography Department in the Field*. Washington, DC: Government Printing Office, 1918

—— General Headquarters. Armies of the North and Northeast. *Study and Exploitation of Aerial Photographs*. Translated and Issued by the Division of Military Aeronautics, U.S. Army. Washington: Government Printing Office, July 1918. Copy in US Army Military History Institute

—— General Headquarters. Armies of the North and Northeast. *Study and Use of Aerial Photography*. November 1921. Copy in in Part XLIII, Section 5, in American Mission with the Commanding General, Allied Forces of Occupation, Mainz, Germany, 'A Study of the Organization of the French Army.' National Archives and Records Administration, Record Group 120, Box 819

—— Grand Quartier Général des Armées du Nord et du Nord-Est. État-Major. *Bulletin du Service de la Photographie Aérienne aux Armées*. Paris: Imprimerie Nationale, 7 January 1916

—— Grand Quartier Général des Armées du Nord et du Nord-Est. État-Major. *Notes sur l'Interprétation des Photographies Aériennes*. Paris: Imprimerie Nationale, 30 December 1916

—— Grand Quartier Général des Armées du Nord et du Nord-Est. État-Major. *Étude et Exploitation des Photographies Aériennes*. Paris: Imprimerie Nationale, 1918

Fraser-Tytler, Neil. *With Lancashire Lads and Field Guns in France, 1915-1918*. Manchester, UK: John Heywood, 1922

French, Field-Marshal Viscount French of Ypres. *1914*. Boston: Houghton Mifflin Company 1919

French Trench Warfare 1917-1918: A Reference Manual. French General Staff. Reprint of translation of Manual of the Chief of Platoon of Infantry, 1918. London: Imperial War Museum and Nashville, TN: The Battery Press, Inc., 2002

Fuller, Alvan Tufts. 'Mysteries of Camouflage,' *Army and Navy Register* (16 August 1919): 222-223

Gilchrist, John W. Stuart. *An Aerial Observer in World War I*. Richmond, VA: Privately printed, 1966

Gleichen, A. E. W. Edward, Count, editor. *Chronology of the Great War, 1914-1918*. 3 volumes London: Ministry of Information, 1918-20; reprint, 3 volumes in 1, London: Greenhill Books, 2000

Goddard, George W., Brigadier General. *Overview: A Life-Long Adventure in Aerial Photography*. Garden City, NY: Doubleday & Company, 1969

Goussot, C. M., Lieutenant. *Précis de Photographie Aérienne*. Paris: Librairie Aéronautique, 1923

Gouvieux, Marc. *Notes d'un Officier Observateur en Avion*. Paris: Pierre Lafitte, 1916

Graffin, Laurence. *Carnets de Guerre, 1914-1918.* Paris: Herscher, 1996

Grahame-White, Claude. *The Aeroplane in War.* London: T. Werner Laurie, 1912

Great Britain. General Staff (Intelligence). General Headquarters. *Notes on the Interpretation of Aeroplane Photographs.* 1st edition (S.S. 445). France: Army Printing and Stationery Services, November 1916

—— *Notes on the Interpretation of Aeroplane Photographs.* 2nd edition (S.S. 550). France: Army Printing and Stationery Services, March 1917. Copy in Combined Arms Research Library, Fort Leavenworth, Kansas

—— *Notes on the Interpretation of Aeroplane Photographs.* 3rd edition (S.S. 631). France: Army Printing and Stationery Services, February 1918. Copy in National Air and Space Museum Archives, A30.2/29. Reprint, edited by Captain Einar W. Chester, U.S. Army Division of Military Aeronautics. Washington, DC: Government Printing Office, 1918

—— *Illustrations to Accompany Notes on the Interpretation of Aeroplane Photographs.* 1st edition (S.S. 445a). France: Army Printing and Stationery Services, November 1916

—— *Illustrations to Accompany Notes on the Interpretation of Aeroplane Photographs.* 2nd edition (S.S. 550a). France: Army Printing and Stationery Services, March 1917. Copy in Combined Arms Research Library, Fort Leavenworth, Kansas

—— *Illustrations to Accompany Notes on the Interpretation of Aeroplane Photographs.* 3rd edition (S.S. 631a). France: Army Printing and Stationery Services, February 1918. Copy in Princeton University Library

Greenhous, Brereton, editor. *A Rattle of Pebbles: The First World War Diaries of Two Canadian Airmen.* Ottawa: Canadian Government Publishing Centre, 1987

Grinnell-Milne, Duncan. *Wind in the Wires.* New York: Ace Publishing Corporation, 1968

Hahn, J. E., Major. *The Intelligence Service within the Canadian Corps.* Toronto: The Macmillan Company of Canada, 1930

Haig, Sir Douglas. *Sir Douglas Haig's Despatches.* Edited by Lieutenant Colonel J.H. Boraston London: J. M. Dent & Sons, Ltd., and New York: E. P. Dutton & Co., 1920

Hanson, Joseph M., Major, editor. *The World War through the Telebinocular: A Visualized, Vitalized History of the Greatest Conflict of All the Ages.* 5th edition. Meadville, PA: Keystone View Company, 1928

Haslett, Elmer. *Luck on the Wing: Thirteen Stories of a Sky Spy.* New York: E. P. Dutton & Company. 1920

Henderson, Sir David, Brigadier-General. *The Art of Reconnaissance.* 3rd edition. London: John Murray, 1914

The History of the 6th Garde Infantry Regiment. n.p., 1931

Höfl, Colonel. *The History of the Royal Bavarian 20th Infantry Regiment, 'Prinz Franz.'* Munich: n.p., 1929

Innes, John R. *Flash Spotters and Sound Rangers: How They Lived, Worked and Fought in the Great War.* London: George Allen & Unwin, Ltd., 1935; reprint, Trowbridge, Wiltshire, UK: Redwood Books, Ltd., 1997

Insall, A. J. *Observer: Memoirs of the R.F.C., 1915-1918.* London: William Kimber, 1970

Instruction sur le Ballon Captif Allonge Type R. Washington, DC: Government Printing Office, 1918

Ives, Herbert E. *Airplane Photography.* Philadelphia: J.B. Lippincott Company, 1920

Jauneaud, Marcel, Commandant. *L'Aviation Militaire et la Guerre Aérienne.* Paris: Ernest Flammarion, 1923

Johnson, Douglas Wilson. *Topography and Strategy in the War.* New York: Henry Holt and Company, 1917

—— *Battlefields of the World War, Western and Southern Fronts: A Study in Military Geography.* New York: Oxford University Press, 1921

Joubert de la Ferte, Sir Philip, Air Chief Marshal. *The Fated Sky, An Autobiography.* London: Hutchinson & Co., 1952

Jünger, Ernst. *The Storm of Steel: From the Diary of a German Storm-Troop Officer on the Western Front.* Translated by Basil Creighton. London: Chatto and Windus, 1929; reprint, New York: Howard Fertig, 1996

Kluck, Alexander von, General. *The March on Paris and the Battle of the Marne 1914.* London: Edward Arnold, 1920

Knappen, Theodore M. *Wings of War: An Account of the Important Contribution of the United States to Aircraft Invention, Engineering, Development and Production during the World War.* London and New York: G. P. Putnam's Sons, 1920

Le Goffic, Charles. *Général Foch at the Marne: An Account of the Fighting in and near the Marshes of Saint-Gond.* New York: E. P. Dutton and Company, 1918

Lewis, Cecil. *Sagittarius Rising [Reminiscences of Flying in the Great War].* London: Peter Davies, 1936; reprint,

London: The Folio Society. 1998

Liddell Hart, Basil H. *Foch: The Man of Orleans*. Boston: Little, Brown, and Company, 1932

'Lord Brabazon of Tara Dies; First Briton to Fly in England; Aviator had Been Early Auto Racer and Minister of Transport for Churchill,' Special to the *New York Times*, May 18, 1964. ProQuest Historical Newspapers, *The New York Times* (1851-2001)

Ludendorff, Erich von, Quartermaster General of the German Army. *Ludendorff's Own Story, August 1914-November 1918: The Great War from the Siege of Liege to the Signing of the Armistice as Viewed from the Grand Headquarters of the German Army*. 2 volumes. New York: Harper & Brothers Publishers, 1919

Macmillan, Norman. *Sir Sefton Brancker*. London, Toronto: William Heinemann Ltd., 1935

Martel, René. *L'Aviation Française de Bombardment (des Origines au 11 Novembre 1918)* Paris: Paul Hartmann, 1939

Maurer, Maurer, editor. *The U.S. Air Service in World War I*. 4 volumes. Washington, DC: Government Printing Office, 1978

Maxwell, W. N. *A Psychological Retrospect of the Great War*. New York: The MacMillan Company, 1923

The Means of Communication between Aeroplanes and the Ground. Edited at the Army War College. Washington: Government Printing Office, 1917

Middleton, Edgar. *The Great War in the Air*. 4 volumes. London: The Waverley Book Co., Ltd., 1920

Miller, Leonard. *The Chronicles of 55 Squadron, R.F.C. and R.A.F.* London: Unwin Brothers, Ltd., 1919

Mitchell, William. *Memoirs of World War I: From Start to Finish of Our Greatest War*. New York: Random House, 1960

Moore-Brabazon, J. T. C., Lord. *The Brabazon Story*. London: William Heinemann, Ltd., 1956

Neumann, Georg Paul. *Die deutschen Luftstreitkraste im Weltkriege*. Berlin: Ernst Siegfried Mittler und Sohn. 1920

Notes on Anti-Aircraft Guns. Compiled at the Army War College from the Latest Available Information. April, 1917. Washington: Government Printing Office, 1917

'The Paris Aero Salon 1913,' *FLIGHT*, January 10, 1914, 32

Pépin, Eugène. *Chinôn*. Paris: Henri Laurens, 1924

Pétain, Henri Philippe, Marshal of France. *Verdun*. New York: The Dial Press, 1930

'The Photographic Eye and Memory of the Military Airman.' *Scientific American* (27 July 1918): 61, 76

Porter, Harold E. *Aerial Observation: The Airplane Observer, The Balloon Observer, and the Army Corps Pilot*. New York: Harper & Brother Publishers, 1921

Powell, E. Alexander, Major. *The Army behind the Army*. New York: Charles Scribner's Sons, 1919

Pulitzer, Ralph. *Over the Front in an Aeroplane and Scenes Inside the French and Flemish Trenches*. New York and London: Harper & Brothers Publishers, 1915

Raleigh, Sir Walter A. and H. A. Jones, *The War in the Air, Being the Story of the Part Played in the Great War by the Royal Air Force*. 6 volumes and Appendices. Oxford: The Clarendon Press, 1922-37; reprint, Nashville, TN: The Battery Press and The Imperial War Museum, 1998

Rickenbacker, Edward V., Captain. *Fighting the Flying Circus*. New York: Frederick A. Stokes Co., [1919]; reprint Chicago: R. R. Donnelley & Sons Company, 1997

Smart, C.D. 'The Front Cockpit: The Diary of Lt. C.D. Smart, No. 5 Sqdn., RFC.' *Cross and Cockade, Journal of the Society of World War I Aero Historians* 10, No. 1 (Spring 1969): 1–33

Smart, Lawrence L., First Lieutenant. *The Hawks that Guided the Guns*. N.p.: Privately printed, 1968

Smith, Neil Leybourne. *History of 3 Squadron, AFC, RAAF*

URL: <http://www.3squadron.org.au/history.htm>. Accessed 10 January 2005

Solomon, Solomon J. *Strategic Camouflage*. New York: E. P. Dutton and Company, 1920

Spears, Sir Edward, Major-General. *Liaison 1914*, 2nd edition. London: Cassell & Co., 1968

Squier, George Owen. *Aeronautics in the United States at the Signing of the Armistice*. New York: n.p., 1919

Steichen, Edward. *A Life in Photography. Published in collaboration with the Museum of Modern Art*. Garden City, NY: Doubleday & Company, Inc., 1963

Study and Utilization of Aerial Photographs. Paris: Imprimerie Nationale, July 1918. Translated Headquarters American Expeditionary Forces, France, July 1918. Copy in National Archives and Records Administration, Record Group 120, Box 6202

Sweetser, Arthur. *The American Air Service: A Record of Its Problems, Its Difficulties, Its Failures, and Its Final Achievements*. New York: D. Appleton and Co., 1919

Sykes, Sir Frederick, Major-General. *Aviation in Peace and War.* London: Edward Arnold & Company, 1922
—— *From Many Angles, An Autobiography.* London, Toronto, Bombay, Sidney: George G. Harrap & Company Ltd., 1942
Toulmin, Henry A., Jr. *Air Service, American Expeditionary Force, 1918.* New York: D. Van Nostrand Company, 1927
Townshend, Philip B., *Eye in the Sky 1918: Reflections of a World War One Pilot on Artillery and Infantry Co-operation Duties.* Croydon, Surrey, UK: E. Hallet & Co. Ltd, 1986
Turner, Charles C. Major. *The Struggle in the Air, 1914-1918.* New York: Longmans, Green & Co., and London: Edward Arnold, 1919
United Kingdom. War Office. General Staff. *British Trench Warfare, 1917-1918: A Reference Manual.* [London?]: n.p., 1918. Reprinted, London: Imperial War Museum, and Nashville, TN: The Battery Press, Inc., 1997
United States Army. Office of Military History. *United States Army in the World War, 1917-1919.* 17 volumes. Washington, DC: Government Printing Office, 1948; reprint, Washington, DC: U.S. Army Center for Military History, 1988-92
United States War Department. 'Report of the Director of Military Aeronautics,' 3 November 1918. *Annual Report of the Secretary of War for the Fiscal Year 1918.* 2 volumes. Washington, DC: Government Printing Office, 1918. I, 1381-1390
Utilization and Role of Artillery Aviators in Trench Warfare. Washington, DC: Government Printing Office, 1917
Vaughan, Edwin C. *Some Desperate Glory: The World War I Diary of a British Officer.* New York: Simon and Schuster, Inc., and Touchstone Books, 1981
Warran, John W., Captain. *Cooperation of Division Aviation with the Field Artillery.* Maxwell Field, Alabama: Air Corps Tactical School, 6 April 1938
Winchester, Clarence, Wills, F.I. *Aerial Photography, A Comprehensive Survey of its Practices & Development.* Boston: American Photographic Puplishing Co., 1925
White, D. editor. *Shorter Contributions to General Geology, 1920.* Washington, DC: Government Printing Office, 1921
Willcox, Cornelis De Witt. *A French-English Military Technical Dictionary.* Washington, DC: Government Printing Office, 1917
Winslow, Carroll D. *With the French Flying Corps.* New York: Charles Scribner's Sons, 1917
Woglom, Gilbert T. *Parakites.* New York: G. P. Putnam's Sons, 1896
Wood, Eric Fisher. *The Note-Book of an Intelligence Officer.* New York: Century Co., 1917

SECONDARY SOURCES

Alegi, Gregory, Ansaldo SVA 5, Windsock Datafile 40. Berkhamsted, Hertfordshire, UK: Albatross Ltd, 1993
Asbrey, Robert B. *The German High Command at War: Hindenburg and Ludendorff Conduct World War I.* New York: Willam Morrow and Company, 1991
Ash, Eric. *Sir Frederick Sykes and the Air Revolution 1912-1918.* London: Frank Cass, 1999
Asprey, Robert B. *The First Battle of the Marne.* Philadelphia, PA: J. B. Lippincott Company, 1962
Babington-Smith, Constance. *Air Spy: The Story of Photo Intelligence in World War II.* New York: Ballantine Books, 1957
Bailey, Frank W. and Cony, Christophe. *French Air Service War Chronology 1914-1918.* London: Grub Street, 2001.
Bailey, J. B. A. *Field Artillery and Firepower.* Oxford, UK: The Military Press, 1989
Barker, Ralph. *The Royal Flying Corps in World War I.* London: Constable & Robinson Ltd, 1995
Beckett, Ian F. W. *The First World War: The Essential Guide to Sources in the UK* National Archives. Kew, Surrey, UK: Public Record Office, 2002
Bickers, Richard Townshend. *The First Great Air War.* London: Hodder & Stoughton, 1988
Biographie P.L. Weiller, *Service Historique de la Défense (SHD),* Vincennes, France
Blond, Georges. *The Marne.* Harrisburg, PA: The Stackpole Company, 1965
Bowen, Ezra. *Knights of the Air.* Alexandria, VA: Time-Life Books. 1980
Brayer, Elizabeth. *George Eastman, A Biography.* Baltimore: The Johns Hopkins University Press, 1996
The British Army and Signals Intelligence during the First World War. Edited by John Ferris Wolfeboro Falls, NH: Alan Sutton Publishing, Inc., 1992

Brown, Malcom. *The Imperial War Museum Book of 1918, Year of Victory.* London: Sidgwick & Jackson, 1998
—— *The Imperial War Museum Book of the Western Front.* London: Sidgwick & Jackson, 1993
Bruce, J. M. *The Aeroplanes of the Royal Flying Corps (Military Wing).* London: Putnam, 1982
—— *The American D.H.4.* London: Profile Publications, 1966
—— *The de Havilland D.H.4.* London: Profile Publications, 1965
—— *The Sopwith 1½ Strutter.* London: Profile Publications, 1967
—— *War Planes of the First World War, Fighters, Great Britain.* 2 volumes. Garden City, NY: Doubleday and Company, Inc., 1968
—— and Jean Noel. *The Breguet 14.* Leatherhead, Surrey, UK: Profile Publications, 1967
Boyle, Andrew. *Trenchard.* New York: W.W. Norton & Company Inc., 1962
Cain, Anthony C. *The Forgotten Air Force: French Air Doctrine in the 1930s.* Washington, DC: Smithsonian Institution Press, 2002
Chamier, J. A. *The Birth of the Royal Air Force.* London: Sir Issac Pitman & Sons, Ltd., 1943
Castex, Jean, Louis Laspalles and Jose Barès. *Le Général Barès, Créateur et Inspirateur de l'Aviation.* Paris: Nouvelles Editions Latines, 1994
Chasseaud, Peter. *Artillery's Astrologers: A History of British Survey and Mapping on the Western Front 1914-1918.* Lewes, UK: Mapbooks, 1999
—— 'British Western Front Mapping 1914-18, Summary of Geodetic Situation with Reference to the Operations Maps Actually in Use in 1918.' In The Imperial War Museum Trench Map Archives CD-ROM. London: The Naval & Military Press and The Imperial War Museum, 2000
—— *Topography at Armageddon: A British Trench Map Atlas of the Western Front, 1914-1918.* Lewes, UK: Mapbooks, 1991
Christienne, Charles and Pierre Lissarague. *A History of French Military Aviation.* Washington, DC: Smithsonian Institution Press, 1986
Collier, Basil. *Heavenly Adventurer: Sefton Brancker and the Dawn of British Aviation.* London: Secker & Warburg, 1959
Cooke, James. *The U.S. Air Service in the Great War, 1917-1919.* Westport, CT: Praeger. 1996
Cooksley, Peter. *Sopwith Fighters in Action.* Carrollton, TX: Squadron/Signal Publications, Inc., 1991
Crary, Jonathan. *Techniques of the Observer: On Vision and Modernity in the Nineteenth Century.* Cambridge, MA: MIT Press, 1991
Davilla, James J. and Arthur M. Soltan. *French Aircraft of the First World War.* Boulder, CO: Flying Machine Press, 1997
Distel, Anne. *Dunoyer de Segonzac.* New York: Crown Publishers, 1980
Dockrill, Michael and David French. *Strategy and Intelligence: British Policy during the First World War.* London: The Hambledon Press, 1996
Douglas, Sholto, *Marshal of the Royal Air Force, Lord Douglas of Kirtleside, Years of Combat.* London: Collins, 1963
Durkota, Alan, Thomas Darcey and Victor Kulikov. *The Imperial Russian Air Service, Famous Pilots & Aircraft of World War One.* Mountain View, CA: Flying Machine Press, 1995
Farrar-Hockley, Anthony. *Death of an Army.* London: Arthur Barker, Ltd., 1967
Ferris, John. 'The British Army and Signals Intelligence in the Field during the First World War.' *Intelligence and National Security* 3, No. 4 (October 1988): 23-48
Finnegan, Terrence J. 'Military Intelligence at the Front, 1914-18.' *Studies in Intelligence* 53, No. 4 (December 2009): 25-40
—— 'Brigadier-General Sir David Henderson's Royal Flying Corps – British Aerial Reconnaissance in the Opening Salvos.' *Over the Front* 25 No. 4 (December 2010): 292-300
Fischer, William E., Jr. *The Development of Military Night Aviation to 1919.* Maxwell AFB, AL: Air University Press, 1998
Frandsen, Bert. *Hat in the Ring: The Birth of American Air Power in the Great War.* Washington, DC: Smithsonian Books, 2003
Frank, Sam Hager. *American Air Service Observation in World War I.* Doctoral Dissertation. Gainesville, FL: University of Florida, 1961

Frankel, Lory, editor. *Steichen at War.* New York: Harry N. Abrams, 1981

French, David, 'Failures of Intelligence: The Retreat to the Hindenburg Line and the March 1918 Offensive.' In *Strategy and Intelligence, British Policy during the First World War.* Michael Dockrill and David French, editors. London: The Hambledon Press, 1996

Gardner, Nikolas. *Trial by Fire: Command and the British Expeditionary Force in 1914.* Westport, CT: Praeger, 2003

Gill, Bob. 'Over the Front in a Balloon Observer's Gondola,' *Journal of the League of World War 1 Aviation Historians* 20, No. 1 (Spring 2005): 4-17

Gliddon, Gerald. *The Battle of the Somme.* Phoenix Mill, UK: Sutton Publishing, 1987

Guttman, Jon. *Caudron G.3.* Berkhamsted, Hertfordshire, UK: Albatros Productions, Ltd., 2002

Herwig, Holger H. *The Marne, 1914.* New York: Random House, 2009

Hodeir, Marcellin. 'La Photographie Aérienne: 'de la Marne à la Somme,' 1914-1916.' *Révue Historique des Armées,* no. 2 (1996): 97-108

—— 'Une Integration Difficile: La Photographie Aérienne comme Élément de Renseignement et de Cartographie Militaire (1890-1919).' Conference: 'Le Terrain du Militaire: Perceptions et Representations.' Château de Vincennes, France, 11-12 September 2002

Hogg, Ian V. *The Guns 1914-18.* New York: Ballantine Books, 1971

Hood, Robert *E. 12 at War, Great Photographers Under Fire.* New York: G.P. Putnam's Sons, 1967

Hooton, Edward R. *War Over the Trenches.* Hersham, Surrey, UK: Ian Allan Publishing Ltd, 2010

Jackson, A. J. *De Havilland Aircraft since 1909.* London: Pitnam, 1978

Jean, David, *Rohrbacher, Georges-Didier, Palmieri, Bernard and Service historique de l'armée de l'air France.* Les escadrilles de l'aéronautique militaire française: Symbolique et histoire 1912-1920. Paris: Sepeg International, 2004

Jenbrau, Hélène and Eugène Pépin. *The Glory of the Loire.* New York: Viking Press, 1971

Joubert de la Ferte, Philip. Sir, Air Chief Marshal. *The Forgotten Ones, The Story of the Ground Crews.* London: Hutchinson & Co. 1961

Judge, Arthur W. *Stereoscopic Photography: Its Application to Science, Industry and Education.* London: Chapman & Hall, Ltd., 1950

Kahn, David. *The Codebreakers: The Story of Secret Writing.* New York: The Macmillan Company, 1967

Kennett, Lee. *The First Air War, 1914-1918.* New York: The Free Press, 1991

Kern, Stephen. *The Culture of Time and Space 1880-1918.* Cambridge, MA: Harvard University Press, 1983

Kilduff, Peter. 'Kenneth P. Littauer: From the Lafayette Flying Corps to the 88th Aero Squadron,' *Journal of the Society of World War I Aero Historians* 13, No. 1 (Spring 1972): 22-45

Lamberton, W. M. *Reconnaissance & Bomber Aircraft of the 1914-1918 War.* Edited by E.F. Cheesman. Letchworth, UK: Harleyford Publications Ltd., 1962

Liddle, Peter H. *The Airman's War 1914-18.* New York: Blandford Press. 1987

Macdonald, Lyn. 1914: *The Days of Hope.* New York: Atheneum, 1987

—— *To the Last Man: Spring 1918.* New York: Carroll & Graf Publishers, Inc., 1999

Martel, Francis. *Pétain: Verdun to Vichy.* New York: E. P. Dutton & Company, Inc., 1943

Martino, Basilio Di. *Ali Sulle Trincee, ricognizione tattica ed osservazione aerea nell'avazione ilaliana durante la Grande Guerra.* Roma: aeronautica militare, Ufficio strorico, 1999

Maurer, Maurer. *Aviation in the U.S. Army 1919-1938.* Washington, DC: Office of Air Force History, 1987

Mead, Peter. *The Eye in the Air: History of Air Observation and Reconnaissance for the Army, 1785-1945.* London: Her Majesty's Stationery Office, 1986

Michelin Road Atlas France. 9th edition. London: Paul Hamlyn, 1996

Miller, Henry W. *The Paris Gun: The Bombardment of Paris by the German Long Range Guns and the Great German Offensives of 1918.* New York: Jonathan Cape & Harrison Smith, 1930

Morgan, William A. 'Invasion on the Ether: Radio Intelligence at the Battle of St. Mihiel, September 1918.' *Military Affairs* 51, No. 2 (April 1987): 57-61

Morris, Alan. *Bloody April.* London: Jarrolds Publishers, 1967

Mousseau, Jacques. *Le siècle de Paul-Louis Weiller, 1893-1993.* Paris: Éditions Stock, 1998

Morrow, John H., Jr. *The Great War in the Air: Military Aviation from 1909 to 1921.* Washington, DC, and London: Smithsonian Institution Press, 1993

Munson, Kenneth. *Aircraft of World War I*. London: Ian Allan, 1968

—— *Bombers: Patrol and Reconnaissance Aircraft, 1914-1919*. New York: Macmillan Publishing Co., Inc., 1968

Neiberg, Michael S. *Fighting the Great War: A Global History*. Cambridge, MA: Harvard University Press. 2005

Nesbit, Roy C. *Eyes of the RAF, A History of Photo-Reconnaissance*. Phoenix Mill, UK: Alan Sutton Publishing, Ltd., 1996

Niven, Penelope. *Steichen*. New York: Clarkson Potter/Publishers, 1997

Occleshaw, Michael. *Armour Against Fate: British Military Intelligence in the First World War*. London: Columbus Books, 1989

O'Connor, Mike. *Airfields and Airmen, Ypres*. London: Leo Cooper, 2001

Oldham, Peter. *The Hindenburg Line*. London: Leo Cooper, 2001

'The Origins of the Italian Air Force,' http://www.finn.it/regia/html/origini.htm, accessed on 19 April 2010

Owers, Colin. De Havilland Aircraft of World War I. Volume I: D.H.1 – D.H. 4. Boulder, CO: Flying Machines Press, 2001

—— Jon S. Guttman, and James J. Davilla. Salmson Aircraft of World War I. Boulder, CO: Flying Machine Press, 2001

Parry, David. 'Descriptions of Aerial Reconnaissance Cameras Used, or Evaluated for Use by the Royal Flying Corps and Royal Air Force.' In *Aerial Archeology* E-Mail Newsletter. 2003

Pépin, Eugène, Dr. *History of the International Institute of Space Law of the International Astronautical Federation (1958-1982)*. New York: American Institute of Aeronautics and Astronautics, Inc., 1982

—— *The Legal Status of the Airspace in the Light of Progress in Aviation and Astronautics*. Institute of International Air Law. Montreal: McGill University, 1957

Phillips, Christopher. *Steichen at War*. New York: H.N. Abrams, 1981

Pieters, Walter M. *The Belgian Air Service in the First World War*. Marceline, Missouri: Walsworth Publishing, 2010

Pitt, Barrie. *Zeebrugge*. New York: Ballantine Books, 1959

Porch, Douglas. *The French Secret Services, From the Dreyfus Affair to the Gulf War*. New York: Farrar, Straus and Giroux, 1995

Pyner, Alf. *Air Cameras R.A.F. & U.S.A.A.F., 1915-1945*. Crouch, UK: Privately published, 1988

Showalter, Dennis E. *Tannenberg: Clash of Empires*. Hamden, CT: Archon Books, 1991

Skelton, M. L. 'Major H.D. Harvey-Kelly, Commanding Officer, No. 19 Squadron.' *Journal of the Society of World War 1 Aero Historians* 16, No. 4 (Winter 1975): 375

Sumner, Ian. *The French Army, 1914-18*. London: Osprey Publishing, 1995

Talbert, Roy, Jr. *Negative Intelligence: The Army and the American Left, 1917-1941*. Jackson, MS: University Press of Mississippi, 1991

Tanks & Weapons of World War I. Edited by Bernard Fitzsimons. New York: Beekman House. 1973

Terraine, John. *Mons: The Retreat to Victory*. New York: The Macmillan Company, 1960

Toelle, Alan D. *Breguet 14*. Berkhamsted, UK: Albatros Productions, Ltd., 2003

Trench Fortifications, 1914-1918: A Reference Manual. London: The Imperial War Museum, and Nashville, TN: The Battery Press, Inc. 1998

Tuchman, Barbara W. *The Guns of August*. New York: Bantam Book, 1976

Underwood, James R. and Peter L. Guth. *Military Geology in War and Peace*. Boulder, CO: The Geological Society of America, Inc., 1998

Venzon, Anne Cipriano, editor. *The United States in the First World War: An Encyclopedia*. New York: Garland Publishing, Inc., 1995

Villèle, Marie-Anne de, Agnès Beylot, and Alain Morgat, editors. *Du Paysage à La Carte: Trois Siècles de Cartographie Militaire de la France*. Vincennes, FR: Ministère de la Défense, Services Historiques des Armées, 2002

Vues d'en Haut: La Photographie Aérienne pendant la Guerre de 1914-18. Paris: Musée de l'Armée, Musée d'Histoire Contemporaine – Bibliothèque de Documentation Internationale Contemporaine. 1989

Watkis, Nicholas C. *The Western Front from the Air*. Phoenix Mill, UK: Sutton Publishing Ltd., 1999

Williams, George K. *Biplanes and Bombsights, British Bombing in World War I*. Maxwell Air Force Base, AL: Air University Press, 1999

INDEX

Page references in *italic* refer to captions. [Colour] indicates the colour section.